Mineral and Metal Neurotoxicology

Edited by

Masayuki Yasui

Michael J. Strong

Kiichiro Ota

M. Anthony Verity

CRC Press

Boca Raton   New York

Acquiring Editor:	Marsha Baker
Project Editor:	Debbie Didier
Marketing Manager:	Susie Carlisle
Direct Marketing:	Becky McEldowney
Cover Design:	Jason Toemmes
Manufacturing Assistant:	Sheri Schwartz

Library of Congress Cataloging-in-Publication Data

Mineral and metal neurotoxicology / edited by Masayuki Yasui ... [et
  al.].
      p. cm.
    Includes bibliographical references and index.
    ISBN 0-8493-7664-5
    1. Neurotoxicology.   2. Metals--Toxicology.   I. Yasui, Masayuki.
  RC347.5.M56   1996
  615.9′2--dc20

96-5735
CIP

Foreword

Evert Nieboer and John D. Turnbull

Exposures to neurotoxic metals and metalloids have occurred in various contexts. The neurotoxicology of lead has been the focus of extensive laboratory, clinical, and epidemiologic research because of widespread exposure through the use of leaded gasoline; similarly, a number of major environmental poisoning episodes have involved alkyl mercury compounds. Medicinal uses and/or accidental or suicidal poisonings first revealed the neurologic dysfunctions associated with arsenic, organotin, and thallium intake; illnesses resulting from occupational or environmental exposures have confirmed these observations. Effects of manganese on the central nervous system appear to be discerned only in occupational settings, while iatrogenic experiences have identified aluminum as neurotoxic. Deficiency or excess of essential metals can also be responsible for neurotoxic effects. This is illustrated by the recent discovery that mutations in two closely related genes, both of which encode proteins involved in copper transport, are responsible for the neurologic deterioration in Menkes disease (copper deficiency) or Wilson disease (copper toxicity due to excess).[1]

Establishing cause-and-effect relationships for neurologic disorders is difficult and, in the context of occupational or environmental exposures, application of rules of evidence is helpful. Guidelines for establishing etiology or causation are common in epidemiology, and these are relevant to toxicology. Critical appraisal criteria include assessment of the following in causal decision making: the strength of the association (magnitude of the observed effect); consistency (has the effect been observed by others elsewhere); temporality (timing and duration of exposure); gradient (evidence for a dose-response relationship); and sense (have competing explanations been ruled out; biological credibility; coherence with other knowledge about the disease; uniqueness of the outcome).[2,3] Biological plausibility is considerably strengthened by positive animal studies, although response differences among species can sometimes confuse this issue. Similarly, a beneficial clinical response to a specific therapy based on cause-and-effect arguments reinforces the causal link.

The importance of subjecting neurotoxicological models to the scrutiny of rules of evidence is easily demonstrated by exploring the sense criterion. A fundamental question concerns chemical or biological sense. Does, for example, a research paradigm of nonisomorphous replacement of Zn^{2+}, Mg^{2+}, or Ca^{2+} conform with the known biochemistry or physiology of these cations?[4] Or can support be drawn from established determinants of reactivity for both the essential and toxic metal ions such as ionic or hydrated size, geometric preferences, donor-atom preferences, complex stability, or kinetic properties?[5] Another example pertains to Alzheimer's disease (AD) and aluminum. Assignment of causation based on biochemical studies without cognizance of the strong genetic contribution to both early and late onset forms of AD[6,7] unduly restricts research scope and experimental design. Lack of coherence indicates a need for continuous feedback and paradigm adjustment. For example, laboratory research supports the notion that tissue damage by free radical pathways may be central to the development of Parkinson's disease (PD). However, clear demonstration that antioxidant therapies are beneficial in PD is still lacking,[8] and other factors may be involved. Finally, when considering causality, it is important to recognize that many diseases are heterogeneous and thus different processes may produce the same phenotypic expression.

Speciation may be defined as the occurrence of an element in separate identifiable forms (i.e., chemical, physical, or morphological state.)[9] Specific forms of an element often determine uptake, the distribution within an organism, and its excretion. Toxic responses can also be strongly speciation dependent.[5] As illustrated in the text, it is imperative that species-specific analytical methods and conceptual models are developed in support of both human and experimental metal neurotoxicology.

From these brief comments, it is evident that the neurotoxicology of metals and metalloids is a topic of broad scope requiring interdisciplinary cooperation for its effective development. Clearly, lessons learned from neurotoxicological research enhances our fundamental understanding of disease mechanisms. The nearly exponential increase in research articles is a daunting challenge for all of us. Even though computer-based retrieval has greatly facilitated literature searches, subject-specific reviews such as those featured in this volume that summarize and synthesize current knowledge constitute an important research aid.

REFERENCES

1. Bull PC and Cox DW: Wilson disease and Menkes disease: New handles on heavy-metal transport. *TIG*, 1994; 10:246–252.
2. Hill AB: The environment and disease: Association or causation? *Proc. R. Soc. Med.*, 1965; 58:295–300.
3. Sackett DL, Haynes RB, Gyatt GH, and Tugwell P: *Clinical Epidemiology. A Basic Science for Clinical Medicine, 2nd Ed.* Boston, Little Brown, 1991, pp 283–302.
4. da Silva JJRF and Williams RJP: *The Biological Chemistry of the Elements.* Oxford, Clarendon Press, 1991.
5. Nieboer E and Fletcher GG: Determinants of reactivity in metal toxicology. In Chang, LW (ed): *Toxicology of Metals.* Boca Raton, FL, CRC Press, 1996, pp 111–130.
6. Nieboer E, Gibson BL, Oxman AD, and Kramer JR: Health effects of aluminum: A critical review with emphasis on aluminum in drinking water. *Environ. Rev.*, 1995; 3:29–81.
7. Barinaga M: Missing Alzheimer's gene found. *Science*, 1995; 269:917–918.
8. The Parkinson Study Group: Effects of tocopherol and deprenyl on the progression of disability in early Parkinson's disease. *N. Eng. J. Med.*, 1993; 328:176–183.
9. Nieboer E and Thomassen Y: The second international symposium on speciation of elements in toxicology and in environmental and biological sciences. *Analyst*, 1995; 120:30N.

Preface

The recognition that our environment, both in its natural state and as it has been modified by ourselves as inhabitants of this planet, can be toxic to our nervous system is gaining increasing acceptance. New to this recognition is the awareness that trace metals and minerals can be as neurotoxic as they can be integral to the function of the nervous system. Although minerals such as zinc, copper, and magnesium have conventionally been held to be integral to the human system, only recently has the awareness of the role of elements such as manganese, cadmium, and selenium increased. Moreover, these elements do not operate independently of each other. For example, magnesium exists in intracellular compartments and its deficit provokes degeneration of the CNS due to a decline in the synthesis and secretion of neurotransmitters. It also triggers the action of neurotoxic metals such as aluminum and mercury. In contrast, although CNS degeneration is induced by the interaction of minerals and metals, both magnesium and zinc have protective effects. This increasing awareness of the complex interactions of metals and minerals, and the diseases induced by alterations in their homeostasis, has prompted the genesis of this text.

In formulating this text, we recognized that disturbances in trace-metal or mineral homeostasis can have broad effects, giving rise to very specific, readily recognized disease phenotypes. In contrast, more subtle effects, at the molecular level may produce profound biochemical defects that are not manifest until cellular damage accumulates. Understanding the nature of these defects also requires an understanding of the risk factors for disease induction, including both the geographic milieu of the affected host and the genetic make-up of the host. For some minerals and metals, the interrelationship amongst these factors is clearly defined — for others it is less so. Our purpose in this text was to provide in a single volume a systematic treatment of areas of new and ongoing research in mineral and metal neurotoxicology, to describe how methods and techniques of mineral and metal neurotoxicology are particularly appropriate for acquiring data on and testing hypotheses about mineral and metal neurotoxicity, to explore the role the neurotoxicologist should take in exploring the relationship between human and environmental factors, and to develop hypotheses based on neurotoxicological data to guide future studies and treatment regimens.

We have also recognized that each mineral or metal addressed within this text could justifiably be the focus of a text unto itself. In the interest of space, therefore, we asked each author to review the major features of his/her area of research in a succinct fashion. Something is always lost in such an endeavor — in this case the breadth of references necessary to accomplish such a task was sacrificed to the desire for text. In many instances, these authors have expressed a gracious willingness to provide additional references to interested readers upon request. We would encourage readers to do so.

No text of this size comes to fruition without significant input from a number of individuals. To the authors who provided timely and succinct reviews of difficult areas of research, we express our appreciation. To Ralph M. Garruto of the National Institutes of Health and Leonard Lundmark of Wakayama University, we owe a special debt of gratitude for their constant advice and assistance. To our secretaries, who maintained the flow of information, a debt of gratitude is acknowledged. The timely advice and assistance of Marsha Baker at CRC Press is acknowledged.

The Editors

Masayuki Yasui, M.D., Ph.D. is an associate professor in the Division of Neurological Diseases at Wakayama Medical College (Wakayama, Japan). He is a specialist of both Neurology and Internal Medicine accredited by the Neurological Society of Japan and the Japanese Society of Internal Medicine.

Dr. Yasui obtained his M.D. degree in 1966 and his Dr. Med. Sci. degree in 1979 from Wakayama Medical College. He was a guest researcher of the National Center of Neurology (Tokyo, Japan) from 1985 to 1988 and at the National Institutes of Neurological Disorders and Stroke, National Institutes of Health (Bethesda, Maryland) in 1990 under the supervision of Dr. Ralph M. Garruto in the laboratories of Dr. D. Carleton Gajdusek, Nobel Laureate.

Dr. Yasui is a member of the Neurological Society of Japan, the Japanese Society of Internal Medicine, the Japanese Society of Stroke, the Japanese Society for Magnesium Research, Japan Society for Biomedical Research on Trace Elements, and Japan Society of Electroencephalography and Electromyography.

Dr. Yasui has presented more than 50 invited lectures at various meetings and seminars of Japan Medical Association. He has published more than 160 research papers and books on electroencephalograms and neurological examination. His primary area of research has been on the interaction among minerals in neurological diseases.

Michael J. Strong, M.D., FRCPC is an associate professor of Neurology in the Department of Clinical Neurological Sciences, Faculty of Medicine at the University of Western Ontario (London, Ontario, Canada) and a research scientist at the John P. Robarts Research Institute, University of Western Ontario in the Neurodegeneration Research Group. He is also a consulting Neurologist at the London Health Sciences Centre (London, Ontario) from 1990 to present.

Dr. Strong obtained his M.D. degree in 1982 from Queen's University (Kingston, Ontario) and his fellowship training in neurology from the University of Western Ontario (London, Ontario) in 1987. From 1987 to 1990, he received fellowship training at the Laboratory of Central Nervous Systems Studies, National Institutes of Neurological Disorders and Stroke, National Institutes of Health (Bethesda, Maryland) under the supervision of Dr. Ralph M. Garruto in the Laboratories of Dr. D. Carleton Gajdusek, Nobel Laureate.

Dr. Strong is a member of the Canadian Neurological Society, Society for Neurosciences, and the American Academy of Neurology. He is Medical Advisor to the ALS Society of Ontario and a member of the Medical Advisory Board of the Muscular Dystrophy Association of Canada. He is also a scholar of the Medical Research Council of Canada.

Dr. Strong's primary area of research has been on the mechanisms of aluminum neurotoxicity, specifically addressing the mechanisms by which aluminum induces intraneuronal inclusions of neurofilament. He has published approximately 28 papers, largely focused on this area of research and its application to the understanding of the human neurodegenerative process, amyotrophic lateral sclerosis.

Kiichiro Ota, M.D., Ph.D. is an associate professor of the Departments of Laboratory Medicine and Blood Transfusion Medicine at Wakayama Medical College (Wakayama, Japan).

Dr. Ota obtained his M.D. degree in 1965 and his Ph.D. in 1975 from Wakayama Medical College. He researched Calcium Metabolism and Parathyroid Hormone as an associate researcher in the Department of Endocrinology and Metabolism at the Michael Reese Hospital, Chicago University, Illinois, USA.

Dr. Ota is a member of Japanese Internal Medicine, Endocrinology, Bone and Mineral Metabolism, Gerontology, Hematology and Blood Transfusion Medicine. His research area has been focused on the implications of minerals and metals on soft tissues, especially central nervous system tissue and bone.

M. Anthony Verity, M.D. received his doctor of medicine degree at St. Mary's Hospital, London, England. He subsequently served internship and residency positions in surgery and medicine, including senior house-officer clinical pathologist at United Bristol Hospitals, England. He completed a residency and fellowship in pathology and neuropathology at UCLA. Specialized training in neurochemistry and neuropathology has been obtained at Oxford University, Bristol University, as honorary consultant in neuropathology to the University of Leeds, and as deputy coroner in neuropathology, Los Angeles County. He was a visiting fellow at the University of Bristol and a Burroughs-Wellcome senior fellow at the University of Newcastle, England. Dr. Verity is currently Emeritus Professor and Chief, Division of Neuropathology, UCLA. He has served on the editorial board of numerous journals including *NeuroToxicology, Neurochemical Pathology*, and *Alzheimer Disease and Allied Conditions* and held numerous counsel and review group appointments in the U.S. Environmental Protection Agency, Los Angles Society of Neurology and Psychiatry, and NIEHS. He has received numerous NIH competitive grants and other grants from foundations and special sources. Research interests have included the neurotoxicology of heavy metals with special reference to mercury and lead experimental neuropathology, dementia, and diagnostic histochemistry and biochemistry of human muscle disease. Dr. Verity is the author of more than 200 publications including books, chapters, and proceedings.

Contributors

Zeev B. Alfassi, Ph.D.
Department of Nuclear Engineering
Ben-Gurion University of the Negev
Beer-Sheva 84 105
P.O.B. 653
Israel

Allen C. Alfrey, M.D.
Renal Section
Department of Veterans Affairs Medical Center
Denver, Colorado 80220
U.S.A.

Björn Arvidson, M.D., Ph.D.
Department of Neurology
University of Uppsala
S-751 85 Uppsala
Sweden

Pierre Bac, M.D., Ph.D.
Maître de Conférences
Laboratoire de Pharmacólogie
Faculté de Pharmacie
5 Rue J.B. Clément
F 92290 Chatenay-Malabry,
France

Michel Bara, Ph.D.
Maître de Conférences
Laboratory of Physiopathology
 of Development
University Pierre et Marie Curie
75252-Paris Cedex 05
France

Joseph S. Beckman, Ph.D.
Departments of Anesthesiology and
 Biochemistry
University of Alabama at Birmingham
Birmingham, Alabama 35233-6810
U.S.A.

David C. Bellinger, Ph.D.
Neuroepidemiology Unit
Harvard Medical School
Children's Hospital
Boston, Massachusetts 02115
U.S.A.

George J. Brewer, M.D.
Department of Human Genetics
University of Michigan
Ann Arbor, Michigan 48109-0618
U.S.A.

Alex W.K. Chan, Ph.D.
DuPont Medical Products
Newtown, Connecticut 06470-5509
U.S.A.

Louis W. Chang, Ph.D.
Departments of Pathology, Pharmacology,
 and Toxicology
University of Arkansas for Medical Sciences
4301 W. Markham, Slot 517
Little Rock, Arkansas 72205-7199
U.S.A.

Jean Constantinidis, M.D., Ph.D.
University of Geneva
Biological Research in Neuropsychiatry
Grange-Falquet 32-34
Chene-Bougeries
CH-1224-Geneva
Switzerland

John P. Crow, Ph.D.
Department of Anesthesiology
University of Alabama at Birmingham
Birmingham, Alabama 35233-6810
U.S.A.

J. Michael Davis, Ph.D.
National Center for Environmental
  Assessment—RTP
U.S. Environmental Protection Agency (MD-52)
Research Triangle Park, North Carolina 27711
U.S.A.

Roger Deloncle, Ph.D.
Institut des Xenobiotiques, E.A.1223
34 Rue du Jardin des Plantes, B.P. 199
86005 Poitiers
France

Robert J. DeLorenzo, M.D., Ph.D., M.P.H.
Department of Neurology
Medical College of Virginia
Virginia Commonwealth University
Richmond, Virginia 23298-0599
U.S.A.

David T. Dexter, B.Sc., Ph.D.
Department of Pharmacology
Charing Cross and Westminster Medical School
London W6 8RF
United Kingdom

Jean Durlach, M.D., Ph.D.
President, International Society for the
  Development of Research on Mg and
Editor-in-Chief, Mg Research
64 rue de Longchamp
F 92200 Neuilly-sur-Seine,
France

Robert W. Elias, Ph.D.
National Center for Environmental
  Assessment
U.S. Environmental Protection Agency
Research Triangle Park, North Carolina 27711
U.S.A.

Hakån Eriksson, Ph.D.
Astra Arcus AB
S-151 85
Sodertalje
Sweden

Trond P. Flaten, Ph.D.
Department of Chemistry
Norwegian University of Science and
Technology
N-7055 Dragvoll
Norway

Donald A. Fox, Ph.D.
College of Optometry
Department of Biochemical and Biophysical
  Sciences
University of Houston
4901 Calhoun
Houston, Texas 77204-6052
U.S.A.

Ralph M. Garruto, Ph.D.
Laboratory of Central Nervous System Studies
National Institutes of Health
Bethesda, MD 20892-4158
U.S.A.

Claire E. Gavin
Astra Arcus USA
P.O. Box 20890
Rochester, New York 14602
U.S.A.

Paul F. Good, Ph.D.
Division of Neuropathology
Mount Sinai School of Medicine
New York, New York 10029
U.S.A.

Lester D. Grant, Ph.D.
National Center for Environmental
  Assessment
U.S. Environmental Protection Agency
Research Triangle Park, North Carolina 27711
U.S.A.

Thomas E. Gunter, Ph.D.
Department of Biophysics
University of Rochester Medical Center
Rochester, New York 14642
U.S.A.

Masazumi Harada, M.D., Ph.D.
Institute of Molecular Embryology and
  Genetics
Kumamoto University School of Medicine
Kuhonji, 4-24-1, Kumamoto 862
Japan

Mohammed I. Hasham, M.Sc.
Division of Neurology
Department of Medicine
University of British Columbia
Vancouver, Canada V6T 1Z3

Alfred R. Haug, Ph.D.
Department of Microbiology
Michigan State University
East Lansing, Michigan 48824
U.S.A.

Harald Hefter, M.D., Ph.D.
Department of Neurology
University of Düsseldorf
D-40225 Düsseldorf
Germany

Edith Heilbronn, Ph.D., M.D.h.c.
Department of Neurochemistry and
  Neurotoxicology
Stockholm University
S-106 91 Stockholm
Sweden

**Peter Jenner, B.Pharm., Ph.D., D.Sc.,
F.R.Pharm.S.**
Neurodegenerative Disease Research Centre
Pharmacology Group
King's College
London SW3 6LX
United Kingdom

Kiyotaro Kondo, M.D., Ph.D.
Department of Public Health
Hokkaido University
School of Medicine
N15, W7, Sapporo
060 Japan

Charles Krieger, M.D., Ph.D.
Division of Neurology
Department of Medicine
University of British Columbia
Vancouver, Canada V6T 1Z3

James C.K. Lai, Ph.D.
Department of Pharmaceutical Sciences
College of Pharmacy
Idaho State University
Pocatello, Idaho 83209-8334
U.S.A.

Louis Lim, Ph.D.
Department of Neurochemistry
Institute of Neurology
University of London, Queen Square
London WC1N 3BG
United Kingdom

Walter J. Lukiw, Ph.D.
Molecular Neurobiology
LSU Eye and Neuroscience Center
Louisiana State University School of Medicine
New Orleans, Louisiana 70112-2234
U.S.A.

R. Bruce Martin, Ph.D.
Department of Chemistry
University of Virginia
Charlottesville, Virginia 22903
U.S.A.

John H. Menkes, M.D.
Departments of Neurology and Pediatrics
University of California Los Angeles
9320 Wilshire Boulevard
Beverly Hills, California 90212
U.S.A.

Margaret J. Minski, Ph.D.
Imperial College Reactor Centre
University of London
Wilwood Park, Ascot, Berkshire,
SL5 7TE
United Kingdom

Hiroshi Morita, M.D., Ph.D.
Department of Medicine (Neurology)
School of Medicine
Shinshu University
Asahi 3-1-1, Matsumoto 390
Japan

Vincent A. Murphy, Ph.D.
Procter & Gamble Pharmaceuticals
Miami Valley Laboratories
P.O.Box 538707
Cincinnati, Ohio 45253-8707
U.S.A.

Evert Nieboer, Ph.D.
Professor of Toxicology
Department of Biochemistry
McMaster University
Health Sciences Centre
1200 Main Street West
Hamilton, Ontario, Canada L8N 3Z5

Yukiharu Okamoto
Department of Laboratory Medicine
Wakayama Medical College
7 Bancho, Wakayama 640
Japan

C. Warren Olanow, M.D., FRCPC
Department of Neurology
Mount Sinai School of Medicine
New York, New York 10029
U.S.A.

Kiichiro Ota, M.D., Ph.D.
Department of Laboratory Medicine
Wakayama Medical College
7 Bancho, Wakayama 640
Japan

Nicole Pages, Ph.D.
Laboratoire de Toxicologie
Faculté de Pharmacie
5 Rue J.B. Clément
92296 — Châtenay-Malabry
France

Daniel P. Perl, M.D.
Neuropathology Division
Mount Sinai School of Medicine
New York, New York 10029
U.S.A.

Jacinda B. Sampson, B.A.
Department of Anesthesiology
University of Alabama at Birmingham
Birmingham, Alabama 35233-6810

Morimi Shimada, M.D.
Department of Pediatrics
Shiga University of Medical Science
Otsu 520-21
Japan

Timothy J.B. Simons, Sc.D.
Physiology Group
Biomedical Science Division
King's College London
Strand, London WC2R 2LS
United Kingdom

Devesh Srivastava, Ph.D.
The Wilmer Ophthamological Institute
Johns Hopkins University
  School of Medicine
Baltimore, Maryland 21287-9257
U.S.A.

Michael J. Strong, M.D., FRCPC
Department of Clinical Neurological Sciences
University Campus, London Health Sciences
  Center
Ontario, Canada N6A 5A5

Tetsuya Tamaki, M.D., Ph.D.
Department of Orthopedic Surgery
Wakayama Medical College
27, 7 – Bancho, Wakayma 640
Japan

Yasunori Taniguchi, M.D., Ph.D.
Department of Orthopedic Surgery
Wakayama Medical College
27, 7 – Bancho, Wakayma 640
Japan

John D. Turnbull, M.D., Ph.D.
Department of Medicine
McMaster University
Health Sciences Centre
1200 Main Street West
Hamilton, Ontario, Canada L8N 3Z5

M. Anthony Verity, M.D.
Department of Pathology (Neuropathology)
Center for the Health Sciences
UCLA School of Medicine
University of California at Los Angeles
Los Angeles, CA 90095-1732
U.S.A.

Victor A. Vitorello, Ph.D.
Centro de Energia Nuclear na Agricultura
Universidade de Sao Paulo
13416 — Piracicaba, Sao Paulo
Brasil

Ikuro Wakayama, M.D., Ph.D.
Research Center of Neurological Diseases
Kansai College of Oriental Medicine
2-11-1 Wakaba, Kumatori, Sennan
Osaka 590-04
Japan

Elizabeth J. Waterhouse, M.D.
Department of Neurology
Medical College of Virginia
Virginia Commonwealth University
Richmond, Virginia 23298-0599
U.S.A.

Erik Ch. Wolters, M.D., Ph.D.
Graduate School of Neuroscience
Institute of Neuroscience
Academic Hospital Vrije Universiteit
1007 MB Amsterdam
The Netherlands

Tsunekazu Yamano, M.D.
Department of Pediatrics
Shiga University of Medical Science
Otsu 520-21
Japan

Nobuo Yanagisawa, M.D., Ph.D.
Department of Medicine (Neurology)
Shinshu University School of Medicine
Matsumoto 390
Japan

Masayuki Yasui, M.D., Ph.D.
Division of Neurological Diseases
Wakayama Medical College
Wakayama 640
Japan

Robert A. Yokel, Ph.D.
Division of Pharmacology and
 Experimental Therapeutics
College of Pharmacy
University of Kentucky
Lexington, Kentucky 40536-0082
U.S.A.

Munehito Yoshida, M.D., Ph.D.
Department of Orthopedic Surgery
Wakayama Medical College
27, 7 – Bancho, Wakayma 640
Japan

Contents

PART I

GENERAL ASPECTS OF MINERALS AND METALS

SECTION 1
INTRODUCTION AND BACKGROUND

Chapter 1

Man, Metals, and Minerals

John H. Menkes

CONTENTS

1.1   INTRODUCTION

The composition of the various elements which make up the human body resembles that found in the earth surface and in the surrounding oceans. In the course of evolution, humans, like other forms of life, have had to adapt themselves to changing environmental conditions by acquiring a variety of mechanisms through which the uptake, transport, and storage of the chemical components essential to biological function are regulated by the body. At least eight trace elements have been found to be essential for man. These are iron, copper, zinc, manganese, cobalt, selenium, chromium, and molybdenum. In addition, some six other trace metals are essential for other mammals. Disease can occur as a result of a deficiency or excess of these essential metals. Several other minerals and metals whose presence is not essential for normal biologic function can act as toxins and induce disease in humans as a result of acute or chronic exposure. Currently, the most important environmental metallic toxins are lead, mercury, arsenic, and thallium.

This introductory chapter is intended to provide an overview of the various diseases which result from a disorder in metal and mineral homeostasis.

1.2 ESSENTIAL METALS AND MINERALS

Diseases resulting from defective homeostasis of essential metals and minerals can result from dietary deficiencies, excessive intake, and from genetic defects of homeostatic control.

1.2.1 Iron

Iron plays a vital role in many enzymes involved in oxidative and amino acid metabolism.[1] In humans, the major amount of iron is in the iron porphyrin complexes hemoglobin and myoglobin, and in various heme-containing enzymes. The remainder is stored in the form of a soluble fraction, ferritin, and an insoluble and nonreactive form, hemosiderin. Transferrin is the principal iron-carrying protein of plasma.[2] In brain, iron is present as heme and nonheme iron. The iron content of brain is highest in the globus pallidus, red nucleus, substantia nigra pars reticulata, putamen, and the dentate nucleus. Only trace amounts are present at birth; the accumulation of the metal increases until adult levels are reached in the late teens. Ferritin and transferrin are specifically localized to oligodendrocytes, suggesting that these cells are important in iron storage and mobilization. In addition, transferrin acts as a trophic factor for neurons, astrocytes, and oligodendrocytes.[3]

The role of iron in dopamine D_2 receptor function has been the subject of considerable recent investigation. In experimental animals iron deficiency leads to the subsensitivity of the D_2 receptor, and peripheral iron status plays an important role in neuroleptic-induced dopamine supersensitivity.[4] The mechanism of this interaction is unclear.

The most common disorder resulting from an abnormality in iron homeostasis is iron deficiency — the world's most prevalent nutritional disorder. The amount of iron absorbed by the body depends not only on the dietary content of the metal and its bioavailability, but also on the activity of the absorptive processes which, in turn, are inversely related to the plasma ferritin concentration and thus to the body iron stores.[5] The lower the body iron stores, the larger the percentage of iron retained. The same absorptive mechanisms are involved in the uptake of lead and cadmium. It should therefore not come as a surprise that lead poisoning is more frequently seen in iron-deficient children.[6]

The effects on neurodevelopmental outcome of chronic iron deficiency suffered during the first two years of life have been a matter of some debate. In the Chilean experience of Walter et al.,[7] developmental test performance, particularly on language items, was impaired in children whose hemoglobin values had been below 10.5 g for more than 3 months. Correction of the iron deficiency failed to improve the performance scores. Although similar results have been obtained from other parts of the world, they are confounded by a variety of environmental and socioeconomic factors.[8]

Both acute and chronic forms of iron poisoning have been encountered. The acute form usually is seen in toddlers and results from the accidental ingestion of large quantities of hematinics or vitamins containing ferrous sulfate or other types of ferrous salts. Symptoms are local or systemic. Free ionic iron can induce shock by a direct vasodepressant effect, or can cause hepatic necrosis and increased permeability of the pulmonary capillary bed.

Chronic iron overload due to the intake of alcoholic beverages brewed in iron containers has been reported from Africa, but this condition does not appear to affect brain function.

At least two genetic disorders of iron regulation have been reported. The more common of these is hemochromatosis, a condition marked by increased iron absorption and accumulation of the metal in a variety of tissues. The gene for hemochromatosis has been localized to the short arm of chromosome 6.[9] Clinical features include hepatic cirrhosis, diabetes, arthropathy, hypermelanotic pigmentation of the skin, and heart failure. Biochemically, hemochromatosis is marked by increased serum iron and an increased transferrin saturation. Since chromium is also transported bound to transferrin, the diabetes which accompanies hemochromatosis is believed to result from chromium deficiency. The nature of the metabolic abnormality responsible for the accumulation of excessive amounts of iron is still unknown, but it has been suggested that it is caused by a tissue- or cell-specific defect in ferritin synthesis.[10] There is no increase in iron accumulation in brain, and neurologic symptoms which have occasionally been reported are probably the result of hepatic dysfunction.

Atransferrinemia is a rare disorder of iron homeostasis which results in severe anemia, but is not accompanied by neurologic symptoms.

Increased brain iron has been noted in several neurologic conditions, especially in Hallervorden-Spatz disease and Parkinson's disease.

In Hallervorden-Spatz disease there is hyperpigmentation of the pallidum, the substantia nigra, and less often of the cerebral cortex. The pigment is located within neurons and glial cells and contains iron. It is still unknown whether the disease results from a disorder of lipofuscin storage or an abnormality in axonal metabolism. In either case iron deposition would be a secondary phenomenon. The clinical highlights of the classic form of the disease are progressive extrapyramidal symptoms and mental deterioration. MR imaging studies can demonstrate increased amounts of iron in the globus pallidus.[11]

A substantial increase in the iron content of the pars compacta of the substantia nigra unaccompanied by an increase in ferritin has been demonstrated in the brain of patients affected with Parkinson's disease. The role of iron in the pathogenesis of this disorder is still unknown. Since iron accumulation occurs in other basal ganglia degenerative diseases, this phenomenon is most likely a response to cell death in the substantia nigra.[12]

1.2.2 Copper

Copper is a metal essential to normal brain function, and both reduced and increased brain copper result in severe neurologic symptoms.[13]

The daily dietary intake of copper ranges between 1 and 5 mg. Humans consuming a free diet absorb about 40% of dietary copper. The absorption site is probably in the proximal portion of the gastrointestinal tract. Metallothionein, a low-molecular-weight metal protein, is involved in regulating copper absorption at high copper intakes in the intestinal transport of the metal, and probably in its initial hepatic uptake.[14] There are two major isoforms of the protein, MT-I and MT-II, with closely related but distinct amino acid sequences. At least three forms of MT-I are demonstrable in humans by protein sequencing. The metallothioneins are encoded by multiple genes, of which one cluster is localized on chromosome 16; the others are dispersed to at least 4 other autosomes. The transcription of each of these genes is induced by copper, zinc, cadmium, and by a variety of nonmetal inducers including glucocorticoid hormones.

Following its intestinal uptake, copper enters plasma where it is bound to albumin in the form of cupric ion. Within 2 h the absorbed copper is incorporated into a liver protein. In liver, copper is either excreted into bile, stored in liver lysosomes in what is probably a polymeric form of metallothionein, or combined with apoceruloplasmin to form ceruloplasmin, which then enters the circulation. More than 95% of serum copper is in this form.

The major function of ceruloplasmin is still unclear. Although it is not involved in copper transport from the intestine, it controls the release of iron into plasma from cells, in which the metal is stored in the form of ferritin. Ceruloplasmin is considered to be an important vehicle for the transport of the metal from the liver and to function as a copper donor in the formation of a variety of copper-containing enzymes. It is also the most prominent serum antioxidant, and as such it catalyzes the oxidation of ferrous ion to ferric ion and prevents the oxidation of polyunsaturated fatty acids and similar substances. Finally, it modulates the inflammatory response and can regulate the concentration of various serum biogenic amines.

Several other copper-containing proteins have been isolated from mammalian tissues. Most prominent of these are the enzymes cytochrome c oxidase, dopamine β-hydroxylase, superoxide dismutase, and tyrosinase.

The effects of nutritional copper deficiency in humans are limited to anemia, neutropenia, and osteoporosis; neurologic symptoms such as are seen after fetal copper deficiency in sheep, rats, or pigs have not been encountered.

Copper toxicity resulting from an increased uptake of the metal has occasionally been seen in children and in adults following prolonged exposure to copper-containing eating utensils.[15] In addition to the high copper intake, a genetic defect in copper metabolism probably contributes to the clinical picture both in Indian children (Indian childhood cirrhosis) and in a similar condition seen in Western countries.[16] Liver failure, renal failure, and hemolysis can occur after toxic exposure to the metal; neurologic symptoms have not been encountered.

Copper is the only metal for which specific transport genes have been identified.[13] Two disorders of copper transport, Wilson disease and Menkes disease (kinky hair disease) have been recognized. The genes whose defects are responsible for these conditions have been identified as coding for copper-transporting ATPases, with considerable homology between the two genes. The gene whose defect causes Wilson disease is expressed in the liver and kidney, whereas the gene responsible for Menkes disease is expressed in a variety of tissues with those tissues which express the Menkes disease gene showing little if any expression of the Wilson disease gene.

As a consequence of the defect in the protein which regulates copper excretion, the dynamic turnover of copper is disturbed in Wilson disease, and the rate at which excess copper is excreted into bile and is incorporated into ceruloplasmin is markedly reduced. Most importantly, copper accumulates within the liver. At first it is firmly bound to copper proteins such as ceruloplasmin and superoxide dismutase, or is in the cupric form complexed with metallothionein. When the copper load overwhelms the binding capacity of metallothionein, cytotoxic cupric copper is released, causing damage to hepatocyte mitochondria and peroxisomes. Ultimately, copper leaks from the liver into the blood, whence it is taken up by other tissues, including the brain, which in turn are damaged by copper.[17]

In the brain, the largest proportion of copper is located in the subcellular soluble fraction, where it is bound not only to cerebrocuprein, but also to a number of other normal cerebral proteins. Microscopic studies reveal a loss of neurons, axonal degeneration, and large numbers of protoplasmic astrocytes, including giant forms termed Alzheimer cells. Copper is deposited in the pericapillary area and within astrocytes, but it is uniformly absent from neurons and ground substance.

Lesser degenerative changes are seen in the brainstem, the dentate nucleus, the substantia nigra, and the convolutional white matter. Copper is also found throughout the cornea, particularly the substantia propria, where it is deposited in an alcohol-soluble and probably chelated form. In the periphery, the metal appears in granular clumps close to the endothelial surface of Descemet's membrane. Here the deposits are responsible for the appearance of the Kayser-Fleischer rings.

Clinically, Wilson disease is a progressive condition with a tendency toward temporary clinical improvement and arrest. Disorders in the function of the basal ganglia predominate, although psychiatric symptoms have been seen initially in a significant proportion of cases.

In Menkes disease, the disorder in copper transport makes the metal inaccessible for the synthesis of ceruloplasmin, superoxide dismutase, and several other copper-containing enzymes, notably ascorbic acid oxidase, cytochrome oxidase, dopamine-β-hydroxylase, and lysyl oxidase.

Because of the defective activity of these metalloenzymes, a variety of pathologic changes are set into motion. Arteries are tortuous, with irregular lumens and frayed and split intimal linings. These abnormalities reflect a failure in elastin and collagen cross-linking caused by dysfunction of the key enzyme for this process — copper-dependent lysyl oxidase.

Changes within the brain result from vascular lesions, copper deficiency, or a combination of the two. There is extensive focal degeneration of gray matter with neuronal loss and gliosis and an associated axonal degeneration in white matter. Cellular loss is prominent in the cerebellum. Here Purkinje cells are hard-hit; many are lost, and others show abnormal dendritic arborization ("weeping willow") and perisomatic processes. Focal axonal swellings (torpedoes) are also observed.[18] Electron microscopy often shows a marked increase in the number of mitochondria in the perikaryon of Purkinje cells, and to a lesser degree in the neurons of cerebral cortex and the basal ganglia. Mitochondria are enlarged, and intramitochondrial electron-dense bodies are present. The pathogenesis of these changes is a matter of controversy, but they are believed to result from a reduction in the activity of the mitochondrial copper-containing enzymes.

The major clinical features of Menkes disease are abnormal hair, progressive cerebral degeneration, vascular tortuosities, bony changes, and hypothermia.

Several other extremely rare disorders of copper metabolism have been recognized. The occipital horn syndrome is caused by a mutation in the same copper-transporting ATPase that results in Menkes disease.[19] Aceruloplasminemia is a dominantly transmitted mutation which results in iron deposition in the liver and brain but no storage of copper. The clinical picture is that of dementia and a variety of movement disorders.[20] This condition confirms what has already been suspected from the manifestations of Wilson disease, namely that copper plays an important role in the mobilization of iron, and that the absence of ceruloplasmin results in hemosiderosis.[21]

1.2.3 Zinc

Zinc plays an important role in eukaryotic DNA transcription. It also functions as a modulator for excitatory synaptic transmission mediated by glutamate receptors within the central nervous system. Dietary zinc deficiency reduces the number of functional N-methyl-D-aspartate receptor channels in cortical membranes, and lead inhibits the activation of the N-methyl-D-aspartate (NMDA) receptor complex by interacting with the zinc allosteric sites.[22]

Several reports suggest that maternal zinc deficiency accompanies a variety of neural tube defects in the developing human central nervous system. These include spina bifida, myelomeningocele, and anencephalus.[23] A depression in the activity of thymidine kinase and impaired DNA synthesis may be responsible. Postnatal zinc deficiency is seen in malabsorption states, notably in cystic fibrosis, anorexia nervosa, hepatic cirrhosis, and in subjects who have been receiving total parenteral nutrition. Symptoms include anorexia, reduced taste sensitivity, poor wound healing, and impaired cellular immunity. Although zinc deficiency adversely affects the behavior of experimental animals, the importance of the metal in human mental function has not been demonstrated. In acrodermatitis enteropathica, an autosomal recessive disorder resulting from a defect in zinc absorption, symptoms include severe gastrointestinal disturbances, dermatitis, and alopecia, but the nervous system is spared.

Since zinc inhibits the intestinal absorption of copper, zinc toxicity can result in symptoms of copper deficiency.[24] The importance of an excessive release of endogenous zinc at the central glutamatergic synapses to the pathogenesis of epileptic brain damage has been the subject of extensive investigation.[25]

Zinc appears to induce β-amyloid protein aggregation and could therefore be involved in the reactions which lead to the deposition of β-amyloid protein in Alzheimer's disease. The reduced concentration of zinc in the temporal lobe and hippocampal cortex of patients with Alzheimer's disease further suggests that the metal is not being distributed in a normal manner.[25a]

1.2.4 Manganese

Human manganese deficiency has not been documented, and if it does occur it must be extremely rare. By contrast, there have been numerous reports of manganese toxicity as a consequence of acute or chronic exposure to the metal. Acute exposure results in hepatitis, pneumonitis, and pancreatitis, but does not induce any symptoms referable to the central nervous system. Chronic manganese toxicity is generally the consequence of occupational inhalation of particles in ferromanganese mines and smelting factories. The clinical picture is one of acute hallucination and psychosis, followed by parkinsonism after an interval of months to many years.[26] The cause for the symptoms is unclear. Manganese catalyzes the autooxidation of dopamine to 6-hydroxydopamine, a substance which destroys dopaminergic nerve terminals. The metal also induces a decrease in striatal glutathione peroxidase, suggesting that it facilitates the formation of free radicals in the brain, particularly within the basal ganglia.[27]

1.2.5 Cobalt

The importance of cobalt in human nutrition is related to the relatively common occurrence of cobalamin (vitamin B_{12}) deficiencies. The cobalamins are a series of porphyrin-like compounds which contain trivalent or monovalent cobalt ions. The usual dietary source of cobalamins is meat, particularly liver, and pure dietary cobalamin deficiency is seen in children of strictly vegan mothers. More commonly, there is failure in absorption of cobalamin due to an inadequate supply of an intrinsic factor which regulates absorption. Neurologic symptoms can develop as a result of cobalamin deficiency, with the most characteristic picture being a spongy demyelination of the long tracts of the spinal cord, particularly the posterior columns. In addition, there are at least four genetic disorders of cobalamin absorption and transport, and several genetic disorders of cobalamin utilization.

1.2.6 Selenium

Selenium is required for the activity of glutathione peroxidase, an enzyme which catalyzes the breakdown of hydrogen peroxide, and thus functions as a free radical scavenger.[28] The clinical importance of selenium deficiency is not clear. Although endemic deficiency has been shown to cause a cardiomyopathy (Keshan disease), neurologic symptoms, if any, are vague and nonspecific. The role of selenium in the various neurologic diseases believed to result from excessive free radical formation within the central nervous system is still unclear, but has become an area of intensive research.[29]

1.2.7 Chromium

Mild degrees of chromium deficiency appear to be fairly widespread and are linked to impaired glucose and lipid metabolism. Chromium is used in a variety of industrial processes and both acute and chronic toxicity have been described following exposure to the hexavalent form of the metal, the latter associated with an increased incidence of a variety of malignancies and developmental malformations.[30] The role of chromium in normal brain function is unknown.

1.2.8 Molybdenum

Three enzymes require molybdenum for their function: sulfite oxidase, xanthene dehydrogenase, and aldehyde oxidase. An autosomal recessive disorder marked by severe developmental delay, seizures, and multiple cerebral infarcts results from a deficiency of the molybdenum cofactor.[31] It is believed that the encephalopathy is the result of sulfate oxidase deficiency, possibly the consequence of the accumulation of a toxic metabolite such as sulfite which could induce a disturbance in pyruvate metabolism, or interfere with the integrity of cell membranes.

Dietary molybdenum deficiency is extremely rare, and has not been well documented.

1.3 NONESSENTIAL METALS AND MINERALS

1.3.1 Lead

Lead toxicity is of considerable importance in clinical neurology. One of the major sources of lead continues to be exterior and, to a lesser extent, interior lead-based paint. Lead can also be absorbed by water with a low mineral content as it passes through lead pipes or lead-soldered copper pipes, and can be ingested from roadside dust contaminated with automobile emissions, or by habitual sniffing of leaded gasoline.

The gastrointestinal tract absorbs about 40% of ingested lead; iron deficiency and constipation encourage absorption; diarrhea has the reverse effect. Lead and calcium compete for a common transport mechanism; as a consequence, a deficient calcium intake and decreased serum levels of 25-hydroxy vitamin D_1 result in increased lead absorption. Following absorption, lead is distributed into three compartments: some enters a rapidly exchangeable pool in the liver and kidney and from there binds to a low-molecular-weight protein of erythrocytes; some enters a rapidly exchangeable pool in soft tissue and is loosely bound to bone; and the major portion is tightly bound as the insoluble and nontoxic lead triphosphate in the skeleton, notably in the epiphyseal portion of growing bone.[32] Lead is also deposited in the hair and nails, and crosses the placenta. Only a small amount of lead enters the brain; the major portion is found in gray matter and the basal ganglia. High phosphate and high vitamin D intakes favor skeletal storage of lead, whereas parathyroid hormone, low phosphate intake, and acidosis promote the release of lead into the bloodstream. Almost all the absorbed lead is excreted in urine; fecal lead represents the unabsorbed fraction. The half life of lead in blood is 1 to 2 months; it is 20 to 30 years in bone. Excretion is so slow that even a slight increase in the average daily intake above 0.3 mg/day, the maximum permissible level for adults, can ultimately induce poisoning.

Almost all organs are affected by lead poisoning, with damage caused by heavy metal inhibition of numerous sulfhydryl enzymes. One of the earliest manifestations of lead intoxication is anemia, induced by a disturbance in heme synthesis and by shortening of the red-cell life span. Lead impairs heme synthesis at several points, with globin synthesis also being disrupted at higher lead levels.

Within the central nervous system lead acts primarily on capillary endothelium; it causes extravasation of plasma followed by endothelial, microglial, and astrocytic proliferation and widespread interstitial edema.[33] Widespread neuronal degeneration occurs, with the neocortex and cerebellum being affected most prominently; neocortical degeneration can also reflect agonal anoxia rather than a direct effect of lead. Experimental work suggests that the immature cerebral vasculature is particularly sensitive to lead. Lead also interrupts the cellular calcium pump in the synaptic complexes, inhibiting acetylcholine release and enhancing the release of dopamine and norepinephrine.[34]

Neurologic symptoms of lead poisoning form a continuum from the most minor to the most severe. They include irritability, seizures, depression of consciousness, and permanent neurobehavioral impairment. From the current state of knowledge we are unable to state whether low-level lead exposure affects the performance or behavior of children. When the effects of environmental lead will ultimately be found they are likely to be small, and overshadowed by the effects that result from the child's socioeconomic status, the quality of home environment, and maternal intelligence.[35]

1.3.2 Mercury

Mercury poisoning can result from exposure to elemental mercury in the form of its vapor, inorganic mercury salts, and organic mercury, most commonly as methylmercury. All forms of intoxication are unusual.

Intoxication with organic mercury compounds derived from the fungicides or methylmercury was responsible for outbreaks of poisoning in Iraq, and for Minamata disease, an encephalopathy encountered along Minamata Bay in Japan. It resulted from the ingestion of methylmercury-contaminated fish. The neurologic picture is complex, and almost every level of the neuraxis is affected.[36] Methylmercury compounds have also been held responsible for congenital mercury poisoning. Ethylmercury intoxication is less common and less severe than intoxication incurred from methylmercury. The neuropathology of organic mercury poisoning consists of focal cerebral and cerebellar atrophy with the axonal retraction bulbs (torpedoes) on Purkinje cells resembling those seen in Menkes disease.[37]

1.3.3 Arsenic

At present, arsenic poisoning most commonly results from the accidental ingestion of pesticides, notably ant poison. The pathologic changes in the brain include generalized edema and pericapillary hemorrhages within the white matter, with foci of demyelination.[38] Nerve biopsies show axonal degeneration. Arsenic toxicity is believed to be due to the ability of the trivalent form to inactivate a number of sulfhydryl enzymes, notably the pyruvate dehydrogenase complex. In addition, the pentavalent form of arsenic uncouples mitochondrial oxidative phosphorylation.

Symptoms of arsenic poisoning are more likely to be gastrointestinal than neurologic. Neurologic symptoms are mainly those of a peripheral sensory neuritis; numbness and severe paresthesia of the distal portion of the extremities predominate.

1.3.4 Thallium

Thallium intoxication is most commonly the result of the accidental ingestion of pesticides, or the inhalation of metal dust from pyrite burners. The clinical picture is marked by a painful peripheral neuropathy, alopecia, and mental disorders.[39] How the metal damages the nervous system is still unknown. In fatal cases one sees white matter edema and degenerative changes in the nerve cells of the cerebral and cerebellar cortex, hypothalamus, the olivary nuclei, and the corpus striatum. In peripheral nerves there is a loss of myelinated fibers and degeneration of the central axons of sensory ganglion cells.[40]

1.3.5 Aluminum

The role of aluminum neurotoxicity in causing dialysis dementia has been well established, although the mechanism for this complication is unclear. The iron-binding protein transferrin is also the major aluminum-binding protein in the plasma, and through its binding to transferrin aluminum can enter the brain.[41] There is evidence that aluminum can act synergistically with iron to increase lipid peroxide formation in the brain, and to increase iron-stimulated brain peroxidation.[42] Whether these synergisms are of importance in the pathogenesis of Alzheimer disease or other neurodegenerative diseases remains to be elucidated.[43]

REFERENCES

1. Sachdev P: The neuropsychiatry of brain iron. *J. Neuropsychiatry Clin. Neurosci.*, 1993; 5:18–29.
2. Hoeck A, Demmel U, Schicha H, et al: Trace element concentration in human brain. *Brain*, 1975; 98:49–64.
3. Morris CM, Candy JM, Bloxham CA, and Edwardson JA: Immunocytochemical localisation of transferrin in the human brain. *Acta Anat. (Basel)*, 1992; 143:14–8.
4. Ben-Shachar D, Livne E, Spanier I, Zuk R, et al: Iron modulates neuroleptic-induced effects related to the dopaminergic system. *Isr. J. Med. Sci.*, 1993; 29:587–92.
5. Conrad ME: Regulation of iron absorption. In Prasad, AS (ed): *Essential and Toxic Trace Elements in Human Health and Disease: An Update*. New York, Wiley-Liss, 1993, pp 203–219.
6. Watson WS, Morrison J, Bethel MIF, et al: Food iron and lead absorption in humans. *Am. J. Clin. Nutr.*, 1986; 44:248–256.
7. Walter T, De Andraca I, Chadud P, and Perales CG: Iron deficiency anemia: Adverse effects on infant psychomotor development. *Pediatrics*, 1989; 84:7–17.
8. Lozoff B and Brittenham GM: Behavioral aspects of iron deficiency. *Prog. Hematol.*, 1985; 14:23–25.
9. Jazwinska EC, Lee SC, Webb SI, et al: Localization of the hemochromatosis gene close to D6S105. *Am. J. Hum. Genet.*, 1993; 53:347–352.
10. Casalgrandi G, Quaglino D, Gualdi R, Conte D, et al: Duodenal ferritin synthesis in genetic hemochromatosis. *Gastroenterology*, 1995; 108:208–17.
11. Schaffert DA, Johnsen SD, Johnson PC, and Drayer BP: Magnetic resonance imaging in pathologically proven Hallervorden-Spatz disease. *Neurology*, 1989; 39:440–442.
12. Jenner P: Altered mitochondrial function, iron metabolism and glutathione levels in Parkinson's disease. *Acta Neurol. Scand. Suppl.*, 1993; 146:6–13.

13. Cox DW: Genes of the copper pathway. *Am. J. Hum. Genet.*, 1995; 56:828–834.

14. Nartey NO, Frei JV, and Cherian MG: Hepatic copper and metallothionein distribution in Wilson disease (hepatolenticular degeneration). *Lab. Invest.*, 1987; 57:397–401.

15. Sethi S, Grover S, and Khodaskar MB: Role of copper in Indian childhood cirrhosis. *Ann. Trop. Paediatr.*, 1993; 13:3–5.

16. Müller T, Feichtinger H, Berger H, and Müller W: Endemic Tyrolean infantile cirrhosis: an ecogenetic disorder. *Lancet*, 1996; 347:877–880.

17. Scheinberg IH and Sternlieb I: *Wilson's Disease*. Philadelphia, W.B. Saunders, 1984.

18. Menkes JH, Alter M, Steigleder GK, et al: A sex-linked recessive disorder with growth retardation, peculiar hair, and focal cerebral and cerebellar degeneration. *Pediatrics*, 1962; 29:764–779.

19. Das S, Levinson B, Vulpe C, et al: Similar splicing mutation of the Menkes/mottled copper-transporting ATPase gene in occipital horn syndrome. *Am. J. Hum. Genet.*, 1995; 56:570–576.

20. Yoshida K, Furihata K, Takeda S, et al: A mutation in the ceruloplasmin gene is associated with systemic hemosiderosis in humans. *Nature Genet.*, 1995; 9:267–272.

21. Logan JL, Harveyson KB, Wisdom GB, et al: Hereditary caeruloplasmin deficiency, dementia and diabetes mellitus. *Q. J. Med.*, 1994; 87:663–670.

22. Guilarte TR, Miceli RC, and Jett DA: Biochemical evidence of an interaction of lead at the zinc allosteric sites of the NMDA receptor complex: effects of neuronal development. *Neurotoxicology*, 1995; 16:63–71.

23. Dreosti IE: Zinc in brain development. In Prasad, AS (ed): *Essential and Toxic Trace Elements in Human Health and Disease: An Update*. New York, Wiley-Liss, 1993, pp 81–90.

24. Sandstead HH: Requirements and toxicity of essential trace elements, illustrated by zinc and copper. *Am. J. Clin. Nutr.*, 1995; 61(Suppl. 3): 621S–624S.

25. Koh JY and Choi DW: Zinc toxicity on cultured cortical neurons: involvement of N-methyl-D-aspartate receptors. *Neuroscience*, 1994; 60:1049–57.

25a. Bush AI, Moir RD, Rosenkranz KM, and Tanzi RE: Zinc and Alzheimer's disease. *Science*, 1995; 268:1921–1923.

26. Huang C, Lu C, Chu N, et al: Progression after chronic manganese exposure. Neurology 1993; 43:1479–1483.

27. Korc M: Manganese as a modulator of signal transduction pathways. In Prasad, AS (ed): *Essential and Toxic Trace Elements in Human Health and Disease: An Update*. New York, Wiley-Liss, 1993, pp 235–255.

28. Prohaska JR: Functions of trace elements in brain metabolism. *Physiol. Rev.*, 1987; 67:858–901.

29. Chazot G and Broussolle E: Alterations in trace elements during brain aging and in Alzheimer's dementia. *Prog. Clin. Biol. Res.*, 1993; 380:269–81.

30. Cohen MD, Kargacin B, Klein CB, and Costa M: Mechanisms of chromium carcinogenicity and toxicity. *Crit. Rev. Toxicol.*, 1993; 23:255–281.

31. Endres W, Shin YS, Gunther R, et al: Report on a new patient with combined deficiency of sulphite oxidase and xanthine dehydrogenase due to molybdenum cofactor deficiency. *Eur. J. Pediatr.*, 1988; 148:246–249.

32. Raghavan SRV and Gonick HC: Isolation of a low-molecular-weight lead-binding protein from human erythrocytes. *Proc. Soc. Exp. Biol. Med.*, 1977; 155:164–167.

33. Pentschew A: Morphology and morphogenesis of lead encephalopathy. *Acta Neuropathol.*, 1965; 5:133–160.

34. Winder C, Garten LL, and Lewis PD: The morphological effects of lead on the developing central nervous system. *Neuropathol. Appl. Neurobiol.*, 1983; 9:87–108.

35. Smith M: The effects of low-level lead exposure on children. In: Smith MA, Grant LD, Sors AI, (eds): *Lead Exposure and Child Development: An International Assessment*. Boston, Kluwer Academic, 1989, pp 3–47.

36. Elhassani SB: The many faces of methyl mercury poisoning. *J. Toxicol. Clin. Toxicol.*, 1983; 19:875–906.

37. Takeuchi T, Eto N, and Eto K: Neuropathology of childhood cases of methylmercury poisoning (Minimata Disease) with prolonged symptoms, with particular reference to the decortication syndrome. *Neurotoxicology*, 1979; 1:1–20.

38. Russell DS: Changes in the central nervous system following arsphenamine medication. *J. Pathol.*, 1937; 45:357–366.

39. Herrero F, Fernandez E, Gomez J, et al: Thallium poisoning presenting with abdominal colic, paresthesia, and irritability. *J. Toxicol. Clin. Toxicol.*, 1995; 33:261–264.

40. Jacobs JM and Le Quesne PM: Toxic disorders. In Adams JH, Duchen LW (eds): *Greenfield's Neuropathology*, 5th ed., New York, Oxford University Press, 1992; pp 881–987.

41. Roskama AJ and Connor JR: Aluminum access to the brain: A role for transferrin and its receptor. *Proc. Natl. Acad. Sci. U.S.A.*, 1990; 87:9024–9027.

42. Ohtawa MM, Seko M, and Takayama F: Effect of aluminum ingestion on lipid peroxidation in rats. *Chem. Pharm. Bull.*, 1983; 31:1415–1418.

43. Gutteridge JMC: Hydroxyl radicals, iron, oxidative stress, and neurodegeneration. *Ann. N.Y. Acad. Sci.*, 1994; 738:201–213.

SECTION 2

ENVIRONMENTAL ASPECTS

Chapter 2

Neurotoxic Metals in the Environment: Some General Aspects

Trond Peder Flaten

CONTENTS

2.1 INTRODUCTION

There is widespread fear in our society of toxic substances lurking in the environment — substances that may be present in low concentrations or may even be unknown to science, but still capable of producing disease after years of exposure. This public fear is generally related to problems of pollution and has focused primarily on carcinogens, but potentially neurotoxic compounds will probably receive increasing attention in the years to come. People expressing such fears are often also notoriously skeptical towards scientific "authorities". This represents an extra challenge for scientists to precisely quantify the risks involved and to communicate their results to the general public. The purpose of this chapter is to discuss some general aspects of inorganic neurotoxicants in water, soil, and outdoor air, and how these affect humans. Individual neurotoxic elements are dealt with in depth in other chapters of this book and therefore specific elements, where discussed, are used to illustrate general principles.

The environment contains a large number of potentially neurotoxic substances, most of which are organic compounds that will not be discussed here. A fundamental difference between metals and organic neurotoxins is that while organic compounds can be "destroyed" in the environment in the sense that their chemical structure can be broken down, metals are indestructible. However, metals can and do react with different ligands, which may greatly alter their environmental behavior and toxicological properties. The best known example is methylation of mercury, which increases its neurotoxicity and also induces bioamplification in food chains (see Section 2.6 below).

2.1.1 The Occupational Environment

The recognition that neurological disorders could be due to environmental exposure to metals was first made in occupational settings. Occupational toxicity to lead was described by Hippocrates in Greece around the year 370 B.C.[1] Mercury neurotoxicity in the working environment has also been known for a long time, notably the "mad hatter syndrome" resulting from chronic exposure to inorganic mercury compounds and "erethism" and "micromercurialism" from elemental mercury vapor in mining and in the chloralkali industry.[2,3] In 1837, just 20 years after James Parkinson first described the disease that now bears his name, a parkinsonian syndrome was described among Scottish factory workers handling powdered manganese dioxide.[4] Although exposure to harmful levels of neurotoxic metals certainly still occurs in the workplace, the number of people with overt neurological damage due to extreme occupational exposure is limited today. Therefore, the rest of this chapter will focus on the nonoccupational environment.

2.2 MINAMATA AND ACUTE VS. CHRONIC EXPOSURE

Outside occupational settings, the best-known large-scale occurrences of neurological disease caused by environmental exposure to metals are the methylmercury epidemics in the fishery village of Minamata in Japan[5] and in rural Iraq.[6] The Minamata incident, especially, clearly demonstrated that metals in the environment can cause neurologic injury, but even here, despite high exposure levels and a rapidly growing number of victims with clear-cut disease, it took 12 years from the time the first case was reported in 1953 to the first actions taken to reduce methylmercury discharges from the Chisso chemical plant to Minamata Bay.[5] In the bay, the methylmercury had bioaccumulated (see below) in the food chain to very high levels in fish and shellfish, and the incidence of acute and subacute Minamata disease started decreasing around 1960 when people largely ceased eating the local seafood. Several years later, investigators began to realize that in addition to the acute and subacute cases, new cases of a clinically different syndrome termed "chronic Minamata disease" continued to turn up, probably as a result of long-term consumption of seafood with methylmercury levels well below those necessary to cause the acute form of the disease.[5] Because seafood is by far the most important dietary source of methylmercury, several fish-eating populations around the world have been studied for possible neurological effects, and their blood levels of methylmercury sometimes reach levels associated with symptoms of poisoning in Minamata and Iraq.[7]

One of the many important lessons to be learned from the Minamata tragedy is the importance of distinguishing between acute and chronic exposure situations. Acute neurotoxicity due to environmental metal exposure involves relatively few people, but chronic exposure to the low levels normally found in the environment represents a potentially much greater threat. In addition, there is a growing consensus that many of the neurodegenerative disorders that plague humankind in old age, notably Alzheimer's disease, Parkinson's disease, and amyotrophic lateral sclerosis, are multifactorial.[8] Thus, it is possible that some fractions of these diseases may reflect long-term, low-level exposure to environmental neurotoxicants, analogous to the situation in cancer and heart disease.

2.3 THE CASE OF LEAD

Lead constitutes the best-known and best-documented case of neurotoxicity due to long-term, low-level environmental exposure to metals. The last few decades have witnessed a gradual and severalfold reduction in the threshold level for lead in blood below which no adverse effects are anticipated.[9] The current level of blood lead warranting concern is 10 µg/dl, but recent studies have indicated subtle cognitive and behavioral deficits in newborn and very young infants in the 10 to 20 µg/dl range and possibly even below 10 µg/dl.[10] Children are more susceptible to lead neurotoxicity because of their more effective gastrointestinal absorption and higher lead intake per kilogram of body weight, and the developing fetus may suffer an extra risk due to an immature blood-brain barrier. Lead is a major cause of mental subnormality in developed industrial countries; in the U.S. alone, it has been estimated that more than 3 million children are exposed to environmental sources of lead that place them at risk to lead toxicity.[11]

Flegal and Smith[12] have estimated, from extrapolation of the linear relation between bone lead and blood lead concentrations in contemporary humans to the bone lead levels of preindustrial era humans, that the natural blood lead concentration in humans is about 0.016 µg/dl. This is more than two orders of magnitude lower than current "background" levels, which are close to the current 10 µg Pb/dl "level of concern" in most industrial countries. This "level of concern" is only ten times lower than levels associated with overt encephalopathy and death, so the gap between what is typical and what is neurotoxic is uncomfortably narrow for lead.

There are several toxicologically relevant sources of lead: notably dust, soil, paint chips, food, and drinking water. Thus, the case of lead clearly illustrates the necessity in public health programs to pinpoint the relative importance of different exposure sources and to keep an integrative approach to preventive measures. For example, the major exposure source for inner-city children in the U.S. is lead paint chips flaking off the surface of walls, furniture, etc. This source is far less important in most European countries, where leaded paint for interior uses has been banned or heavily regulated since the 1920s and 1930s.[13] Where corrosive water is supplied through lead-containing pipes and plumbing, as is particularly common in Scotland,[14] drinking water may be the dominant source of lead.[15] In other urban areas, lead in air from vehicle exhausts has been the dominant lead source, and a gradual reduction of the content of lead in gasoline has been closely followed by striking reductions in population levels of lead in blood.[16,17] Thus, the introduction of lead-free gasoline represents a success story in preventive environmental medicine. Likewise, raising drinking water pH has resulted in clinically important reductions of blood lead among postpartal women in Scotland.[14] Regarding lead in food, the removal of lead from soldered cans may also have contributed to the observed decline in population lead levels.[17] Globally, the most important (but also by far the most expensive) current preventive measure is to reduce the exposure from lead-based paints.

Lead, like mercury and cadmium, tends to accumulate in the body over time. Note that although cadmium is neurotoxic, it is far less relevant as an environmental neurotoxicant than lead or mercury. This is partly because the critical effect of cadmium, that is, the first biological effect observed when the dose is gradually increased, is on the kidneys and not on the nervous system. For elements that accumulate in the body and for which the mean population intake approaches the health effect threshold, it is important to realize that every increase in exposure may be of clinical relevance. For example, Figure 2.1 shows that a 20% increase in the intake of every individual in a population can lead to a doubling in the number of symptomatic persons. Needleman[18] makes a related point, using data from epidemiological studies to show that a decrease in median IQ score of only 6 points is associated with a 4-fold increase in the rate of severely impaired children (IQ below 80), and prevents 5% of children from achieving superior function (IQ above 125). Thus, a source of toxicant or an IQ effect that might seem trivial for the individual, might be important from a public health point of view.

2.4 THE EXTENT OF THE PROBLEM

The degree of exposure to environmental neurotoxicants is not well known. Even for lead and mercury, where there is no doubt that neurotoxicity due to environmental exposure is a large problem, too few data exist to accurately quantify the problem. For other potentially neurotoxic metals in the environment, like manganese, aluminum, and zinc, the extent of the problem can only be speculated upon. The largest uncertainties involved in quantifying the problem may be of a diagnostic nature. The changes related to chronic, low-level exposure would be expected to be slowly progressive, subtle, and nonspecific, even subclinical, and represent formidable challenges to toxicologists and epidemiologists. Subtle cognitive and behavioral changes are very difficult to evaluate, especially in children, and such changes will generally not be specific to a particular

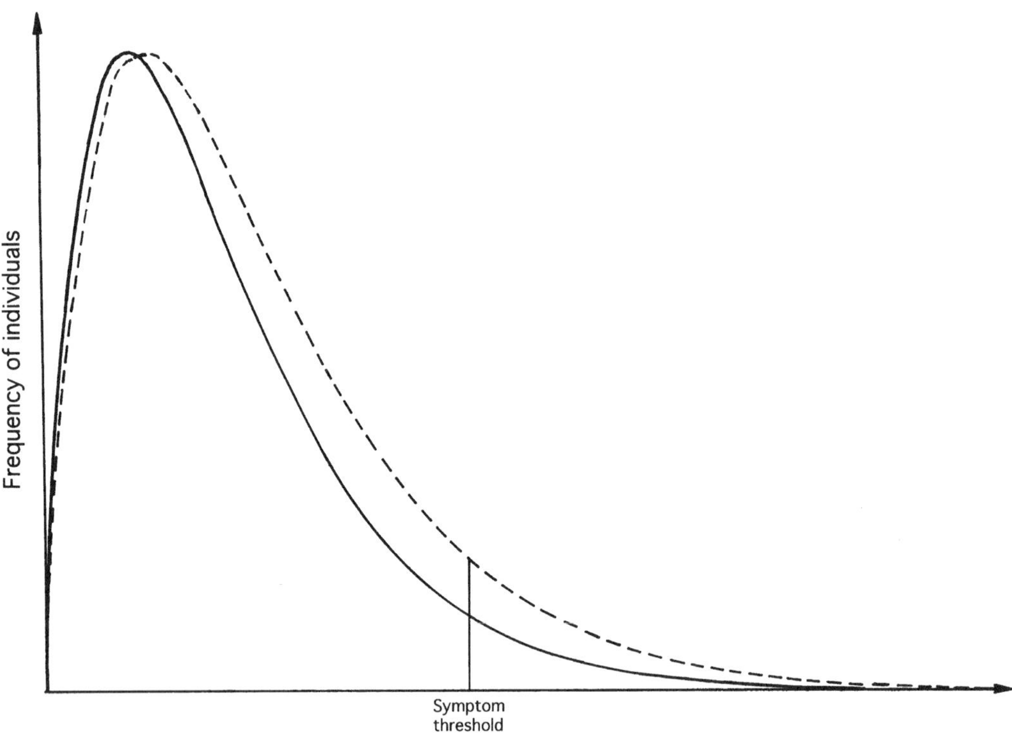

FIGURE 2.1 Hypothetical frequency distribution in a population of the intake of a toxic element or the concentration of this element in a body compartment (solid line). A skewed distribution, as this one, is typical for elemental intake or for concentrations of toxic, nonessential elements in tissues, e.g., for lead in blood; 5% of the individuals in this population have intakes high enough to cause a defined symptom. If every individual was to increase intake by 20% (dashed line), this would result in an approximate doubling in the number of people showing symptoms (the area under the curve). Thus, a small increase in the average intake may result in a substantial increase in the number of people that cross the symptom threshold.

neurotoxicant. In addition, the latent time until a disease is diagnosable is likely to be long in many cases, which makes it notoriously difficult to define the exposure retrospectively, as is necessary in epidemiological case-control studies. Finally, it is worth noting that a recent report on environmental neurotoxicity by the U.S. National Research Council stated that "…it would not be surprising if the direct and indirect societal costs of subclinical neurological losses or deficits — such as reduction in intelligence, diminution in achievement, and waste of opportunity — might equal or exceed those of neurotoxic effects that are clinically recognized."[19]

2.5. ELEMENT SPECIATION AND BIOAVAILABILITY

The physicochemical forms of an element, that is, the actual species found in the exposure media and in the different body compartments, may be instrumental in determining bioavailability and toxicity. For a particular element, the most important variables governing speciation are (1) particulate vs. dissolved state in aqueous solution, (2) variations in oxidation state, and (3) the ligand(s) to which the element is chemically bound. Aluminum and mercury serve to illustrate some general relationships between the fundamental chemical properties of different elements and their respective biological effects. Notably, aluminum and mercury are on the extreme ends of the scale from class A metals which largely bind to oxygen-containing ligands, to class B metals which seek sulfur- and nitrogen-containing ligands.[20] Mercury, being the archetypal class B metal, binds strongly to the ubiquitous sulfhydryl groups on proteins and enzymes, while sulfhydryl groups are totally irrelevant as targets for aluminum toxicity. Further, while variation in oxidation state is a major determinant of the biological properties of mercury (0 in metallic, gaseous mercury, +1 or +2 in inorganic salts, and +2 in methylmercury), +3 is the only biologically relevant oxidation state for aluminum. For aluminum, on the other hand, many different potential ligands are available in the environment and in biological systems, and sorting out which ligands are relevant under different conditions has proved to be a very difficult task.

For mercury, environmental pathways, exposure routes, absorption, metabolism, and toxicological properties vary widely with speciation.[2,3] Mercury is dissipated in the environment largely by transport through the atmosphere, where it exists almost exclusively in the elemental form (Hg^0), as a gas or adsorbed to the surface of small particles. This elemental mercury may then be oxidized to the more water-soluble divalent form (Hg^{2+}) dissolved in raindrops and deposited from the atmosphere. In soil and water, Hg^{2+} can be (1) reduced to gaseous Hg^0, thereby returning mercury to the atmosphere, or (2) methylated by microorganisms. The resultant methylmercury unit CH_3Hg^+ is kinetically remarkably inert towards decomposition, so it may persist in the environment or in an organism even though thermodynamic conditions strongly favor demethylation. The electrically neutral methylmercury species, CH_3HgCl, $(CH_3)_2Hg$, etc. are quite lipophilic, and can thus readily pass biological membranes and be concentrated along ecological food chains — a process known as bioamplification. Any substance that is more soluble in the tissues of simple organisms than in the surrounding water will be a more concentrated component of the diet of the more complex species which feed on the simpler ones. Methylmercury has a very strong tendency for bioamplification, which is why the fish in Minamata Bay could attain such toxic levels. Aluminum (like most other metals), on the other hand, is not bioamplified. Rather, concentrations in plants generally seem to be higher than in the animals that eat them, probably because aluminum is very poorly absorbed.

Because metallic mercury is volatile at ambient temperatures, most human exposure is by inhalation, and the lipid-soluble gaseous mercury is readily absorbed in the lung. In the gut, however, the absorption of metallic mercury is negligible (about 0.01%), while the absorption of inorganic mercury species is about 10%, and as much as 90 to 95% of ingested methylmercury is absorbed.[3] Inside the human body, the fate of the different mercury species is also very different due to their different chemical properties.[2,3]

The bioavailability of an element depends not only on the ingested species of the element, but also on other simultaneously ingested compounds. For aluminum, it is well documented that concomitant intake of low-molecular-weight organic ligands like the common dietary constituents citrate or maltolate greatly enhances intestinal absorption. For example, orange juice, which contains citric acid, increases severalfold the absorption of aluminum from an aluminum-containing antacid.[21] Therefore, citrate increases the potential for aluminum neurotoxicity.[22] On the other hand, citrate complexation clearly decreases aluminum toxicity in plants[23] and in fish,[24] probably because the mechanism of toxicity involves binding of aluminum to exterior cell membranes.[25,26] Thus, in fish and plants it is quite possible that organic ligands reduce the acute toxicity of aluminum while at the same time increasing its uptake.

Finally, it is important to realize that the speciation of an element may change several times from its source through cycling in the environment and on its way through different compartments of the organism to the final target site. For example, acidification of the environment leads to mobilization of aluminum from otherwise very poorly soluble minerals like feldspars and kaolinite in soils and bedrock. In the subsequent transport of dissolved aluminum through soil water, groundwater, rivers, and lakes, aluminum may change ligands several times depending on the water chemistry. Relevant ligands in natural waters include OH^-, F^-, H_4SiO_4, HSO_4^- and a host of naturally occurring organic compounds. It is conceivable that some natural waters may contain small amounts of organic aluminum complexes with properties similar to aluminum maltolate: stable enough to pass through the stomach and accumulate in brain and bone.[27] After uptake from the gastrointestinal tract, aluminum probably largely follows the iron pathway in being bound to transferrin in blood extracellularly, while the slightly more acid intracellular environment favors binding to phosphate-containing ligands like inositol or membrane phosphate groups.[28]

2.6 ELEMENT-ELEMENT INTERACTIONS

The subject of interactions between different elements is closely related to the previous one of speciation and bioavailability. The basis of such interactions often is similarity in chemical structure, for example when elements interact by competing for a common receptor. Thus, nontoxic elements may indirectly be relevant to neurotoxicity.[29] For example, lead and cadmium absorption is reduced by calcium supplementation, and lead may "displace" zinc, iron, and copper, inducing "deficiency syndromes" of these essential elements.[29] A related mechanism has been termed "molecular mimicry", meaning that elements may form compounds with chemical structure similar to essential metabolites, like when the arsenate oxyanion replaces phosphate in the process of ATP synthesis.[29]

The hyperendemic foci of amyotrophic lateral sclerosis and parkinsonism-dementia in the Western Pacific have been intensively studied for almost 50 years in search of etiological factors.[30] These massive research efforts have not resulted in a definitive answer, but the most plausible hypothesis is that a basic, perhaps genetic, defect in mineral metabolism induces a form of secondary hyperparathyroidism, provoked by chronic nutritional deficiencies of calcium and magnesium. This leads to enhanced gastrointestinal absorption of aluminum and the observed codeposition of very high levels of aluminum, calcium, and silicon as aluminosilicates or hydroxyapatites in neurones.[30]

While calcium protects against lead toxicity by reducing gastrointestinal absorption and tissue accumulation of lead, selenium counteracts mercury toxicity without reducing tissue concentrations of mercury. This is probably related to the very strong binding of selenium to mercury and the resulting diversion of mercury from a small protein to a much larger selenium-containing protein.[29] The selenium-mercury interaction has led to innovative approaches in environmental remediation and protection: in Sweden, adding selenium to lakes has resulted in severalfold reductions of the mercury levels in fish,[31] and ampules with selenium placed on top of coffins during cremations have reduced the emission of mercury to the air from the amalgam fillings of the corpses.[32]

In ruminants, a high molybdenum intake may induce copper deficiency.[33] With this in mind, molybdenum was tried as a therapeutic agent in pilot studies in Wilson disease,[34,35] which results from an inborn error in copper metabolism (see Chapter 41). The lack of success in these initial trials was probably related to metabolic differences between ruminants and humans: in the rumen, molybdenum is converted to thiomolybdate derivatives, a conversion that does not take place in humans. Taking this difference into account, Brewer and co-workers[36] were able to control the acute neurological symptoms of Wilson disease safely and effectively by oral administration of preformed thiomolybdate.

In general, the importance of trace elements is better recognized in veterinary than in human medicine. Analogous to the copper-molybdenum antagonism in ruminants inspiring trials of molybdenum therapy in Wilson disease, I believe that there are probably other such natural models of neurotoxicants in animals, representing an important untapped resource for knowledge relevant to human neurological diseases. Perhaps one fruitful strategy is to systematically search among the known trace element-health relationships in animals for ideas about analogous relationships that might be of importance in humans.

2.7 ESSENTIALITY-TOXICITY DUALISM AND THE "TOTAL DOSE-RESPONSE CURVE"

One of the oldest dogmas in toxicology is that "the dose makes the poison", which is often visualized in the so-called "total dose-response curve" (Figure 2.2). From this point of view, a rigorous classification of elements as either "essential" or "toxic" is rather arbitrary, because if the dose is high enough all essential elements are toxic. Conversely, even toxic elements like arsenic and lead may in fact be essential, but if so, practically all individuals have a more than adequate intake of these elements. In any case, even for the most toxic elements there is an exposure threshold below which toxicity does not occur. For zinc, iron, copper, and perhaps manganese, which are all potential neurotoxicants, low-level exposure may actually be beneficial because substantial parts of the population may have a suboptimal intake. Finally, it should be remembered that although the focus of this chapter is on toxicity, trace element deficiency may also cause neurological disease: cretinism and other more subtle neurological deficits resulting from iodine deficiency are clearly a major public health problem worldwide.[37]

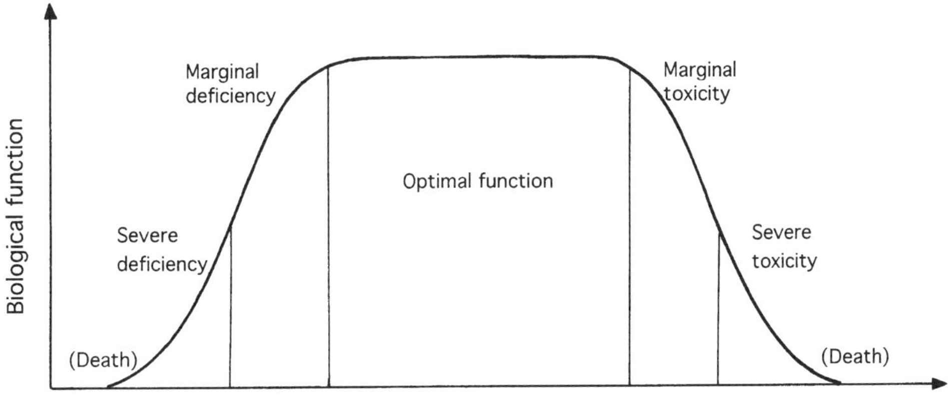

FIGURE 2.2 Total biological dose-response curve for essential trace elements. The vertical dividing lines may be at different intakes for different individuals depending on genetic variables, age, sex, pregnancy or lactation, and intake of interfering dietary factors.

2.8 RESEARCH NEEDS AND FUTURE PRIORITIES

The existing risk assessment techniques, which have been developed largely in response to worries about low-level exposure to carcinogens, need to be tuned towards environmental neurotoxicants.[19,38] To be able to judge such risks quantitatively, it is instrumental to know the cellular and molecular mechanisms of neurotoxicity; this is perhaps the most pressing field of future research needs. Such mechanistic knowledge would also help develop new sensitive biological markers for neurotoxicants and neurologic disease.[19] Biological markers are important to detect potentially harmful exposure to neurotoxicants preclinically, before irreversible damage occurs. Biomarkers are also valuable as indicators of internal dose in epidemiological studies.

Well-designed epidemiological studies are of utmost importance to define possible risks. Because of the long latent periods involved, especially in the chronic neurodegenerative diseases of the aged, it is often difficult to define exposure retrospectively in case-control studies. Longitudinal studies need to be very large to study rare complications of low-dose exposure. Therefore, in special cases when defined groups of people have had elevated exposures to environmental neurotoxicants, they should be followed for a long time in prospective epidemiological studies, together with suitable control groups. Aspects of epidemiological design potentially useful in studies of environmental neurotoxicants include identification of high-risk individuals based on current etiological hypotheses: twin, family, and household studies; and identification of high-incidence foci, perhaps in non-Western anthropological populations where confounding factors are less likely to occur compared with large, genetically diverse cosmopolitan societies.[30] It is important to recognize that knowledge of toxicity mechanisms will often be helpful in improving the design of epidemiological studies. Finally, our knowledge of the prevalence and incidence of neurological diseases in defined geographic areas is limited, making general epidemiologic surveillance a prime need in the evaluation of population risk from environmental neurotoxicants.

In dealing with health effects from low-level exposure to neurotoxicants, the most important public health strategies are general environmental contamination control and exposure prevention, rather than traditional medical intervention strategies which focus on individual poisoning cases. Further, with special reference to lead, one should adopt the view that health is not merely the absence of overt disease, but the presence of maximally achievable function. Finally, it is important to learn to communicate the risk of potentially harmful environmental neurotoxicants to the public without creating unnecessary worry and panic.

REFERENCES

1. Nriagu JO: *Lead and Lead Poisoning in Antiquity.* New York, John Wiley & Sons, 1983.
2. Berlin M: Mercury. In Friberg L, Nordberg GF, Vouk V (Eds): *Handbook on the Toxicology of Metals,* Volume 2, 2nd edn. Amsterdam, Elsevier, 1986, pp 387–445.
3. Goyer RA: Toxic effects of metals. In Klaassen CD, Amdur MO, Doull J (Eds): *Casarett and Doull's Toxicology: the Basic Science of Poisons,* 5th edn. New York, McGraw-Hill, 1996, pp 691–736.
4. Couper J: On the effects of black oxide of manganese when inhaled into the lungs. *Br. Ann. Med. Pharmacol.,* 1837; 1: 41–42.
5. Harada M: Minamata disease: methylmercury poisoning in Japan caused by environmental pollution. *Crit. Rev. Toxicol.,* 1995; 25: 1–24.
6. Bakir F, Damluji SF, Amin-Zaki L, et al: Methylmercury poisoning in Iraq: an interuniversity report. *Science,* 1973; 181: 230–241.
7. World Health Organization: Environmental Health Criteria 101. Methylmercury. Geneva, WHO, 1990.
8. Calne DB, Eisen A, McGeer E, and Spencer P: Alzheimer's disease, Parkinson's disease, and motoneurone disease: abiotropic interaction between ageing and environment? *Lancet,* 1986; 2: 1067–1070.
9. Mushak P: Defining lead as the premiere environmental health issue for children in America: criteria and their quantitative application. *Environ. Res.,* 1992; 59: 281–309.
10. Schwartz J: Low-level lead exposure and children's IQ: a meta-analysis and search for a threshold. *Environ. Res.,* 1994; 65: 42–55.
11. Agency for Toxic Substances and Disease Registry: The Nature and Extent of Lead Poisoning in Children in the United States: A Report to Congress. Atlanta, GA, U.S. Department of Health and Human Services, 1988.
12. Flegal AR and Smith DR: Lead levels in preindustrial humans. *N. Engl. J. Med.,* 1992; 326: 1293–1294.
13. Rabin R: Warnings unheeded: a history of child lead poisoning. *Am. J. Public Health* 1989; 79: 1668–1674.
14. Moore MR, Richards WN, and Sherlock JG: Successful abatement of lead exposure from water supplies in the west of Scotland. *Environ. Res.,* 1985; 38: 67–76.
15. Elwood PC, Gallacher JEJ, Phillips KM, et al: Greater contribution to blood lead from water than from air. *Nature,* 1984; 310: 138–140.
16. Ducoffre G, Claeys F, and Bruaux P: Lowering time trend of blood lead levels in Belgium since 1978. *Environ. Res.,* 1990; 51: 25–34.
17. Pirkle JL, Brody DJ, Gunter EW, et al: The decline in blood lead levels in the United States: the National Health and Nutrition Examination Surveys (NHANES). *J. Am. Med. Assoc.,* 1994; 272: 284–291.
18. Needleman HL: The future challenge of lead toxicity. *Environ. Health Perspect.,* 1990; 89: 85–89.
19. National Research Council Committee on Neurotoxicology and Models for Assessing Risk: Environmental Neurotoxicology. Washington, D.C., National Academy Press, 1992.
20. Nieboer E and Richardson DHS: The replacement of the nondescript term 'heavy metals' by a biologically and chemically significant classification of metal ions. *Environ. Pollut. [B],* 1980; 1: 3–26.
21. Weberg R and Berstad A: Gastrointestinal absorption of aluminum from single doses of aluminum containing antacids in man. *Eur. J. Clin. Invest.,* 1986; 16: 428–432.
22. Molitoris BA, Froment DH, Mackenzie TA, et al: Citrate: a major factor in the toxicity of orally administered aluminum compounds. *Kidney Int.,* 1989; 36: 949–953.
23. Ownby JD and Popham HR: Citrate reverses the inhibition of wheat root growth caused by aluminum. *J. Plant Physiol.,* 1989; 135: 588–591.
24. Lacroix GL, Peterson RH, Belfry CS, and Martin-Robichaud DJ: Aluminum dynamics on gills of Atlantic salmon fry in the presence of citrate and effects on integrity of gill structures. *Aquat. Toxicol.,* 1993; 27: 373–402.
25. Exley C, Chappell JS, and Birchall JD: A mechanism for acute aluminium toxicity in fish. *J. Theor. Biol.,* 1991; 151: 417–428.
26. Flaten TP and Garruto RM: Polynuclear ions in aluminum toxicity. *J. Theor. Biol.,* 1992; 156: 129–132.
27. van Ginkel MF, van der Voet GB, D'Haese PC, et al: Effect of citric acid and maltol on the accumulation of aluminum in rat brain and bone. *J. Lab. Clin. Med.,* 1993; 121: 453–460.

28. Birchall JD and Chappell JS: Aluminium, chemical physiology, and Alzheimer's disease. *Lancet*, 1988; 2: 1008–1010.

29. Chang LW: The concept of direct and indirect neurotoxicity and the concept of toxic metal/essential element interactions as a common biomechanism underlying metal toxicity. In Isaacson RL, Jensen KF (Eds): *The Vulnerable Brain and Environmental Risks*, Vol. 2: Toxins in Food. New York, Plenum Press, 1992, pp 61–82.

30. Garruto RM: Pacific paradigms of environmentally-induced neurological disorders: clinical, epidemiological and molecular perspectives. *Neurotoxicology*, 1991; 12: 347–377.

31. Paulsson K and Lundbergh K: The selenium method for treatment of lakes for elevated levels of mercury in fish. *Sci. Total Environ.*, 1989; 87/88: 495–507.

32. Hogland WKH: Usefulness of selenium for the reduction of mercury emission from crematoria. *J. Environ. Qual.*, 1994; 23: 1364–1366.

33. Mills CF and Davis GK: Molybdenum. In Mertz W (Ed): *Trace Elements in Human and Animal Nutrition*, Volume 1, 5th edn. San Diego, Academic Press, 1987, pp 429–463.

34. Bickel H, Neale FC, and Hall G: A clinical and biochemical study of hepatolenticular degeneration (Wilson disease). *Q. J. Med.*, 1957; 26: 527–558.

35. Walshe JM: Tetrathiomolybdate (MoS_4) as an 'anti-copper' agent in man. In Scheinberg IH, Walshe JM (Eds): *Orphan Diseases and Orphan Drugs*. Manchester, Manchester University Press, 1986, pp 76–85.

36. Brewer GJ, Dick RD, Johnson V, et al: Treatment of Wilson disease with ammonium tetrathiomolybdate. I. Initial therapy in 17 neurologically affected patients. *Arch. Neurol.*, 1994; 51: 545–554.

37. Hetzel BS and Dunn JT: The iodine deficiency disorders: their nature and prevention. *Annu. Rev. Nutr.*, 1989; 9: 21–38.

38. Rees DC and Glowa JR: Extrapolation to humans for neurotoxicants: issues and challenges. In Isaacson RL, Jensen KF (Eds): *The Vulnerable Brain and Environmental Risks*, Volume 3: Toxins in Air and Water. New York, Plenum Press, 1994, pp 207–230.

Chapter 3

Effects of Acid Rain: A Natural Model of Aluminum Toxicity in Fish

Ralph M. Garruto, Trond Peder Flaten, and Ikuro Wakayama

CONTENTS

3.1 INTRODUCTION

The search for a fundamental understanding of causation and pathogenic mechanisms of disease has many approaches, often with serendipity, opportunism, creativity, and research design playing a significant part in scientific discovery. In today's world of biological reductionism and tunnel-vision approaches to find "the root cause" of a disease of unknown etiology, we fail to frame our research within a broader cross-disciplinary context that often allows for a much better understanding of both causation and process. Such is the case in the study of neurodegenerative disorders and emerging neurotoxicants.

Aluminum as a potent neurotoxicant dates back 100 years to Siem's and Döllken's studies of aluminum neurotoxicity in experimental animals.[1] Yet the study of the effects of aluminum and aluminum compounds on biological systems has only come to the research forefront during the last half-dozen years. A retrieval of relevant references from the literature database *Toxline* lists some 4000 articles since 1990. Acceptance that aluminum can be toxic to biological systems varies widely between different scientific disciplines: from agriculture, where it has been accepted at least since 1918,[2] to neurology, where the metal is still perceived as a "red herring" and often unworthy of further biomedical study.

Aluminum toxicity in plants is considered to be a global problem (see Chapter 4). It has been estimated that 40% of arable soils of the world and up to 70% of potentially usable land for food and/or biomass production are acidic and consequently subject to aluminum toxicity.[3] In acid soils aluminum interferes with cell division in plant roots, decreases plant rooting depths and use of subsoil nutrients, and increases susceptibility of plants to drought. The molecular basis for aluminum toxicity in plants is not known in detail. In some plants, aluminum toxicity appears as an induced calcium deficiency or as reduced calcium transport problems.[4] In addition, biochemical, cellular, and/or molecular mechanisms of aluminum toxicity vary widely between species, due to differences in genetic susceptibility or by the ability of plants to alter their microenvironment. In some plant species, aluminum may even be beneficial by increasing plant growth, possibly indirectly by increasing iron solubility and availability.[4]

In certain regions of the world, forests have been in dramatic decline and among the trees most affected are the evergreens, particularly spruce and fir. In addition to the direct toxic effects of aluminum on plant root systems, aluminum may interact in yet another way, by inhibiting calcium uptake and causing a decline in cambial growth.[5] This may be induced by the preferential binding of aluminum to the calcium regulatory protein calmodulin,[6] although it has been argued on a thermodynamical basis that calmodulin cannot bind very strongly to aluminum.[7] In any case, this process is particularly damaging in older trees that have an increased demand for calcium; they lose sapwood and become increasingly vulnerable to insects and various pathogenic microorganisms.

Such natural environmental models may provide an important foundation and frame of reference for understanding issues of toxicity in more complex, mobile organisms where environmental conditions and exposure rates are more difficult to evaluate. The more sedentary (or geographically restricted) an organism, the better we are able to determine its life history events, including the kind and degree of exposure to toxic elements. This is as true for natural paradigms in humans as it is for flora and other fauna. For example, exposure to chemical toxicants in lake or river water is often well known for a nonmigratory aquatic species living in an intensely monitored lake, as is the case for the fish studied in the present work.

3.2 NATURAL PARADIGMS: A GEOGRAPHICAL AND ECOSYSTEM APPROACH

For more than three decades our laboratory has been involved in the study of natural models of neurodegenerative disease that occur among geographically isolated and genetically and culturally homogeneous, small, technologically simple populations, often living in close association with flora and fauna. Among them are toxic and deficiency diseases, such as hyperendemic goiter and cretinism with an associated spectrum of deafness, mutism, and mental subnormality as a result of simple iodine deficiency, and aluminum-induced Pacific amyotrophic lateral sclerosis (ALS) and parkinsonism-dementia (PD) among three genetically, culturally, and geographically distinct populations: the Chamorros of the Mariana Islands, the Auyu and Jakai people of West New Guinea, and the Japanese from the Kii peninsula of Honshu Island.[8,9]

In an attempt to search for new models in nature that mimic the natural environmental conditions we see in these Pacific foci, we recently began a research effort to understand how aluminum alters the nervous system under conditions of chronic environmental exposure. We are studying a number of species of fish from lakes affected by acid precipitation in the Adirondack Mountains in the eastern U.S., comparing them to fish from a pH-neutral lake in the same area. Such models are attempts to understand not only how aluminum alters the nervous system, but also attempts to understand species-specific thresholds of aluminum toxicity under different water chemistry conditions, to identify the bioavailable forms of aluminum, be they mononuclear or polynuclear[10] (Figure 3.1), and to further understand the cellular and molecular mechanisms of aluminum intoxication. Important contributions towards understanding these mechanisms have already come from impressive studies of both domestic and wild species of plants.

3.3 FISH: A NATURAL MODEL OF CHRONIC ALUMINUM INTOXICATION

In the animal world, it is now well established that aluminum is the main toxicant leading to fish extinction in areas where acid rain has leached the metal from the watershed soil.[11-14] Biological organisms in such watersheds are vulnerable mainly because of a lack of easily soluble minerals and other buffering agents that are able to neutralize the acid precipitation. The main target organ in mature fish is the gill, where aluminum interferes with iono- and osmoregulation and also leads to gill irritation, sometimes with associated thickening of the mucous layer, respiratory dysfunction,

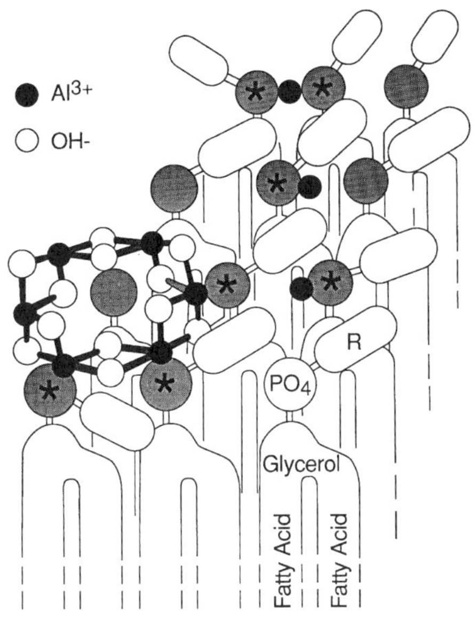

FIGURE 3.1 Hypothetical binding mechanisms of mononuclear aluminum (exemplified by Al³⁺) and polynuclear aluminum [exemplified by $Al_6(OH)_{12}^{6+}$] to phosphate groups (shaded) on a phospholipid cell membrane. R denotes the side-chain on the polar head group. The negatively charged phosphate groups form strong bonds with A-type cations; aluminum forms particularly strong bonds due to its high charge and low ionic radius. The large polynuclear aluminum ion may form strong bonds with several negatively charged membrane phosphate groups (phosphate groups bound to aluminum are indicated by asterisks), thereby increasing membrane rigidity more effectively than mononuclear aluminum, which may bind to more than one phosphate group but only if they are close together (top). The membrane changes induced by aluminum binding may increase aluminum entry into cells where it can exert secondary toxic effects.[14] (Redrawn from Flaten, T.P. and Garruto, R.M., *J. Theor. Biol.*, 1992; 156, 129-132.)

and ultimately death. The toxicity is mitigated when the concentration of calcium and, to a lesser degree sodium, in natural waters is increased, probably due to nonspecific competition by calcium for aluminum binding sites on the gills. Aluminum complexing agents, notably fluoride, silicic acid, humic molecules, and citrate also ameliorate aluminum toxicity, probably by preventing aluminum binding to the gill membrane.[14] Previously, the nervous system of fish had not been examined for aluminum-induced neuropathological changes. In acid lakes, not only fish gills but also primary receptor neurons in the olfactory neuropathway are in direct and continuous contact with water containing up to several hundred micrograms of soluble aluminum per liter, levels that when found in the blood of humans with chronic renal failure rapidly cause dialysis encephalopathy.

During several field trips to the western Adirondack Mountains, with the assistance of wildlife biologists in the New York State Department of Environmental Conservation and the Adirondack Lakes Survey Corporation, we selected a series of lakes in a watershed affected by acid precipitation where lake water chemistries have been intensely monitored over the past decade (Table 3.1). A simple open-air field laboratory was set up from portable components at each of the lake sites and various species of fish were caught live using gill nets. The nets were set out, monitored, and retrieved every hour in a labor intensive way using one or two separate boat crews.

Seven species of live fish in good condition were deeply anesthetized with 100 mg/l of tricaine methanesulfonate (MS 222) dissolved in a container of local lake water. Once anesthetized, fish were perfusion-fixed using a mixture of 4% paraformaldehyde and 0.5% glutaraldehyde in 0.1 M phosphate buffer after administration of sodium heparin in physiological saline through the ventricle of the heart. Dissections were performed 1 to 2 h postperfusion and the brain and spinal cord along with other tissues were removed and immersed in 10% buffered formalin or 4% paraformaldehyde for 7 days before final processing for light and electron microscopy. Also, fresh tissue from other fish of the same species were dissected, embedded in O.C.T. compound, and quick-frozen on dry ice for aluminum staining. The age of each fish was determined from its scales. The research protocol has been approved by the NINDS Animal Care and Use Committee at the National Institutes of Health.

Our results suggest that the central nervous systems of fish from lakes affected by acid precipitation have some neuropathological changes consistent with those found in experimental aluminum intoxication and in Pacific ALS.[15] Specifically, the neuropathological lesions seen include chromatolysis, neurons with displaced nuclei, perikaryal and neuritic inclusions, and most interesting plaque-like lesions that were strongly immunoreactive to antibodies (SMI 31 and 34) against

Table 3.1   Water Chemistry of the Four Lakes Studied in the Western Adirondack Mountain Region of New York State, an Area Heavily Affected by Acid Precipitation

Water chemistry	Eighth Lake	Moss Lake	Sagamore Lake	Big Moose Lake
Surface area	123 ha	46 ha	68 ha	515 ha
Elevation	546 m	536 m	580 m	556 m
Mean depth	11.9 m	5.7 m	10.5 m	—
Field pH	6.5–7.0	5.7–6.1	5.3–5.7	4.6–5.5
Dissolved organic carbon	3.5–4.6 mg/l	2.9–4.7 mg/l	6.3–8.4 mg/l	1.5–4.5 mg/l
Total dissolved aluminum	0.02–0.03 mg/l	0.07–0.10 mg/l	0.27–0.35 mg/l	0.32–0.60 mg/l
Silicon	1.0–1.8 mg/l	2.5–2.9 mg/l	2.8–3.6 mg/l	—
Calcium	6.5–6.7 mg/l	2.6–3.0 mg/l	2.4–2.5 mg/l	2.0 mg/l (avg.)
Magnesium	0.9–1.0 mg/l	0.5–0.6 mg/l	0.6 mg/l	0.4 mg/l (avg.)
Sulfate	6.4–6.8 mg/l	6.4–6.5 mg/l	3.7–6.6 mg/l	7.2 mg/l (avg.)

Note: The lakes were selected to represent a spectrum of water chemistries: Eighth Lake is a control lake with neutral pH and very low aluminum concentrations. Moss Lake is distinctly acid with aluminum levels that are elevated, but not to the extent that they would generally cause significant reductions in fish populations. Big Moose and Sagamore Lakes both have aluminum concentrations known to be highly toxic to fish, but Sagamore Lake has a higher pH, and is also higher in dissolved organic carbon, which is known to reduce acute aluminum toxicity, probably by complexing aluminum, thereby preventing binding of aluminum to the gill epithelial membrane. The levels of calcium and silicon in Sagamore Lake may also be high enough to significantly protect fish against acute aluminum toxicity. The lakes have been monitored chemically for several years, and the chemical composition has stayed quite stable over time. Thus, knowing the age of the fish from their scales, the dose received by each fish can be calculated with quite good precision.

phosphorylated epitopes of heavy and medium molecular weight neurofilament and microtubule-associated protein tau (MAP-tau) (Figure 3.2). The lesions appear in the electron microscope on toluidine blue-stained thin sections both with and without a dense core. The lesions are Congo Red/birefringent negative under polarized light and are not immunoreactive to antibodies against amyloid β/A4 protein, microtubule-associated protein 2 (MAP-2), or glial fibrillary acidic protein (GFAP).

Finally, in solochrome azurine-stained sections of unfixed frozen cryocut tissue from fish chronically exposed to aluminum we observed the focal deposition of aluminum (blue staining) in the secondary lamellar epithelium of the gills (Figure 3.3), in the sensory layer of the olfactory epithelium, and also partly in the liver. In one fish, we observed diffuse accumulation of aluminum in the olfactory glomeruli layer of the olfactory bulb where axons from primary sensory neurons terminate and make synaptic connections with neuritic processes of secondary neurons. This latter result is particularly interesting in light of the well-documented olfactory deficits occurring in several neurological disorders, including Guamanian PD, Parkinson's disease, motor neuron disease, and Alzheimer's disease.[16–21]

As an interesting technical development, our studies demonstrated that the standard immersion fixation process leaches aluminum from gill tissues in fish. Also, fixation of fresh-frozen cryocut sections of gill for only 30 min with periodate-lysin-paraformaldehyde greatly reduces the extent of aluminum staining relative to the staining found when frozen cryocut sections were mounted directly on glass slides and air dried. Likewise, in fish of the same species from the same lake, aluminum staining was always much more intense in sections from frozen cryocut gills than from sections of gill tissue that were perfused with 4% paraformaldehyde. Similar problems were noted earlier in our development of models of chronic aluminum intoxication in rabbits,[22] where significant leaching of aluminum from formalin- or paraformaldehyde-fixed tissues from the left hemisphere of the brain occurred compared to frozen tissue from the right hemisphere of the brain of the same rabbit (Garruto, unpublished data).

3.4 OLFACTORY NEURAL PATHWAYS: AN ENVIRONMENTAL LINK

Environmental agents have been suggested to predispose individuals to Alzheimer's disease through the olfactory neural pathways.[23] Such agents are thought to spread transneuronally from sensory

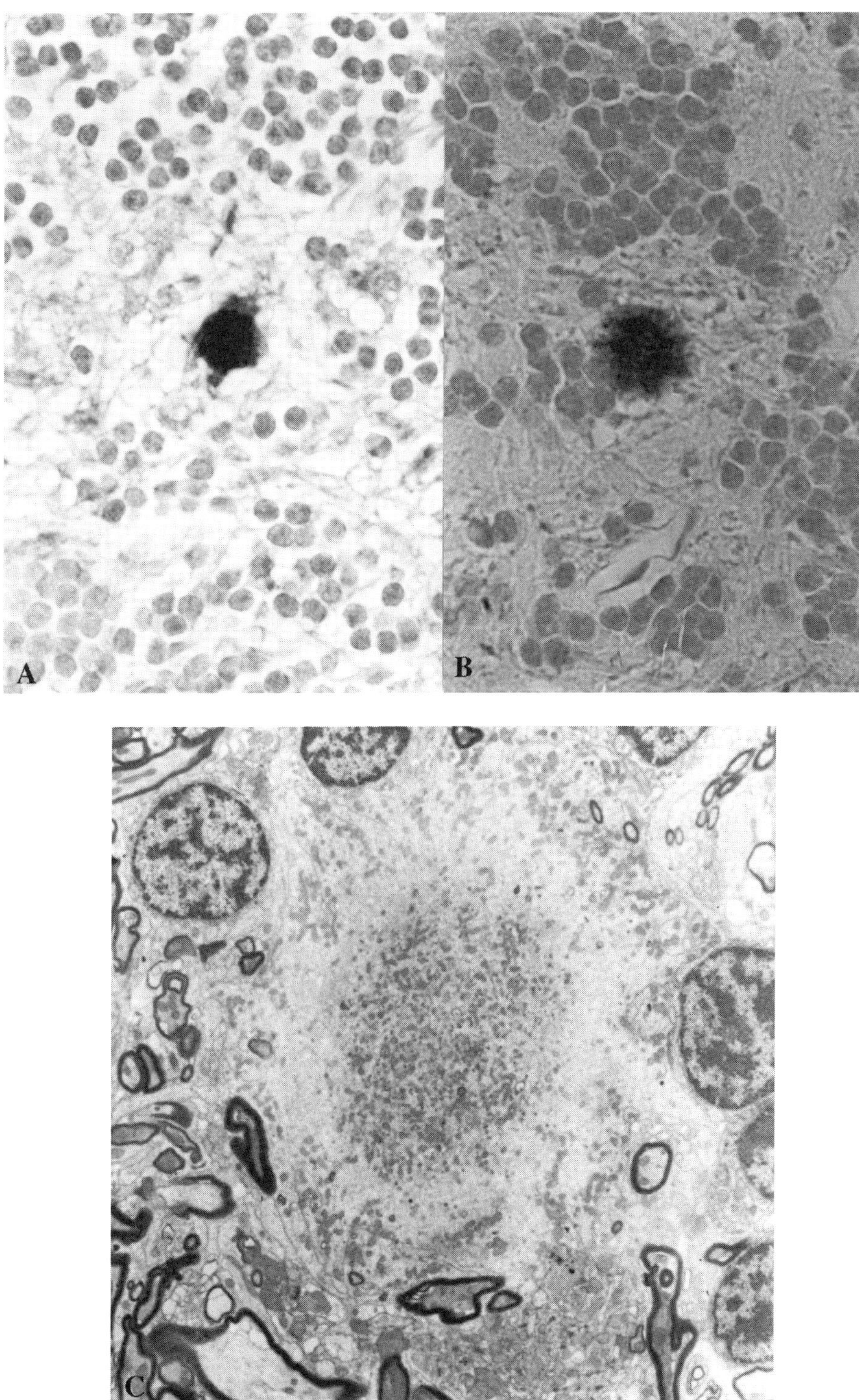

FIGURE 3.2 (a) Plaque-like lesions in the cerebellar granular layer of an 11-year-old lake trout from the high-aluminum Sagamore Lake were immunoreactive to antibodies against phosphorylated epitopes of heavy- and medium-weight neurofilament subunit proteins. (b) The same plaque-like lesions in the same fish were immunoreactive to antibodies against microtubule-associated protein tau. (c) Electron micrograph of plaque-like lesion in the same fish. The structure has a membrane in the outer layer, indicating that this is an intracytoplasmic layer. On higher magnification aggregation of mitochrondria are seen in the center and fine fibrillary structures and/or granules are seen in the periphery (× 1500).

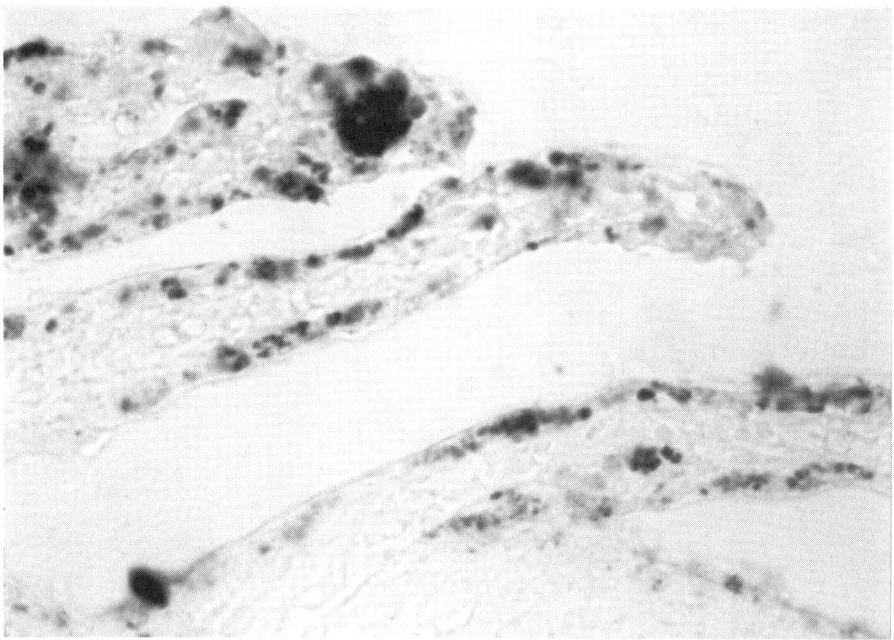

FIGURE 3.3 Prominent dark globular staining of aluminum in the secondary lamellae of the gill from a yellow perch from Big Moose Lake. The tissue was embedded in O.C.T. compound and quick-frozen on dry ice at the lake site. Cryocut sections were stained for aluminum with solochrome azurine (dark) and lightly counterstained with eosin.

neurons in the olfactory epithelium by transsynaptic or retrograde degeneration to olfactory-related areas of the brain (e.g., nucleus basalis of Meynert, locus ceruleus, and brainstem) and ultimately to the cerebral cortex,[21,23] all of which are severely affected in Alzheimer's disease.

The existence of olfactory deficits in Alzheimer's disease and Parkinson's disease is well established.[16,17,19] In Alzheimer's disease the changes include a marked impairment of odor identification and discrimination that is not attributable to the patient's poor memory, since these changes occur early in the disease. Recent studies of the neuropathology of the olfactory system in Alzheimer's patients demonstrated the presence of abnormally appearing olfactory tissue[24] and pathological changes in olfactory neurons, including the presence of abnormal dystrophic neurites that are morphologically identical to those found in the cerebral cortex of Alzheimer's disease patients.[25] Immunohistochemically, these abnormal neurites are reactive to antibodies against MAP-tau, ubiquitin, and phosphorylated epitopes of the heavy neurofilament subunit protein.[25]

Recently, olfactory impairment has been found in Guamanian PD,[18] thus strengthening the clinical, neuropathological, immunohistochemical, and cellular and molecular association with Alzheimer's disease. Olfactory dysfunction in Guamanian PD is among the earliest of signs and of the same magnitude of dysfunction as seen in Alzheimer's disease and in idiopathic Parkinson's disease.[18] The neurodegenerative connection is further linked by recent findings of olfactory impairment in motor neuron disease.[20]

The olfactory organ is the only vertebrate organ in which nerve cells are directly exposed to the environment. Experimentally, there is evidence of direct uptake of aluminum via the nasal-olfactory pathway after intranasal application of aluminum salts in rabbits.[26] In this study, aluminum produced granulomas in the olfactory bulb, cerebral cortex, and hippocampus, and cytoplasmic granules of macrophages in the granulomas contained significant amounts of aluminum. Neurofilamentous lesions, however, were not observed. In another study, olfactory deficits were reported in rabbits after intraventricular administration of aluminum chloride.[27] Aluminum deposition and pathology in the olfactory epithelium have also been noted in fish naturally[15] and experimentally[28,29] exposed to aluminum.

3.5 SUMMARY

The neuropathological changes we see in fish from acid-rain lakes are variable in number, distribution, and type. This variability is likely to be age related, species specific, and water chemistry associated, particularly with regard to the amount of dissolved aluminum, water pH, and other factors such as calcium, fluoride, silicon, and organic ligands. The olfactory system contains nerve cells which are directly and continuously exposed to the external environment and probably represents a major pathway for aluminum intoxication of the nervous system. It is likely that the wide variability in lesions we see in the nervous system is also a result of the high capacity of the fish CNS to regenerate. Thus in fish, unlike in mammals, chronic continuous neuronal degeneration may not be cumulatively observed and, as such, represents a weakness of the model. In nature, clearly we have a series of potential models of aluminum intoxication and calcium/aluminum interaction with striking parallels to human disorders such as ALS and PD in the Western Pacific and Alzheimer's disease. We believe these and related studies hold considerable promise for helping us to identify and understand what particular aluminum species are bioavailable, how aluminum alters the nervous system through chronic intoxication, and how to develop further insights into the pathogenesis of Pacific ALS and PD and related disorders with a hallmark pathology of cytoskeletal pooling and neurofibrillary degeneration.

ACKNOWLEDGMENTS

We wish to thank M. J. Sturm, W. Kretser, S. Capone, T. Dudones, D. W. Bath, and J. Gallagher for their assistance on the project.

REFERENCES

1. Döllken V: Über die Wirkung des Aluminiums mit besonderer Berücksichtigung der durch das Aluminium verursachten Läsionen im Zentralnervensystem. *Arch. Exp. Pathol. Pharmacol.*, 1897; 98–120.
2. Hartwell BL and Pember FR: The presence of aluminum as a reason for the difference in the effect of so-called acid soil on barley and rye. *Soil Sci.*, 1918; 6:259–279
3. Osmond CB, Bjorkman O, and Anderson, DJ: *Physiological Processes in Plant Ecology.* Berlin, Springer-Verlag, 1980.
4. Foy CD, Chaney RL, and White MC: The physiology of metal toxicity in plants. *Annu. Rev. Plant Physiol.*, 1978; 29:511–566
5. Shortle WC and Smith KT: Aluminum-induced calcium deficiency syndrome in declining red spruce. *Science*, 1988; 240:1017–1018.
6. Siegel N and Haug A: Aluminum interaction with calmodulin: evidence for altered structure and function from optical and enzymatic studies. *Biochim. Biophys. Acta*, 1983; 744:36–45.
7. Martin RB: Bioinorganic chemistry of aluminum. In Sigel H and Sigel A (eds): *Metal Ions in Biological Systems, Vol. 4: Aluminum and Its Role in Biology.* New York, Marcel Dekker, 1988, pp 1–57.
8. Garruto RM: Pacific paradigms of environmentally-induced neurological disorders: clinical, epidemiological and molecular perspectives. *Neurotoxicology*, 1991; 12:347–378.
9. Garruto RM and Yase Y: Neurodegenerative disorders of the western Pacific: the search for mechanisms of pathogenesis. *Trends Neurosci.*, 1986; 9:368–374.
10. Flaten TP and Garruto RM: Polynuclear ions in aluminum toxicity. *J. Theor. Biol.*, 1992; 156:129–132.
11. Driscoll CT, Baker JP, Bisogni JJ, and Schofield CL: Effect of aluminium speciation on fish in dilute acidified waters. *Nature*, 1980; 284:161–164.
12. Schofield CL and Trojnar RJ: Aluminum toxicity to brook trout (*Salvelinus fontinalis*) in acidified waters. In Toribara TY, Miller MW, and Morrow PE (eds): *Polluted Rain.* New York, Plenum Press, 1980, pp 341–366.
13. Rosseland BO, Blakar IA, Bulger A, et al: The mixing zone between limed and acidic river waters: complex aluminium chemistry and extreme toxicity for salmonids. *Environ. Pollut.*, 1992; 78:3–8.
14. Exley C, Chappell JS, and Birchall JD: A mechanism for acute aluminium toxicity in fish. *J. Theor. Biol.*, 1991; 151:417–428.
15. Flaten TP, Wakayama I, Sturm MJ, et al: Natural models of aluminium neurotoxicity in fish from lakes influenced by acidic precipitation. In Allan RJ and Nriagu JO (eds): *Heavy Metals in the Environment - 9th International Conference*, Vol. 2. Edinburgh, CEP Consultants Ltd., 1993, pp 285–288.
16. Doty RL, Reyes PF, and Gregor T: Presence of both odor identification and detection deficits in Alzheimer's disease. *Brain Res. Bull.*, 1987; 18: 597–600.

17. Doty RL, Deems DA, and Stellar S: Olfactory dysfunction in parkinsonism: a general deficit unrelated to neurologic signs, disease stage, or disease duration. *Neurology*, 1988; 38: 1237–1244.

18. Doty RL, Perl DP, Steele JC, et al: Odor identification deficit of the parkinsonism-dementia complex of Guam: equivalence to that of Alzheimer's and idiopathic Parkinson's disease. *Neurology*, 1991; 41 (Suppl. 2): 77–81.

19. Reyes PF, Deems DA, and Suarez MG: Olfactory-related changes in Alzheimer's disease: a quantitative neuropathologic study. *Brain Res. Bull.*, 1993; 32:1–5.

20. Elian M: Olfactory impairment in motor neuron disease: a pilot study. *J. Neurol. Neurosurg. Psychiatry*, 1991; 54: 927–928.

21. Harrison PJ: Neurodegeneration and the nose. *Clin. Otolaryngol.*, 1990; 15: 289–291.

22. Strong MJ, Wolff AV, Wakayama I, and Garruto RM: Aluminum-induced chronic myelopathy in rabbits. *Neurotoxicology*, 1991; 12:9–22.

23. Roberts E: Alzheimer's disease may begin in the nose and may be caused by aluminosilicates. *Neurobiol. Aging*, 1986; 7:561–567.

24. Johnson EW, Jafek BW, Eller PM, et al: Abnormalities in olfactory biopsies of Alzheimer's patients. *Soc. Neurosci. Abstr.*, 1991; 17: 351.

25. Tabaton M, Cammarata S, Mancardi GL, et al: Abnormal tau-reactive filaments in olfactory mucosa in biopsy specimens of patients with probable Alzheimer's disease. *Neurology*, 1991; 41:391–394.

26. Perl DP and Good PF: Uptake of aluminium into central nervous system along nasal-olfactory pathways. *Lancet*, 1987; i:1028.

27. Polonskaia EL, Butikova VI, and Zhirov SV: Pathological changes in the central nervous system studied on the model of aluminum encephalopathy. *Zh. Nevropatol. Psikhiatr. Im. S.S. Korsakova*, 1990; 90:71–74.

28. Klaprat DA, Brown SB, and Hara TJ: The effect of low pH and aluminum on the olfactory organ of rainbow trout, *Salmo gairdneri*. *Environ. Biol. Fishes*, 1988; 22:69–77.

29. Norrgren L, Glynn AW, and Malmborg O: Accumulation and effects of aluminium in the minnow (*Phoxinus phoxinus* L.) at different pH levels. *J. Fish Biol.*, 1991; 39:833–847.

Chapter 4

Cellular Aspects of Aluminum Toxicity in Plants

Alfred R. Haug and Victor Vitorello

CONTENTS

4.1 INTRODUCTION

Soil acidity is a major factor limiting plant growth and biomass production. In the heterogeneous soil system the pH of the aqueous phase plays a fundamental role in properties like the solubility and hydrolysis of ions, the electric charges of soil colloids, and the uptake of nutrients by plant roots. As the pH decreases, the solubility of ions like Al(III), Mn(II), and Cu(II) generally increases, eventually reaching levels toxic for many crop plants.

Regarding metal toxicity, that caused by aluminum is by far the major factor for plant growth limitation on naturally acidic soils (pH <5.0). For example, 70% of the soils in tropical America are affected by aluminum toxicity. Industrial activities and fertilizer usage have also contributed to the acidification and thus to the levels of free aluminum in terrestrial and aquatic environments previously not afflicted with aluminum toxicity.[1–3]

Despite the long history of studies on aluminum-related limitation of plant growth, there still exists a dearth of knowledge concerning the molecular basis for aluminum toxicity and the impact of low pH on plant root biochemistry.[1–3] In this chapter we are therefore focusing our attention primarily on molecular aspects of aluminum toxicity in plant cells rather than on phenomenological descriptions of the anatomy/physiology of plants under aluminum stress. Regarding mechanisms of aluminum toxicity in plant cells, information will also be provided, and conclusions drawn, from appropriate experiments conducted on mammalian cells. We hope that such a strategy will stimulate interest in studies directed towards elucidating primary mechanisms of aluminum toxicity manifested in both plant and mammalian cells.

4.2 Al(III) CHEMISTRY

In soils, aluminum is predominantly found as aluminosilicate and clay minerals. In the environment, chemical reactions of aluminum are mostly occurring in water. For these reasons, and also because the water content of a typical biological cell is approximately 80% of the cellular weight, it is necessary to understand reactions of aluminum in aqueous solution. Concerning details about aluminum properties (thermodynamic data, kinetics, chelation) in minerals and in solution, informative articles are available.[4–7]

In short, in aqueous solution the charged Al(III) exists as a hexahydrated species, $Al(H_2O)_6^{3+}$, at low pH. As the pH increases, in particular in the range from 5 to 6.2, the hydrated metal ion has a strong tendency to form various hydrolytic species. A series of polymeric aluminum species is also generated, particularly at higher pH values. Besides the pH value and Al(III) concentration, the extent of hydrolysis is strongly dependent on the presence of chelating ligands. The hydrated Al(III) ion has a considerable affinity to oxygen-containing negatively charged ligands such as carboxyl, hydroxyl, carbonyl, and phosphate groups associated with organic molecules. Apart from binding to hydroxyl, strong inorganic complexes are formed with fluoride, sulfate, and phosphate ligands. The mode of metal uptake across the plant plasma membrane may depend upon a chelate formed by Al(III). Following interiorization, the movement of Al(III) to biochemically critical targets in the cytosol and/or membranes is presumably also dependent upon the types of ligands available. Participation of Al(III) in chemical reactions is in part governed by the slow exchange rate of hydration water, a consequence of the small ionic radius (0.053 nm) of Al(III).

4.3 Al(III) AND PLANT GROWTH

A pH value of 5.0 is the approximate threshold below which aluminum toxicity becomes manifest in whole plants. This threshold is rather ill-defined since it also depends on factors such as the presence of ions [Ca(II) and phosphate], chelators, and the plant genotype.[1,3] In a "typical" acidic soil, the pH is around 4.5 and the concentration of mobile aluminum may vary between 10 and 40 µM (1 ppm Al(III) equals 37 µM). However, in acidic forest ecosystems displaying forest decline, pH values around 3.8 and mobile aluminum levels of about 1 mM have been reported.[8] Toxicity has been tentatively ascribed to the hexahydrated, trivalent Al(III) species, rather than to its hydrolysis products with lower electric charge.[1–3]

Al(III)-stressed plants show inhibition of growth, mainly that of the roots; the root apex is apparently a critical tissue.[9] Al(III)-related derangements of root metabolism reportedly impact transport of essential nutrients between roots and plant tissues above ground, thus reducing plant biomass accumulation.[1–3] Symptoms on roots include root thickening and discoloration. At the cellular level, perturbations of plasma membrane structure, vacuolation of root cap cells, and changes in nuclear structure are observable.[3,9] Since primary target(s) of Al(III) action on cells have not been clearly identified, some or all of these symptoms may reflect secondary injuries to the cell.

Regarding growth of Al(III)-stressed plants, considerable attention has been paid to interactions between Al(III) and Ca(II). Not attributable to ionic strength effects, elevated Ca(II) or Mg(II) levels in the root medium have been found to ameliorate the effects of toxic Al(III) on plant growth.[1] An attractive feature of these observations is that an increase in soil acidity is generally accompanied by a loss of exchangeable cations like Ca(II) and Mg(II).[8] Findings derived from these types of amelioration studies led to the hypothesis that toxic Al(III) ions are displacing Ca(II) from the membrane surface, thus impairing the functioning of surface ligands and transport channels critical for proper root growth.[1–3] Recent experiments on wheat discredited this type of Ca(II)-displacement hypothesis for explaining Al(III)-related rhizotoxicity.[10] Rather than unspecifically residing at the cell surface, the initial Al(III)-induced lesion seems to be associated with regulatory processes which are in part dependent upon signal-transducing elements of the plasma membrane, at least in mammalian cells.[11,12]

4.4 GENETICS AND ALUMINUM TOLERANCE

Among plant species and genotypes of a given species there exist wide variations regarding tolerance to Al(III).[1] In the acidic cerrado region of central Brazil, for example, Al(III)-accumulating tree species have been found which apparently even require Al(III) in the growth medium for normal growth.[13] However, the biochemical and genetic control mechanisms underlying tolerance to Al(III) are poorly understood. At this time, hard evidence is lacking whether tolerance to Al(III) is associated with processes outside the cell (exclusion) and/or inside the plant cell (internal detoxification). Given this dearth of knowledge, and findings that rhizotoxicity seems to be a pronounced feature of toxic Al(III) on plants,[1] parameters like root length or root elongation of Al(III)-stressed plants are generally selected as indicators when studying the inheritance of Al(III) tolerance in crop plants. Based on these selection techniques, there is evidence that a single incomplete dominant gene is responsible for the inheritance of aluminum tolerance in wheat.[14] Al(III) resistance of a tobacco variant was also attributed to a single dominant gene mutation when tobacco cells were selected for Al(III) resistance in cell cultures.[15] Furthermore, by selecting biochemical parameters as indicators for Al(III) sensitivity, experiments on two wheat varieties suggested that the effects of Al(III) on K(I) uptake are not primary factors associated with sensitivity towards toxic Al(III).[16] Despite the paucity of information on the molecular basis for Al(III) resistance, significant progress has been achieved, solely relying on conventional plant breeding programs, in selecting and releasing Al(III)-tolerant plants for crop production on acidic soils.[1]

4.5 MOLECULAR FEATURES

4.5.1 Al(III) Uptake Across the Plasma Membrane

Numerous studies have been conducted concerning Al(III) uptake and distribution in intact plants or in plant tissue.[1,17] However, considerable ambiguity remains concerning the relationships between the measured uptake kinetics and putative Al(III)-binding compartments. Indeed, few data are available on the effects of Al(III) at the cellular and molecular levels, such as mode(s) of Al(III) uptake across the plasma membrane, role of metal speciation in membrane traversal, and initial targets critical for cellular physiology. Outside the plant cell, in the Donnan free space of the root cell, major aluminum-binding sites are negatively charged carboxyl groups of pectic substances.[1–3] However, we believe that these surface-exposed, unspecific binding sites can be dismissed as primary aluminum targets. Rather, likely primary targets for aluminum-triggered rhizotoxicity are expected to reside in the cytosol and/or in the plasma membrane of the plant cell,[18] i.e., Al(III) must be internalized by the cell. This notion is also supported by findings on mammalian cells, where key elements of phosphoinositide-associated signal transduction pathways, coupled with Ca(II) regulation, were implicated as potential primary targets for toxic Al(III) ions.[11,12]

In oak root cortex cells, application of Al(III) (pH 4; 37 µM to 3.7 mM) enhanced membrane permeability to nonelectrolytes (methyl urea, urea) and lowered permeability to water.[19] The plasma membrane was also implicated as a putative Al(III) target in studies on *Amaranthus* protoplasts, where $^{45}Ca^{2+}$ uptake was inhibited within minutes by Al(III) application (pH 4.5; 10 and 50 µM).[20] These kinds of permeability changes are presumably related to pronounced Al(III)-triggered membrane lipid phase changes, *in vivo* and in membrane vesicles isolated from *Thermoplasma acidophilum* (pH 2).[2] Moreover, compared with nonstressed mycelia, the proportion of less-ordered membrane domains was decreased when spin-labeled fungal mycelia were exposed to Al(III) stress (1 to 10 mM).[21] Al(III)-induced changes in lipid packing were also observed in response to Al(III)-binding (20 µM) to the erythrocyte membrane.[22] The phosphodiester group was found to be the primary binding site of Al(III) with lipids.[23] Furthermore, based on findings that Al(III) neutralizes the surface charge of phosphatidyl choline vesicles, this type of interaction was implicated in the nonspecific, Al(III)-related inhibition of cation uptake by root cells.[24] In the presence of Al(III) (20 to 50 µM), changes in the membrane surface charge are generated by displacing virtually all bivalent cations, such as Ca(II), present at bulk concentrations as high as 1 mM.[25] To reiterate, Al(III)-related Ca(II) displacement at the membrane surface does not seem to be the initial lesion for rhizotoxicity of plants.[10] Nevertheless, it is reasonable to assume that nonspecific Al(III) binding to the membrane surface, followed by traversal of the membrane, are sequential events required for Al(III) interiorization by the cell. Membrane traversal, perhaps involving some sort of carrier,[26] may in part depend on the physical state of the membrane.

4.5.2 Interiorized Al(III)

Once interiorized, Al(III) may interact with membrane-bound, cytosolic, and nuclear constituents of the cell. Given the slow solvent exchange rate of Al(III),[7] it must be kept in mind, however, that initially established interactions with certain sites may gradually change until a final equilibrium has been reached where Al(III) is irreversibly coordinated to a ligand, perhaps different from the initial one.

Forming the interface between the cell and its microenvironment, the plasma membrane detects external stimuli at the surface, processes these signals across the membrane, then converts them into a cascade of reactions involving second messengers like Ca(II). In animal cells, chief among cellular communication channels is the phosphoinositide signaling pathway. A major product of this pathway is a second messenger, inositol-1,4,5-triphosphate (IP_3) which, in turn, plays a crucial role in mobilizing intracellular Ca(II).[27] In plants, the elements of this pathway are present but evidence is lacking as to the operation of this system in response to a specific stimulus.[28] Regarding aluminum toxicity, short-term studies on neuroblastoma cells demonstrated that application of Al(III) (pH 6.8, 2 µM to 1 mM) inhibited intracellular Ca(II) mobilization which, in turn, coincided with a decrease in inositol phosphate production. Being crucial for IP_3 formation, phosphatidylinositol-4,5-diphosphate-specific phospholipase C probably harbors an Al(III) target. A second Al(III) target seems to reside on molecular switching elements of the phosphoinositide signaling pathway, viz., on Mg(II)/guanine nucleotide binding (Gp) proteins. It was suggested that Al(III) is occupying a site at or near the GTP binding center on the Gp protein.[11,12] Besides direct interactions with proteins, Al(III) may perturb the protein's lipid milieu, especially at higher [Al(III)]. For example, the hydrolyzing efficacy of phospholipases is in part dependent on the physical state of their lipid environment.[12] Evidently further studies are necessary to identify Al(III) targets on elements of the signal transduction pathway in eukaryotic cells.

Concerning Al(III) interactions with membrane-bound enzymes, studies on barley root plasma membranes indicated that Al(III) ions are capable of inhibiting a membrane-associated Ca(II)-ATPase activity. The Al(III)-induced lower activity could in part be restored by applying massive amounts (50 mM) of monovalent cations. As a result of increased ionic strength, Al(III) was presumably displaced from lipids in the boundary lipid region of the enzyme, thus restoring the motional requirements of the intact protein.[29] There is also the possibility that ATPase activity is diminished resulting from the formation of an Al(III)-ATP chelate when millimolar $AlCl_3$ levels are employed.[30] Furthermore, applying Al(III) to vesicles enriched in right-side-out plasma membranes isolated from maize roots, Al(III) (pH 6.0; 40 µM) inhibited NADH-dependent electron transfer processes,[31] i.e., reactions generally associated with the uptake of iron by the plant root.[32]

Following internalization by the wheat root cell, Al(III) was distributed in the cytosolic fraction [generally 55% of total Al(III)], nuclear fraction (around 30%), and the mitochondrial fraction (around 13%).[33] Interestingly, a roughly similar distribution of Al(III) was found in neuroblastoma cells.[26] In the cytosol there exists a multitude of ligands capable of binding Al(III), e.g., ATP,[30] citric acid,[6,7] various organic acids,[7] and acidic polypeptides.[2] However, evidence is lacking whether intracellular chelators are instrumental in intrinsic Al(III) tolerance mechanisms of plant cells. On the other hand, like in mammalian cells,[26] this large chelator pool may contribute to the high and rapid accumulation of Al(III) by plant cells. By maintaining a low intracellular-free [Al(III)] the equilibrium is shifted in favor of Al(III) influx.

Al(III) ions reportedly also interact with mitochondrial membranes. Experiments on isolated mitochondria from untreated wheat roots indicated that Al(III) ions (pH 7; 12 to 75 μM) interacted with the mitochondrial respiratory pathway, presumably by interfering with the oxidation of substrates that donate electrons to Complexes I and II.[34] Also involved in energy-generating processes are NADP-dependent dehydrogenases, which were strongly inhibited by Al(III) (pH 6.85; 2 μM AlCl$_3$) with respect to the substrate, isocitrate.[35]

In the cytosol, interiorized Al(III) may bind to calmodulin,[2] a highly conserved protein (about 17 kDa) crucial for Ca(II)-mediated regulation in eukaryotic cells. Responding to altered intracellular Ca(II) levels, e.g., triggered via signaling pathways, this protein is activated by undergoing conformational changes generated by specifically binding four Ca(II) per protein. The activated protein can then associate with and stimulate certain calmodulin-dependent partner enzymes.[36] Concerning Al(III)-related alterations of calmodulin, only *in vitro* experiments are available. Presumably binding only weakly to calmodulin,[37] the protein undergoes helix-coil transitions in the presence of Al(III), thus changing the protein's internal dynamics and its efficacy in correctly docking with partner proteins.[38–40] Al(III)-induced changes on calmodulin are instrumental in malfunctions of calmodulin-dependent reactions, e.g., that of 3′,5′-cyclic nucleotide phosphodiesterase, certain plant ATPases,[2] and NAD kinase activity in wheat root tips.[41] The latter region harbors high levels of calmodulin,[42] and has been implicated as being a site for Al(III) injury.[9] At this time evidence is lacking whether Al(III) inactivates calmodulin in a cell. Abundantly present, cytosolic chelators are possibly protecting this important regulatory protein from Al(III) injury.[43]

The nucleus may be an important sink for interiorized Al(III), especially in cells under long-term Al(III) stress. For example, when onion root tips were treated with AlCl$_3$ (pH 5.5, 1 mM, several hours), cell division was inhibited.[44] Various findings indicate that Al(III) accumulates on DNA,[44] where the phosphodiester group is apparently a binding site.[45] Under cytosolic conditions, Al(III) binding to the DNA is presumably weak.[46] Given the condensed state of DNA in the nucleus of a cell, interstrand DNA cross-linking via Al(III) bridges is highly probable.[47] Such cross-linking would be consistent with observations that Al(III)-treated pea chromatin is more heat stable.[44] Moreover, Al(III)-induced changes in DNA structure might be expected to disrupt the physiological processes of gene expression. It is noteworthy that the mRNA levels of calmodulin and tubulin were considerably diminished in Al(III)-treated rabbit brains.[48] Compared with Al(III)-induced derangements of membrane structure and signaling processes (detectable within minutes),[11,12] Al(III) interactions at the DNA, and numerous toxicity responses (growth, cell secretion, callose formation, etc.) are long-term effects.[1,2,3,49] Whether some of these long-term responses are directly caused by Al(III) is questionable.

4.6 CONCLUDING REMARKS

Only rudimentary information is currently available on primary targets and the initial phases in the temporal progression of aluminum toxicity, both in plant and animal cells. What, at present, appears to be a multitude of Al(III) effects in cells, may eventually originate from a single (or a few) primary target(s) critical for the temporal development of the aluminum toxicity syndrome. Given the significance of an intact plasma membrane and of functional communication pathways for cellular survival, any Al(III)-triggered change therein is expected to have severe repercussions on the cell's ability to respond to external/internal signals. By focusing on identifying initial key interaction sites for toxic Al(III), we may also contribute to, and in fact simplify, our endeavors of developing Al(III)-resistant varieties through biotechnological means.

ACKNOWLEDGMENT

V.V. was supported in part by a fellowship, grant Nr. 200892/91-6, from the Conselho Nacional de Desenvolvimento Cientifico e Tecnologico (Brasil).

REFERENCES

1. Foy CD, Chaney RL, and White MC: The physiology of metal toxicity in plants. *Annu. Rev. Plant Physiol.*, 1978; 29: 511–566.
2. Haug A: Molecular aspects of aluminum toxicity. *CRC Crit. Rev. Plant Sci.*, 1984; 1: 345–373.
3. Taylor GJ: The physiology of aluminum phytotoxicity. In Sigel H (ed): *Metal Ions in Biological Systems*, Vol. 24. New York, Marcel Dekker, 1988, pp 123–163.
4. Sposito G (ed): *The Environmental Chemistry of Aluminum*. Boca Raton, CRC Press, 1989.
5. Akitt JW: Multinuclear studies of aluminum compounds. *Prog. Nucl. Magn. Reson. Spectrosc.*, 1989; 21: 1–149.
6. Martin RB: The chemistry of aluminum as related to biology and medicine. *Clin. Chem.*, 1986; 32: 1797–1806.
7. Orvig C: The aqueous coordination chemistry of aluminum. In Robinson GH (ed): *Coordination Chemistry of Aluminum*. New York, VCH Publishers, 1993, pp 85–121.
8. Godbold DL, Fritz E, and Hüttermann A: Aluminum toxicity and forest decline. *Proc. Natl. Acad. Sci. U.S.A.*, 1988; 85: 3888–3892.
9. Bennet RJ, and Breen CM: Aluminium toxicity: Towards an understanding of how plant roots react to the physical environment. In Randall PJ, Delhaize E, Richards RA, and Munns R (eds): *Genetic Aspects of Plant Mineral Nutrition*. Dordrecht, Kluwer Academic, 1993, pp 103–116.
10. Kinraide TB, Ryan PR, and Kochian LV: Al^{3+}-Ca^{2+} interactions in aluminum rhizotoxicity. II. Evaluating the Ca^{2+}-displacement hypothesis. *Planta*, 1994; 192: 104–109.
11. Shi B, Chou K, and Haug A: Aluminium impacts elements of the phosphoinositide signalling pathway in neuroblastoma cells. *Mol. Cell. Biochem.*, 1993; 121: 109–118.
12. Haug A, Shi B, and Vitorello V: Aluminum interaction with phosphoinositide-associated signal transduction. *Arch. Toxicol.*, 1994; 68: 1–7.
13. Haridasan M, and de Araujo GM: Aluminium-accumulating species in two forest communities in the cerrado region of central Brazil. *Forest Ecol. Manage.*, 1988; 24: 15–26.
14. Wheeler DM, Edmeades DC, Christie RA, and Gardner R: Comparison of techniques for determining the effect of aluminium on the growth of, and the inheritance of aluminium tolerance in wheat. In Randall PJ, Delhaize E, Richards RA, and Munns R (eds): *Genetic Aspects of Plant Mineral Nutrition*. Dordrecht, Kluwer Academic, 1993, pp 9–16.
15. Conner AJ, and Meredith CP: Strategies for the selection and characterization of aluminum-resistant variants from cell cultures of *Nicotiana plumbaginifolia*. *Planta*, 1985; 166: 466–473.
16. Pettersson S, and Strid H: Effects of aluminium on growth and kinetics of $K^+(^{86}Rb^+)$ uptake in two cultivars of wheat with different sensitivity to aluminium. *Physiol. Plant*, 1989; 76: 255–261.
17. Zhang G, and Taylor GJ: Effects of biological inhibitors on kinetics of aluminium uptake by excised roots and purified cell wall material of aluminium-tolerant and aluminium-sensitive cultivars of *Triticum aestivum L. J. Plant Physiol.*, 1991; 138: 533–539.
18. Haug A, and Shi B: Biochemical basis of aluminium tolerance in plant cells. In Wright RJ, Baligar VC, and Murrmann RP (eds): *Plant-Soil Interactions at Low pH*. Dordrecht, Kluwer Academic, 1991, pp 839–850.
19. Zhao XJ, Sucoff E, and Stadelmann EJ: Al^{3+} and Ca^{2+} alteration of membrane permeability of *Quercus rubra* root cortical cells. *Plant Physiol.*, 1987; 83: 159–162.
20. Rengel Z, and Elliott DC: Aluminium inhibits $^{45}Ca^{2+}$ uptake by *Amaranthus* protoplasts. *Biochem. Physiol. Pflanz*, 1992; 188: 177-186
21. Zel J, Svetek J, Crne H, and Schara M: Effects of aluminum on membrane fluidity of the mycorrhizal fungus *Amanita muscaria*. *Physiol. Plant*, 1992; 89: 172–176.
22. Weis C, and Haug A: Aluminum-altered membrane dynamics in human red blood cell white ghosts. *Thromb. Res.*, 1989; 54: 141–149.
23. Hauser H, and Phillips MC: Interactions of the polar groups of phospholipid bilayer membranes. In Cadenhead DA, and Danielli JF (eds): *Progress in Surface and Membrane Science*. Vol. 13. New York, Academic Press, 1979, pp 297–413.
24. Akeson MA, Munns DN, and Burau RG: Adsorption of Al^{3+} to phosphatidylcholine vesicles. *Biochim. Biophys. Acta*, 1989; 986: 33–40.
25. Deleers M: Cationic atmosphere and cation competition binding at negatively charged membranes: Pathological implications of aluminum. *Res. Commun. Chem. Pathol. Pharmacol.*, 1985; 49: 277–294.
26. Shi B, and Haug A: Aluminum uptake by neuroblastoma cells. *J. Neurochem.*, 1990; 55: 551–558.
27. Berridge MJ: Inositol triphosphate and calcium signalling. *Nature*, 1993; 361: 315–325.
28. Coté GG, and Crain RC: Biochemistry of phosphoinositides. *Annu. Rev. Plant Physiol.*, 1993; 44: 333–356.
29. Caldwell CR, and Haug A: Divalent cation inhibition of barley root plasma membrane-bound Ca^{2+}-ATPase activity and its reversal by monovalent cations. *Physiol. Plant*, 1982; 54: 112–118.
30. Panchalingam K, Sachedina S, Pettegrew JW, and Glonek T: Al-ATP as an intracellular carrier of Al(III) ion. *Int. J. Biochem.*, 1991; 23: 1453–1469.
31. Loper M, Brauer D, Patterson D, and Tu SI: Aluminum inhibition of NADH-linked electron transfer by corn root plasma membranes. *J. Plant Nutr.*, 1993; 16: 507–514.

32. Holden MJ, Luster DG, and Chaney RL: Enzymatic iron reduction at the root plasma membrane: Partial purification of the NADH-Fe chelate reductase. In Manthey JA, Crowley DE, and Luster DG (eds): *Biochemistry of Metal Micronutrients in the Rhizosphere*. Boca Raton, Lewis Publishers, 1994, pp 285–294.

33. Niedziela G, and Aniol A: Subcellular distribution of aluminium in wheat roots. *Acta Biochim. Pol.*, 1983; 30: 99–105.

34. de Lima ML, and Copeland L: The effect of aluminium on respiration of wheat roots. *Physiol. Plant*, 1994; 90: 51–58.

35. Yoshino M, Yamada Y, and Murakami K: Inhibition by aluminum ion of NAD- and NADP-dependent isocitrate dehydrogenases from yeast. *Int. J. Biochem.*, 1992; 24: 1615–1618.

36. Cohen P, and Klee CB (eds): *Calmodulin*. Amsterdam, Elsevier, 1988.

37. Martin RB: Bioinorganic chemistry of aluminum. In Sigel H, and Sigel A (eds): *Metal Ions in Biological Systems*, Vol 24. New York, Marcel Dekker, 1988, pp 1–57.

38. Yuan S, and Haug A: Frictional resistance to motions of bimane-labelled spinach calmodulin in response to ligand binding. *FEBS Lett.*, 1988; 234: 218–223.

39. Weis C, and Haug A: Aluminum-altered conformational changes in calmodulin alter the dynamics of interaction with melittin. *Arch. Biochem. Biophys.*, 1987; 254: 304–312.

40. Haug A, and Vitorello V: Aluminium coordination to calmodulin, thermodynamic and kinetic aspects. *Coord. Chem. Rev.*, in press.

41. Slaski JJ: Effect of aluminium on calmodulin-dependent and calmodulin-independent NAD kinase activity in wheat root tips. *J. Plant Physiol.*, 1989; 133: 696–701.

42. Allan E, and Trewavas A: Quantitative changes in calmodulin and NAD kinase during early cell development in the root apex of *Pisum sativum L. Planta*, 1985; 165: 493–501.

43. Suhayda CG, and Haug A: Citrate chelation as a potential mechanism against aluminum toxicity in cells: the role of calmodulin. *Can. J. Biochem. Cell Biol.*, 1985; 63: 1167–1175.

44. Matsumoto H: Biochemical mechanism of the toxicity of aluminium and the sequestration of aluminium in plant cells. In Wright RJ, Baligar VC, and Murrmann RP (eds): *Plant-Soil Interactions at Low pH*. Dordrecht, Kluwer Academic, 1991, pp 825–838.

45. Dyrssen D, Haraldsson C, Nyberg E, and Wedborg M: Complexation of aluminum with DNA. *J. Inorg. Biochem.*, 1987; 29: 67–75.

46. Martin RB: Aluminium speciation in biology. In *Aluminium in Biology and Medicine* (Ciba Foundation Symposium no. 169). Chichester, John Wiley & Sons, 1992, pp 5–25.

47. Karlik SJ, and Eichhorn GL: Polynucleotide crosslinking by aluminum. *J. Inorg. Biochem.*, 1989; 37: 259–269.

48. Crapper McLachlan DR: Aluminum neurotoxicity: Criteria for assigning a role in Alzheimer's disease. In Lewis TE (ed): *Environmental Chemistry and Toxicology of Aluminum*. Boca Raton, Lewis Publishers, 1989, pp 299–315.

49. Basu A, Basu U, and Taylor GJ: Induction of microsomal membrane proteins in roots of an aluminum-resistant cultivar of *Triticum aestivum L.* under conditions of aluminum stress. *Plant Physiol.*, 1994; 104: 1007–1013.

SECTION 3

RECENT CONCEPT OF MINERALS/METALS RESEARCH

Determination of Trace Elements in Biological Samples — Neutron Activation Analysis

Zeev B. Alfassi and Masayuki Yasui

CONTENTS

5.1 INTRODUCTION

Various instrumental methods have been developed for multielement determination of trace elements in biological samples.[1,2] All of these methods have advantages and disadvantages, and the choice of which method to use depends on the special problem, i.e., the analyzed element (or elements), its concentration, the interfering elements, and the availability of the appropriate equipment. In this chapter a short description of the main methods will be given, followed by a more detailed description of one method — neutron activation analysis.

5.2 INDUCTIVELY COUPLED PLASMA (ICP)[3]

In this method the analyzed sample must be in solution. Biological samples are dissolved usually by application of concentrated acids.[4] In the instrument the solution is evaporated and turned into plasma consisting of excited atoms and atomic ions. The atoms or ions can be detected either by their emitted light when returning to their ground state (ICP-AES stands for Atomic Emission Spectrometry), or the ions can be determined with a mass spectrometer (ICP-MS). ICP-MS is more sensitive by about two orders of magnitude than ICP-AES, and the sensitivity of most of the elements is about the same (0.02 to 0.2 ppb). However, for masses lower than 80 Da there are interferences due to cluster ions and compounds ions such as MO^+, MCl^+, MOH^+, and MOH_2^+. Thus, for example, Mg is more sensitive by ICP-AES than by ICP-MS, due to the unusually high sensitivity of Mg to ICP-AES (0.1 ppb compared to 1 to 250 ppb for other elements), and the low sensitivity of Mg to ICP-MS (0.7 ppb compared to 0.002 to 0.1 ppb for most elements). ICP-MS is much more expensive than ICP-AES. The use of ICP-MS enlarges considerably the number of elements which can be directly determined in body fluids. In serum, ICP-AES allows the direct determination of only Na, Ca, Mg, Fe, Cu, and Zn, whereas with ICP-MS many other elements can be determined, e.g., Pb, Cd, and Hg.

The main disadvantage of the ICP is the need to dissolve the sample, which is laborious and may lead to contamination from the dissolving agents, the solvent, and the atmosphere as well as the high cost of the equipment.

ICP can be used to simultaneously measure many elements, as all of them form excited ions and atoms. In the case of ICP-AES it is done either by using a diode array for detection or by fast scanning of the monochromator. In the ICP-MS it is done by scanning.

Table 5.1 gives the detection limit of various elements for ICP-AES and ICP-MS together with detection limits for other methods.

5.3 ATOMIC ABSORPTION SPECTROMETRY (AAS)[7,8]

AAS is a method for elemental analysis in solution. It is a sensitive method which can detect different elements in the range of few parts per million (when atomization is done by flame) or few parts per billion (for electrothermal furnace atomization). The temperature required is lower than in ICP, and hence the atoms are not excited. While in ICP the element is determined by its emission (ICP-AES), in AAS it is determined by absorption of light emitted from a hollow cathode lamp. Consequently, AAS is used to determine a single element at one time since each element has its own cathode lamp, although there are lamps which can be used to determine two or three metals. Although ICP is more sensitive (for most elements) and can simultaneously determine many elements, AAS is more common in analytical chemistry use due to its lower price. The graphite furnace technique is more sensitive than the flame atomization (usually two to three orders of magnitude) but the analysis takes much longer (may reach 2 to 3 min per element/sample). In flame atomization the solution is continuously fed while in the graphite furnace instrument a small amount of solution is injected, dried, and evaporated. The furnace version uses much smaller samples.

5.4 ELECTROCHEMICAL METHODS[9]

Electrochemical methods of analysis, which can be used for analysis of below micromolar levels, are mainly voltammetric methods. The analyte must be as a solute in a solution. All the methods include two steps. The first step is a preconcentration step in which the analyte is accumulated at the electrode by either faradic or nonfaradic processes. The next step, which is the step forming the analytical signal, is the subsequent stripping of the analyte from the electrode. This is the reason that these methods are called stripping voltammetry analyses (anodic, cathodic, or adsorptive).

The electrode used is usually mercury, either as a drop electrode or thin-film electrode deposited on a substrate (usually glassy carbon).

5.5 X-RAY EMISSION SPECTROMETRY (XRES)[10–12]

XRES is a multielement nondestructive technique in which thin samples (a few micrometers) or the surface of thicker samples are analyzed for their elemental concentration.

In XRES one of the electrons in the inner shells (either the innermost K shell, or the next one-L shell) is ejected out of the atom. This vacancy in the electron shell is filled immediately by electrons from outer shells. The energy difference between the binding energy of the electrons in

TABLE 5.1 Determination Limits for Single Element Standards: A Comparison of Neutron Activation With Other Analytical Techniques

Element	FAAS[5] (mg dm^{-3})	ICP-AES[5] (mg dm^{-3})	ICP-MS[5,6] (mg dm^{-3})	INAA (mg kg^{-1})
Al	0.06	0.046	0.00097	0.001
Sb	0.12	0.064	0.00028	0.00005
As	0.003	0.11	0.0028	0.00001
Ba	0.06	0.0026	0.0002	0.001
Bi	0.12	0.272	0.00001	0.010
B	2.1	0.0096	0.00038	—
Cd	0.003	0.0050	0.00057	0.0005
Ca	0.003	0.020	0.010	0.10
Ce	—	0.10	0.0001	0.0001
Cs	0.06	83	0.00009	0.0001
Cr	0.009	0.014	0.00011	0.003
Co	0.015	0.012	0.00034	0.00001
Cu	0.009	0.010	0.00023	0.00001
Dy	0.15	0.054	0.0001	0.000001
Eu	0.09	0.0054	0.00006	0.00001
Gd	6.0	0.050	0.0001	0.0001
Ga	0.21	0.092	0.00098	0.00005
Ge	0.3	0.096	0.00096	0.010
Au	0.03	0.034	0.00011	0.0000001
In	0.12	0.13	0.00026	0.0001
Ir	1.5	0.056	0.00008	0.000001
Fe	0.015	0.012	0.001	0.10
La	6.0	0.020	0.00008	0.00001
Pb	0.06	0.084	0.0003	1.0
Li	0.003	0.045	0.00029	—
Mg	0.0006	0.060	0.00094	0.010
Mn	0.006	0.0028	0.00029	0.00001
Hg	0.51	0.050	0.00023	0.0001
Mo	0.06	0.016	0.0007	0.001
Nd	3.0	0.015	0.0002	0.0005
Ni	0.015	0.030	0.004	0.010
Nb	6.0	0.072	0.00003	0.001
Os	0.3	0.0007	0.00044	0.001
Pd	0.045	0.088	0.0041	0.00005
P	120	0.15	0.032	—
Pt	0.21	0.11	0.00041	0.0005
K	0.003	12	0.004	0.0002
Pr	15	0.094	0.00009	0.00001
Re	2.4	0.012	0.004	0.00001
Rh	0.015	0.088	0.002	0.0001
Rb	0.015	75	0.00036	0.0001
Ru	0.9	0.060	0.00044	0.001
Sm	3.0	0.086	0.0002	0.000005
Sc	0.15	0.0030	0.00006	0.00001
Se	0.6	0.15	0.00028	0.005
Si	0.45	0.024	0.033	0.010
Ag	0.006	0.00032	0.00032	0.001
Na	0.001	0.058	0.003	0.0001
Sr	0.015	0.00084	0.0001	0.010
Ta	4.5	0.050	0.00001	0.00005
Te	0.15	0.082	0.003	0.010
Tb	3.0	0.056	0.00003	0.0001
Tl	0.06	0.080	0.00003	0.010
Th	—	0.166	0.00001	0.0001
Sn	0.3	0.09	0.00043	0.001
Ti	0.21	0.0076	0.0024	0.003
W	3.6	0.060	0.00001	0.00001

TABLE 5.1 (continued) Determination Limits for Single Element Standards: A
Comparison of Neutron Activation With Other Analytical Techniques

Element	FAAS[5] (mg dm^{-3})	ICP-AES[5] (mg dm^{-3})	ICP-MS[5,6] (mg dm^{-3})	INAA (mg kg^{-1})
U	90	0.50	0.00002	0.00001
V	0.15	0.015	0.00016	0.00001
Yb	0.12	0.0036	0.00006	0.00001
Y	0.6	0.0070	0.0001	0.0001
Zn	0.003	0.0036	0.00059	0.010
Zr	4.5	0.014	0.00011	0.10

the two shells is compensated for by emission of a photon of electromagnetic radiation in the X-ray
wavelength region. The energy of the X-ray is characteristic of the atom (element) from which the
electron was ejected, and the number of photons emitted (for the specific energy) is proportional
to the concentration of this element. The energy of the photons is determined either by using a
crystal monochromator, measuring one wavelength at a time due to its angle of scattering (wavelength dispersion), or using an Si(Li) detector (or a Ge detector) that measure simultaneously all
the photons according to their energy (energy dispersion). The energy-dispersive detectors have
the advantage of simultaneous measurement of the whole spectra rather than measurement of a
single line at one time and thus they are the detectors used in many analyses when the concentration
of several elements is to be determined.

As the penetration of X-rays in materials is quite low, XRES can be used only to measure the
concentration on the surface. The intensity of the characteristic X-rays emitted from a sample varies
with the sample thickness.

The wide use of XRES analysis is based on two important characteristics of X-rays: (1) the
small numbers of lines in the characteristic X-ray spectrum of each element, which makes the
interpretation easier, and (2) the direct correlation between the wavelength (energy) and the atomic
number by Mosley's law.

Excitation of the samples can be induced by a radioactive source emitting X-rays, X-ray tubes,
accelerated electrons, and accelerated positively charged particles H^+ or He^{+2}. Using charged
particles (either electrons or positive ions) allows the beam to be focused to very small sizes,
enabling determination of the elemental composition of very small samples or the measurement of
the variation of the concentration as a function of the geometry (referred to as scanning).

Most systems using X-rays for excitation (the method is then called X-ray fluorescence [XRF],
as higher-energy X-rays induce lower-energy X-ray emission) use X-ray tubes due to their higher
intensity of X-rays. The use of X-ray-emitting radioisotopes is due to one of the following factors:
(1) lower price of the radioisotope source, (2) few numbers of lines, which can be used to increase
the selectivity of some elements, and (3) smaller size of the radioisotope source.

In electron probe X-ray spectroscopy, or as it is usually called, microprobe analysis, an electron
beam of a moderate energy of 10 to 50 keV, similar to that of an electron microscope, is focused
to a minute area of several micrometers in diameter in order to eject inner electrons from the surface
(up to a few micrometers in depth) atoms. Most modern electron probes have the capability of
sweeping the electron beam across the sample in order to obtain the concentration across the sample
surface. The common instrument — SEM (scanning electron microscope) — can give the sample
image by either back-scattered or transmitted electrons together with the elemental composition
by energy-dispersion X-rays measurement.

Proton-induced X-ray emission (PIXE) analysis is done with a small accelerator (commonly
accelerating protons to 2.5 MeV) attached to a vacuum chamber with a sample changer. An X-ray
energy-dispersive detector views the sample at a short distance. The detector output signal is fed
to a multichannel analyzer, hosted by a computer, via a pulse processing technique. For a millimeter-
sized proton beam PIXE, the absolute detection limit is in the picogram range. PIXE is more
sensitive than XRF for thin samples as the XRF absolute detection limit is in the nanogram range.
However, they have about the same sensitivity for thick samples. The probability for electron

ejection (ionization cross-section) decreases with increasing atomic number for positive ion excitation, whereas for photon excitation (XRF) it increases with increasing Z. This seems to favor PIXE for low Z, however the same effect also appears in the background, which largely controls the detection limit.

The lowest detection limits are obtained for matrices consisting mainly of light elements (Z <10), making biological samples good candidates for PIXE analysis. The most intensively studied biological applications are investigations of elemental distribution within biological cells and tissues done with nuclear microprobes.[11]

New scanning nuclear microscope instruments can measure both the X-ray emission and the ions transmitted through and scattered from the sample, in order to supply more information on the elemental composition.[11] For example, Watt and Landsberg[13] used scanning transmission ion microscopy (STIM) to measure the image, the Rutherford backward scattering (RBS) of the protons, and the PIXE spectra to prove that aluminum is not involved in the etiology of Alzheimer's disease, and previous suggestions[14] were probably due to contamination of tissue by aluminosilicates present in most reagents.

5.6 NEUTRON ACTIVATION ANALYSIS (NAA)

Chemical analysis by nuclear activation is an elemental analysis, i.e., it determines the contents of the various elements, but cannot tell in what chemical form (compounds, valence states) they are. The analysis is based on a reaction of the analyzed element with nuclear projectiles (neutrons or accelerated small charged particles as for example protons, deuterons, ^{3}He, or α particles, or gamma photons). The reaction can be written in the same form as a chemical reaction.

Target + projectile $\rightarrow$ light product + heavy product

or in the more concise form of writing of nuclear physics:

Target (projectile, light product) heavy product.

Thus, for example, the first production of artificial radionuclide by Joliot and Curie was done by the reaction of alpha particles with aluminum metal.

^{27}Al + α $\rightarrow$ ^{30}P+n (α is the ^{4}He nucleus)

or, in the physics notation, ^{27}Al $(\alpha, n)^{30}$P.

The light product is similar to the projectile: a neutron, a small charged particle, or a photon. The basis of the nuclear activation method for chemical analysis is the measurement of the amount of either the light or the heavy product produced in a known flux of projectiles for a known length of time. The amount of the products is proportional to the number of the target atoms, and hence the measurement of the amount of the product yields the amount of the target atoms. The amount of the products formed is too small to be measured chemically (except in very rare cases),[1] and the only way to measure them is by the nuclear physics method of pulse counting. The light product can be measured due to its energy if it is a photon or high kinetic energy charged particles, or due to its reaction if it is a neutron. The common denominator of all these processes is that they must be done a very short time after the formation of the product, otherwise the light product will lose either its energy or its identity by reaction with the surrounding media. Consequently, the measurement of the light particle must be done during the bombardment of the target with the projectiles. This kind of measurement is called **Prompt Activation Analysis**.

In the case where the heavy product is radioactive, its amount can be measured by its radioactivity. This is the more common case, and since the radioactivity can be measured after the end of the interaction between the target and the beam of the projectiles, this method of analysis is called **Delayed Activation Analysis**. Since the heavy product must be radioactive, not all elements can be measured with each projectile. However, by the choice of the correct bombarding projectile, every element can be determined. Since many elements are activated simultaneously the radioactivity measurement must yield both the identity of the radioactive nuclides and their amounts. The sometimes used term "radiosotope" is wrong, since when saying "isotope" we should name the element. **Nuclide** is the term in nuclear physics which is parallel to atom in chemistry. It consists of a nucleus (composed of neutrons and photons) and electrons around it. As the identity of radionuclides (short term for radioactive nuclides) in a mixture cannot be determined from β-emission (unless very few are in the mixture), γ-ray emitters are almost exclusively used in activation analysis. The radionuclides used for identification are called indicator radionuclides and abbreviated IRN. Only a few β-emitting IRNs such as ^{32}P are used in activation analysis. In this case the phosphorus should be separated from the other radionuclides in the activated sample. The analysis by nuclear activation with activity measurements only after chemical separation, is named **Radiochemical Activation Analysis (RAA)**, whereas analysis by nuclear activation with direct measurement of the nuclear activity is named **Instrumental Activation Analysis (IAA)**. RAA is used not only for β⁻- and β⁺-emitters, but also for γ-emitting IRNs when they are masked by the more active nuclides in the sample.

The measurement of both the activity and the identity of the radionuclide can be done only for measurement of gamma rays but not for beta particles. The γ photons emitted by a radionuclide have a discrete specific energy (or sometimes several specific energies). Thus, the identity of the emitting nuclide can be determined by the energy of the gamma rays, whereas its amount is measured by the number of counts due to this energy. β-Particles, on the other hand, do not have discrete energies and the various β-emitting radionuclides in a mixture cannot be separated instrumentally. The amounts of various γ-emitters in a mixture can be determined by using a γ-ray spectrometer system. This system consists of a detector — a germanium crystal doped with Li in old models, which are named Ge(Li) detectors, or in more modern versions a very pure germanium designated as HPGe. This crystal, working as an ionization detector, sends an electric pulse every time a γ photon interacts within it. The voltage of the pulse is proportional to the energy transferred to the crystal by the photon and hence the pulse voltage is proportional to the energy of the γ photons. The pulse voltage is amplified (by a preamplifier and an amplifier) and sorted according to its voltage by a multichannel analyzer (MCA). In most modern systems, the MCA is on an electronic card inserted inside a PC, which can be used simultaneously to analyze the data.

Figure 5.1 shows an example of the γ-ray spectrum obtained from a liver sample. In this figure the number of pulses in each channel is plotted against the number of the channel. The transformation of the number of the channel to the energy of the photon is done by standard γ-ray sources, usually ^{60}Co, ^{137}Cs, and a few others. The varying half-lives of the different radionuclides formed by irradiation dictates, in order to measure as many elements as possible, that the sample will be usually irradiated twice. First the sample is irradiated for a short time (2 to 5 min) followed by a short counting (5 to 20 min). After several hours of decay, the sample is reirradiated for longer times (20 min to 8 h) followed by two countings, one for 20 to 60 min after a decay of 0.5 to 1 h and one after a decay of about 10 h for 10 to 30 h. If interest is in the concentration of only one element, one irradiation and one counting is usually sufficient.

The most common activation analysis is done with thermal neutrons as projectiles and using delayed analysis, i.e., measuring the induced radioactivity after the end of irradiation. The use of thermal neutrons is due to several factors: (1) relative large cross-sections (probability of reactions), (2) sources with large fluxes of neutrons (nuclear reactors), (3) large penetration of the neutrons, which allows the measurement of the trace elements in the bulk and not only on the surface, and (4) the reaction with thermal neutrons leads almost always to only one type of reaction (n,γ), and thus there is one-to-one correlation between the measured product and the target.

The term thermal neutrons means neutrons with kinetic energy equal to that in equilibrium at room temperature. In all nuclear reactions the emitted neutrons have high kinetic energy, and the neutrons must be slowed down (moderated) before hitting the target. Fast neutrons have the

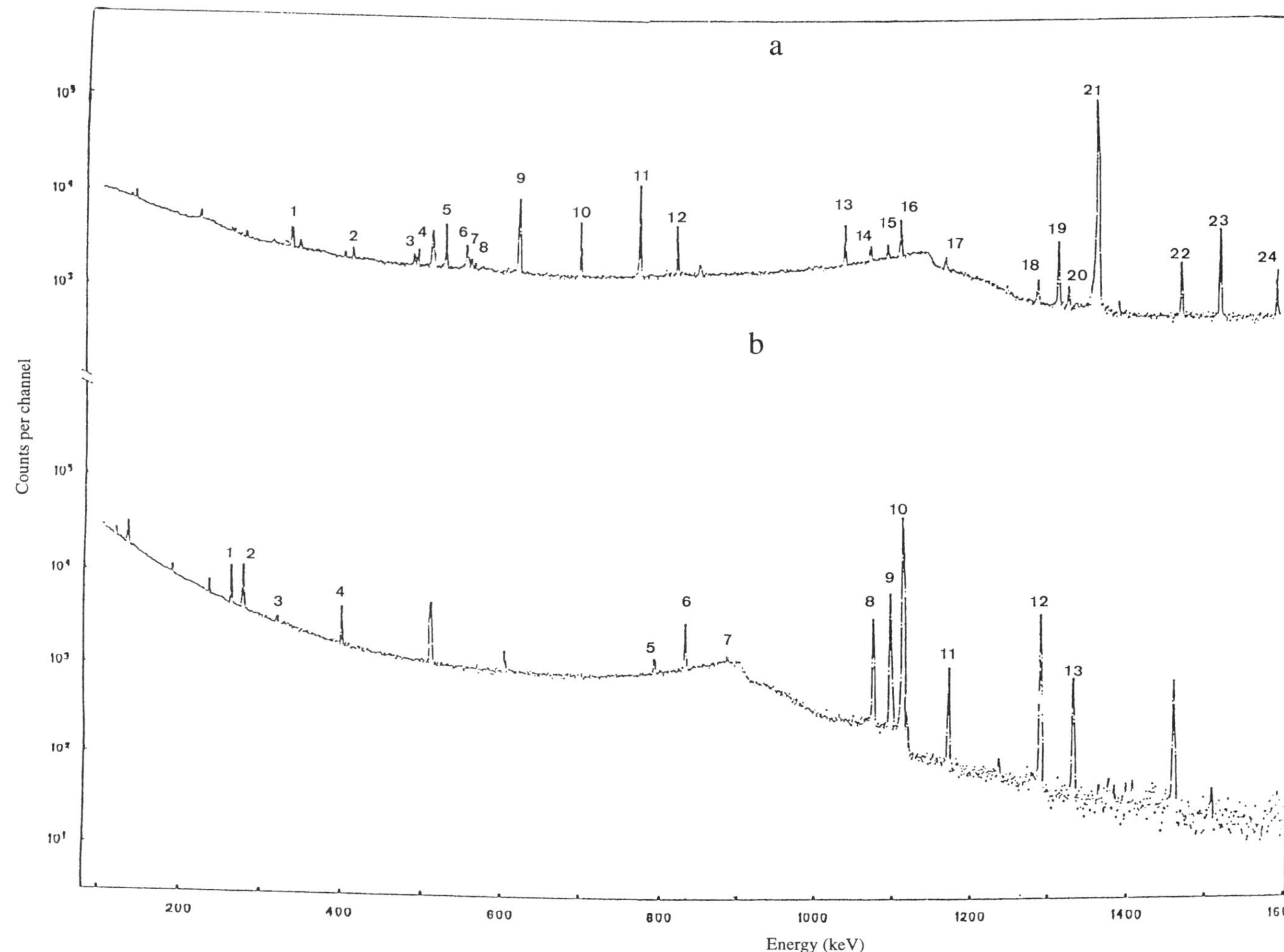

FIGURE 5.1 The γ-ray spectrum of reactor neutrons activated liver sample, irradiated for 3 h with flux of 8×10^{12} n $\bullet$ cm^{-2} s^{-1}. (a) Decay time 15 h, counting time 1 h (b) decay time 50 d, counting time 20 h. The numbers on the peaks refer to the following γ lines: (a) 1. 336 keV — ^{115}Cd, 2. 412 keV — ^{198}Au, 3. 487 keV — ^{140}La, 4. 492 keV — ^{115}Cd, 5. 528 keV — ^{115}Cd, 6. 554 keV — ^{82}Br, 7. 559 keV — ^{76}As, 8. 564 keV — ^{122}Sb, 9. 619 keV — ^{82}Br, 10. 698 keV — ^{82}Br, 11. 776 keV — ^{82}Br, 12. 827 keV — ^{82}Br, 13. 1044 keV — ^{82}Br, 14. 1077 keV — ^{86}Rb, 15. 1099 keV — ^{59}Fe, 16. 1116 keV — ^{65}Zn, 17. 1173 keV — ^{60}Co, 18. 1292 keV — ^{59}Fe, 19. 1317 keV — ^{82}Br, 20. 1332 keV — ^{60}Co, 21. 1369 keV — ^{24}Na, 22. 1474 keV — ^{82}Br, 23. 1525 keV — ^{42}K, 24. 1569 keV — ^{140}La. (b) 1. 265 keV — ^{75}Se, 2. 279 keV — ^{75}Se + ^{203}Hg, 3. 320 keV — ^{51}Cr, 4. 401 keV — ^{75}Se, 5. 796 keV — ^{134}Cs, 6. 835 keV — ^{54}Mn, 7. 889 keV — ^{46}Sc, 8. 1077 keV — ^{86}Rb, 9. 1099 keV — ^{59}Fe, 10. 1116 keV — ^{65}Zn, 11. 1173 keV — ^{60}Co, 12. 1292 keV — ^{59}Fe, 13. 1332 keV — ^{60}Co.

disadvantages of lower cross sections and more than one type of reaction. The main source for neutrons is the nuclear reactor.

Not all the elements can be detected by thermal neutron activation analysis (NAA). Thermal neutrons form by reaction with some of the element nuclides which are either stable or have only β-emission without γ-emission, and cannot be measured by radioactivity decay techniques. Other elements form too short-lived or too long-lived radionuclides, which renders the measurement impractical.

The technique of neutron activation analysis involves the insertion of the analyte (the target sample) into a nuclear reactor (either inside the reactor core, or in the surrounding moderator) for a period of time t_i. After some delay time t_d, the irradiated sample is counted on a γ-spectrometer — a Ge-γ-detector connected to a multichannel analyzer — for a period of time t_c (the subscripts stand for irradiation, delay or decay, and counting). The number of counts due to each radionuclide is given by the equation:

$$C = N_A \cdot \frac{m}{M} \cdot f \cdot \varphi \cdot \sigma \cdot I_\gamma \cdot \varepsilon \, \frac{1 - e^{-\lambda t_i}}{\lambda} \cdot e^{-\lambda t_d} (1 - e^{-\lambda t_c}), \tag{5.1}$$

where N_A is Avogadro's number, m is the mass of the element responsible for the formation of this radionuclide, M is its atomic mass, and f is the natural abundance of the isotope leading to the radionuclide used for measurement — the IRN. For example, chlorine has two natural isotopes — ^{35}Cl (the left-side superscript denotes the atomic mass, the sum of the number of protons and neutrons) and ^{37}Cl. Irradiation with thermal neutrons leads to ^{36}Cl and ^{38}Cl by $^{35}Cl(n,\gamma)$ ^{36}Cl and $^{37}Cl(n,\gamma)^{38}Cl$. Although both ^{36}Cl and ^{38}Cl are radioactive, only ^{38}Cl can be used in NAA since ^{36}Cl is both a pure β^--emitter and very long-lived ($t_{1/2} = 3 \times 10^5$ years). Each of these reasons is sufficient to prevent the measurement of the decay of ^{36}Cl. Consequently, the concentration of chlorine is determined via the decay of ^{38}Cl alone, and it is only the amount of ^{37}Cl which is important. Computation of m $\cdot$ f, where m is the mass of chlorine and f is the natural abundance of the ^{37}Cl isotope, gives the mass of ^{37}Cl in the sample. The symbol ϕ is the flux of the neutrons, i.e., the number of neutrons passing through an area unit (cm^2) in a time unit (s), and σ is the cross section (geometrical probability) that the (n,γ) reaction will occur. The unit of σ is usually the barn (1 b $= 10^{-24}$ cm^2). I_γ is the intensity of the specific γ line used for measurement (i.e., the number of photons emitted per disintegration of one nuclide); ε is the efficiency of the detection of this γ line by the detection system and λ is the decay constant of the radionuclide, which is related to the half life by the equation ($\lambda = \ln 2 / t_{1/2}$).

Equation 5.1 shows that the number of counts measured, C, is proportional to the mass of the element, and hence can be used in order to calculate it. If the physical parameters ϕ and σ are known accurately, m can be calculated from the measured C. However, the flux of the neutrons sometimes fluctuates and its spectral composition (how much the neutrons are really thermalized) also can fluctuate. Thus the simultaneous monitoring of $\phi \cdot \sigma$ is mandatory. In the past samples were irradiated (for the long irradiation) together with a multielement standard. However, more common today is to measure the sample together with one elemental standard or only a two-component standard. The two components are used for calculation of the average thermal flux and the average epithermal neutron flux. The condition for the two components is that they will have very different ratios of activation cross section in the thermal and the epithermal regions. The two components can be different compounds but preferably they are two isotopes of one element[15] or two components of one simple compound.[16]

The simplest form of NAA is the instrumental one, i.e., instrumental neutron activation analysis (INAA). In this method the sample is irradiated in a nuclear reactor. After irradiation, the sample is counted on a Ge detector without any treatment. Biological samples are irradiated in polyethylene capsules. Counting is done sometimes in the polyethylene capsule or after removing the sample from the capsule — depending on the elements to be determined, the trace elements in the polyethylene, and the irradiation site. The samples are introduced to the reactor mainly by pneumatic devices (called "rabbits"). However for open pool-type reactors and for long irradiations, the samples can be introduced from above, either by ropes or by vertical tubes.

This method of analysis is advantageous over the destructive methods (ICP or AAS) since it saves the work involved with dissolution of the samples and, even more important, it avoids the possibility of contamination by the reagents in the solubilization process. However, in cases of very minute amounts or in cases where the gamma photons from the major elements interfere with the measurement of the counts originating from the required elements, a chemical separation should be done before counting. NAA with chemical separation before counting is referred to as radio-chemical neutron activation analysis (RNAA).[17] RNAA suffers from the same disadvantage as the other nondestructive methods — the work involved in the solubilization process. In the case of RNAA this process is even more tedious and time-consuming since it involves radioactive materials and requires very careful work behind a lead shield and sometimes even remote-controlled operation with manipulators. However, RNAA has one advantage of which none of the other destructive methods can boast. The dissolution is done after irradiation, and since the measurement is done by counting the radioactive products, the impurities in the reagents and in the room do not influence the measurement. Thus, the use of RNAA does not require the use of clean rooms or special clean reagents, and can be used to measure smaller concentrations of very abundant elements.

Some elements cannot be measured by INAA with thermal neutrons due to counting interferences from the Compton effect of the major elements in biological tissues — Na and Cl. However, fortunately the major elements do not have a high cross section for activation with epithermal neutrons (neutrons with energies above normal). Consequently, elements which have a high cross section for epithermal neutrons, e.g., Br, I, and Cd, can be measured better by activation with epithermal neutrons.[18]

An advantage of INAA over the nondestructive multielement analysis method of XRES is due to the fact that both neutrons and γ photons have large penetration. Consequently, INAA can determine the bulk of a thick sample, whereas XRES methods measure only the concentrations on the surface. Biological tissues are not homogeneous and hence the surface concentrations are not always indicative of the whole tissue.

Due to large variations of the half-lives of the IRNs (indicator radioactive nuclide — the radionuclide produced by the activation which is used for determination of the concentration of the element) different times of irradiation, decay, and counting are required for various elements. Sato[20] suggested four different schemes of timings for determination of various elements in biological samples (three times of irradiations and the last one being counted twice, after different decay times). His suggestion for the various elements is given in Table 5.2. Table 5.3 gives three examples from the literature for irradiation of biological samples together with their timings and the limit of detection of some elements via INAA.

5.6.1 Interferences in NAA

There are two main kinds of interferences in the calculation of trace element concentration by INAA. The first one is the formation of the same radionuclide from two different elements. This would be impossible if all neutrons were thermal ones, since then the only possible reaction is (n,γ). However, all reactors have some higher-energy neutrons which can induce also other reactions. Examples of this case are the determination of Mg in the presence of Al, Cr in the presence of Fe, or Al in the presence of Si or P. Mg can be determined only through its less abundant isotope $^{26}Mg(n,\gamma)^{27}Mg$. The cross section for this reaction is rather low, making this measurement not very sensitive. It has interference, since ^{27}Mg can be formed from nonthermal neutrons reacting with Al in an (n,p) reaction $^{27}Al(n,p)^{27}Mg$. Similarly, the determination of chromium by $^{50}Cr(n,\gamma)^{51}Cr$ is interfered with by the nonthermal neutron reaction $^{54}Fe(n,\alpha)^{51}Cr$. Al is determined by ^{28}Al, $^{27}Al(n,\gamma)^{28}Al$, which can also be produced by the reactions $^{31}P(n,\alpha)^{28}Al$ and $^{28}Si(n,p)^{28}Al$. In flesh samples it was found that the contribution of P to ^{28}Al is more than ten times higher than the ^{28}Al from Al, in most reactors.[21]

Table 5.2 INAA Timings for Various Elements in Biological Samples

Element	IRN	Half-life of IRN	γ-ray (keV) of IRN	Timings system[c]	Interfering reactions
Na	^{24}Na	15.02 h	1369	2, 3	^{24}Mg(n,p), ^{27}Al(n,α)
Mg	^{27}Mg	9.46 min	1014	1	^{27}Al(n,p)
Al	^{28}Al	2.24 min	1779	1	^{28}Si(n,p), ^{31}P(n,α)
Si	^{31}Si	2.62 h	1266	2	^{31}P(n,p), ^{34}S(n,α)
S	^{37}S	5.0 min	3103	1	
Cl	^{38}Cl	37.3 min	1642	1, 2	
K	^{42}K	12.36 h	1525	2	^{42}Ca(n,p)
Ca	^{49}Ca	8.72 min	3084	1	
Sc	^{46}Sc	83.8 d	889	4	^{46}Ti(n,p)
V	^{52}V	3.76 min	1434	1	
Cr[a]	^{51}Cr	27.70 d	320	4	^{54}Fe(n,α)
Mn[a]	^{56}Mn	2.58 h	847	2	^{56}Fe(n,p)
Fe	^{59}Fe	44.6 d	1099	4	
Co	^{60}Co	5.27 years	1173	4	^{63}Cu(n,α)
Ni[b]	^{65}Ni	2.52 h	1482	2	^{65}Cu(n,p), ^{68}Zn(n,α)
Ni[a]	^{58}Co	70.8 d	811	4	
Cu[a]	^{66}Cu	5.10 min	1039	1	^{66}Zn(n,p)
Zn	^{65}Zn	244.1 d	1116	4	
Ga[b]	^{72}Ga	14.10 h	834	3	
Ge[b]	^{77}Ge	11.30 h	265	3	
As[a]	^{76}As	26.3 h	559	3	^{79}Br(n,α)
Se[a]	^{75}Se	118.5 d	265	4	
Br	^{82}Br	35.34 h	776	3	
Rb	^{86}Rb	18.8 d	1077	4	
Sr[a]	^{87m}Sr	2.80 h	388	2	
Mo[a]	^{99}Mo-^{99m}Tc	66.02 h	141	3	
Cd[a]	^{115}Cd-^{115}In	53.4 h	336	3	
Sb[a]	^{122}Sb	2.68 d	564	3	
Sb	^{124}Sb	60.20 d	1691	4	
I[a]	^{128}I	24.99 min	443	2	
Cs	^{134}Cs	2.06 years	796	4	
Ba[a]	^{139}Ba	82.9 min	166	2	
Au[b]	^{198}Au	2.70 d	412	3	
Hg[a]	^{203}Hg	46.8 d	279	4	

[a] Radiochemical is not indispensable but is desirable.
[b] Radiochemical separation is necessary.
[c] The timing systems are:

	system:	1	2	3	4
t_i:		1–10 min	2–20 min	10 h	10 h
t_d:		1–2 min	1 h	5–8 d	30 d
t_c:		2–5 min	10 min	1–2 h	3–5 h

Another kind of interference is from two radionuclides having very close γ lines. Examples of this kind of interference are the 846.8 keV line of ^{56}Mg and the 843.8 keV line of ^{27}Mg. Nowadays, good spectrometers are able to separate these two peaks, but this was impossible 10 or 15 years ago when germanium detectors had lower resolution. However, if they cannot be separated by the spectrometer there are two ways to overcome this interference:

1. Since the half life of ^{56}Mn (2.56 h) is longer than that of ^{27}Mg (9.45 m), a decay of 2 h will leave practically only ^{56}Mn (the activity of ^{27}Mg will decrease by a factor of 6647, whereas ^{56}Mn activity will be decreased by less than a factor of 2). Thus, measurements of the sample after the irradiation and again after a decay of 2 to 3 h will give the activities of both ^{56}Mn (from the delayed measurement) and ^{27}Mg (by subtracting the corrected activity of ^{56}Mn from the first measurement).

Table 5.3 Detection Limits Found for INAA of Various Biological Samples

Biological sample	Irradiation time[a]	Delay time	Counting time	Detection limits (ppb)
1. Human blood[b]	5 d	7 mon	2 h	Co(4), Cs(4), Fe(10,000), Se(20)
	24h, E	21 d	2 h	Br(1000), Fe(1500), Rb(300), Se(100), Zn(500)
2. Hair[c]	30 s	10 s	20 s	Ag(140), Cl(13,000), F(23,000), Se(160)
	10 m	5 m	5 m	Al(1600), Ba(7600), Ca(52,000), Cu(3800), I(260) K(16,000), Mn(750), Na(4100), S(0.28%), Zn(3500)
	16 h	2 d	50 m	As(46), Au(2), Sb(45)
	16 h	21 d	50 m	Co(97), Hg(200), Ag(140), Ba(7600), Se(160), Zn(3500)
3. Meat, fish and[d] poultry	15 s	2 m	10 m	Ca(82,000), Cl(2400), Cu(8300), K(0.1%), Mg(0.013%), Mn(1900), Na(260), V(81)
	4 h	5–28 d	7 h	Ag(120), As(650), Br(130), Cd(170), Co(23), Cr(350), Cs(39), Eu(3.7), Fe(6800), Rb(430), Sb(32), Sc(1.4), Se(300), Zn(460).

[a] d = days; h = hours; s = seconds; m = minutes; mon = months; E = irradiation with epithermal neutrons.
[b] F. Chisela and P. Bratter, *Anal. Chim. Acta*, 188, 85 (1986).
[c] A. Chatt, M. Sayjad, K.N. De Silva, and C.A. Secord in *IAEA Health-Related Monitoring of Trace Element Pollutants Using Nuclear Techniques*, IAEA TECDOC - 330, IAEA, Vienna 1985, p. 33.
[d] W.C. Cunningham and W.B. Stroube Jr, *Sci. Tot. Environ.*, 63, 29 (1987).

2. The activities of both radionuclides can be calculated using the fact that ^{27}Mg also has another γ line at 1014 keV. The ratio of the activities of the two lines can be measured in a pure sample of Mg, and thus the 843.8 keV activity of ^{27}Mg can be calculated from the 1014 keV activity. Subtracting this activity from the combined (843.8 + 846.8) keV peak, yields the activity of the ^{56}Mn 846.8 keV peak.

5.7 CHARGED PARTICLES ACTIVATION ANALYSIS (CPAA)[22,23]

CPA Analysis, i.e., activation analysis in which the radionuclides are formed by reaction with high-energy charged particles (mainly H$^+$, D$^+$, He^{+2}, and ^{3}He^{+2}) from accelerators is less popular than NAA due to: (1) lower cross section, (2) more than one product for almost every element, (3) measurement of elements on the surface only, due to the lower penetration of high-energy charged particles, and (4) the dependence of the cross section on the energy and the slowing down of the charged particles moving through the matter leads to errors in measurement of relatively thick inhomogeneous samples. However, CPAA allows the measurement of elements which cannot be measured by almost any other method. In biological samples the main use for CPAA is determination of carbon, nitrogen, and oxygen. The three elements are positron-producing emitters, which can be determined after radiochemical separation.[24]

5.8 APPLICATIONS OF NAA TO CENTRAL NERVOUS SYSTEM (CNS) TISSUE SAMPLES

5.8.1 Sensitivity of NAA to Trace Elements and Its Limitations

NAA is more sensitive in the determination of trace elements in biological samples including CNS tissues than other analytical methods such as atomic absorption spectrometry because it is not necessary to use a large sample (more than 10^2 mg dry weight) but rather only a minute one (more than 1 mg dry weight). NAA can analyze many trace elements in CNS tissues, but it is important to remember that it is the medical doctor's purpose to focus on the pathological aspects of the disease and the biological researcher's task to focus on the physiological aspects of trace element analysis. Thus we need to select our target from among the trace elements, depending on the neurological disease or type of neurotoxicity.

Besides the extreme sensitivity of NAA, it also produces quick and accurate results of multiple short-lived species (trace elements) simultaneously, by using several factors such as sample weight, length of activation (bombardment) time, and by varying the elapsed time from the end of bombardment to the time of radioactivity measurement.

It is the purpose of this section to provide information about the concentration of trace elements in CNS samples of neurological diseases compared to controls, and to show that Ca, Al, Mn, Fe, Zn, and Cu are important trace metals related to neurological disease and neurotoxicity. Table 5.1 shows the comparative sensitivity of various methods of trace element analysis. It should be noted that NAA can give particularly sensitive results for Al and Mn.

The chief limitation of NAA is the small number of nuclear reactors with NAA facilities. The method's extreme sensitivity can also pose problems if the sample size is large because of strong radioactivity after neutron activation. Hence it is profitable to measure short-lived species such as Ca, Al, Mn, and Cu by NAA because only a small amount of the sample is needed (about 1 mg dry weight). Also, NAA can analyze element levels in small regions of the CNS.

5.8.2 Normal Mineral Range of Controls

It is important to compare the mineral content in the CNS in cases of neurological disease with those of controls basically living in the same environment as the disease cases, because drinking water, garden soil, air, and many factors including minerals influence the human body. Moreover, analysis of metals such as Al needs age-matched samples because Al concentration in the brain increases with age.[25] Table 5.4(a) shows trace element concentrations in the CNS by NAA (with the exception of MG which is not receptive to NAA and was analyzed by ICP). Although exact relationships are generally unclear at this time, we know that correlation between elements (for example, between Ca and Mg) is important in disease pathogenesis.[26]

Table 5.4 (a)　Elemental Concentration (µg/g dry weight) in 26 Anatomical Regions in CNS Tissue of Normal Controls by NAA (Mg by ICP)

Ca	Al	Mn	Fe	Zn	Cu	Mg
518–700	18.0–28.2	1.45–2.43	GM:169–206 WM:213–257	GM: 73.8–81.6 GM: 49.1–55.7	13.7–38.1	440–665

Note: GM: Gray Matter; WM: White Matter of frontal and occipital lobes.

The normal range of Ca, Al, Mn, and Cu concentrations are determined in 26 anatomical CNS regions of neurologically normal controls: precentral, postcentral, frontal, parietal, temporal, and occipital gyri; insula; hippocampus; thalamus; caudate nucleus; capsula interna; globus pallidus; putamen; crus cerebri; substantia nigra; red nucleus; pons; olivary nucleus; medulla; cervical segment of spinal cord; white and gray matter cerebellum; and white matter of the frontal, parietal, temporal, and occipital cortex.[26,27] Fe and Zn content as long-lived species show those levels of gray and white matter of the frontal and occipital regions, respectively.[28]

Wet mass weight of each sample ranged between 50 to 200 mg. The samples were dried at 105°C to a constant weight (10 to 50 mg). The samples were irradiated in the thermal neutron flux of 2.3×10^{13} neutrons/cm²s for 2 min for Ca, Al, Mn, and Cu, and for 60 min for Fe and Cu. The radioactivity of ^{49}Ca, ^{28}Al, ^{56}Mn, ^{64}Cu, ^{59}Fe, and ^{69}Zn in the samples was measured by a germanium detector. The intra- and interassay coefficients of variation for the measurement of those elements by NAA were 5 to 7% and 8 to 12%, respectively. Results were expressed as micrograms per gram dry weight (mean +/– SD).

5.8.3 Usefulness of NAA as a Research/Diagnostic Tool

The most noteworthy disease from an elemental point of view is Wilson disease (hepatocerebral disease), caused by the toxic accumulation of Cu in the liver, brain, and other organs. Table 5.4(b) shows the extremely high content of Cu in the CNS tissues of Wilson disease patients. From the biopsied liver or from autopsied tissues with Wilson disease, we got a high Cu content of more than 200 µg/g dry weight compared to controls of less than 50 µg/g (see Chapter 43 in Section 13: Copper).

Amyotrophic lateral sclerosis (ALS) cases from the Western Pacific presented high Al and/or Ca deposits, and elevated Ca/Mg ratios in the CNS.[26] A reduction of Ca intake leads to the mobilization of Ca from the bones and deposition in soft tissues, in particular, in the tissues of the CNS. Mg and Zn also play important roles in activating neurons; they are mobilized from bones in order to sustain their physiological levels in the CNS.[29] Experimental as well as clinico-epidemiological studies in high-incidence foci in the western Pacific strongly implicate a low content of Ca and Mg and high levels of Al in garden soil and drinking water.[30] This demonstrates the probability that CNS degeneration is induced by a long-term low Ca and Mg intake together with excessive intake of Al and manganese (Mn), and is implicated in the pathogenesis of ALS and parkinsonism-dementia (PD).[26–28,31] Thus through using NAA, ALS and PD have been reported as mineral-related neurological diseases with regard to Ca, Al, Mn, Fe, and Zn (Table 5.4(b)).

Table 5.4 (b) Elemental Concentration in Neurologically Diseased CNS Tissue

CNS disease	Ca	Al	Mn	Fe	Zn	Cu	Mg
Wilson disease		→	→			↑↑	
ALS	↑	↑	↑	↑	↓	→	↓
Parkinson's disease	→	↑	↑	↑	→	→	↓
Parkinsonism-dementia	↑	↑	↑	↑	↓	→	↓
Alzheimer's disease	↑	↑	↑		↓	→	
Multiple sclerosis	↑↓→	→	→	→	↓	→	↓

Note: ↑: increased; ↓: decreased; →: unchanged

Some reports[32–34] have been published concerning iron (Fe), Mn, and Al using atomic absorption spectrometry or X-ray microanalysis of mineral metabolism in Parkinson's disease. As for NAA, the Ca concentration was unchanged in all 26 anatomic regions of Parkinson's-disease brains compared to control brains. The Al concentration in the substantia nigra, caudate nucleus, and globus pallidus of pathological foci responsible for Parkinson's disease was higher in Parkinson's brains and significantly higher in gray matter and basal ganglia by NAA.[35] Also, the Mg concentration was lower in the cortex, white matter, basal ganglia, and brain stem of the same Parkinson's-disease brains compared to controls by ICP.[35]

The relationship between Al and Alzheimer's disease has been well examined by other methods such as atomic absorption and discussed.[36] Using NAA, Japanese Alzheimer's disease cases revealed a diffusely high content of Ca and a slight increase in Al content in the ten analyzed brain sites, with a significant positive correlation between Ca and Al, and Ca and Mn.[37]

Multiple sclerosis (MS) is characterized by the presence of demyelinated plaques in the white matter of CNS tissues. Although the pathogenesis of MS remains unknown, recent studies indicate that low concentration of minerals, including Mg and Zn, might be involved in the pathogenesis;[38,39] Zn content is reduced along with Mg in the CNS. In the 26 anatomical regions of the CNS (whole CNS), the average Ca concentration was significantly higher in MS patients compared to controls, but the Ca content was not significantly different in gray matter, basal ganglia, brain stem, or spinal cord of MS patients compared to controls, although the Ca content was significantly lower in white matter by the NAA method.[40] The finding of lower Ca and Mg contents in white matter suggests that cholesterol synthesis was inhibited in fat-rich white matter owing to MS-induced demyelination and resulted in a decrease in Ca content in the white matter of MS patients.

REFERENCES

1. Alfassi ZB: *Determination of Trace Elements*, VCH, Weinheim, 1994.
2. Alfassi ZB: *Chemical Analysis by Nuclear Methods*, John Wiley & Sons, Chichester, 1994.
3. Broekart JAC: Plasma optical emission and mass spectrometry, *Determination of Trace Elements*, VCH, Weinheim, 1994, 191–251.
4. Chu CC, Chen PY, Yang MH, et al: Determination of trace impurities in silicone and chlorosilanes by inductively coupled plasma-atomic emission spectrometry and neutron activation analysis, *Analyst*, 1990, 115, 29–33.
5. Potts PJ: *A Handbook of Silicate Rock Analysis*, Blackie, Glasgow, 1987.
6. Jarvis KE and Willams JG: The analysis of geological samples by slurry nebulization inductively coupled plasma-mass spectrometry (ICP-MS) *Chem. Geol.*, 1989, 77, 56–82.
7. Date AR and Gray AL: Determination of trace element in geological samples by inductively coupled plasma source mass spectrometry, *Spectrochim. Acta*, 1985, 40B, 115–137.
8. Pelly IZ: Determination of trace elements by atomic absorption spectrometry, *Determination of Trace Elements*, VCH, Weinheim, 1994, 145–190.
9. Von Wandruszka R: Trace element determination by electrochemical methods, *Determination of Trace Elements*, VCH, Weinheim, 1994, 393–424.
10. Tapper UAS, Przybylowicz WJ, and Annegarn HJ: Particle-induced x-ray emission, *Chemical Analysis by Nuclear Methods*, John Wiley & Sons, Chichester, 1994, 323–360.
11. Lindh U: Use of microprobes, *Chemical Analysis by Nuclear Methods*, John Wiley & Sons, Chichester, 1994, 361–390.
12. Biran-Izak T and Mantel M: X ray fluorescence analysis with radioactive sources, *Chemical Analysis by Nuclear Methods*, John Wiley & Sons, Chichester, 1994, 391–416.
13. Watt F and Landsberg J: Nuclear microscopy: biomedical applications, *Nucl. Instrum. Methods Phys. Res.*, 1993, B77, 249–260.
14. Crapper DR, Krishnan SS, and Dalton AJ: Brain aluminum distribution in Alzheimer disease and experimental neurofibrillary degeneration, *Science*, 1973, 180, 511–514.
15. De Corte F, Speecke A, and Hoste J: Reactor neutron activation analysis by a triple comparator method, *J. Radioanal. Chem.*, 1969, 3, 205–218.
16. Alfassi ZB and Lavi N: Use of salts for calibration of the neutron flux, *J. Radioanal. Nucl. Chem.*, 1991, 150, 183–187.
17. Alfassi ZB: Radiochemical neutron activation analysis, *Determination of Trace Elements*, VCH, Weinheim, 1994, 309–357.
18. Alfassi ZB: Epithermal neutron activation analysis, *J. Radioanal. Nucl. Chem.*, 1985, 90, 151–165.
19. Alfassi ZB: Activation with nuclear reactors, in *Activation Analysis*, Vol. II, Alfassi ZB, Ed., CRC Press, Boca Raton, FL, 1990, 3–73.
20. Sato T: Activation analysis of biological materials, in *Activation Analysis*, Vol. II, Alfassi ZB, Ed., CRC Press, Boca Raton, FL, 1990, 323–358.
21. Alfassi ZB and Rietz B: Determination of Aluminum by INAA in biological samples, with special reference to NBS SRM-1577, Bovine liver, *Analyst*, 1994, 119, 2407–2410.
22. Alfassi ZB and Peisach M: *Elemental Analysis by Particle Accelerators*, CRC Press, Boca Raton, FL, 1991.
23. Sellschop JPH and Annegarn HJ: Charged particle activation analysis, *Elemental Analysis by Particle Accelerators*, CRC Press, Boca Raton, FL, 1991, 75–150.
24. Alfassi ZB: Radiochemical activation analysis by accelerated charged particles, *Elemental Analysis by Particle Accelerators*, CRC Press, Boca Raton, FL, 1991, 151–178.
25. McDermott JR, Smith AL, Iqbal K, et al: Brain aluminum aging and Alzheimer disease. *Neurology*, 1979; 29:809-814.
26. Yasui M, Yase Y, Kihira T, et al: Magnesium and calcium contents in CNS tissues of amyotrophic lateral sclerosis patients from the Kii Peninsula, *Jpn. Eur. Neurol.*, 1992; 32:95–98.
27. Yasui M, Yase Y, Ota K, et al: Aluminum deposition in the central nervous system of patients with amyotrophic lateral sclerosis. *Neurotoxicology*, 1991; 12:615–620.
28. Yasui M, Ota K, and Garruto RM: Concentration of zinc and iron in the brain of Guamanian patients with amyotrophic lateral sclerosis and parkinsonism-dementia. *Neurotoxicology*, 1993; 14:445–450.
29. Yasui M, Yano I, Yase Y, et al: Distribution of magnesium in central nervous system tissue, trabecular and cortical bone in rats fed with unbalanced diets of minerals. *J. Neurol. Sci.*, 1990; 99:177–183.
30. Iwata S: Study of the effects of environmental factors on the local incidence of amyotrophic lateral sclerosis. *Ecotoxicol. Environ. Safety*, 1977; 1:297–303.
31. Garruto RM, Fukatsu R, Yangihara R, et al: Imaging of calcium and aluminum in neurofibrillary tangle-bearing neurons in parkinsonism-dementia of Guam. *Proc. Natl. Acad. Sci. U.S.A.*, 1984; 81:1875–1879.

32. Dexter DT, Carayon A, Javoy-Agid F, et al: Alterations in the levels of iron, ferritin and other metals in Parkinson's disease and other neurodegenerative disease affecting the basal ganglia. *Brain*, 1991; 114:1953–1975.

33. Good PF, Olanow CW, and Perl DP: Neuromelanin-containing neurons of the substantia nigra accumulate iron and aluminum in Parkinson's disease: A LAMMA study. *Brain Res.*, 1992; 593:343–346.

34. Riederer P, Sofic E, Rausch WD, et al: Transition metals, ferritin, glutathion, and ascorbic acid in Parkinsonian brains. *J. Neurochem.*, 1989; 52:515–520.

35. Yasui M, Kihira T, and Ota K: Calcium, manganese and aluminum concentration in Parkinson's disease. *Neurotoxicology*, 1992; 13:593–600.

36. Lukiw WJ, Kruck TPA, McLachlan DRC: Alteration in human linker histon-DNA binding in the presence of aluminum salts in vitro and in Alzheimer's disease. *Neurotoxicology*, 1987; 8:291–302.

37. Yoshimasu F, Yasui M, Yase Y, et al: Studies on amytrophic lateral sclerosis by neutron analysis. 2. Comparative study of analytical results on Guam PD, Japanese ALS and Alzheimer disease cases. *Folia Psychiatr. Neurol. Jpn.*, 1980; 34:75–82.

38. Yasui M, Yase Y, Ando K, et al: Magnesium concentration in brains from multiple sclerosis patients. *Acta Neurol. Scand.*, 1990; 80:197–200.

39. Yasui M, Ota K, and Garrutu RM: Aluminum decreases the zinc concentration of soft tissues and bones of rats fed a low calcium-magnesium diet. *Biol. Trace Element Res.*, 1991; 31:293–304.

40. Yasui M, Kihira T, and Ota K: Calcium concentration in brains from multiple sclerosis patients. *Biol. Trace Element Res.*, 1993; 36:251–255.

Chapter 6

Axonal Transport of Neurotoxic Metals

Björn Arvidson

CONTENTS

6.1   INTRODUCTION

Neurons have a characteristic morphology with the axons extending over very long distances from the cell body to the target sites. As the ratio of cell body diameter to axonal length may be as much as $1:10^5$, neurons need efficient transport mechanisms to carry cellular constituents along the axons to the nerve terminals and back to the cell bodies. This process is known as axonal transport and occurs in both anterograde (from cell body to nerve terminals) and retrograde (reverse) direction within the axon. During recent years, it has been shown that certain metals may accumulate in neurons after retrograde transport, but the biological effects of this are incompletely known. In this article, recent findings regarding axonal transport of neurotoxic metals are reviewed, and axonal transport of metals as a possible etiological factor in certain diseases of the nervous system is discussed.

6.2 BASIC PRINCIPLES OF AXONAL TRANSPORT

6.2.1 Functions of Axonal Transport

Because biosynthesis is largely restricted to the region of the cell body and dendrites, there is a need for a constant transport of material from the cell body out into the axon. This process, *anterograde* axonal transport, is divided into a rapid and a slow phase. Rapid anterograde transport involves the movement of predominantly membrane-associated organelles (membrane proteins and lipids, small vesiculotubular structures) along linear arrays of microtubules. The rate of rapid anterograde transport is 100 to 400 mm/d. Slow axonal transport involves the anterograde movement of cytoskeletal structures and cytoplasmic proteins including glycolytic enzymes at two major and one or more minor rate components. Slow axonal transport may serve to replenish worn-out structural proteins and to supply such proteins to growing axons. Materials are also returned from the nerve terminals to the cell body by *retrograde* transport which occurs at a rate approaching that of fast anterograde transport. Among structures involved in retrograde transport are vacuoles, cup-shaped bodies, multivesicular bodies, and cisternal or tubular structures.[1] When it reaches the cell body, transported material may be delivered to the lysosomes for degradation, to the nuclear compartments for regulation of gene expression, or to the Golgi complex for repackaging.[2] Trophic substances and growth factors are returned to the cell body by retrograde transport. These factors assure the survival of the neuron, and may modulate neuronal gene expression. Retrograde transport of exogenous substances provides a pathway for viruses and certain toxins to enter the central nervous system.[1] Certain toxins transported to the cell body may cause degenerative cell lesioning and even cell necrosis. Examples of such toxins are doxorubicin and toxic lectins.

6.2.2 Molecular Mechanism of Anterograde and Retrograde Axonal Transport

During the last decade, biochemical research has generated substantial data contributing to our understanding of the mechanism of fast axonal transport. Two separate motor molecules, kinesin and cytoplasmic dynein, have been implicated in the movement of organelles along axonal micro-tubules.[2,3] Kinesin generates movement in the anterograde direction, whereas cytoplasmic dynein is thought to generate force in the retrograde direction (Figure 6.1). The cellular and molecular biology of the probable retrograde motor protein, cytoplasmic dynein, has not been as thoroughly characterized as has that of its anterograde counterpart, kinesin.[3] Some force-producing motor molecule must also be involved in the transport of slow-component constituents, but the identity of this molecule remains unknown.

6.3 METHODS FOR LOCALIZING METALS IN NERVOUS TISSUE IN CONNECTION WITH STUDIES ON AXONAL TRANSPORT

Using autoradiography, certain radioactively labeled metals can be localized in nervous tissue after axonal transport. Examples are ^{109}Cd, ^{203}Hg, and ^{54}Mn. Microautoradiography can be used to enhance resolution. The accumulation of a radioactive metal in a nerve or a group of neurons can also be quantified by gamma spectrometry. Autometallography is a histochemical process by which certain metals and metal sulfides or selenides are visualized by catalyzing the reduction of silver ions on their surfaces.[4] The autometallographic technique can be used to localize gold, silver, zinc, and mercury at light and electron microscopy levels. The classic procedure used a developer with a rather low pH, silver nitrate as silver ion donor, and a protecting colloid, gum arabic. The method has since

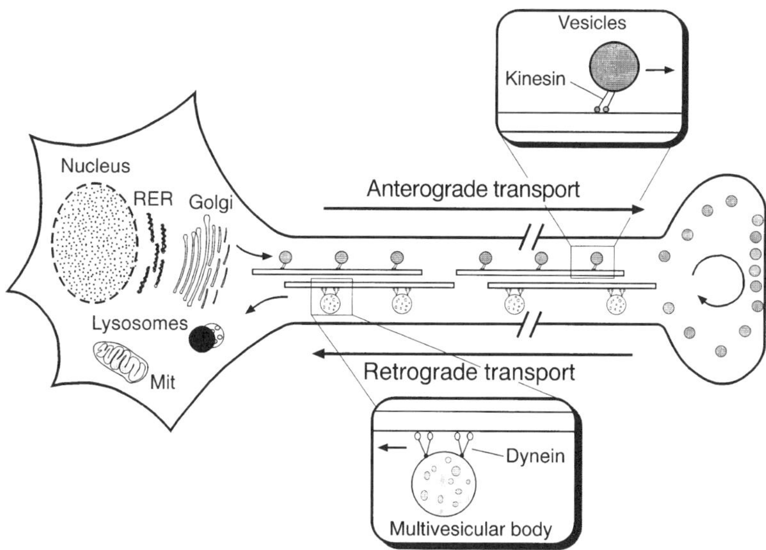

FIGURE 6.1 Schematic illustration of the movement of organelles within the axon in fast axonal transport. The organelles move along axonal microtubules in both the anterograde and retrograde directions. Membrane-associated proteins are synthesized on rough endoplasmic reticulum and assembled in the cell body after passage through the Golgi apparatus. Polypeptides are packed into vesicular organelles, and the vesicles are then transported anterogradely at rates of 150 to 400 mm/d. Movement in the anterograde direction is believed to be generated by the molecular motor kinesin. Retrograde transport of a multivesicular body is also illustrated. The retrograde motor, cytoplasmic dynein, moves organelles from the nerve terminal to the cell body at a rate approaching that of fast anterograde transport. Mit, mitochondrion; RER rough endoplasmic reticulum. (Modified from Hammerschlag, R., Cyr, J.L., and Brady, S.T., *Basic Neurochemistry: Molecular, Cellular and Medical Aspects*, 5th ed., Siegal, G.J., et al., Eds., Raven Press, New York, 1994. With permission.)

been modified and improved, with either silver lactate or silver acetate now being used as the silver ion source, and with smaller amounts of the reduction molecule hydroquinone in the medium.[4]

Perl's Prussian blue reaction has been used to visualize iron deposits in neurons after retrograde axonal transport.[5] Owing to their inherent electron opacity, a few metals (e.g., colloidal gold and the iron content of ferritin) can be visualized directly in the electron microscope.[6]

6.4 NEUROTOXIC METALS FOR WHICH AXONAL TRANSPORT HAS BEEN DEMONSTRATED

6.4.1 Lead

After injection of radiolabeled lead into the triceps surae muscle of rats, a wavelike pattern of retrograde axonal transport was observed in the sciatic nerve.[7] The transport rate was about 10 mm/d, which is slower than that of most proteins and growth factors. A second smaller peak of radioactivity was observed in the distal sciatic nerve, suggesting a wave of ^{203}Pb that traveled at an even slower rate.

In a recent study on the uptake of lead into motor axons, mice were injected with a 5% lead nitrate solution in the right tibialis anterior muscle. Deposits of lead were observed in the neuromuscular junction, and in the axoplasm and mitochondria of terminal and preterminal motor axons.[8]

6.4.2 Cadmium

Axonal transport of cadmium was first demonstrated in the hypoglossal nerve of rats.[9] Injection of a few microliters of radiolabeled cadmium in the tongue of rats was followed by an accumulation of cadmium in both hypoglossal nuclei, as revealed by autoradiography of freeze-dried sections from the lower brainstem. When one of the hypoglossal nerves was sectioned before injection, only the contralateral nucleus was labeled, strongly suggesting that retrograde axonal transport was the mechanism involved. Axonal transport of cadmium has also been demonstrated in the olfactory nerve of rats and fish. Evans and Hastings[10] instilled a solution of radiolabeled $CdCl_2$ intranasally on one side in rats, and found the Cd level in the ipsilateral olfactory bulb to be nearly 40 times higher than that in the contralateral bulb. This finding strongly supports the idea that cadmium is transported from the olfactory primary sensory neurons to the bulb by anterograde axonal transport (Figure 6.2). In this way, the olfactory primary neurons may serve as a route of entry for cadmium and other toxic agents into the CNS.[10]

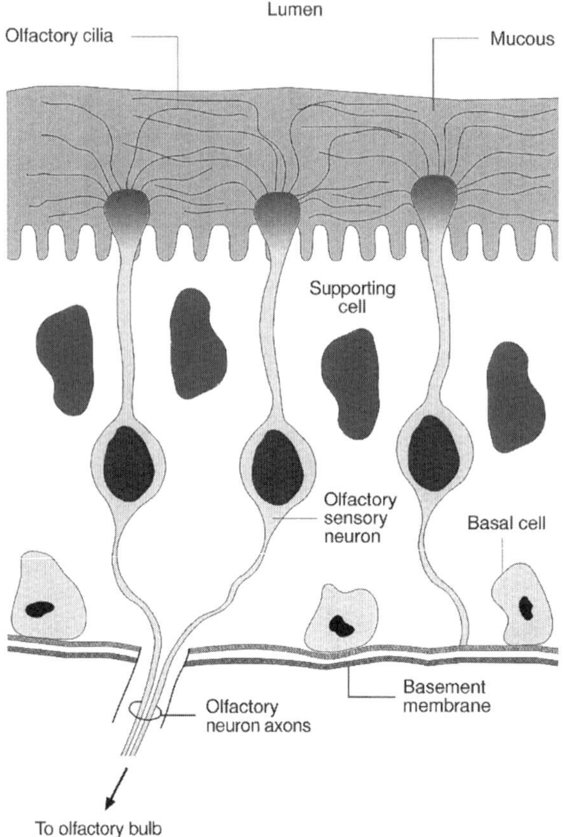

FIGURE 6.2 A schematic illustration of the olfactory epithelium. The olfactory neuron is bipolar, and a single dendrite extends to the epithelial surface. From the dendrites, fine olfactory cilia extend into a layer of mucus that lines the nasal lumen. Each olfactory neuron projects a single axon to the olfactory bulb. Experimental studies in rats have shown that after nasal instillation, cadmium is taken up into the olfactory cilia and transported anterogradely in the axons to the olfactory bulb. (Reproduced from Buck, L.B. et al., *Basic Neurochemistry: Molecular, Cellular and Medical Aspects*, 5th ed., Siegel, G.J., et al., Eds., Raven Press, New York, 1994. With permission.)

In a study on the distribution of [109]Cd in the brown trout, it was noted that [109]Cd, dissolved in the aquarial water, accumulated in the epithelium of the olfactory rosette, the olfactory nerve, and the anterior part of the olfactory bulb. This distribution of cadmium suggests an uptake of cadmium in the nerve cells of the olfactory epithelium with subsequent axonal transport to the olfactory bulb.[11] Axonal transport of cadmium has also been demonstrated in the olfactory nerve of the pike. Radiolabeled cadmium was applied in the olfactory chambers of pike, and the dynamics of the axoplasmic flow of cadmium in the olfactory nerve analyzed with gamma spectrometry and autoradiography.

Cadmium was transported at a rate of 2.18 +− 0.05 mm/h at 10°C along the olfactory nerves, accumulating predominantly in the anterior parts of the olfactory bulbs.[12]

6.4.3 Mercury

Retrograde transport of inorganic mercury was first observed in the hypoglossal nerve of mice. There was selective labeling of the hypoglossal nuclei of the lower brainstem following injection of the mercury isotope ^{203}Hg in the tongue.[13] In a few later studies, transport of inorganic mercury from the injection site to the parent motor or sensory neurons has been observed. The injection of $HgCl_2$ in the vibrissae region of rats caused accumulation of mercury deposits in the innervating facial motor and trigeminal sensory neurons, as well as in the primary sensory neurons of the trigeminal mesencephalic nucleus.[14] After injection of mercuric chloride into the tooth pulp of the first upper molar of rats, the metal could be localized to the innervating sensory neurons in the trigeminal ganglion.[15] Injection of mercuric chloride in the triceps surae muscle of rats caused accumulation of mercury deposits in ipsilateral ventral horn motoneurons and dorsal root ganglion cells.[16]

Motor neuron uptake of mercuric chloride from the circulation was found in mice after parenteral administration of an aqueous solution of the salt.[17] Mercury was distributed to the muscles by the general circulation, and gained access to the neuromuscular junctions which are situated outside the blood-brain barrier. After uptake of the metal into the nerve terminals, it was transported to motor neurons in the brain stem and spinal cord (Figure 6.3). A general labeling of motor neurons in the spinal cord and brain stem has also been observed after parenteral administration of horse-radish peroxidase, fluorescent tracer molecules, and the binding fragment of tetanus toxin.[1]

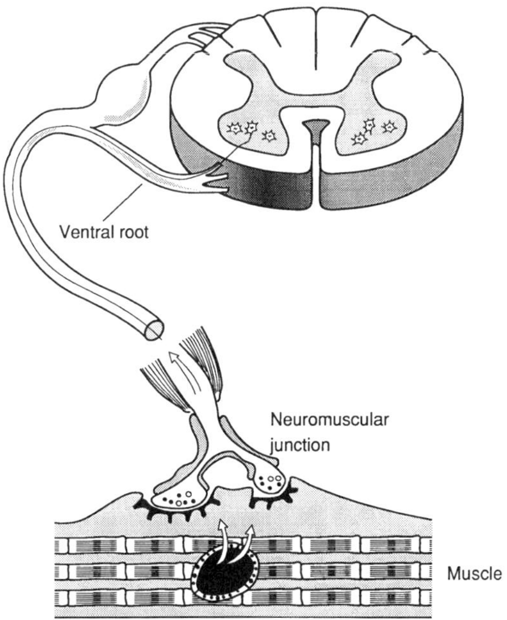

FIGURE 6.3 The putative mechanism for the accumulation of inorganic mercury in spinal motoneurons. Mercuric chloride, administered by injection or in the drinking water, is resorbed to blood and is present in plasma as a metal-protein complex. The protein-bound metal (depicted in black) leaks from capillaries in skeletal muscle into the synaptic clefts at the neuromuscular junctions. The metal is subsequently taken up into vesicles in the nerve terminals, probably by fluid-phase endocytosis, and is transported retrogradely in axons to the anterior horn motoneurons. (From Arvidson, B., *Muscle and Nerve*, 15, 1089–1094, 1992. With permission.)

6.4.4 Iron

Iron-dextran and ferritin were transported in a retrograde fashion in the hypoglossal nerves of mice after repeated injections of these substances in the tongue. In neurons of the hypoglossal nucleus, iron deposits were localized with Perl's Prussian blue reaction.[5] Transport of iron from the vibrissae muscles to the facial nerve nucleus was demonstrated by Olsson and Kristensson.[6] A comparative study was performed on the transport of native ferritin (NF) and cationized ferritin (CF) in the facial nerve of mice. Lower doses of CF than NF were required to demonstrate the presence of iron deposits in cell bodies of the facial neurons using Perl's Prussian blue reaction. This difference is explained by the fact that, in contrast to NF, CF is adsorbed on the surface of the axon terminal due to charge interaction. In this way, the uptake of CF into the axon terminal and the subsequent axonal transport is increased.[6]

Retrograde transport of iron has also been demonstrated within the central nervous system. After injection of iron-dextran into the rat corpus striatum, the metal was transported to neurons of the substantia nigra. In nigral cell bodies, iron was localized to lysosomes and mitochondria.[18] Double labeling of neuronal cell bodies with iron-dextran and horseradish peroxidase can be used to trace neuronal collateral projections.[19]

6.4.5 Thallium

The thallium isotope ^{204}Tl$^+$ was found to be transported in both anterograde and retrograde directions in the frog sciatic nerve, though retrograde transport was somewhat greater. The transport rate was about 30 mm/d at 18°C in both directions. As transport was depressed by either vinblastin, 2,4,-dinitrophenol, or low temperature, an active microtubule-dependent component is probably involved.[20]

6.4.6 Manganese

The axoplasmic flow of radiolabeled manganese was studied by autoradiography and gamma spectrometry in the olfactory nerves of pike. Manganese accumulated in the olfactory bulbs and also migrated into large areas of the telencephalon, probably via the secondary olfactory axons present in the medial olfactory tract. In addition, manganese labeled the diencephalon down to the hypothalamus. The conclusion drawn from this study was that, in contrast to cadmium, manganese can pass transsynaptically from primary to secondary olfactory neurons in the olfactory bulbs, and that the metal then migrates along secondary olfactory pathways into the telencephalon and diencephalon.[21]

6.5 MECHANISMS INVOLVED IN THE UPTAKE OF METALS IN NERVE TERMINALS

The uptake of macromolecules at the synapse may occur by different mechanisms. Certain molecules such as nerve growth factor, cholera toxin, and ricin bind to specific binding sites on the nerve terminal membrane before uptake. This mechanism has been named *specific adsorptive uptake* (or receptor-mediated uptake). Specific adsorptive uptake is very efficient and retrograde transport can be demonstrated even when concentrations of the substance present around the nerve terminals are very low.[1] Substances that do not bind to the nerve terminal membrane, such as native ferritin, horseradish peroxidase, or thorium dioxide, are taken up by a process of nonspecific or *fluid phase endocytosis*. This is a passive process by means of which small amounts of the substance are trapped along with extracellular fluid within endocytotic vesicles produced as part of a normal, ongoing process of synaptic plasma membrane turnover. An uptake mechanism which is intermediate in efficiency between nonspecific fluid-phase uptake and specific adsorptive uptake is the uptake of substances which bind to the axon terminal by charge interactions, for example, cationized ferritin or basic horseradish peroxidase isoenzymes. This form of uptake has been named *nonspecific adsorptive endocytosis.*

The mechanism by which different metals are taken up into nerve terminals for subsequent retrograde axonal transport is not known in detail. Many metals will bind to proteins in plasma after parenteral administration, forming a metal-protein complex. It is likely that these metals are internalized as metal-protein complexes into nerve terminals by fluid-phase endocytosis. Another mechanism for uptake of metals might be passage of the ionic form of the metal through ion channels of the nerve terminal membrane. *In vitro* experiments on frog muscles have suggested that divalent mercury ions may enter the motor nerve terminal through sodium and calcium channels. Evidence has also been presented that lead might enter the motor terminal in the tibialis muscle of mice by passage through calcium channels.[22]

6.6 ACCUMULATION OF METALS IN THE CELL BODY OF NEURONS AFTER AXONAL TRANSPORT

6.6.1 Localization of Metals Within the Nerve Cell

After injection of iron-dextran into the vibrissae muscles of mice, the metal was visualized with Perl's Prussian blue reaction as a granular reaction product in the cytoplasm of facial motoneurons. At the ultrastructural level, iron was seen as a finely granular electron-dense material within membrane-limited organelles, presumably representing lysosomes. Iron could no longer be detected in neurons from day 25 after the injection and onwards.[6] Nguyen-Legros et al.[18] studied axonal transport of iron to neurons in the substantia nigra of rats. Iron deposits were seen mainly within lysosomes but also within the matrix of mitochondria. The neurons were still labeled 10 days after injection of iron-dextran, but the labeling then gradually decreased. The electron microscopy findings in this study suggest that the iron lost by neurons is taken up by surrounding glial cells.

In studies on axonal transport of inorganic mercury, mercury deposits appear at the light microscopy level as fine granules in the cytoplasm of neurons. At the ultrastructural level, mercury has been localized exclusively to lysosomes.[16,17] Mercury deposits can be visualized in neurons for long periods after administration of mercury. Thus, after intramuscular injection of mercuric chloride in mice, mercury was still present in spinal motoneurons 4 months after the injection.[17]

6.6.2 Morphological and Functional Effects of Metals on the Nerve Cell Following Retrograde Transport

Investigations of nerve cell morphology after retrograde transport of iron or mercury have so far yielded no evidence of degenerative lesioning or cell necrosis. In a study of the effects of retrogradely transported iron on neurons of the facial nucleus, immature and young mice were given repeated injections of iron-dextran around the vibrissae muscles which caused a marked iron load in Schwann cells and nerve cell bodies. No morphological changes were found in neurons even after 265 days, and there was no increase in the amount of lipopigment in nerve cell bodies as compared to controls. An analysis of axon diameter in relation to myelin sheath thickness in facial nerves 223 days after repeated injections of iron-dextran around the vibrissae muscles in suckling mice showed no difference from controls, indicating facial nerve development to be normal.[6] Nguyen-Legros et al.[18] found no evidence of nerve cell degeneration in their study of iron transport in the nigrostriatal pathway of rats.

Studies on the localization of inorganic mercury in neurons after retrograde transport have shown the metal to be localized exclusively to lysosomes. Experiments *in vitro* have demonstrated that accumulation of inorganic mercury in a lysosome may cause labilization of the lysosomal membrane, and it has also been reported that mercuric chloride may impair the ability of renal lyosomal enzymes to digest substrate proteins. In electron microscopy studies of neurons accumulating mercury after retrograde transport, the lysosomes have been found to be intact and to manifest no vacuolization or membrane disruption. The nerve cells have a normal appearance without any signs of degeneration.[16,17] A possible explanation of the absence of demonstrable nerve cell damage after accumulation of inorganic mercury in lysosomes might be that the mercury is bound to selenium, which reduces the toxicity of mercuric chloride in animals. In this context, it is interesting that Aoi et al.[23] reported mercury in lysosomes of renal tubular cells from two patients intoxicated with inorganic mercury to occur in association with selenium. To our knowledge, the ability of nerve cell lysosomes containing mercury to digest substrate proteins has not been investigated.

6.7 NEUROLOGIC DYSFUNCTION POSSIBLY RELATED TO AXONAL TRANSPORT OF METALS

6.7.1 Anosmia and Cadmium Exposure

A relationship between cadmium exposure and olfactory impairment was first reported in 1948 by Friberg, who described proteinuria, emphysema, and anosmia in alkaline battery workers. Adams and Crabtree[24] reported impairment of olfaction in 106 alkaline battery workers exposed to cadmium and nickel dust, and postulated this to be due to an irritant effect of metal dust on the nasal mucosa. Later studies have confirmed that chronic occupational cadmium exposure is associated with olfactory impairment though the exact mechanism responsible for this effect of cadmium has not been clarified. Histologic findings in a worker exposed to cadmium oxide dust for 16 years were reported by Baader.[25] The nasal mucosa was atrophic, and the olfactory bulbs were stained bright yellow. The fact that the olfactory bulbs were affected suggests that the underlying mechanism might be axonal transport of cadmium from the olfactory receptor cells to the bulbs, where cadmium may, by some unknown mechanism, disturb bulb function. Experiments in rats have shown that intranasally instilled cadmium is transported from the olfactory primary sensory neurons through the cribriform plate to the olfactory bulbs.[10] Although olfactory impairment is a common finding in cadmium-exposed workers, no effect of cadmium on olfaction was observed in one experimental study. Rats exposed to cadmium oxide dust for 20 weeks did not develop anosmia, nor was there a significant shift in the detection threshold despite increased levels of cadmium in the olfactory epithelium and bulbs. The explanation might be that there is a long latency period between cadmium exposure and the development of anosmia.[26]

Transport of cadmium has been demonstrated in the olfactory nerves of trout and pike.[11,12] Cadmium has a noxious effect on olfactory epithelium in fish and is a strong inhibitor of olfactory function. It has therefore been suggested that low concentrations of cadmium in the water may be injurious to olfactory sense in fishes.[11] As olfaction is considered important for fish migration, this in turn may also be disturbed.

6.7.2 Amyotrophic Lateral Sclerosis

Neurotoxic metals have long been discussed as a pathogenetic factor in the development of amyotrophic lateral sclerosis (ALS). Increased exposure to heavy metals in ALS has been reported by some investigators,[27] but in other studies no such correlation was found.[28] A few case-reports have been published in which exposure to lead or mercury was followed by a clinical picture resembling ALS.[29] To clarify the pathogenetic relation of lead and mercury to ALS, concentrations of these metals in various tissues and fluids of patients with motor neuron disease have been analyzed, but the results are conflicting. In a study of muscle tissue of patients with different forms of motor neuron disease, no lead was detected in terminal axons or at the neuromuscular junctions.[30] It has recently been demonstrated in experimental animals that mercury in blood is taken up by motor nerve terminals and transported back to all motor neurons in the spinal cord and brain stem.[17] This mechanism might be of importance for the pathogenesis of ALS, because the metal will accumulate in a selective manner in lower motoneurons, i.e., those that are preferentially damaged in ALS. It is not known if there is also a transsynaptic transfer of the metal, but if this mechanism exists for mercury it may explain the damage which occurs to upper motoneurons in ALS. Experimental studies have failed to disclose any degenerative changes in motoneurons accumulating mercury by retrograde transport, but this might be due to a long latency period between the accumulation of mercury in motoneurons and the onset of nerve cell degeneration.

6.7.3 Neurotoxic Effects of Manganese

A recent study on the transport of manganese in the olfactory neurons of the pike showed that manganese accumulated in the olfactory bulbs after application of $^{54}Mn^{2+}$ in the olfactory chambers. In addition, the metal migrated into large areas of the telencephalon and diencephalon, presumably due to a transsynaptic passage to the secondary olfactory neurons and their axons in the medial olfactory tract. Based on these results, a theory has been proposed that uptake of manganese into the brain via olfactory neurons might underlie the extrapyramidal motor disorder which occurs in workers with occupational exposure to manganese-containing dusts or fumes.[21] The results are interesting, but additional experimental studies in mammals are necessary before this theory can be accepted or rejected.

6.8 SUMMARY AND CONCLUSIONS

There is a constant flow of material from the region of the cell body and dendrites of neurons out into the axon, a process known as anterograde axonal transport. Similarly, materials are returned to the cell body by retrograde transport. Experimental work in recent years has shown that neurotoxic metals such as lead, mercury, and cadmium may accumulate in neurons after retrograde transport. Metals may also be transported in the anterograde direction. The mechanism for uptake of metals at the neuromuscular junction prior to retrograde transport is not known in detail. Uptake of metal-protein complexes into recycling vesicles of the nerve terminal and passage of the ionic form of the metal through ion channels of the nerve terminal membrane are probably important mechanisms. The distribution of iron and inorganic mercury within the nerve cell after retrograde transport has been studied in rodents with electron microscopy. With histochemical techniques, inorganic mercury has been localized exclusively to lysosomes, and iron to both lysosomes and mitochondria. Neither mercuric mercury nor iron has been shown to induce any degenerative changes in neurons after retrograde transport. The long-term effects of lysosomal accumulation of these metals in neurons are still incompletely known, however. Studies in rats and certain fish (trout, pike) have demonstrated that cadmium is transported from the olfactory primary sensory neurons to the olfactory bulb. Based on these experimental studies, it has been suggested that retrograde transport of cadmium from the olfactory epithelium to the bulbs is a mechanism by which cadmium induces olfactory dysfunction in humans. Inorganic mercury in blood leaks from capillaries in muscle, and by retrograde transport from nerve terminals in muscle accumulates selectively in lower motor neurons of the spinal cord and brain stem in rodents. This mechanism is probably relevant for other neurotoxic metals as well. By this mechanism, mercury, and perhaps other metals which undergo retrograde transport, may play a part in the pathogenesis of amyotrophic lateral sclerosis.

REFERENCES

1. Schwab M and Thoenen H: Retrograde axonal transport. In Lajtha A (ed) *Handbook of Neurochemistry*, vol 5. New York, Plenum, 1983, pp 381–404.
2. Hammerschlag R, Cyr JL, and Brady ST: Axonal transport and the neuronal cytoskeleton. In Siegel GJ, Agranoff BW, Wayne Albers R, et al (eds): *Basic Neurochemistry: Molecular, Cellular and Medical Aspects*. 5th Ed. New York, Raven Press, 1994, pp 545–571.
3. Cyr JL and Brady ST: Molecular motors in axonal transport. Cellular and molecular biology of kinesin. *Neurobiology*, 1992; 6: 137–155.
4. Danscher G: Applications of autometallography to heavy metal toxicology. *Pharmacol. Toxicol.*, 1991; 69: 414–423.

5. Malmgren L, Olsson Y, Olsson T et al: Uptake and retrograde axonal transport of various exogenous macromolecules in normal and crushed hypoglossal nerves. *Brain Res.*, 1978; 153: 477–493.

6. Olsson T and Kristensson K: Neuronal uptake of iron: somatopetal axonal transport and fate of cationized and native ferritin, and iron-dextran after intramuscular injections. *Neuropathol. Appl. Neurobiol.*, 1981; 7: 87–95.

7. Baruah JK, Rasool CG, Bradley WG, et al: Retrograde axonal transport of lead in rat sciatic nerve. *Neurology*, 1981; 31: 612–616.

8. Pamphlett R and Bayliss A: Lead uptake in motor axons. *Muscle and Nerve*, 1992; 15: 620–625.

9. Arvidson B: Retrograde axonal transport of cadmium in the rat hypoglossal nerve. *Neurosci. Lett.*, 1985; 62: 45–49.

10. Evans J and Hastings L: Accumulation of Cd (II) in the CNS depending on the route of administration: intraperitoneal, intratracheal or intranasal. *Fundam. Appl. Toxicol.*, 1992;19:275–278.

11. Tjälve H, Gottofrey J, and Björklund I: Tissue disposition of $^{109}Cd^{2+}$ in the brown trout (*Salmo Trutta*) studied by autoradiography and impulse counting. *Toxicol. Environ. Chem.*, 1986; 12: 31–38.

12. Gottofrey J and Tjälve H: Axonal transport of cadmium in the olfactory nerve of the pike. *Pharmacol. Toxicol.*, 1991; 69: 242–252.

13. Arvidson B: Retrograde axonal transport of mercury. *Exp. Neurol.*, 1987; 98: 198–203.

14. Arvidson B: Accumulation of mercury in brainstem nuclei of mice after retrograde axonal transport. *Acta Neurol. Scand.*, 1990; 82: 234–237.

15. Arvidson B and Arvidsson J: Retrograde axonal transport of mercury in primary sensory neurons innervating the tooth pulp in the rat. *Neurosci. Lett.*, 1990; 115: 29–32.

16. Schiönning J: Retrograde axonal transport of mercury in rat sciatic nerve. *Toxicol. Appl. Pharmacol.*, 1993; 121: 43–49.

17. Arvidson B: Inorganic mercury is transported from muscular nerve terminals to spinal and brainstem motoneurons. *Muscle and Nerve*, 1992; 15: 1089–1094.

18. Nguyen-Legros J, Cesaro P, Gay M, et al: Evolution du lieu de stockage du dextran-fer transporté par le flux axonal rétrograde dans le système nerveux central du rat. *Acta Neuropathol.*, 1981; 54: 101–112.

19. Olsson T and Kristensson K: A simple histochemical method for double labelling of neurons by retrograde axonal transport. *Neurosci. Lett.*, 1978; 8: 265–268.

20. Bergquist J-E, Edström A, and Hansson PA: Bidirectional axonal transport of thallium in frog sciatic nerve. *Acta Physiol. Scand.*, 1983; 117: 513–518.

21. Tjälve H, Mejàre C, and Borg-Neczak K: Uptake and transport of manganese in primary and secondary olfactory neurons in pike. *Pharmacol. Toxicol.*, 1995; 77: 23–31.

22. Pamphlett R and Bayliss A: The effect of nerve crush and botulinum toxin on lead uptake in motor axons. *Acta Neuropathol.*, 1992; 84: 89–93.

23. Aoi T, Higuchi T, Kidokoro R, et al: An association of mercury with selenium in inorganic mercury intoxication. *Hum. Toxicol.*, 1985; 4: 637–642.

24. Adams RG and Crabtree N: Anosmia in alkaline battery workers. *Br. J. Ind. Med.*, 1961; 18: 216–221.

25. Baader EW: Chronic cadmium poisoning. *Ind. Med. Surg.*, 1952; 21: 427.

26. Hastings L: Sensory neurotoxicology: use of the olfactory system in the assessment of toxicity. *Neurotoxicol. Teratol.*, 1990; 12: 455–459.

27. Currier RD and Haerer AF: Amyotrophic lateral sclerosis and metallic toxins. *Arch. Environ. Health*, 1968; 17: 712–719.

28. Gresham LS, Molgaard CA, Golbeck AL, et al: Amyotrophic lateral sclerosis and occupational heavy metal exposure: a case-control study. *Neuroepidemiology*, 1986; 5: 29–38.

29. Adams CR, Ziegler DK, and Lin JT: Mercury intoxication simulating amyotrophic lateral sclerosis. *J. Am. Med. Assoc.*, 1983; 250: 642–643.

30. Pamphlett R: Looking for lead at endplates in motor neuron disease. *Med. J. Austr.*, 1991; 154: 637.

PART II

SPECIFIC MINERALS AND METALS RELATED TO EXPERIMENTAL AND CLINICAL NEUROTOXICOLOGY

SECTION 4

ALUMINUM

Chapter 7

Chemistry of Aluminum in the Central Nervous System

R. Bruce Martin

CONTENTS

7.1 CHEMISTRY OF Al^{3+} AND COMPARISON WITH OTHER METAL IONS

To understand the roles of an element we need to know not only the gross amount present, but also the locale and complexes or compounds into which the element enters. The chemistry of aluminum is relatively simple. Its hydroxide is much more soluble than that of Fe^{3+}, and it exhibits only one oxidation state in biological systems, Al^{3+}. Metallic Al is too reactive to be found free in nature, and the metal is won from its ores only with difficulty.[1] Thus there is no oxidation-reduction chemistry to Al^{3+} in biology. (We employ Al(III) as a generic representation of the 3+ ion when a specific form is not indicated.)[1]

Al^{3+} is a small ion: the effective ionic radius of Al^{3+} in 6-fold coordination is only 54 pm. By way of comparison, other values are Ga^{3+}, 62; Fe^{3+}, 65; Mg^{2+}, 72; Zn^{2+}, 74; Fe^{2+}, 78; and Ca^{2+}, 100 pm.[2] On the basis of the radii, Al^{3+} is closest in size to Fe^{3+} and Mg^{2+}, and it is to these ions that we compare Al^{3+}. Ca^{2+} is much larger, and in its favored 8-fold coordination exhibits a radius of 112 pm, yielding a volume 9 times that of Al^{3+}. In the mixed crystal Ca$_3$Al$_2$(OH)$_{12}$, each hexacoordinate Al^{3+} is surrounded by six hydroxide ions and each cubic Ca^{2+} by eight hydroxide ions. Each metal ion adopts its own favored coordination number. The Al–O distances are 192 pm and the average of the Ca–O distances is 250 pm.[3] The difference of 58 pm agrees exactly with the difference of ionic radii quoted above between six-coordinate Al^{3+} and eight-coordinate Ca^{2+}. Thus the Al^{3+} and Ca^{2+} sites are distinctly different; one metal ion does not substitute for the other. For these reasons it is unlikely that Al^{3+} binds strongly to the Ca^{2+} sites of calmodulin.[1,4] Only weak binding of Al^{3+} to calmodulin has been found by some investigators.[5,6] With one-quarter of its amino acid residues bearing carboxylate side chains, calmodulin is an acidic protein that should bind multiply charged ions as a polyelectrolyte. When it does so, physical changes upon addition of Al^{3+} are merely those of denaturation. It is, however, likely that Al^{3+} interacts with calmodulin-regulated proteins that involve phosphate groups. By this route calmodulin-dependent reactions may exhibit an Al^{3+} dependence.[1,4]

We have argued that in biological systems Al^{3+} will be more competitive with Mg^{2+} than with Ca^{2+}.[1,7] In both mineralogy and biology comparable ionic radii frequently outweigh charge in determining behavior. More Al^{3+} is accumulated by central nervous system tissue when the Mg^{2+} concentration is low.[8] Both Al^{3+} and Mg^{2+} favor oxygen donor ligands, especially phosphate groups.[9] Al^{3+} is 10^7 times more effective than Mg^{2+} in promoting polymerization of tubulin to microtubules.[10] In this study the free Al^{3+} concentration was controlled near 10^{-12} M with nitrilotriacetate (NTA). Wherever there is a process involving Mg^{2+}, there exists an opportunity for interference by Al^{3+}.

The most likely Al^{3+} binding sites are oxygen atoms, especially if they are negatively charged. Carboxylate, deprotonated hydroxy (as in catecholates and serine and threonine), and phosphate groups are the strongest Al^{3+} binders. These binding characteristics differ sharply from those of the heavy metal ions that bind to sulfhydryl and amine groups. Even when part of a potential chelate ring, sulfhydryl groups do not bind Al^{3+}. Amines bind Al^{3+} strongly only as part of multidentate ligand systems as in NTA and EDTA. Amino acids are weak binders barely competing with metal ion hydrolysis.[11] The nitrogenous bases of DNA and RNA do not bind Al^{3+} strongly.[1,4] The weakly basic phosphate group of RNA and DNA also bind Al^{3+} weakly,[12] while the basic and chelating phosphate groups of nucleoside di- and triphosphates do bind Al^{3+} strongly.[13] Within cells Al^{3+} is likely bound to nucleoside di- and triphosphates.[13]

7.2 EXCHANGE

In addition to stability of metal ion complexes, an important and often overlooked feature is the rate of ligand exchange out of and into the metal ion coordination sphere. Ligand exchange rates take on special importance for Al^{3+} because they are slow and systems may not be at equilibrium. The rate for exchange of inner sphere water with solvent water is known for many metal ions, and the order of increasing rate constants in acidic solutions is given by

$$Al^{3+} \ll Fe^{3+} < Ga^{3+},\ Be^{2+} \ll Mg^{2+} < Fe^{2+} < Zn^{2+},\ Mn^{2+} < Ca^{2+},\ Cd^{2+} < Cu^{2+},\ Hg^{2+},\ Pb^{2+}$$

Included in this listing are metal ions discussed in other chapters of this volume. Each inequality sign indicates an approximate 10-fold increase in rate constant from 1.3 s^{-1} for Al^{3+} and increasing through 9 powers of 10 to $>10^9$ s^{-1} for Pb^{2+} at 25°C. Though these specific rate constants refer to water exchange in aquo metal ions, they also reflect relative rates of exchange of other ligands. Contrary to popular opinion, the strongly binding heavy metal ions Cd^{2+}, Hg^{2+}, and Pb^{2+} also undergo rapid exchange, a feature that promotes relatively rapid distribution of these toxic metal ions throughout the body.[2] Reducing Fe^{3+} to Fe^{2+} gains a 10^4-fold rate increase. Chelated ligands exchange more slowly, but the order remains. The slow ligand exchange rate for Al^{3+} makes it useless as a metal ion engaged in enzyme active site reactions. The 10^5 times faster rate for Mg^{2+} furnishes enough reason for Al^{3+} inhibition of enzymes with Mg^{2+} cofactors. Processes involving rapid Ca^{2+} exchange would be thwarted by substitution of the 10^8-fold slower Al^{3+}.

7.3 Al^{3+} HYDROLYSIS

Whatever ligands may be present, understanding the state of Al(III) in any aqueous system demands awareness of the species that Al(III) forms with the components of water at different pH values. In solutions more acid than pH <5, Al(III) exists as an octahedral hexahydrate, $Al(H_2O)_6^{3+}$, usually abbreviated as Al^{3+}. As a solution becomes less acidic, $Al(H_2O)_6^{3+}$ undergoes successive deprotonations to yield $Al(OH)^{2+}$, $Al(OH)_2^{+}$, and a soluble $Al(OH)_3$, with a decreasing and variable number of water molecules.[1,14] Neutral solutions give an $Al(OH)_3$ precipitate that redissolves owing to formation of tetrahedral aluminate, $Al(OH)_4^{-}$, the primary soluble Al(III) species at pH >6.2. Aqueous polynuclear complexes are unlikely at less than 10 µM total Al(III).[15]

The four successive deprotonations from $Al(H_2O)_6^{3+}$ to yield $Al(OH)_4^-$ squeeze into an unusually narrow pH range of less than one log unit, with pK_a values of 5.5, 5.8, 6.0, and 6.2.[16] In contrast, for Fe^{3+} the range is more than six log units. The narrow span for Al^{3+} is explained by the cooperative nature of the successive deprotonations due to a concomitant decrease in coordination number from six to four.[14] Thus only two species dominate over the entire pH range, the octahedral hexahydrate $Al(H_2O)_6^{3+}$ at pH <5.5, and the tetrahedral $Al(OH)_4^-$ at pH >6.2, while there is a mixture of hydrolyzed species and coordination numbers between 5.5 < pH < 6.2 (distribution curves appear in the references).[11,14,17,18] If in addition other ligands are incapable of holding Al(III) in solution, it becomes necessary to include the solubility equilibrium.[1,11,14] To avoid precipitation, Al(III) may be administered internally as a soluble citrate complex, using at least a 2:1 citrate-to-metal ion mole ratio. Maltol forms a weaker complex and allows a higher free Al^{3+} concentration.[11]

7.4 Al^{3+} IN THE PLASMA AND CEREBROSPINAL FLUID

The best starting point for discussion of the chemistry of aluminum in the cerebrospinal fluid (CSF) is to review the chemistry of aluminum in the blood plasma, where there is good agreement between experimental measurements and elementary considerations. By the latter we mean the distribution of Al^{3+} from knowledge of the ligands present and their known stability and solubility product constants with Al^{3+} and other metal ions.

Table 7.1 compares those components of the blood plasma and the CSF that have been considered important in setting the level of free Al^{3+}. There have been inconsistencies in reporting concentrations in the CSF; the results in Table 7.1 are taken from a single compilation. The ratio of CSF to plasma contents listed in the last column of Table 7.1 indicates that except for transferrin (and Zn^{2+}) the two fluids contain comparable concentrations of the listed components. Thus, consideration of extensive work on the plasma provides an excellent starting point for discussing the CSF.

Table 7.1 Comparison of Blood Plasma and Cerebrospinal Fluid (CSF)

	Plasma[a]	CSF[b]	CSF / Plasma
pH	7.4	7.33	
Inorganic phosphate	1.1 mM	0.49 mM	0.44
Citrate	0.10 mM	0.18 mM	1.8
Transferrin, free sites	50 μM	< 0.25 μM	< 0.005
Citrate / phosphate	0.09	0.37	4.1
Citrate / transferrin	2.0	> 720	> 360
Lactate	0.9 mM	1.4 mM	1.6
Amino acids	4.0 mM	1.8 mM	0.5
Catecholamines	5 nM	1.6 nM	0.3
Mg^{2+} (free)	0.9 mM	1.0 mM	1.1
Ca^{2+} (free)	1.2 mM	0.9 mM	0.7
Zn(II) (total)	14 μM	0.5 μM	0.04
pAl, 1 μM total Al(III)	14.6	14.0	
pX (free Al^{3+})	8.6	8.0	

[a] From *Geigy Scientific Tables*, 8th ed., Vol. 3, Ciba-Geigy Limited, Basel, 1984.

[b] From *Geigy Scientific Tables*, 8th ed., Vol. 1, Ciba-Geigy Limited, Basel, 1981.

Citrate is the main small molecule binder of Al^{3+} in the plasma.[4,12,19] Even though much of the citrate in plasma occurs as a Ca^{2+} or Mg^{2+} complex, Al^{3+} easily displaces Ca^{2+} from citrate. Including consideration of alkaline earth cations in the plasma, there is an almost 10^8 mole ratio of citrate-bound to -unbound Al^{3+}.[11] Lactate and amino acids (including aspartate and glutamate[17]) are weak binders, barely competitive with Al^{3+} hydrolysis to $Al(OH)_4^-$ in neutral solutions.[11] It is unlikely that any process involving Al^{3+} occurs via an amino acid complex as there are more strongly binding ligands in a biological milieu. Though the stability constants for binding to catecholamines are high, there is competition from protons in neutral solutions, and unless they occur at significant concentrations they are not apt to be determining Al^{3+} binders.[20] In the plasma, insoluble and soluble phosphate complexes are not competitive with those of citrate in sequestering Al^{3+}.[19]

Transferrin is the main protein binder of Al^{3+} in the plasma; albumin exhibits little ability to bind Al^{3+}.[4,21] Though not able to displace the much more strongly binding Fe^{3+}, with only 30% of its sites occupied with Fe^{3+}, unoccupied transferrin sites occur at 50 μM in the plasma. In the blood plasma about 90% of Al^{3+} is bound to transferrin and most of the remaining 10% to citrate.[19] Thus mainly transferrin sets the free Al^{3+} concentration allowed in the plasma.

In the CSF the transferrin concentration is about 0.5% of that in the plasma (Table 7.1). Moreover, if most of the transferrin sites in the CSF are occupied by the much more strongly binding Fe^{3+}, then this protein is not a factor in Al^{3+} binding. A report of twice the transferrin concentration of 0.5 μM in the CSF[22] is still so low that it does not alter the conclusion that transferrin seems unlikely to set the level of free Al^{3+} in the CSF.

On the other hand, the citrate concentration is 1.8 times greater in the CSF than in the plasma (Table 7.1). The citrate/transferrin site ratio is 2.0 in the plasma and >720 in the CSF (Table 7.1). A 2.4 times greater citrate concentration of 0.44 mM in the CSF has been found by NMR.[23] Thus citrate appears as the most likely candidate to set the free Al^{3+} concentration in the CSF.

7.5 pAl = −log [Al³⁺]

In addition to the hydrolysis features considered previously in this chapter (Section 7.3), the amount of free, aqueous Al^{3+} in solution depends upon several variables: ligands present, their stability constants with Al^{3+}, and total Al(III) to total ligand mole ratio. For ligands with protons competing with metal ion for binding sites in the pH range of interest, the pH is also a variable. Thus, instead of simple association of metal ion and basic ligand, $Al^{3+} + L \rightarrow AlL$, the relevant reaction may become displacement of a proton from the acidic ligand by the metal ion, $Al^{3+} + HL \rightarrow AlL + H^+$. For ligands containing amino, phenolate, and catecholate groups, the amount of free, aqueous Al^{3+} in neutral solutions becomes pH dependent. Thus for these ligands, listed stability constants overstate effective binding strengths and need to be lowered to reflect competition of the proton with the metal ion for basic binding sites. The most practical method to allow for proton-metal ion competition at a ligand is to calculate conditional stability constants applicable to a single pH.[1,11] Conditional stability constants may also allow for deprotonation of metal ion coordinated water that yields more stable complexes with increasing pH in some Al^{3+} complexes, such as with citrate, nitrilotriacetate, and EDTA.

Results from quantitative evaluation of conditional stability constants are revealingly expressed as the negative logarithm of the free Al^{3+} concentration, $-\log [Al^{3+}] = pAl$. Analogous to pH, higher pAl values represent lesser amounts of free Al^{3+}. Calculated values of pAl for a variety of biological ligands appear in three articles.[11,12,17] In calculating the values in Table 7.1, competitive binding by Mg^{2+} and Ca^{2+} with citrate is taken into account.

The second to last row in Table 7.1 shows the allowed free Al^{3+} molar concentration as $10^{-14.6}$ M in the plasma and $10^{-14.0}$ M in the CSF for a 1 μM total Al(III) concentration. Alternatively, we may express the result as the mole fraction of free Al^{3+}, $X = [Al^{3+}]/[Al(III)]$, or

$$pX = -\log(\text{mole fraction free } Al^{3+}) = pAl + \log[Al(III)]$$

where the last term in brackets is the total Al(III) molar concentration. The values for pX appear as the last entry in Table 7.1.

The conclusions presented in the last two lines of Table 7.1 suggest that for the same total Al(III) concentration, the free Al^{3+} concentration is four times greater in the CSF than in the blood plasma. If the greater CSF citrate concentration of 0.44 mM is used, then for the CSF the last two entries in Table 7.1 become pAl = 14.4 and pX = 8.4, so that the difference with the plasma is narrowed to a factor of less than two.

Therefore, we conclude that the higher citrate concentration in the CSF than in the plasma holds the allowed free Al^{3+} concentration in the CSF to about two to four times that of the plasma, where the level is set mainly by unoccupied transferrin sites. In both fluids the allowed level of free Al^{3+} is very low: less than 10^{-14} M for 1 μM total Al(III).

Though the allowed free Al^{3+} concentration in the CSF is only two to four times greater than that in the plasma, there may be a much greater *rate* of reaction of citrate-bound Al^{3+} in the CSF compared to the nearly inert transferrin bound Al^{3+} in the plasma.

7.6 Al^{3+} BINDING TO PHOSPHORYLATED PROTEINS

A low level of free Al^{3+} in the CSF is consistent with the finding of Al(III) associated with brain proteins. That calmodulin binds Al^{3+} only weakly has been mentioned previously in this chapter (Section 7.1). Phosphorylation and dephosphorylation reactions normally accompany cellular processes. The phosphate groups of any phosphorylated protein provide the requisite basicity, and in conjunction with juxtaposed carboxylate or other phosphate groups, become strong Al^{3+} binding sites. (The substituted malonate function furnished by gamma-carboxyglutamate side chains should also provide strong Al^{3+} binding sites.) Abnormally phosphorylated proteins have been found in Alzheimer's-disease brains.[24,25] Al(III) induces covalent incorporation of phosphate into human tau protein.[26] Al^{3+} aggregates highly phosphorylated brain cytoskeletal proteins[27] and induces conformational changes in phosphorylated neurofilament peptides that are irreversible to added citrate.[28] High Al(III) contents have been found associated with increased linker histones in the nuclear region of Alzheimer's-disease brains.[29] Al(III) induces neurofibrillary tangles in the perikaryon of neurons.[30] Ternary Al^{3+} complexes have received little study, and Al(III) has been used as a tanning or cross-linking reagent. Al^{3+} seems capable of cross-linking proteins, and proteins and nucleic acids.

More examples and details of Al^{3+} interactions with proteins appear in the other chapters in Section 4 of this volume.

REFERENCES

1. Martin RB: Bioinorganic chemistry of aluminum. *Metal Ions Biol. Syst.*, 1988; 24: 1–57.
2. Martin RB: Bioinorganic chemistry of metal ion toxicity. *Metal Ions Biol. Syst.*, 1986; 20: 21–65.
3. Weiss R and Grandjean D: Structure de l'aluminate tricalcique hydrate. *Acta Crystallogr.*, 1964; 17: 1329–30.
4. Martin RB: The chemistry of aluminum as related to biology and medicine. *Clin. Chem.*, 1986; 32: 1797–1806.
5. Richardt G, Federolf G, and Habermann E: The interaction of aluminum and other metal ions with calcium-calmodulin-dependent phosphodiesterase. *Arch. Toxicol.*, 1985; 57: 257–259.
6. You G and Nelson DJ: Al^{3+} versus Ca^{2+} ion binding to methionine and tyrosine spin-labeled bovine brain calmodulin. *J. Inorg. Biochem.*, 1991; 41: 283–291.
7. Macdonald TL and Martin RB: Aluminum ion in biological systems. *Trends Biochem. Sci.*, 1988; 13: 15–19.
8. Mitani K: Relationship between neurological diseases and aluminium load, especially amyotrophic lateral sclerosis, and magnesium status. *Magnesium Res.*, 1992; 5: 203–213.
9. Martin RB: Bioinorganic chemistry of magnesium. *Metal Ions Biol. Syst.*, 1990; 26: 1–13.
10. Macdonald TL, Humphreys WG, and Martin RB: Promotion of tubulin assembly by aluminum ion in vitro. *Science*, 1987; 236: 183–186.
11. Martin RB: Aluminum in biological systems. In Nicolini M, Zatta PF, Corain B (eds): *Aluminium in Chemistry, Biology, and Medicine.* New York, Raven Press, 1991, pp 3–20.
12. Martin RB: Aluminum: a neurotoxic product of acid rain. *Acc. Chem. Res.*, 1994; 27: 204–210.
13. Kiss T, Sovago I, and Martin RB: Al^{3+} binding by adenosine-5′-phosphates: AMP, ADP, and ATP. *Inorg. Chem.*, 1991; 30: 2130–2132.
14. Martin RB: Fe^{3+} and Al^{3+} hydrolysis equilibria. Cooperativity in Al^{3+} hydrolysis reactions. *J. Inorg. Biochem.*, 1991; 44: 141–147.
15. Furrer G, Trusch B, and Muller C: The formation of polynuclear Al-13 under simulated natural conditions. *Geochim. Cosmochim. Acta*, 1992; 56: 3831-3838.
16. Ohman L: Stable and metastable complexes in the system $H^+–Al^{3+}$–citric acid. *Inorg. Chem.*, 1988; 27: 2565–2570.

17. Martin RB: Aluminium speciation in biology. In Ciba Foundation Symposium 169: *Aluminium in Biology and Medicine*. Chichester, John Wiley & Sons, 1992, pp 5–25 and 104–108.

18. Martin RB: The elements in the environment. In Downs AJ (ed): *Chemistry of Aluminium, Gallium, Indium and Thallium*. London, Blackie, 1993, pp 474–490.

19. Ohman L and Martin RB: Citrate as the main small molecule binding Al^{3+} in serum. *Clin. Chem.*, 1994; 40: 598–601.

20. Kiss T, Sovago I, and Martin RB: Al^{3+} binding to catecholamines and tiron. *J. Am. Chem. Soc.*, 1989; 111: 3611–3614.

21. Martin RB, Savory J, Brown S, et al.: Transferrin binding of Al^{3+} and Fe^{3+}. *Clin. Chem.*, 1987; 33: 405–407.

22. Elovaara I, Palo J, Erkinjuntti T, and Sulkava R: Serum and cerebrospinal fluid proteins. *Eur. Neurol.*, 1987; 26: 229-234.

23. Bell JD, Brown JCC, Sadler PJ, et al.: High resolution proton nuclear magnetic resonance studies of human cerebrospinal fluid. *Clin. Sci.*, 1987; 72: 563–570.

24. Sternberger NH, Sternberger LA, and Ulrich J: Aberrant neurofilament phosphorylation in Alzheimer disease. *Proc. Natl. Acad. Sci. U.S.A.*, 1985; 82: 4274–4276.

25. Grundke-Iqbal I, Iqbal K, Tung Y, et al.: Abnormal phosphorylation of the microtubule associated protein tau in Alzheimer cytoskeletal pathology. *Proc. Natl. Acad. Sci. U.S.A.*, 1986; 83: 4913–4917.

26. Abdel-Ghany M, El-Sebae AK, and Shalloway D: Aluminum-induced nonenzymatic phospho-incorporation into human tau and other proteins. *J. Biol. Chem.*, 1993; 268: 11976–11981.

27. Diaz-Nido J and Avila J: Aluminum induces the in vitro aggregation of bovine brain cytoskeletal proteins. *Neurosci. Lett.*, 1990; 110: 221–226.

28. Hollosi M, Shen ZM, Perczel A, and Fasman GD: Stable intrachain and interchain complexes of neurofilament peptides. *Proc. Natl. Acad. Sci. U.S.A.*, 1994; 91: 4902–4906.

29. Lukiw WJ, Krishnan B, Wong L, et al.: Nuclear compartmentalization of aluminum in Alzheimer's disease. *Neurobiol. Aging*, 1991; 13: 115–121.

30. Leterrier JF, Langui D, Probst A, and Ulrich J: A molecular mechanism for the induction of neurofilament bundling by aluminum ions. *J. Neurochem.*, 1992; 58: 2060–2070.

Chapter 8

The Metabolism and Toxicokinetics of Aluminum Relevant to Neurotoxicity

Robert A. Yokel

CONTENTS

8.1　INTRODUCTION

This chapter focuses on the routes and mechanisms of aluminum (Al) absorption, its distribution and retention, and its routes and mechanisms of elimination. The toxicokinetics of Al are discussed when relevant to its neurotoxicity.

Table 8.1 shows common sources and normal and elevated exposure to, and absorption of, Al. Diet is a major source of Al exposure under normal conditions. The Al body burden in normal adult humans is about 25 to 60 mg and can be greater than 1 g in end-stage renal disease (ESRD) patients (Figure 8.1).

8.2　ALUMINUM QUANTITATION AND DETECTION

The accurate quantitation of brain Al is difficult due to low concentrations and the ubiquitous presence of Al. This contributes to discrepancies among studies. The use of an Al tracer, when possible, could avoid this problem. Although used as an Al tracer, ^{67}Ga does not faithfully model Al.[1,2] The isotopic tracer [26]Al can be measured using accelerator mass spectrometry (AMS) which

TABLE 8.1 Main Sources of Al, Al Concentration in the Source, Exposure to Source, Daily Al Exposure From the Source, Estimated Percentage Absorbed, and Calculated Amount Absorbed Daily

Source	[Al]	Exposure to source	Daily Al exposure	Estimated percentage absorbed	Al absorbed daily
Normal exposure					
Air — rural	$0.2\ \mu g/m^3$	$15\ m^3/d$	$3\ \mu g$	? nose to brain	?
				1 from lungs	$0.03\ \mu g$
				0.1 from gut	$0.003\ \mu g$
Air — urban	$1\ \mu g/m^3$	$15\ m^3/d$	$15\ \mu g$	? nose to brain	?
				1 from lungs	$0.15\ \mu g$
				0.1 from gut	$0.015\ \mu g$
Water	$50\ \mu g/l$	$2\ l/d$	$100\ \mu g$	0.2	$0.2\ \mu g$
Food			14 mg	0.05	$7\ \mu g$
Antiperspirant	$5\text{–}7.5\%^a$	1 g/d	50–75 mg	? topically	?
Elevated exposure					
Industrial air	$25\text{–}2500\ \mu g/m^3$	$10\ m^3/shift$	0.25–25 mg	1.5	$4\text{–}375\ \mu g$
Antacid/phosphate binders			Up to 5000 mg	0.1	$5000\ \mu g$
Dialysate	If tap water $50\ \mu g/l$	110 l/dialysis 3 times weekly	$2360\ \mu g$	25	$590\ \mu g$
Total parenteral nutrition solution	Neonatal/pediatric $110\text{–}270\ \mu g/l^b$	0.1 l/kg/d	$11\text{–}27\ \mu g/kg$	100	$11\text{–}27\ \mu g/kg$
	Adult $40\text{–}135\ \mu g/l^b$	2 l/d	$80\text{–}270\ \mu g$	100	$80\text{–}270\ \mu g$

[a] Based on 20% Al zirconium glycine complex or 25% Al chlorohydrate in a topical product, which are typical concentrations.[34]

[b] Based on maintenance fluids and normal neonatal/pediatric or adult electrolyte supplementation.

has a detection limit of $\approx 4 \times 10^{-18}$ g ^{26}Al. This method has been recently utilized in Al metabolism studies and holds promise to resolve some major issues, such as the bioavailability of Al from water as a function of Al speciation.

8.3 ORAL ABSORPTION

8.3.1 Site and Mechanism

The percentage of Al taken up from different regions of the gastrointestinal (GI) tract and the rate of absorption suggest that this process occurs primarily in the jejunum and duodenum. The acidic pH of the stomach may solubilize Al from insoluble species such as $Al(OH)_3$, facilitating absorption. On the other hand, food may complex Al as nonabsorbable forms. Further work is needed to understand the influence of diet on the bioavailability of Al.

Several mechanisms probably mediate GI Al absorption. Saturable uptake has been claimed, although this might be an experimental artifact due to formation of large nonabsorbable Al species at high concentrations. An energy-dependent process has been shown, perhaps involving calcium channels. Considerable evidence also suggests an energy-independent process, probably via the paracellular pathway. Aluminum may be taken up into mucosal cells, which may provide a barrier to its absorption. Small amounts of Al may then be slowly released into circulation, as seen with iron.[3] The clearance of considerable Al in the first pass through the liver[4] may protect the brain from immediate exposure to oral Al when the liver is functioning well.

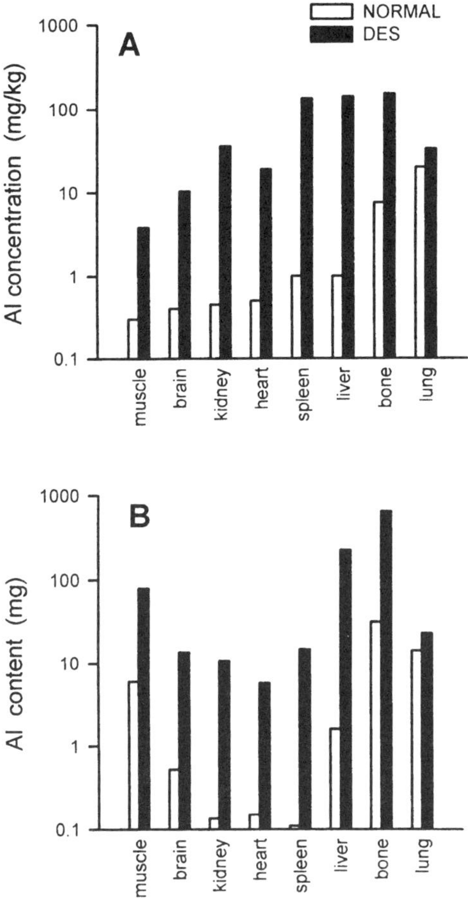

Figure 8.1   Aluminum concentration (mg/kg wet weight), Panel A, and Al content (mg), Panel B, in the human tissues that contain the majority of the Al body burden. Values are averages from normal humans from several studies and from those dying with symptoms of the dialysis encephalopathy syndrome (DES).[31,35] Aluminum content is based on normal organ weights of adults at autopsy. Note that the Y axis is on a log scale.

8.3.2 Percentage Absorbed

The percentage of Al absorbed by animals and humans after oral administration of various Al species using ^{27}Al and ^{26}Al was generally <1%.[5,6] There are no good estimates of the percentage of Al absorbed from drinking water. Two weeks after a single oral dose of ^{26}Al in drinking water, rat brains contained ^{26}Al. However, the variability of brain ^{26}Al (200- to 300-fold among the 8 treated rats) was considerable.[7] Further study of this issue is needed.

8.3.3 Oral Al Intake and Alzheimer's Disease

The relationship between drinking water Al and the incidence of Alzheimer's disease (AD) has been assessed in several epidemiological studies. Some found a weak correlation.[8] A few small studies of the relationship between Al-containing antacid use and AD have not provided convincing evidence for an association.[8]

8.3.4 Influence of Small Molecular Weight Species

The percentage of Al absorbed is dependent on the species of Al consumed and produced in the gut. In both the rabbit and rat, a greater percentage of soluble than insoluble Al species was absorbed.[9,10] Small molecular weight ligands, such as citrate, silicate, and fluoride, can alter Al absorption and retention, and thereby its toxicity.

Numerous studies in normal and renal-impaired humans and animals have shown that some carboxylic acids and carbonate can enhance absorption and brain Al accumulation. Citric and ascorbic acids generally produced a greater effect than gluconic, lactic, malic, oxalic, succinic, and tartaric acids.[11] Several deaths in chronic renal failure patients were attributed to concurrent $Al(OH)_3$ and citrate intake. Citrate may facilitate Al absorption by formation of soluble Al species as well as its ability to chelate Ca to increase paracellular pathway permeability. However, the results of many of these studies are questionable due to the use of single blood samples (which cannot define the time course or extent of absorption), urinary Al output (which cannot account for retained Al), or tissue Al as an indication of Al absorption.

Silicon, as monomeric silicic acid, forms stable, soluble hydroxyaluminosilicate which may reduce GI Al absorption and facilitate Al excretion by reducing its reabsorption in the kidney.[12] Addition of silicon to the diet was claimed to reduce brain Al accumulation in 28-, but not 23-month-old rats.[13] The basis of the age difference is unknown. Silicon reduced the absorption of ^{26}Al (in orange juice) by 85%.[6] Increased urinary Al and Si excretion were seen after renal transplantation.[12] Increasing dietary silicic acid reduced tissue Al accumulation. Beer increased the output of Al in urine, after Al citrate dosing, presumably due to its silicon content.[12] The neurotoxic β-sheet conformation of β-amyloid-(1-42) produced by Al was reversed by orthosilicate.[14] The above results encourage further investigation of the ability of silicate to reduce Al accumulation and toxicity. However, reports that Al and Si are colocalized in senile plaques and that neuronal cytotoxicity results from Al silicate exposure raise the possibility that Si could increase Al-induced neurotoxicity, which must also be evaluated.

Fluoride may inhibit Al absorption. A few animal studies suggest a slight reduction of brain Al accumulation when F is added to the diet.[15] Epidemiological studies suggest increased fluoride in the water is associated with less mental impairment.[16] However, fluoride increased Al-induced behavioral toxicity in an animal study.[15] Furthermore, no obvious benefit was observed from fluoride given daily for 1 year to patients in the early stages of AD (Shore, D., personal communication, 1992). The ability of fluoride to reduce Al neurotoxicity is not encouraging.

Iron status negatively correlates with Al absorption and brain Al accumulation. Although GI Al uptake was suggested to involve a transferrin-mediated process, transferrin failed to facilitate Al uptake into Caco-2 cells. The mechanism of the iron-Al interaction is unclear.

Dietary calcium deficiency increased the rate and extent of Al absorption, tissue Al accumulation, and Al-induced neuropathology in rats. Calcium decreased Al uptake by both the rat everted gut sac and perfused small intestine, suggesting a common uptake mechanism for Al and Ca uptake. However, addition of Ca to $Al(OH)_3$ did not affect Al absorption in humans with normal renal function who presumably had normal Ca status.

In summary, the potential for silicon, iron, and calcium to protect the brain from Al requires further study.

8.3.5 Influence of Other Factors

A parathyroid hormone (PTH)-induced increase of serum and brain Al in rats was shown in early studies, consistent with the effects obtained with dietary Ca deficiency and a low-Ca-induced hyperparathyroidism. Subsequent studies failed to show an appreciable effect of PTH on Al absorption, although one study found PTH-induced elevated brain Al. Because PTH stimulates $1,25\text{-}(OH)_2$-vitamin D, a PTH effect could be mediated by vitamin D. There is evidence for vitamin D dependence of Al absorption in renal-intact, but not uremic, animals. Although vitamin D has been shown to increase Al in serum and in some peripheral tissues, it decreased brain Al in most studies, suggesting that possible PTH effects are not mediated by vitamin D.

Greater Al absorption has been shown in uremic than in normal humans and animals. Uremia may increase Al uptake through the paracellular pathway.[17]

A study in rats showed that concurrent gastric intubation of Al and ethanol in the drinking water nonsignificantly increased blood and liver Al, and significantly increased Al-induced neurotransmitter changes.[18] This apparent facilitation of Al absorption and toxicity by ethanol should be further studied.

8.4 NONORAL ABSORPTION

Inhalation exposure to Al chlorohydrate, the active ingredient of aerosol antiperspirants, did not increase rodent brain Al after exposure to doses that reduced body weight. Occupational exposure to Al fumes, dusts, and flakes can elevate serum, urine, and bone Al. Brain Al has not been studied. It is not known if absorption occurs across the lungs, from the GI tract after mucociliary clearance, or via the olfactory tract. Pulmonary Al absorption is suggested by the elevated serum and urine Al following exposure to the respirable particles in welding fumes and the better correlation between respirable than total Al with urinary Al output.[19] Although there has been no good determination of the fractional absorption of Al from lung to blood, results from two studies of Al industry workers suggest ≈1 to 2% absorption.[19,20] Fractional absorption was greater at lower, and less at higher, air Al concentrations (Gitelman, H.J., personal communication, 1995). Neurocognitive deficits in some workers exposed to Al dusts and fumes have been attributed to Al exposure.[8] The route and percentage of Al absorbed after inhalation and as a function of particle composition and size should be investigated, as well as the ability of this absorbed Al to enter the brain.

Aluminum may directly enter the brain via the olfactory tract, as shown with manganese. Instillation of Al into the nasal recess of rabbits produced pathological changes in the olfactory bulb and cerebral cortex, and elevated Al in cortical macrophages.[21] The olfactory route could enable Al entry into the hippocampus, which is affected in AD. Nasal exposure to Al acetylacetonate produced Al accumulation in brain foci, including the cerebellum.[22] Aluminum could have distributed through the olfactory bulb and then the cerebrospinal fluid (CSF) to the cerebellum. In this case, the CSF may expose all brain surfaces to Al absorbed from nasal mucosa. This potential route of Al absorption should be studied further.

Aluminum-contaminated intravenous and total parenteral nutrition solutions produce obvious Al dosing. The Al in dialysate solutions effectively transfers into plasma, due to extensive transferrin-Al binding. Dialysis encephalopathy syndrome has resulted from such exposure, as discussed by Alfrey in Chapter 13.

Application of Al chlorohydrate to shaved dorsal skin of mice for 130 days raised brain, serum, and urine Al concentrations,[23] refuting the assumption that Al is not transdermally absorbed. The total Al applied during the 130 days (0.5 to 2 mg/kg body weight) is comparable to the daily Al exposure of humans using topical antiperspirants (Table 8.1). Al uptake through mouse skin was also shown,[23] supporting the whole-mouse observations.

Continuous urinary bladder alum instillation to control hemorrhage has produced elevated serum Al and an acute encephalopathy in at least eight patients.[24] This illustrates the potential of acute neurotoxicity after exposure to massive amounts of soluble Al.

8.5 DISTRIBUTION

The initial volume of Al distribution in animals[5] is consistent with blood volume. Aluminum is unequally distributed (Figure 8.1). The high concentration in lung may reflect inhalation exposure and the elevated concentrations in bone, liver, and spleen may reflect sequestration. The elevated brain and bone Al in dialysis encephalopathy syndrome victims (Figure 8.1) has been modeled in renal-intact animals.

Plasma Al is ≈80 to 90% bound to proteins[5] in normal renal function (see Chapter 7). Most recent studies concluded that transferrin is the only plasma protein that binds Al, although some suggested Al binding to albumin. Binding of Al to albumin would be nonspecific and much weaker than to transferrin. Nonprotein-bound plasma Al is believed to be Al citrate (see Chapter 7).

Coadministration of citrate and Al to renally intact animals results in formation of Al citrate, which readily distributes out of plasma and is renally eliminated. Most single- and short-term dosing studies failed to show citrate-enhanced Al tissue accumulation. Some studies of repeat Al citrate dosing or prolonged exposure showed elevated bone, muscle, and brain Al; others did not. Although citrate in the presence of renal function may facilitate Al elimination, it is unclear if it could enhance Al excretion and reduce tissue Al accumulation in renal-impaired subjects.

Lower transferrin binding of ^{67}Ga, used as a tracer for Al, was reported in serum from AD subjects.[25] The authors suggested that reduced transferrin Al binding could generate a higher percentage of nonprotein-bound Al species (perhaps Al citrate) that might gain access to the brain more easily than transferrin-bound Al, enhancing neurotoxicity. However, the Al species entering the brain have not been identified. There is considerable suggestive evidence that transferrin can transport Al across the blood-brain barrier, presumably causing release of free Al in brain extracellular fluid (ECF). When high Al doses are given to animals, Al rapidly enters brain ECF and CSF.[26,27] Under these conditions, Al probably enters ECF and CSF as a small-molecular-weight, nontransferrin-bound species, suggesting a second mechanism of Al brain entry. Decreased transferrin Al binding, e.g., produced by uremia, should favor greater formation of small-molecular-weight Al species such as Al citrate. This may increase brain Al distribution if these species enter the brain, but should decrease brain entry of transferrin-bound Al. An energy-dependent mechanism transports Al out of brain ECF,[28] apparently into blood. Transferrin, but not citrate, has been shown *in vitro* to enhance Al uptake into neuroblastoma cells and oligodendroglia, but not astrocytes. However, the CSF, and presumably ECF, transferrin concentration is very low (see Chapter 7), raising doubt concerning the role of transferrin in brain cell Al uptake. Until the mechanisms of Al blood-brain barrier permeability and brain cell Al entry are determined, the impacts of altered transferrin Al binding, citrate concentration, and uremia on brain Al entry cannot be predicted.

Brain Al concentrations are lower than in many other tissues (Figure 8.1). Increases of 4- to 6-fold in brain Al concentrations in rabbits and 10- to 20-fold increases in victims of dialysis encephalopathy syndrome (Figure 8.1) were associated with neurotoxicity and death, respectively. Some studies of AD brain show an increase of Al concentration in bulk brain samples (of a fewfold) and at subcellular sites (up to 500 mg/kg). Other studies found no increase. The inconsistent findings of a small increase in bulk brain Al in AD in relation to the small increase that is sufficient to produce neurotoxicity have hindered the resolution of the role of Al in AD. If Al is elevated in AD brain, it is not reflected in CSF Al, which has generally not been found to be elevated.

Spinal cord Al distribution has been most studied in animals given cisternal medullary Al injections. The predominant effects are in the upper spinal cord, as might be predicted considering CSF circulation.

Aluminum is highly concentrated in the lysosomes of Kupffer's cells, hepatocytes, mesangial cells, and neurons. Intraneuronal Al accumulation has also been reported in the nucleus, neurofibrillary tangles, senile plaques, and cytoplasm, and in association with lipofuscin.[29] Chromatin-associated Al may alter DNA transcription and increase replication errors. It has been debated whether Al in neurofibrillary tangles and senile plaques is a cause or effect of the formation of these pathological structures. Lipofuscin granule Al may have derived from neuronal lysosomal Al. In the dialysis encephalopathy syndrome, Al accumulates in choroidal epithelial inclusions, glia to a greater extent than neurons, and in neuronal cytoplasm. This is very different than seen in AD.[30]

8.6 ELIMINATION RATES AND TISSUE RETENTION

The concentration of Al in brain and other tissues was elevated in nondialyzed uremic patients,[31] suggesting uremia-facilitated Al retention. Partially nephrectomized animals showed up to a fivefold increase in whole-body Al concentration, and elevated Al in bone and, in some studies, brain, heart, and muscle, when compared to controls. This increase is not due to reduced Al clearance in partial renal failure, but may result from formation of small-molecular-weight Al species, perhaps Al citrate, that have a different distribution.

The Al elimination half-life $(t_{1/2})$ has been determined in animals and humans.[5] The $t_{1/2}$ was estimated to be 13 and 85 days in 2 studies in renal-impaired humans after discontinuation of Al-containing medications. The apparent $t_{1/2}$ increased with increased duration of sampling after acute Al loading of rabbits, suggesting compartments with very long $t_{1/2}$s. Further evidence of multiple compartments of Al storage is provided by the positive correlation between $t_{1/2}$ and exposure duration. Elimination $t_{1/2}$s of hours, weeks, and years were seen after termination of short-term inhalation exposure, <1-year exposure, and upon retirement, respectively.[32] Furthermore, a study in one human who received i.v. ^{26}Al suggested that 85 to 90% of the Al was eliminated in <24 h but that the terminal $t_{1/2}$ increased with time — up to 7 years at the latest estimate.[2] This unusual kinetic behavior might result from retention of an Al species other than that administered, creating a depot, probably in bone, from which the Al is slowly eliminated. Slow elimination coupled with continued exposure may produce an increasing body Al burden with age. There are reports of increased brain, blood, liver, and kidney Al with age; however, other studies failed to find an association between age and brain, blood, or hair Al.

There are no determinations in humans of Al retention time in specific tissues. Elevated brain Al concentrations in previously dialyzed humans who had successful renal transplants years prior to death suggest lack of rapid removal of brain Al.[33] Brain Al slightly increased from 5 to 35 days after i.p. ^{26}Al injection to rats, supporting the persistence of brain Al over time.

8.7 EXCRETION

The primary organ of Al elimination is the kidney. Urinary Al may be a good index of recent Al exposure, but not body, and presumably brain, Al burden. The rate of Al clearance is consistent with GFR,[5] although proximal tubular Al reabsorption and Al excretion in the distal nephron have been suggested. Several animal studies suggested a decrease in Al clearance and an increase in $t_{1/2}$ with increased Al concentration. These results may be due to the high Al concentrations studied that probably formed unfilterable Al "complexes", reducing the plasma filterable Al fraction.[9]

Biliary Al excretion, which accounts for 0.1 to 3% of total Al elimination in several species, including humans,[2] is not appreciably influenced by uremia. Enterohepatic circulation of Al has not been investigated.

8.8 SUMMARY

Brain Al is influenced by the amount, chemical species, routes and duration of Al exposure, and ability of the kidney to eliminate Al. Aluminum may enter the brain by transferrin receptor-mediated endocytosis and other mechanisms. It is actively removed from the brain. The mechanisms of Al entry into brain cells are not understood. Study of the metabolism of Al in the brain has been difficult but may now be conducted more reliably and at physiological concentrations by using ^{26}Al and accelerator mass spectrometry.

ACKNOWLEDGMENT

Partially supported by NIH grant ES 4640.

REFERENCES

1. Greger JL, Chang MM, and MacNeil GG: Tissue turnover of aluminum and Ga-67: Effect of iron status. *Proc. Soc. Exp. Biol. Med.*, 1994; 207:89–96.

2. Priest ND, Newton D, Day JP, et al: Human metabolism of aluminum-26 and gallium-67 injected as citrates. *Hum. Exp. Toxicol.*, 1995; 14:287–293.

3. van der Voet GB: Intestinal absorption of aluminum. Relation to neurotoxicity. In Isaacson RL and Jensen KF (eds): *The Vulnerable Brain and Environmental Risks*, Vol. 2: Toxins in Food, New York, Plenum Press, 1992, pp 35–47.

4. Xu Z-X, Tang J-P, Badr M, et al: Kinetics of aluminum in rats. III. Effect of route of administration. *J. Pharm. Sci.*, 1992; 81:160–163.

5. Wilhelm M, Jager DE, and Ohnesorge FK: Aluminum toxicokinetics. *Pharmacol. Toxicol.*, 1990; 66:4–9.

6. Edwardson JA, Moore PB, Ferrier IN, et al: Effect of silicon on gastrointestinal absorption of aluminum. *Lancet*, 1993; 342:211–212.

7. Walton J, Tuniz C, Fink D, et al: Uptake of trace amounts of aluminum into the brain from drinking water. *Neurotoxicology*, 1995; 16:187–190.

8. Doll R: Review: Alzheimer's disease and environmental aluminum. *Age Ageing*, 1993; 22:138–153.

9. Yokel RA and McNamara PJ: Influence of renal impairment, chemical form and serum protein binding on intravenous and oral aluminum kinetics in the rabbit. *Toxicol. Appl. Pharmacol.*, 1988; 95:32–43.

10. Froment DH, Buddington B, Miller NL, and Alfrey AC: Effect of solubility on the gastrointestinal absorption of aluminum from various aluminum compounds in the rat. *J. Lab. Clin. Med.*, 1989; 114:237–242.

11. Domingo JL, Gomez M, Llobet JM, and Corbella J: Influence of some dietary constituents on aluminum absorption and retention in rats. *Kidney Int.*, 1991; 39:598–601.

12. Birchall JD, Bellia JP, and Roberts NB: On the mechanisms underlying the essentiality of silicon interactions with aluminum and copper. *Coord. Chem. Rev.*, 1996; in press.

13. Carlisle EM and Curran MJ: Effect of dietary silicon and aluminum on silicon and aluminum levels in rat brain. *Alzheimer Dis. Assoc. Disord.*, 1987; 1:83–89.

14. Fasman GD, Perczel A, and Moore CD: Solubilization of β-amyloid-(1-42)-peptide: Reversing the β-sheet conformation induced by aluminum with silicates. *Proc. Natl. Acad. Sci. U.S.A.*, 1995; 92:369–371.

15. Yokel RA: Aluminum chelation: Chemistry, clinical and experimental studies and the search for alternatives to desferrioxamine. *J. Toxicol. Environ. Health*, 1994; 41:131–174.

16. Kraus AS and Forbes WF: Aluminum, fluoride and the prevention of Alzheimer's disease. *Can. J. Public Health*, 1992; 83:97–100.

17. Ittel TH, Paulus C-P, Handt S, et al: Induction of intestinal mucosal atrophy by difluoromethylornithine: A nonuremic model of enhanced aluminum absorption. *Miner. Electrolyte Metab.*, 1992; 18:15–23.

18. Flora SJS, Dhawan M, and Tandon SK: Effects of combined exposure to aluminum and ethanol on aluminum body burden and some neuronal, hepatic and haematopoietic biochemical variables in the rat. *Vet. Hum. Toxicol.*, 1991; 10:45–48.

19. Gitelman HJ, Alderman FR, Kurs-Lasky M, and Rockette HE: Serum and urinary aluminum levels of workers in the aluminum industry. *Ann. Occup. Hyg.*, 1995; 39:181–191.

20. Pierre F, Baruthio F, Diebold F, and Biette P: Effect of different exposure compounds on urinary kinetics of aluminum and fluoride in industrially exposed workers. *Occup. Environ. Med.*, 1995; 52:396–403.

21. Perl DP and Good PF: Uptake of aluminum into central nervous system along nasal-olfactory pathways. *Lancet*, 1987; i:1028.

22. Zatta, P, Favarato, M, and Nicolini, M: Deposition of aluminum in brain tissues of rats exposed to inhalation of aluminum acetylacetonate. *Neuroreport*, 1993; 4:1119–1122.

23. Anane, R, Bonini, M, Grafeille, J-M, et al: Bioaccumulation of water soluble aluminum chloride in the hippocampus after transdermal uptake in mice. *Arch. Toxicol.*, 1995; 69:568–571.

24. Perazella M and Brown E: Acute aluminum toxicity and alum bladder irrigation in patients with renal failure. *Am. J. Kidney Dis.*, 1993; 21:44–46.

25. Farrar G, Altmann P, Welch S, et al: Defective gallium-transferrin binding in Alzheimer disease and Down syndrome: possible mechanism for accumulation of aluminum in brain. *Lancet*, 1990; 335:747–750.

26. Yokel RA, Lidums V, McNamara PJ, et al: Aluminum distribution into brain and liver of rats and rabbits following intravenous aluminum lactate or citrate: A microdialysis study. *Toxicol. Appl. Pharmacol.*, 1991; 107:153–163.

27. Xu Z-C, Tang J-P, Xu Z-X, et al: Kinetics of aluminum in rats. IV. Blood and cerebrospinal fluid kinetics. *Toxicol. Lett.*, 1992; 63:7–12.

28. Allen DD, Orvig C, and Yokel RA: Evidence for energy-dependent transport of aluminum out of brain extracellular fluid. *Toxicology*, 1995; 98:31–39

29. Schuurmans Stekhoven JH, Renkawek K, Otte-Holler I, et al: Exogenous aluminum accumulates in the lysosomes of cultured rat cortical neurons. *Neurosci. Lett.*, 1990; 119:71–74.

30. Reusche E and Seydel U: Dialysis-associated encephalopathy: light and electron microscopic morphology and topography with evidence of aluminum by laser microprobe mass analysis. *Acta Neuropathol. (Berl)*, 1993; 86:249–258.

31. Alfrey AC, Hegg A, and Craswell P: Metabolism and toxicity of aluminum in renal failure. *Am. J. Clin. Nutr.*, 1980; 33:1509–1516.

32. Ljunggren KG, Lidums V, and Sjogren, B: Blood and urine concentrations of aluminum among workers exposed to aluminum flake powders. *Br. J. Ind. Med.*, 1991; 48:106–109.

33. McDermott JR, Smith AI, Ward MK, et al: Brain-aluminum concentration in dialysis encephalopathy. *Lancet*, 1978; i:901–904.

34. POISINDEX Information System, Micromedex, Inc. Englewood, CO.

35. Flendrig JA, Kruis H, and Das HA: Aluminum and dialysis dementia. *Lancet*, 1976; i:1235.

Chapter 9

Aluminum: On Both Sides of the Blood-Brain Barrier

Roger Deloncle and Nicole Pages

CONTENTS

9.1 INTRODUCTION

Aluminum is the third most abundant element in the earth's crust and many toxicological effects of this metal have been described. While implication of aluminum in the etiology of Alzheimer's disease has been controversial, the role of this metal has clearly been established in dialysis encephalopathy and in amyotrophic lateral sclerosis.

To participate in neurodegenerative processes Al has to traverse the blood-brain barrier (BBB), and the mechanism for that crossing as yet remains unclear. Nevertheless, aluminum-related toxicological effects can be noticed on both sides of the BBB.

9.2 ALUMINUM BEFORE PASSING THE BLOOD-BRAIN BARRIER

At physiological pH, aluminum salts are almost insoluble and poorly absorbed, but aluminum may be found in colloïdal form or as soluble complexes subsequent to reactions with chelating agents such as citric or lactic acid. These chemical forms are thought to facilitate transfer through the gastrointestinal barrier.

Upon entering the blood, the main fraction of aluminum is found in the plasma, bound to transferrin. Quantitative spectroscopic determination under blood-plasma conditions of Al^{3+} binding to the two sites of transferrin yields successive stability constants of log K_1 = 12.9 and log K_2 = 12.3.[1] This protein, involving brain transferrin receptors, has been implicated in aluminum transfer to the brain.[2] On the other hand, aluminum can form a complex with small biological molecules such as citric, lactic, aspartic, or glutamic acids. For these two dicarboxylic amino acids (AH_2), the formation of $Al(OH)HA^+$ and $Al(HA)^{2+}$ complexes, taking into account the hydrolysis of aluminum ion in solution, have been reported with respective constants of log β_1 = 7.97 and log β_2 = 11.81 for aspartic acid and log β_1 = 8.10 and log β_2 = 11.62 for glutamic acid. With these values, calculations for the equilibrium model speciation of aluminum in serum indicate that aluminum-aspartate or -glutamate complexes are not major blood constituents under physiological conditions.[3] However, we have already demonstrated in sodium L-glutamate-overloaded rats that aluminum may be transferred through the erythrocyte membrane and the blood-brain barrier as a glutamate complex.[4] In our *in vivo* experiment, blood levels — about 5.10^{-3} mol l^{-1} and 70.10^{-3} mol l^{-1} for aluminum and glutamate, respectively — were 1000 times higher than the levels in physiological conditions. Such concentrations can explain the *in situ* formation of an aluminum-glutamate complex and its transfer through biological membranes and the BBB.

In this context it is noteworthy that in aluminum intoxication as well as in Alzheimer patients showing aluminum brain deposits, this metal is mainly found in the brain cortex, hippocampus, and amygdala. These brain areas are rich in glutaminergic neurons as well as in transferrin receptors.

Aluminum which is blood-transported is partially eliminated by kidney filtration, but the remaining aluminum fraction may interfere with the physiological metabolism of calcium and magnesium. Indeed, experiments on the influence of calcium on aluminum accumulation by rat jejunal slices suggest that Al does interact with the gastrointestinal calcium transporting system.[5] As with calcium, aluminum metabolism depends in part on vitamin D and on parathyroid hormone function, and it has been shown in rats that subcutaneous injections of 1,25-dihydroxyvitamin D_3 resulted in increased intestinal aluminum absorption from the diet and in increased aluminum concentrations in serum, the kidney, and the bone.[6] In studying calcium transport in the adult rat brain by measuring Ca-ATPase activity on rat brain microsomes and synaptosomes, Mundy et al.[7] obtained, with $AlCl_3$, a stimulation in Ca^{2+}-ATPase activity in a concentration-dependent manner but to a greater extent than with Mg^{2+}. They concluded that aluminum acted at multiple sites to displace both Mg^{2+} and Ca^{2+}, increasing the activity of ATPase but disrupting transport of calcium.

Moreover, high levels of intravenous aluminum increased total plasma calcium, and *in vitro* studies showed that aluminum alters calcium influx and efflux in bone.[8] In the same way, *in vivo* experiments on rats and monkeys maintained on calcium-deficient diets demonstrated that bone Al levels were significantly increased in rats fed diets deficient in Ca alone or deficient in Ca and Mg with or without added Al. Reduced levels of serum Ca and Mg, parathyroid hormone, and alkaline phosphatase were observed in monkeys fed a low Ca-Mg diet with added Al.[9]

In various organs such as in bone and articular cartilage, substitutions of calcium by aluminum induced disturbances such as osteomalacia, osteopenia, or arthropathy. The resulting weakened bone matrix could lead to multiple fractures as observed in dialysis and in Alzheimer's disease patients.[10]

In muscle, calcium is associated with actin and myosin where the cation plays a role in the contraction process. Again, calcium replacement by aluminum leads to disorders in muscle contraction. This type of substitution could be related to findings of aluminum deposits in myocardic cells and in dialysis patients afflicted with cardiomyopathy.[11] It has also been demonstrated that aluminum alters the electrical and mechanical properties of atrial muscle in the frog.[12] Toxicological experiments with liposome-transported aluminum or with lipophilic aluminum compounds, such as acetylacetonate, caused myocardial infarcts in rabbits resembling those in humans.[13]

Alkaline earth elements are not the only metals which can be displaced by aluminum. In particular, interferences with iron are expected since ionic radii are in the same order of magnitude for Al^{3+} and Fe^{3+}. In fact, metabolism of these elements is closely linked, and aluminum is able to bind to specific iron proteins such as ferritin or transferrin, the latter molecule allowing aluminum blood transport. It has also been demonstrated that aluminum accumulation in serum, liver, and spleen is increased in Fe-depleted rats[14] and that aluminum chemistry with chelating agents such as desferrioxamine is analogous to that of iron.

Similarly, aluminum displacement of iron in bone marrow would explain dysfunctions in hemoglobin synthesis and microcytic anemia observed in aluminum-intoxicated patients. Moreover, Garbossa et al.[15] demonstrated, with mouse bone marrow cell cultures stimulated with erythropoietin, that aluminum chloride or citrate at concentrations as low as 0.37 µmol/l inhibit erythropoiesis through a mechanism dependent upon the availability of transferrin to bind to aluminum.

9.3 ALUMINUM, BIOLOGICAL MEMBRANES, AND THE BLOOD-BRAIN BARRIER

Al-related modifications of membrane fluidity presumably play a role in neurodegenerative processes such as Alzheimer's disease. Van Rensburg et al.[16] described an increase in platelet membrane fluidity and a decrease in erythrocyte membrane fluidity in Alzheimer's disease patients. When adding aluminum sulfate to platelet and erythrocyte membrane suspensions from healthy volunteers, similar variations in membranous fluidities were obtained and the fluidity was found to change with increasing aluminum concentrations. The mechanism leading to these modifications in membrane fluidity remains unclear. However Al-related peroxidation processes have been implicated in membrane alterations.[17]

Furthermore, by analyzing the membrane-bound IgG and the erythrocyte anion transport with labeled sulfate, Bosman et al.[18] demonstrated membrane changes in erythrocytes from Alzheimer patients. They also found an increased breakdown in integral membrane protein band 3, suggesting that the normal aging process of these cells is disturbed, this breakdown in band 3 being possibly initiated by oxidative damage. Moreover, it has been proven that autopsy samples of Alzheimer's disease cortex showed increased peroxidation *in vitro*.[19]

Aluminum may affect various oxidative processes or enzymatic properties, and even facilitates the oxidation of NADH by the superoxide anion. It appears that Al(III) forms a complex with O^{2-} that is a stronger oxidant than O^{2-} itself, and which may contribute to the adverse biological effects of Al(III).[20] Moreover, it has been demonstrated that aluminum salts can accelerate iron-stimulated peroxidation of membrane lipids.[17] Similarly, by studying *in vitro* effects of Al on lipid peroxidation in mouse brain homogenates and purified brain subcellular fractions, Oteiza[21] demonstrated that this metal has both oxidant and antioxidant effects in mouse brain membranes. In brain microsomes, Al increased TBARS (thiobarbituric acid-reactive substances) production and conjugated dienes formation, both depending on the addition of Fe^{2+}. In brain homogenates, microsomes, and myelin, aluminum has the potential for exhibiting both prooxidant and antioxidant activity, depending on the concentration of Fe^{2+} and Al in the medium and on membrane integrity. The capacity of Al to stimulate Fe^{2+}-induced lipid peroxidation in liposomes was evaluated by measuring the formation of TBARS. Results indicate that Al can stimulate Fe^{2+}-supported lipid peroxidation through binding to the membrane and by promoting rearrangement of membrane lipids.

In Alzheimer's disease, a compositional aberration in membrane lipid has been suggested since membrane lipids from Alzheimer's patients form a stable bilayer at a critical temperature which differs from that of the physiological behavior.[22] Similarly, alterations in the blood-brain barrier for this disease have been suggested.

To demonstrate aluminum-induced alterations in blood-brain barrier permeability, experiments have been carried out with various salts. Using quinidine and aspirin, Jakovljevic et al.[23] demonstrated that aluminum chloride influences the brain kinetics of drugs. $AlCl_3$ subcutaneous pretreatment in mice inhibited the entry into and exit out of the brain of quinidine (a drug that dissociates as a cation), but $AlCl_3$ does not compete for penetration with acetylsalicylic acid (a drug that dissociates as an anion) but slows down its brain elimination. Moreover, Banks and Kastin[24] demonstrated that intraperitoneal injections of $AlCl_3$ increase blood-brain barrier permeability of iodinated N-Tyr delta sleep-inducing peptide (DSIP) and β-endorphin by 60 to 70%. Other experiments by Kim et al.[25] have illustrated that intraperitoneal injections of aluminum chloride and lactate, but not hydroxide, can induce changes in the permeability of the blood-brain barrier to (^{14}C)-sucrose. These modifications were reversible since these alterations, noticeable 2 and 4 h after Al injection, could not be observed 24 h following injection. Favarato et al.[26] performed a similar experiment with (^{14}C)-sucrose on rats intraperitoneally injected with various aluminum salts, such as lactate, acetylacetonate, or maltolate, at pH 7.5. They found an increase in brain (^{14}C) radioactivity for acetylacetonate or maltolate but not for lactate. They concluded that Al(III) alters the BBB function either permanently or transiently depending on the physicochemical properties of the metal coordination sphere.

In order to measure the effects of aluminum on the blood-brain barrier, experiments were also carried out on brain microvessel monolayer endothelial cell cultures (BMEC) which can be considered as an *in vitro* experimental model for the blood-brain barrier. With that model, Audus et al.[27] demonstrated that micromolar concentrations of aluminum increased BMEC permeability and attenuated BMEC pinocytosis. Their results confirmed *in vivo* findings of aluminum-induced increases in BBB permeability to peptides and nonpeptides.

Similarly, employing brain microvascular endothelial cell cultures, Vorbrodt et al.[28] studied the influence of aluminum maltolate on anionic sites localized on cell structures. A 26% reversible decrease, similar to those (28%) obtained for cadmium chloride, a well-known toxic element for these cultures, was observed in the surface density of anionic sites for aluminum maltolate. The authors tried to demonstrate the same phenomenon by chronic aluminum exposure in mice. Using aluminum lactate in drinking water for 6 weeks and for 6 months, the barrier function of the endothelium of central cortex blood microvessels, however, was not noticeably affected and only a few changes in localization of anionic sites were observed.

In the rat, we have already demonstrated that aluminum, in the form of a glutamate complex, can cross the erythrocyte membrane and the blood-brain barrier.[4] For a better understanding of Alzheimer's disease we have suggested a mechanism based on alteration of biological membranes and the blood-brain barrier.[29]

To illustrate aluminum effects on membrane alteration, we carried out experiments on rat chronic intoxication. For 5 weeks, treated animals received daily a subcutaneous aluminum-glutamate injection (Al =10.8 mg/kg/d; L-glu = 588 mg/kg/d) prior to receiving a single intravenous administration of aluminum and glutamate (Al = 10.8 mg/kg, L-glu = 588 mg/kg). A significant increase in aluminum was observed in the hippocampus, the occipitoparietal cortex, the cerebellum, and the striatum. Moreover, half of the aluminum-glutamate-treated animals had neurological disturbances such as trembling, equilibrium difficulties, and convulsions leading to death about 1 h after the intravenous administration. An increased level of glutamic acid in the parietal cortex was measured in these aluminum-glutamate-treated rats ($14.0 \pm 0.9 \ 10^{-9}$ mol/mg) compared with controls receiving only sodium L-glutamate ($10.5 \pm 0.5 \ 10^{-9}$ mol/mg) or physiological saline ($7.5 \pm 0.4 \ 10^{-9}$ mol/mg). These findings demonstrated an alteration of the blood-brain barrier in aluminum-glutamate treated rats, allowing that an increased glutamate transfer to the brain is in part responsible for the neurological disturbances observed.[30]

To test the hypothesis of the peroxidation process in the blood-brain barrier alteration, we treated rats according to a chronic aluminum-glutamate intoxication schedule for 10 weeks. The animals received subcutaneous injections (2 ml/kg) of Al-L-glutamate (Al = 0.1 M ; glu = 1 M) or Na-L-glutamate (glu = 1 M), or isotonic sodium chloride once daily. The animals were killed 24 h after the last injection. Three brain regions were selected, since we had already noticed an aluminum accumulation in the cerebral cortex, the hippocampus, and the striatum. Measurements were performed on thiobarbituric acid-reactive substances (TBARS) and polyunsaturated fatty acids (PUFAs), as markers of the lipid peroxidation; aluminum and iron levels were also measured.

In aluminum-L-glutamate-treated rats, Al blood plasma was increased (28 μmol l^{-1}) and accumulation of this metal was observed in the three brain regions, with higher levels in the striatum. Furthermore, a decrease in iron concentration was observed in blood plasma and in the striatum only. An overproduction of TBARS was observed in the hippocampus but not in the cerebral cortex nor striatum. These results suggest that lipid peroxidation might be involved in the alteration of neuronal membranes in the hippocampus after such an Al treatment. In the striatum, the absence of an increased lipid peroxidation might be related to a decrease in the Fe level, following a larger Al accumulation.[31]

9.4 ALUMINUM TOXICOLOGICAL EFFECTS AFTER PASSING THE BLOOD-BRAIN BARRIER

Roskams and Connor[2] implicated transferrin receptors in aluminum brain transfer. Subsequent to the demonstration in rats that aluminum, as a glutamate complex, was able to cross the BBB, we proposed a mechanism involved in Alzheimer's disease. This mechanism is based on the alteration of permeation across biological membranes and the BBB.[29]

Whatever the process for aluminum transfer to the brain may be, in aluminum intoxication, as well as in Alzheimer's disease patients showing aluminum deposits, aluminum is mainly found in the brain cortex, the hippocampus, and the amygdala. These brain regions are rich in glutaminergic neurons as well as in transferrin receptors.

As previously found in other target organs, upon entering the brain aluminum may displace physiological minerals such as magnesium, calcium, or iron, and induce disturbances in their brain metabolism. For example, replacing calcium in the synaptic area might disturb neurotransmitter release and alter neurotransmitter systems, as observed in Alzheimer's disease and in experimental aluminum intoxications of animals.

Abnormal protein formation in the brain might result from metal complex formation with precursors present in the cell. Substitution of aluminum for magnesium in the neuron might disturb the magnesium-dependent polymerization-depolymerization process of neurotubules. Employing synthetic fragments of the human neurofilament protein mid-sized subunit, Hollosi et al.[32] induced β-pleated sheet formation in the phosphorylated fragment with Al^{3+} and Ca^{2+}. Al^{3+}-induced conformational changes were irreversible to citrate chelation, whereas Ca^{2+}-induced conformational changes were reversible with citric acid. Peptides containing at least one free serine residue were shown to form an intramolecular Al^{3+} complex rather than an intermolecular one, though Ca^{2+} did not show a tendency for intramolecular complexing. These findings may have a bearing on neurofibrillary tangle and plaque formation in Alzheimer's disease patients. Aluminum may also induce amyloid deposition in Alzheimer's disease brains since aluminum involvement in β-amyloid peptide aggregation has been demonstrated.[33]

In neurons, glutamic acid in the form of a stable aluminum-glutamate complex ($\log \beta_1 = 8.10$ and $\log \beta_2 = 11.62$)[3] would no longer be able to detoxify cellular ammonia, thus leading to neuronal death. With regard to ammonia, in Alzheimer's disease patients, a postprandial hyperammonemia has been described and an increase in brain ammonia production has been reported.[34] Furthermore, exposure to aluminum lactate in rat brain astrocyte cultures resulted in glutaminase and glutamine synthetase alterations and decreased intracellular glutamate levels.[35] This may alter the availability of neurotransmitter glutamate in neurons and support the ammonia toxicologic hypothesis we had previously proposed for a better understanding of Alzheimer's disease.[29]

9.5 CONCLUSION

As developed in the previous sections, its appears that aluminum is a highly toxic metal with various harmful effects depending on the target organ where this element is deposited. These toxicological effects may in part be related to substitution of physiological metals, among them calcium, magnesium, and iron. Disturbances by aluminum deposits can occur, e.g., in bone, where multiple fractures can be induced, or in muscle like the myocardium where myocardic infarct has been described. In bone marrow, aluminum deposits may alter hemoglobin synthesis and lead to subsequent anemia.

However, the more serious disturbances occur in the brain. While aluminum involvement in neurodegenerative processes has clearly been established for dialysis encephalopathy or amyotrophic lateral sclerosis, the role for aluminum in Alzheimer's dementia is still controversial. However, the ability of Al to induce neurofibrillary tangle formation, to promote aggregation of β-amyloid peptide, or to disturb neurotransmitter systems are strong arguments for a role of this metal in this disease.

An alteration of the blood-brain barrier would be necessary for aluminum brain transfer, but the mechanism leading to this alteration is still completely unknown, even though aluminum-induced peroxidation is perhaps a process involved therein.

REFERENCES

1. Martin RB: The chemistry of aluminum as related to biology and medicine. *Clin. Chem.*, 1986; 32/10:1797–1808.
2. Roskams AJ and Connor JR: A role for transferrin and its receptor. *Proc. Natl. Acad. Sci. U.S.A.*, 1990; 87:9024–9027.
3. Harris WR: Equilibrium model for speciation of aluminum in serum. *Clin. Chem.*, 1992; 38/9:1809–1818.
4. Deloncle R, Guillard O, Clanet F, et al: Aluminum transfer as glutamate complex through blood-brain barrier. Possible implication in dialysis encephalopathy. *Biol. Trace Elem. Res.*, 1990; 25:39–45.
5. Provan SD and Yokel RA: Influence of calcium on aluminum accumulation by the rat jejunal slice. *Res. Commun. Chem. Pathol. Pharmacol.*, 1988; 59(1):79–92.
6. Burnatowska-Hledin MA, Doyle TM, Eadie MJ, et al: 1,25-Dihydroxyvitamin D3 increases serum and tissue accumulation of aluminum in rats. *J. Lab. Clin. Med.*, 1986; 108(2):96–102.
7. Mundy WR, Kodavanti PRS, Dulchinos VF, et al: Aluminum alters calcium transport in plasma membrane and endoplasmic reticulum from rat brain. *J. Biochem. Toxicol.*, 1994; 9(1); 17–23.
8. Goodman WG and O'Connor J: Aluminum alters calcium influx and efflux from bone in vitro. *Kidney Int.*, 1991; 39:602–607.
9. Yasui M, Yase Y, Ota K, and Garruto RM: Evaluation of magnesium, calcium and aluminum metabolism in rats and monkeys maintained on calcium-deficient diets. *Neurotoxicology*, 1991; 12:603–614.
10. Buchner DM and Larson EG: Falls and fractures in patients with Alzheimer's type dementia. *J. Am. Med. Assoc.*, 1987; 257(11):1492–1495.
11. Timsit F, Galle P, Bourdon R, et al: Cardiomyopathie aluminique chez un hémodialysé chronique. *Néphrol*, 1985; 6:263.
12. Meiri H and Shimoni Y: Effects of aluminum on electrical and mechanical properties of frog atrial muscle. *Br. J. Pharmacol.*, 1991; 102:483–491.
13. Bombi GG, Corain B, Favaratto M, et al: Experimental aluminum pathology in rabbits: effects of hydrophilic and lipophilic compounds. *Environ. Hlth. Perspect.*, 1990; 89:217–223.
14. Brown TS and Schwartz R: Aluminum accumulation in serum liver and spleen of Fe-depleted and Fe-adequate rats. *Biol. Trace Elem. Res.*, 1992; 34:1–10.
15. Garbossa G, Gutnisky A, and Nesse A: The inhibitory action of aluminum on mouse bone marrow cell growth, Evidence for an erythropoietin- and transferrin-mediated mechanism. *Miner. Electrolyte Metab.*, 1994; 20:141–146.
16. Van Rensburg SJ, Carstens ME, Potocnik FCV, et al: Membrane fluidity of platelets and erythrocytes in patients with Alzheimer's disease and the effect of small amounts of aluminium on platelet and erythrocyte membranes. *Neurochem. Res.*, 1992; 17(8):825–829.
17. Gutteridge JMC, Quinlan GJ, Clark I, et al.: Aluminium salts accelerate peroxidation of membrane lipid stimulated by iron salts. *Biochim. Biophys. Acta*, 1985; 835:441–447.
18. Bosman GJCG, Bartholomeus IGP, and De Grip WJ: Alzheimer's disease and cellular aging: membrane-related events as clues to primary mechanism. *Gerontology*, 1991; 37:95–112.
19. Subbarao, KV, Richardson, JS, and Ang, LC: Autopsy samples of Alzheimer's cortex show increased peroxidation in vitro. *J. Neurochem.*, 1990; 55(1):342–345.
20. Kong S, Liochev S, and Fridovitch I: Aluminum (III) facilitates the oxidation of NADH by the superoxide anion. *Free Rad. Biol. Med.*, 1992; 13:79–81.
21. Oteiza I: A mechanism for the stimulatory effect of aluminum on iron-induced lipid peroxidation. *Arch. Biochem. Biophys.*, 1994; 308(2):374–379.
22. Ginsberg L, Atack JR, and Rapoport SI: Evidence for a membrane lipid defect in Alzheimer disease. *Mol. Chem. Neuropathol.*, 1993; 19:37–46.
23. Jakovljevic V, Banic B, and Radunovic A: The effect of aluminum chloride upon the transition of drugs through the blood-brain barrier into the central nervous system. *Eur. J. Drug Metab. Pharmacokinet.*, 1991; 16(3):171–175.
24. Banks WA and Kastin AJ: Aluminium increases permeability of the blood brain barrier to labelled DSIP and β-endorphin: Possible implications for senile and dialysis dementia. *Lancet*, Nov. 26, 1983:1227–1229.
25. Kim YS, Lee MH, and Wisniewski HM: Aluminum induced reversible change in permeability of the blood brain barrier to (^{14}C)sucrose. *Brain Res.*, 1986; 377:286–291.
26. Favarato M, Zatta P, Perazzolo M, et al.: Aluminum III influences the permeability of the blood brain barrier to (^{14}C)sucrose in rats. *Brain Res.*, 1992; 569:330–335.
27. Audus KL, Shinogle JA, Guillot FL, et al: Aluminum effects on brain microvessel endothelial cell monolayer permeability, *Int. J. Pharmacol.*, 1988; 45:249–257.
28. Vorbrodt AW, Dobrogowska DH, and Lossisnky AS: Ultracytochemical studies of the effects of aluminum on the blood-brain barrier of mice. *J. Histochem. Cytochem.*, 1994; 42(2):203–212,
29. Deloncle R and Guillard O: Mechanism of Alzheimer's disease: Arguments for a neurotransmitter-Aluminium complex implication, *Neurochem. Res.*, 1990; 15(12):1239–1245.
30. Deloncle R, Guillard O, Huguet F, et al: Modification of the blood-brain barrier through chronic intoxication by aluminum glutamate. Possible role in the etiology of Alzheimer's disease. *Biol. Trace Elem. Res.*, 1995, 47:227–234.

31. Huguet F, Deloncle R, Guillard O, et al: Effects of chronic treatment with aluminum-glutamate complex on TBARs production and iron level in different rat brain region. Joint Meeting "Structure activity relationships in Toxicology", Société Française de Toxicologie & Societa Italiana di Tossicologia, Marseille, 18–19 Avril 1994.

32. Hollosi M, Shen TM, Perczel A, et al: Stable interchain complexes of neurofilament peptides: A putative link between Al^{3+} and Alzheimer disease. *Proc. Natl. Acad. Sci. U.S.A.*, 1994; 91:4902–4906.

33. Mantyh PW, Ghilardi JR, Rogers S, et al: Aluminum, iron and zinc ions promote aggregation of physiological concentrations of β amyloid peptide. *J. Neurochem.*, 1993; 61(3):1171–1174.

34. Hoyer S, Nitsch R, and Oesterreich K: Ammonia is endogenously generated in the brain in the presence of presumed and verified dementia of Alzheimer type, *Neurosci. Lett.*, 1990; 17(3):358–362.

35. Zielke HR, Jackson MJ, Tildon JT, et al: A glutamatergic mechanism for aluminum toxicity in astrocytes. *Mol. Chem. Neuropathol.*, 1993; 19:219–233.

Chapter 10

Modeling of Acute and Chronic Aluminum Neurotoxicity

Michael J. Strong

CONTENTS

10.1 ALUMINUM AS A NEUROTOXIC AGENT

Whether aluminum can be considered to be a mammalian neurotoxicant is no longer an issue. When administered through a variety of routes in a number of experimental paradigms, aluminum will induce significant biochemical and cytoskeletal pathology. However, the topography and extent of neuropathological degeneration is dependent upon the aluminum compound administered (organic vs. inorganic), the route of administration of the aluminum compound, the host species, and the rate of aluminum exposure (acute vs. chronic). These critical factors interrelate to induce patterns of neurotoxicity that can be manipulated to induce selective disease states.

For similar reasons, identifying a single mechanism of aluminum neurotoxicity is not possible. Although traditional acute *in vivo* aluminum toxicity models induced with massive doses of aluminum are invariably accompanied by fulminant, fatal encephalomyelopathies, such nonspecific models have significantly enhanced our understanding of the cellular mechanisms of aluminum toxicity. Studies utilizing isolated cytoskeletal proteins have allowed a finer molecular approach. In contrast, chronic *in vivo* models, utilizing either sublethal, repetitive inocula or chronic infusions of aluminum, induce patterns of neuronal degeneration that are relatively system-specific and thus find greater parallels to a human disease state. Used in concert, these models have proven invaluable in understanding the pathophysiology of aluminum neurotoxicity, and indirectly, mechanisms of human disease induction.

In practice, two broad paradigms of experimental aluminum neurotoxicity exist — those that are associated with neuronal pathology and those that, for the most part, are not. The clinical and neuropathological features of these two paradigms can be correlated with two patterns of human disease in which aluminum appears to play a role. The former finds parallels in the previously endemic focus of amyotrophic lateral sclerosis (ALS) in the Western Pacific and the latter to dialysis encephalopathy.[1] In the case of ALS, a critical, early neuropathological hallmark is the aggregation of neurofilament (NF) within motor neurons. In dialysis encephalopathy, the clinical phenomenology finds few neuropathological correlates and might be best viewed as the result of a fundamental derangement of neuronal biology. In this chapter, we will review the morphological and biochemical characteristics of these models. The role of aluminum in the development of specific human disease states will be reviewed by others.

10.2 ALUMINUM NEUROTOXICITY MANIFESTED BY CYTOSKELETAL PATHOLOGY

10.2.1 Acute Aluminum Neurotoxicity

10.2.1.1 *In Vivo* Models

Amongst those animals developing neurofilamentous pathology in response to aluminum exposure, the most frequently studied is the New Zealand white rabbit. When administered aluminum by a variety of routes, including oral, intravenous, or intracisternal, rabbits develop a fulminant encephalomyelopathy marked by massive intraneuronal aggregates of NF, often within dendrites (reviewed in Strong et al., 1990).[2] With sufficient aluminum exposure, virtually all neuronal populations become involved although a proclivity for certain neuronal systems exists. For example, motor neurons tend to be remarkably sensitive, as do the anterior thalamic nuclear groups and neurons of the parasubiculum. In contrast, oculomotor nuclei and Purkinje cells are rarely involved.

Clinically, acute aluminum exposure induces impairments in classical conditioning responses (e.g., nictitating membrane response and conditioned avoidance responses). Generalized motor slowing with a loss of tone occurs early and is followed by quadraparesis, spasticity, convulsions, and coma. Impairments in the tonic mobility response are marked. Cerebellar involvement is rare. Unfortunately, with prolonged survival virtually all neuronal populations show some degree of degeneration.

10.2.1.2 *In Vitro* Models

When exposed to aluminum, neuroblastoma cell lines develop intraneuronal argentophilic aggregates that are immunoreactive to phosphorylated NF,[3–5] although human neuroblastoma cells (IMR 32) also develop immunoreactivity to a monoclonal antibody recognizing an epitope of phosphorylated tau protein that is also present in the paired helical filamentous inclusions (PHFs) of Alzheimer's disease.[6] The extent of neuronal degeneration is dependent on the pH of the culture medium[7] and the developmental state of the cells.[8] Utilizing a neuroblastoma × glioma hybridoma (NG108-15) exposed to 2 mM aluminum lactate for 4 days, Singer and colleagues[9] observed a significant increase in cAMP and choline acetyltransferase levels in the absence of either alterations in adenylate cyclase levels or in NF phosphorylation or structure. As will be discussed, this suggests that the increased cAMP activity observed in some *in vitro* models of aluminum neurotoxicity may not be mediated by aluminum activation of G-protein for adenylate cyclase.

Although dorsal root ganglia (DRG) tend to be refractory to the induction of neurofilamentous aggregates following aluminum exposure *in vivo*, *in vitro* studies have tended to utilize these cells.[10] NF aggregates develop in approximately 50% of DRGs exposed to 1 mM aluminum lactate and are accompanied by reductions in the high (NFH), intermediate (NFM), and low (NFL) molecular weight neurofilament subunit proteins and β-tubulin mRNA levels.[11] Following cessation of aluminum exposure, these NF aggregates resolve and mRNA levels return to steady-state levels. Homogenates of rat brain neurons will also develop neurofilamentous aggregates following $AlCl_3$ exposure, although the percentage of neurons involved can be increased from 5 to 20% if concomitantly exposed to maltol.[12,13] This relative insensitivity of cortical and DRG neurons to the *in vitro* induction of pathology is a reflection of neuron-specific thresholds of aluminum susceptibility, and mirrors the existence of *in vivo* neuronal selectivity. In contrast, fetal rabbit motor neuron-enriched monolayer cultures are exquisitely sensitive.[14] Moreover, correlating with their sensitivity *in vivo*, dissociated motor neurons are tenfold more sensitive to the toxicity of $AlCl_3$ *in vitro* than are dissociated hippocampal neurons. Although neuropathological changes do not necessarily correlate with electrophysiological evidence of neuronal degeneration,[15] the striking disparity between thresholds to aluminum toxicity of motor neurons and hippocampal neurons in both *in vivo* and *in vitro* paradigms indicates that the mechanisms of aluminum toxicity are neuronspecific. *In vitro*, the hierarchical sensitivity of aluminum toxicity observed following intravenous aluminum maltolate in rabbits, recapitulated in matrix cultures of midbrain,[16] also supports this concept.

10.2.1.3   Neurochemistry of Acute Aluminum Toxicity

Although conventional models of acute aluminum neurotoxicity yield fulminant encephalomyelopathies which seem to have little correlation with a human disease state, their study has provided considerable information regarding the pathogenesis of neurofilamentous aggregates. From these, it is becoming increasingly clear that aluminum perturbs a number of processes from gene transcription to posttranslational processing of cytoskeletal proteins.

Through increased linker histone binding, aluminum alters DNA transcription by inducing chromatin condensation, and a reduction in the amount of transcribable DNA.[17–19] Chromatin-bound aluminum will inhibit the nuclear binding of corticosterone receptor complexes, hormone-induced chromosomal puffing, and chromatin-dependent RNA synthesis.[20,21] A reduction in the rate of total RNA transcription has been observed.[22] However, although suppressions in total RNA occur concomitantly with the induction of neurofilamentous aggregates, it is not clear that the two are causally related.

Using a model of acute aluminum neurotoxicity in which rabbits were intracisternally inoculated with $AlCl_3$, we demonstrated a dosage-dependent suppression of NFL and NFM mRNA steady-state levels, sparing those of NFH mRNA, that directly correlated with the induction of pathology.[23] However, sciatic axotomy preceding aluminum exposure, reducing the expression of all NF subunits,[24–26] failed to inhibit NF aggregate formation. Although a uniform suppression of NFH, NFM, and NFL mRNA levels occurred ipsilateral to the axotomy, a twofold induction of inclusion formation also occurred on the axotomized side. This suggests that acute alterations in the steady-state levels of NF mRNA are not integral to disease induction. In support of this, we have failed to observe an alteration in the level of NF, actin, or tubulin (T_{26}, $T\alpha 1$) mRNA levels in a model of chronic $AlCl_3$ neurotoxicity. To understand the etiology of aluminum-induced neurofilamentous degeneration, our focus has thus been directed towards posttranslational modifications in NF synthesis.

Primary amongst these issues is that of determining whether aluminum-induced neurofilamentous aggregates are composed of biochemically intact NF. Because these aggregates are composed primarily of phosphorylated NF, interest has focused on the process of NF phosphorylation. Physiologically, NF phosphorylation is a posttranslational modification that occurs extensively within the carboxy-terminal domain of NFH and NFM. Phosphorylation is integral in modulating NF interactions with each other, with other cytoskeletal proteins, and in determining neuritic caliber. Importantly, the majority of NF phosphorylation occurs in a somatofugal fashion such that little phosphorylated NF is normally observed within neuronal perikarya. To determine whether aluminum-induced NF inclusions were composed of excessively phosphorylated NF, we performed time-course dephosphorylation studies of an NF-enriched cytoskeletal fraction isolated from young adult New Zealand white rabbits 48 h following inoculation with a range of $AlCl_3$ amounts. Although both NFM and NFH were resistant to dephosphorylation, we observed no alteration in the net phosphorylation state of either.[27] This suggests a resistance to phosphatase susceptibility rather than an excessive phosphorylation and is in keeping with the observations by Savory et al.[28] It has also been demonstrated that aluminum will inhibit both rat brain synaptosomal protein phosphatase 2A activity[29] and an NF-associated phosphatase that coisolates with the 0.8 M KCl extract of the Triton® X-100-insoluble cytoskeletal fraction of bovine spinal cord.[30] Given these observations, we have proposed that aluminum induces the formation of a pool of phosphatase-resistant, highly phosphorylated NF.[1]

The mechanisms of aluminum-induced NF phosphatase resistance remain to be delineated. In theory, phosphatase resistance could be conferred by interactions of Al^{3+} with the carboxy terminus of NFH or NFM at either phosphate-dependent or -independent sites. *In vitro* studies have shown that Al^{3+} will covalently bind to the negatively charged phosphate residues of both NFH and NFM and cross-link adjacent carboxy-terminus domains. Al^{3+} can also bind to phosphate-independent NFH carboxy-terminus groups[31] and will induce the aggregation of NFL — a phosphate-poor protein.[32] Dephosphorylation of NFH and NFM also does not fully inhibit aluminum-induced aggregation of carboxy-terminus NFH fragments.[33,34] Such phosphate-independent binding may occur at glutamic acid carboxyl groups.[35] As a consequence of this, conformational changes to the NF protein might secondarily affect the accessibility of phosphorylation sites, perhaps via intramolecular cross bridging between Al^{3+} binding sites.[34] Intermolecular interactions with cross-linking between adjacent NF carboxy-terminus domains will also lead to their aggregation in a β-pleated sheet conformation with reduced solubility.[35] The observations of Garruto and colleagues of the colocalization of Al^{3+} with neurofilamentous aggregates, using both solochrome azurine staining and electron beam X-ray microanalysis with wavelength-dispersive spectrometry, lends support to the proposal that Al^{3+}/NF interactions may occur *in vivo*.[36]

The extent to which posttranslational alterations in the NF phosphorylation state also affect NF catabolism remains to be fully explored. Although we have found no alteration in the rate of calpain-mediated NFH or NFM proteolysis following aluminum exposure *in vivo*, *in vivo* it has been observed that aluminum will inhibit calpain-mediated proteolysis of human NF.[37] Cathepsin-D-mediated NF proteolysis is unaffected.

These experiments have led us to propose that the NF aggregation induced by acute $AlCl_3$ neurotoxicity in New Zealand white rabbits reflects perturbations in the posttranslational processing of NF. While selective suppressions of the steady-state levels of individual NF mRNA do occur, these do not appear to be integral to the disease process. Rather, through mechanisms of both phosphatase resistance and phosphatase inhibition, a pool of highly phosphorylated neurofilament resistant to proteolysis is created. Whether cross-linking by Al^{3+} occurs within these inclusions is not certain, but if present, this would be predicted to generate an insoluble proteinaceous aggregate.

10.2.2 Chronic *In Vivo* Neurotoxicity

In spite of the neurochemical insights gained from acute models, these models continue to suffer from the disadvantage of widespread neuronal degeneration that bears little relationship to a human disease state. To circumvent this, models of chronic aluminum exposure have been developed to more closely approximate the topographic specificity of human disease.

10.2.2.1 Nonhuman Primate

Building on the hypothesis that the Western Pacific environment, rich in bioavailable aluminum and deficient in calcium, might induce a secondary form of hyperparathyroidism accompanied by enhanced gastrointestinal absorption of aluminum, a number of investigators have studied the effects of these dietary abnormalities in nonhuman primates. Macaques *(Macaca fuscata)* fed a low calcium and magnesium diet for seven months with aluminum lactate supplementation of the drinking water develop anterior horn cell chromatolysis, argentophilic bodies, and axonal and dendritic swellings — the latter due to an accumulation of interwoven masses of NF.[38] Cynomolgus monkeys (*Macaca fascicularis*), fed an aluminum- and manganese-supplemented diet deficient in calcium for a period of 41 to 46 months, develop chromatolytic, swollen neurons, with neurofibrillary thickening and perikaryal argentophilic masses occurring in some neurons of the anterior horns, trigeminal and hypoglossal nuclei, reticular formation, and substantia nigra.[39] Aluminum- and manganese-supplemented monkeys were most prominently affected. Neuroaxonal spheroids were observed in the anterior and lateral columns and spinocerebellar tracts of the spinal cord. Ultrastructurally and immunohistochemically, both the perikaryal and neuritic inclusions consisted of phosphorylated neurofilamentous material. Because of the complexity of these models, both with regards to the time required to induce disease and the inability to establish consistent clinicopathological correlates, alternate models of chronic aluminum toxicity have been developed.

10.2.2.2   Chronic Aluminum Toxicity in New Zealand White Rabbits

A number of investigators have attempted to develop models of aluminum neurotoxicity with prolonged survival and less fulminant neuropathology.[2] To an extent, these models, utilizing either intracisternal reservoirs of insoluble aluminum compounds or repeated subcutaneous injections, succeeded in producing these more selective changes. Building on these observations, we utilized monthly intracisternal inoculums of 100 µg $AlCl_3$ in young adult New Zealand white rabbits to induce a chronic progressive motor system degeneration.[40] In a time and cumulative dose-dependent manner, these rabbits develop neurofilamentous inclusions in spinal motor neurons and select brainstem nuclei (thought to subserve upper motor neuron functions). As with acute aluminum neurotoxicity, these inclusions contain predominantly highly phosphorylated neurofilament, although immunoreactivity to poorly phosphorylated neurofilament is also present. Inclusions also demonstrate immunoreactivity to a monoclonal antibody (SM1 34) recognizing a distinctive phosphorylation state, not present following acute aluminum exposure. Ubiquitin-protein conjugation also occurs within these inclusions, although only secondarily.

In order to assess the degree to which neurofilamentous aggregates are capable of remodeling, as suggested by Uemura and colleagues,[41,42] we induced a chronic state of aluminum neurotoxicity and then allowed rabbits to recover by removing them from further aluminum exposure.[43] All rabbits showed significant clinical recovery. However, while perikaryal neurofilamentous aggregates largely resolved, the neuroaxonal spheroids did not resolve. Using time-course dephosphorylation studies, we subsequently demonstrated that perikaryal inclusions differed from neuroaxonal spheroids in their phosphatase sensitivity. Neuroaxonal spheroids were uniquely resistant to enzymatic dephosphorylation while perikaryal inclusions could be readily dephosphorylated. While this suggests that the induction of inclusions may differ within regions of the neuron, it also suggests that a failure of phosphatase activity may be sufficient to induce NF inclusion formation. It is interesting thus that NFH/microtubule binding is inhibited when NFH is incompletely dephosphorylated.[44]

10.3 ALUMINUM NEUROTOXICITY IN THE ABSENCE OF CYTOSKELETAL PATHOLOGY

As described in the Section 10.1, the response to aluminum administration is host specific. In contrast to the species described above, rats, while developing significant behavioral and neurochemical changes in response to aluminum exposure, do not develop cytoskeletal pathology. However, as with the above models, the extent of neurobehavioral and neurochemical changes is dependent upon the rate, dose, and speciation of aluminum administered.

Following oral loading of aluminum, a uniformly fatal disorder characterized by lethargy, failure to thrive, and bleeding diathesis occurs in rats. These neurobehavioral changes are accompanied by a number of alterations in cytoskeletal protein phosphorylation and in second messenger dependent pathways, although neurofilamentous aggregates do not form.

Increased *in vivo* phosphorylation has been observed for microtubule-associated protein 2 (MAP-2) and NFH, accompanied by elevated brain cAMP and cGMP levels.[45–48] It has been observed that aluminum inhibits agonist-dependent GTPase activity with prolonged increases in activity in the active GPT α-subunit complex, thereby leading to enhanced cyclic AMP synthesis.[49] Altered intraneuronal Ca^{++} homeostasis has been demonstrated in a number of pathways, and will ultimately affect pathways in phosphorylation. Following chronic aluminum exposure, elevations in the calcium phospholipid-dependent protein kinase C (PKC) activity have been observed as a result of increased intracellular calcium levels stimulating the translocation of PKC to the cell membrane with stabilization of this translocation.[45,50] Altered Ca^{++} homeostasis is also the basis of aluminum-induced inhibition of the fast phase of voltage-dependent uptake of calcium and synaptosomes,[51] altered Ca^{++}/calmodulin binding,[52] inhibition of Ca^{++}/Mg^{++}-ATPase,[53] the displacement of calcium from membrane phospholipid binding sites,[54,55] and the inhibition of PKC-mediated glutamate release.[56] Inhibitions of agonist-stimulated phosphoinositide hydrolysis results in significant reductions in inositol 1,4,5-triphosphate.[57] *In vivo*, micromolar quantities of aluminum inhibits phosphoinositide-mediated calcium signaling in rat hepatocytes.[58]

10.4 SUMMARY

The development of these experimental models of aluminum neurotoxicity has yielded a wealth of information with regards to the induction of neuronal pathology. It is clear that factors such as host specificity, aluminum speciation, solution characteristics, tempo of aluminum administration, and the experimental design being utilized are of importance in understanding these mechanisms. In those species that respond by the induction of cytoskeletal pathology, this is almost exclusively neurofilamentous in nature and appears to be related to significant perturbations in the posttranslational modification of these proteins. However, it is also clear that a number of processes at the level of DNA transcription are impacted upon. The extent to which these processes are critical to the induction of pathology in models distinct from that described above remains to be determined. In the rat, a species that does not develop significant cytoskeletal pathology, but rather a concomitant fulminant neurobehavioral deficit, the impact on calcium-mediated pathways and phosphorylation is profound.

As will be discussed by others within this text, these models of aluminum neurotoxicity find interesting parallels in the development of human neurodegenerative disease. The challenge will be to further apply these to understand the pathophysiology of human neurodegeneration.

ACKNOWLEDGMENTS

M.J.S. is supported by a scholarship from the Medical Research Council of Canada and by the Amyotrophic Lateral Sclerosis Society of Canada. The assistance of Rita Casciano in typing this manuscript is gratefully acknowledged.

REFERENCES

1. Strong MJ: Aluminum-induced neurodegeneration with special reference to amyotrophic lateral sclerosis and dialysis encephalopathy. *Rev. Biochem. Toxicol.*, 1994; 11:75–115.

2. Strong MJ, Yanagihara R, Wolff AV, et al: Experimental neurofilamentous aggregates. Acute and chronic models of aluminum-induced encephalomyelopathy in rabbits. In Rose FC, Norris FH (eds): *Amyotrophic Lateral Sclerosis. New Advances in Toxicology and Epidemiology.* London, Smith-Gordon, 1990, pp 157–173.

3. Shea TB, Clarke JF, Wheelock TR, et al: Aluminum salts induce the accumulation of neurofilaments in perikarya of NB2a/dl neuroblastoma. *Brain Res.*, 1989; 492:53–64.

4. Miller CA and Levine EM: Effects of aluminum salts on cultured neuroblastoma cells. *J. Neurochem.*, 1974; 22:751–758.

5. Cole GM, Wu K, and Timiras PS: A culture model for age-related human neurofibrillary pathology. *Int. J. Dev. Neurosci.*, 1985; 3:23–32.

6. Guy SP, Jones D, Mann DMA, and Itzhaki RF: Human neuroblastoma cells treated with aluminum express an epitope associated with Alzheimer's disease neurofibrillary tangles. *Neurosci. Lett.*, 1991; 121:166–168.

7. Shi B and Haug A: Aluminum uptake by neuroblastoma cells. *J. Neurochem.*, 1990; 55:551–558.

8. Roll M, Banin E, and Meiri H: Differentiated neuroblastoma cells are more susceptible to aluminum toxicity than developing cells. *Arch. Toxicol.*, 1989; 63:231–237.

9. Singer HS, Searles CD, Hahn I-H, et al: The effect of aluminum on markers for synaptic neurotransmission, cyclic AMP, and neurofilaments in a neuroblastoma x glioma hybridoma (NG108-15). *Brain Res.*, 1990; 528:73–79.

10. Seil FJ, Lampert PW, and Klatzo I: Neurofibrillary spheroids induced by aluminum phosphate in dorsal root ganglia neurons in vitro. *J. Neuropathol. Exp. Neurol.*, 1969; 28:74–85.

11. Gilbert MR, Harding BL, Price DL, et al: Aluminum-induced neurofilamentous changes in cultured dorsal root ganglia explants. [Abstract] *J. Neuropathol. Exp. Neurol.*, 1989;48:362.

12. Langui D, Anderton BH, Brion J-P, and Ulrich J: Effects of aluminum chloride on cultured cells from rat brain hemispheres. *Brain Res.*, 1988; 438:67–76.

13. Langui D, Probst A, Anderton B, et al: Aluminum-induced tangles in cultured rat neurones. *Acta Neuropathol.*, 1990; 80:649–655.

14. Strong MJ and Garruto RM: Neuron-specific thresholds of aluminum toxicity in vitro. A comparative analysis of dissociated fetal rabbit hippocampal and motor neuron-enriched cultures. *Lab. Invest.*, 1991; 65:243–249.

15. Farnell BJ, DeBoni U, and Crapper McLachlan DR: Aluminum neurotoxicity in the absence of neurofibrillary degeneration in CA1 hippocampal neurons *in vitro*. *Exp. Neurol.*, 1982; 78:241–258.

16. Hewitt CD, Herman MM, Lopes MBS, et al: Aluminum maltol-induced neurocytoskeletal changes in fetal rabbit midbrain in matrix culture. *Neuropathol. Appl. Neurobiol.*, 1991; 17:47–60.

17. Lukiw WJ, Kruck TPA, and Crapper McLachlan DR: Alterations in human linker histone-DNA binding in the presence of aluminum salts *in vitro* and in Alzheimer's disease. *NeuroToxicology*, 1987; 8:291–302.

18. Lukiw WJ, Kruck TPA, and McLachlan DR: Linker histone-DNA complexes: enhanced stability in the presence of aluminum lactate and implications for Alzheimer's disease. *FEBS Lett.*, 1989; 253:59–62.

19. Lukiw WJ, St. George-Hyslop P, and McLachlan DR: Chromatin structure, gene expression, and nuclear aluminum in Alzheimer's disease. In Harrison PJ (ed): *Regulation of Gene Expression and Brain Function*. New York, Springer-Verlag, 1994, pp 31–45.

20. Sanderson C, Crapper McLachlan DR, and DeBoni U, Inhibition of corticosterone binding in vitro, in rabbit hippocampus, by chromatin bound aluminum. *Acta Neuropathol. (Berl.)*, 1982; 57:249–254.

21. Sanderson CL, Crapper McLachlan DR, and DeBoni U, Altered steroid induced puffing by chromatin bound aluminum in a polytene chromosome of the black fly, *Simulium vittatum*. *Can. J. Genet. Cytol.*, 1982; 24:27–36.

22. Sarkander H-I, Balb G, Schlösser R, et al: Blockade of neuronal brain RNA initiation sites by aluminum: a primary molecular mechanism of aluminum-induced neurofibrillary changes?. In Cervós-Navarro J, Sarkander H-I (eds): *Brain Aging: Neuropathology and Neuropharmacology*. New York, Raven Press, 1983, pp 259–274.

23. Strong MJ, Mao K, Nerurkar VR, et al: Dose-dependent selective suppression of light (NFL) and medium (NFM) but not high (NFH) molecular weight neurofilament gene transcription in acute aluminum neurotoxicity. *Mol. Cell. Neurosci.*, 1994; 5:319–326.

24. Wong J and Oblinger MM: Changes in neurofilament gene expression occur after axotomy of dorsal root ganglion neurons: an in situ hybridization study. *Metab. Brain Dis.*, 1987; 2:291–303.

25. Hoffman PN and Cleveland DW: Neurofilament and tubulin expression recapitulates the developmental program during axonal regeneration. Induction of a specific β-tubulin isotype. *Proc. Natl. Acad. Sci. U.S.A.*, 1988; 85:4530–4533.

26. Hoffman PN, Pollock SC, and Striph GG: Altered gene expression after optic nerve transection: reduced neurofilament expression as a general response to axonal injury. *Exp. Neurol.*, 1993; 119:32–36.

27. Strong MJ and Jakowec DM: 200 kDa and 160 kDa neurofilament protein phosphatase resistance following in vivo aluminum chloride exposure. *NeuroToxicology*, 1994; 15:799–808.

28. Savory J, Herman MM, Hundley JC, et al: Quantitative studies on aluminum deposition and its effects on neurofilament protein expression and phosphorylation, following intraventricular administration of aluminum maltolate to adult rabbits. *NeuroToxicology*, 1993; 14:9–12.

29. Yamamoto H, Saitoh Y, Yasugawa S, and Miyamoto E: Dephosphorylation of t factor by protein phosphatase 2A in synaptosomal cytosol fractions, and inhibition by aluminum. *J. Neurochem.*, 1990; 55:683–690.

30. Shetty KT, Veeranna, and Guru SC: Phosphatase activity against neurofilament proteins from bovine spinal cord: effect of aluminum and neuropsychoactive drugs. *Neurosci. Lett.*, 1992; 137:83–86.

31. Pierson KB and Evenson MA: 200 kDa neurofilament protein binds Al, Cu and Zn. *Biochem. Biophys. Res. Commun.*, 1988; 152:598–605.

32. Troncoso JC, March JL, Häner M, and Aebi U: Effect of aluminum and other multivalent cations on neurofilaments in vitro: an electron microscopic study. *J. Struct. Biol.*, 1990; 103:2–12.

33. Leterrier JF, Langui D, Probst A, and Ulrich J: A molecular mechanism for the induction of neurofilament bundling by aluminum ions. *J. Neurochem.*, 1992; 58:2060–2070.

34. Leterrier JF, Langui D, Probst A, and Ulrich J: A molecular mechanism of aluminum neurotoxicity (reply). *J. Neurochem.*, 1993; 60:385–387.

35. Hollósi M, Ürge L, Perczel A, et al: Metal ion-induced conformational changes of phosphorylated fragments of human neurofilament (NF-M) protein. *J. Mol. Biol.*, 1992; 223:673–682.

36. Garruto RM, Yanigahara R, Shankar SK, et al: Experimental models of metal-induced neurofibrillary degeneration. In Tsubaki T, Yase Y (eds): *Amyotrophic Lateral Sclerosis*. Amsterdam, Elsevier Science, 1988, pp 41–50.

37. Nixon RA, Clarke JF, Logvinenko KB, et al: Aluminum inhibits calpain-mediated proteolysis and induces human neurofilament proteins to form protease-resistant high molecular weight complexes. *J. Neurochem.*, 1990; 55:1950–1959.

38. Yano I, Yoshida S, Uebayashi Y, et al: Degenerative changes in the central nervous system of Japanese monkeys induced by oral administration of aluminum salt. *Biomed. Res.*, 1987; 10:33–41.

39. Garruto RM, Shankar SK, Yanagihara R, et al: Low-calcium, high aluminum diet-induced motor neuron pathology in cynomolgus monkeys. *Acta Neuropathol.*, 1989; 78:210–219.

40. Strong MJ, Wolff AV, Wakayama I, and Garruto RM: Aluminum-induced chronic myelopathy in rabbits. *NeuroToxicology*, 1991; 12:9–22.

41. Uemura E: Intranuclear aluminum accumulation in chronic animals with experimental neurofibrillary changes. *Exp. Neurol.*, 1984; 85:10–18.

42. Uemura E and Ireland WP: Synaptic density in chronic animals with experimental neurofibrillary change. *Exp. Neurol.*, 1984; 85:1–9.

43. Strong MJ, Gaytan-Garcia S, and Jakowec D: Reversibility of neurofilamentous inclusion formation following repeated sublethal intracisternal inoculums of $AlCl_3$ in New Zealand white rabbits. *Acta Neuropathol.*, 1995; 90:57–67.

44. Hisanaga S, Yasugawa S, Yamakawa T, et al: Dephosphorylation of microtubule-binding sites at the neurofilament-H tail domain by alkaline, acid, and protein phosphatases. *J. Biochem.*, 1993; 113:705–709.

45. Johnson GVW, Cogdill KW, and Jope RS: Oral aluminum alters *in vitro* protein phosphorylation and kinase activities in rat brain. *Neurobiol. Aging*, 1990; 11:209–216.

46. Johnson GVW and Jope RS: Phosphorylation of rat brain cytoskeletal proteins is increased after orally administered aluminum. *Brain Res.*, 1988; 456:95–103.

47. Johnson GVW and Jope RS: Aluminum alters cyclic AMP and cyclic GMP levels but not presynaptic cholinergic markers in rat brain *in vivo*. *Brain Res.*, 1987; 403:1–6.

48. Johnson GVW, Watson AL, Jr., Lartius R, et al: Dietary aluminum selectively decreases MAP-2 in brains of developing and adult rats. *NeuroToxicology*, 1992; 13:463–474.

49. Johnson GVW: The effects of aluminum on agonist-induced alterations in cyclic AMP and cyclic GMP concentration in rat brain regions *in vivo*. *Toxicology*, 1988; 51:299–308.

50. Cochran M, Elliott DC, Brennan P, and Chawtur V: Inhibition of protein kinase C activation by low concentrations of aluminum. *Clin. Chim. Acta*, 1990; 194:167–172.

51. Koenig ML and Jope RS: Aluminum inhibits the fast phase of voltage-dependent calcium influx into synaptosomes. *J. Neurochem.*, 1987; 49:316–320.

52. Siegel N and Haug A: Aluminum interaction with calmodulin. Evidence for altered structure and function from optical and enzymatic studies. *Biochim. Biophys. Acta*, 1983; 744:36–45.

53. Suhayda CG and Haug A: Organic acids prevent aluminum-induced conformational changes in calmodulin. *Biochem. Biophys. Res. Commun.*, 1984; 119:376–381.

54. Farnell BJ, Crapper McLachlan DR, Baimbridge K, et al: Calcium metabolism in aluminum encephalopathy. *Exp. Neurol.*, 1985; 88:68–83.

55. Deleers M, Servais J-P, and Wülfert E: Micromolar concentrations of Al^{3+} induce phase separation, aggregation and dye release in phosphatidylserine-containing lipid vesicles. *Biochim. Biophys. Acta*, 1985; 813:195–200.

56. Provan SD and Yokel RA: Aluminum inhibits glutamate release from transverse rat hippocampal slices: role of G proteins, Ca channels and protein kinase C. *NeuroToxicology*, 1992; 13:413–420.

57. McDonald LJ and Mamrack MD: Aluminum affects phosphoinositide hydrolysis by phosphoinositidase C. *Biochem. Biophys. Res. Comm.*, 1988; 155:203–208.

58. Schöfl C, Sanchez-Bueno A, Dixon CJ, et al: Aluminum perturbs oscillatory phosphoinositide-mediated calcium signalling in hormone-stimulated hepatocytes. *Biochem. J.*, 1990; 269:547–550.

Chapter 11

MOTOR NEURON DISEASE

Michael J. Strong and Ralph M. Garruto

CONTENTS

11.1 INTRODUCTION

The evidence linking environmental aluminum exposure to the induction of a human neurodegenerative disease state is strongest for the occurrence of amyotrophic lateral sclerosis (ALS) and parkinsonism-dementia (PD) complex in the Western Pacific. Initially described amongst the Chamorro of Guam,[1–5] subsequent epidemiological studies confirmed the extraordinary occurrence of ALS in this focus; however, 15 years later, a second high-incidence disease focus consisting of parkinsonism-dementia (PD) was identified amongst the same people.[6,7] At approximately the same time, two further high-incidence foci of ALS and PD were discovered amongst the Auyu and Jakai people of Southwest New Guinea [8] and amongst inhabitants of the Kii Peninsula of Honshu Island in Japan.[9] The intensive longitudinal, epidemiological, clinical, and neuropathological study of these closely related high-incidence foci has provided a tremendous insight into the role of altered trace metal metabolism in the induction of an adult-onset neurodegenerative disease state. This analysis, including the recapitulation of the disease in nonhuman primates, will be briefly reviewed in this chapter.

11.2 EPIDEMIOLOGY

11.2.1 Clinical and Neuropathological Phenomenology

Amyotrophic lateral sclerosis (ALS) is an adult-onset, progressive, neurodegenerative disorder which manifests as profound muscle wasting and weakness, ultimately leading to death through respiratory failure. Three clinical variants are recognized, including a classical sporadic variant, an autosomal dominantly inherited variant, and the Western Pacific variant. Apart from inheritance characteristics and geographic considerations, the three variants are clinically indistinguishable.

The onset of symptoms is often with unilateral limb wasting and fasciculations that progress to involve all limb and trunk musculature. Bulbar dysfunction and spasticity are additional key clinical manifestations. Extrapyramidal signs, sensory dysfunction, alterations in bladder function, and dementia are rare occurrences. Overlap with PD may occur — particularly in the Western Pacific variant.[10] The mean age of onset is approximately 59 years and the mean duration is 49 months (Southwestern Ontario ALS clinical demographic data). However, the disease occurrence is age dependent and the duration markedly skewed by long-term survival of greater than 10 years in approximately 20% of patients. The male to female ratio is approximately 1.1:1. While no etiologic agent has been discovered for the common sporadic variant and only 10 to 12% of families with the autosomal dominant variant harbor mutations in the superoxide dismutase gene (SOD1),[11] the epidemiological and longitudinal studies of the Western Pacific variant have provided insights into the etiology of ALS.

Traditionally, the combination of widespread motor neuron loss in a topographically specific pattern with chromatolytic anterior horn cells, atrophic proximal axonal processes, atrophy of ventral spinal roots, and demyelination of the corticospinal tracts establishes the diagnosis of ALS.[12] Although not an invariable feature of all clinically confirmed cases of ALS,[13] corticospinal tract degeneration, when present, is extensive and involves the entirety of the tract with a concomitant loss of Betz cells in the precentral gyrus.[14] Pallor of the spinocerebellar tracts and posterior columns with degeneration of Clarke's column occurs in the familial variant.[15] In the Western Pacific variant, neurofibrillary tangles are a prominent feature.[10]

11.2.2 Epidemiological Analysis

The establishment of a research center on Guam by the National Institutes of Neurologic Disorders and Stroke (1956) marked the beginning of an intensive longitudinal study of the epidemiological, clinical, and neuropathological characteristics of ALS and PD in the Western Pacific. Since then, more than 400 patients from a mean Chamorro population base of 40,000 have been studied. Initially, ALS and PD accounted for more than 20% of all deaths over the age of 25 on Guam.[16–18] This epidemiological analysis also tracked the dramatic changes in incidence rates of the disease on the island in the ensuing four decades. Whereas in the remote high-incidence focus in South-western New Guinea prevalence rates were originally as high as 1,300 per 100,000, in the ensuing 4 decades the disease has rapidly declined and all but disappeared in some villages.[19] Similar trends were observed for the Kii Peninsula and for the Chamorro in Guam where incidence rates were approximately 50-fold that of the worldwide incidence rates. Concomitant with this dramatic decline in the prevalence and incidence of the disease, the mean age of onset of Western Pacific ALS increased in men from 47.6 to 51.9 years and in women from 42.1 to 52.5 years — more closely approximating sporadic ALS.[10] The mean duration increased by 2 years in males and 4 years in females. Sex ratios, originally 2:1, declined to near unity by the early 1980s. Despite these dramatic epidemiological changes, neither the remaining clinical manifestations of the disease or the neuropathological findings changed.

Accompanying these epidemiological changes, dramatic social and economic changes unfolded in the Western Pacific. Following World War II, Guam was converted from a horticultural and fisher-folk subsistence economy to a Westernized culture with a cash economy. Similar trends were observed in the Kii Peninsula and Western New Guinea, leading to the suggestion that the hyper-endemic occurrence of ALS in this region was the result of a cohort effect with exposure to a common environmental factor. Migration studies supported this hypothesis — Chamorro migrants to mainland U.S. continued to exhibit a five- to tenfold increased risk of developing ALS[20] and Filipino migrants to Guam developed ALS and PD on average within two decades following their arrival.[21] In contrast, short-term migrants to Guam, including U.S. military personnel and construc-tion workers and their dependents, did not show an increased risk of developing the disease.[22] These studies suggested that a traditional lifestyle and long-term rural existence on Guam of approximately two decades would increase by five- to tenfold the risk of developing ALS. In addition, a significant genetic component has not been identified by analysis of afflicted families, and linkage to the superoxide dismutase gene mutation,[23,24] recognized in the inherited variant of ALS, has not been documented for the ALS population on Guam.[25] Searches for both conventional and unconventional viral etiologies have similarly proven fruitless[26] and follow-up studies of the potential role of cycad toxicity in the occurrence of Guam ALS and PD have not been substantiated.[27]

11.2.3 Epidemiological Evidence of Toxic and Essential Element Involvement

In 1961 Kimura and colleagues described an ALS-like syndrome in the Kii Peninsula in regions where rivers and drinking water were low in minerals.[9] Following this, Yase and colleagues, working in both the Kii Peninsula and Guam, postulated a mechanism for the involvement of manganese and other metals in the induction of these disorders.[28–30] Subsequently, in all three hyperendemic foci low environmental calcium and high levels of aluminum were reported.[8,31–33] Importantly, the aluminum within the soils of these regions consisted of a highly bioavailable form that would be predicted to be more readily absorbed. In the high-incidence southern region of Guam, the soil was found to be a lateritic form of clay overlying volcanic rock and containing 40% oxides and hydroxides of aluminum, iron, and silicon. In west New Guinea, affected villages lay along waterways that were deficient in calcium and magnesium but high in aluminum and iron.[8] In two separate studies, further evidence of significant calcium deficiency was gained from the observation of biochemical disturbances in calcium and vitamin D metabolism in more than 35% of Guamanian patients with ALS and PD,[34,35] although a third study performed after the prevalence of the disease had dramatically declined did not confirm this observation.[36] The bone mass of Guamanian patients with ALS was found to be significantly decreased when compared to neurologically normal Guamanian patients.[37] Guamanian children and adolescents were also observed to have less bone mass when compared to other populations.[34,38]

Elevated levels of calcium and aluminum have also been observed in spinal cord tissues of ALS cases from the Kii Peninsula,[39,40] and have been supported by similar observations using a number of techniques, including neutron activation analysis, X-ray microanalysis, and graphite furnace atomic absorption spectrometry.[41–43] Elevated tissue aluminum levels were not observed in brain tissues from patients dying with Creutzfeldt-Jacob disease and other neurodegenerative processes, suggesting that brain degeneration alone was not sufficient to account for elevated brain aluminum levels.[44] Subsequently, aluminum was shown to colocalize with calcium and silicon in neurofibrillary tangle-bearing neurons.[45–48] The latter is a pathological hallmark of the Western Pacific ALS variant. Garruto and colleagues, using computer-controlled X-ray microanalysis and wavelength-dispersive spectrometry, produced precise maps of the hippocampus and spinal cord of ALS and PD patients and demonstrated the colocalization of calcium, aluminum, and silicon within neurofibrillary tangle-bearing neurons and the walls of the cerebral blood vessels.[46,47,49,50] Subsequently, a number of investigators using widely varying techniques confirmed these results.[40,43,51–54] Importantly, the abnormal intraneuronal deposition of calcium, aluminum, and silicon was also observed to occur in neurologically normal Guamanians with neurofibrillary tangle formation but not in those free of such lesions.

These systematic clinical, epidemiological, and elemental studies generated the hypothesis that a basic defect in mineral metabolism occurred in these hyperendemic foci of ALS, with chronic nutritional deficiencies of calcium and magnesium provoking a form of secondary hyperparathyroidism. As a consequence of this process, the gastrointestinal absorption of aluminum was enhanced and calcium, aluminum, and silicon were thought to be deposited as hydroxyapatites or aluminosilicates within neurons. The mechanisms by which this fundamental alteration in aluminum absorption might bring about neuronal degeneration is addressed elsewhere in this text.

11.3 NONHUMAN PRIMATE STUDIES

In order to test the above hypothesis, nonhuman primates were administered diets supplemented with aluminum. In the studies of Garruto et al., juvenile Cynomolgus monkeys were fed a low-calcium diet with or without supplemental aluminum for 41 to 46 months.[55] Although the animals clinically appeared healthy without apparent behavioral or neurological deficits, at autopsy widespread neurodegenerative changes were observed. These included central chromatolysis, aberrant perikaryal accumulations of phosphorylated neurofilament, neurofibrillary tangle-like structures, axonal spheroids, and basophilic and hyaline-like inclusions in motor neurons. These were observed in the spinal cord, brainstem, substantia nigra, and cerebrum. By X-ray microanalysis, aluminum was found to be present in spinal cord tissues. Yase and colleagues observed similar changes in Japanese macaques maintained on low calcium diets with supplemental aluminum.[33,56,57]

These studies demonstrated that motor neuron pathology could be induced in nonhuman primates chronically fed low-calcium, aluminum-supplemented diets. The colocalization of aluminum in the spinal cord suggests that such deposition, enhanced by a low-calcium diet, may occur over the course of years and may be a lifelong process.

11.4 SUMMARY

The longitudinal and detailed clinical, epidemiological, and elemental analysis of the hyperendemic foci of ALS and PD in the Western Pacific islands of Guam, Western New Guinea, and the Kii Peninsula have provided evidence for aluminum toxicity as an etiologic process in the Western Pacific variant of ALS. It has been proposed that long-term dietary deficiencies of calcium, rendering a secondary hyperparathyroid state, in the presence of highly bioavailable aluminum compounds, will bring about neuronal degeneration. The supportive evidence for such a hypothesis includes the colocalization of aluminum, calcium, and silicon with degenerating neurons, and the induction of neuropathological features resembling those of motor neuron disease in nonhuman primates chronically fed a low-calcium, aluminum-supplemented diet.

ACKNOWLEDGMENTS

M.J.S. is supported by a research scholarship from the Medical Research Council of Canada, and by research funding by the Amyotrophic Lateral Sclerosis Society of Canada. The authors are grateful to Rita Casciano for her assistance in the preparation of this manuscript.

REFERENCES

1. Zimmerman HM: Monthly report to medical officer in command. U S Naval Medical Research Unit Number 2, 1945.
2. Juergens SM, Kurland LT, Okazaki H, and Mulder DW: ALS in Rochester, Minnesota, 1925–1977. *Neurology*, 1980; 30:463–470.
3. Arnold A, Edgren DC, and Palladino VS: Amyotrophic lateral sclerosis: fifty cases observed on Guam. *J. Nerv. Ment. Dis.*, 1953; 105:490–493.
4. Koerner DR: Amyotrophic lateral sclerosis on Guam: a clinical study and review of the literature. *Ann. Int. Med.*, 1952; 24:1204–1220.
5. Tillema S and Wijnberg CJ: 'Epidemic' amyotrophic lateral sclerosis on Guam: epidemiologic data. *Doc. Med. Geogr. Trop.*, 1953; 5:366–370.
6. Hirano A, Kurland LT, Krooth RS, and Lessell S: Parkinsonism-dementia complex, an endemic disease on the island of Guam. I. Clinical features. *Brain*, 1961; 84:642–661.
7. Hirano A, Malamud N, and Kurland LT: Parkinsonism-dementia complex, an endemic disease on the island of Guam. II. Pathological features. *Brain*, 1961; 84:662–679.
8. Gajdusek DC and Salazar AM: Amyotrophic lateral sclerosis and parkinsonian syndromes in high incidence among the Auyu and Jakai people of West New Guinea. *Neurology*, 1982; 32:107–126.
9. Kimura K, Yase Y, Higashi Y, et al: Epidemiological and geomedical studies on amyotrophic lateral sclerosis. *Dis. Nerv. Syst.*, 1963; 24:155–159.

10. Rodgers-Johnson P, Garruto RM, Yanigahara R, et al: Amyotrophic lateral sclerosis and parkinsonism-dementia on Guam: a 30-year evaluation of clinical and neuropathological trends. *Neurology*, 1986; 36:7–13.

11. Rosen DR, Siddique T, Patterson D, et al: Mutations in Cu/Zn superoxide dismutase gene are associated with familial amyotrophic lateral sclerosis. *Nature*, 1993; 362:59–62.

12. Hirano A: Cytopathology of amyotrophic lateral sclerosis. In Rowland LP (ed): *Advances in Neurology, Amyotrophic Lateral Sclerosis and Other Motor Neuron Disorders*. New York, Raven Press, 1991, pp 91–101.

13. Brownell B, Oppenheimer DR, and Hughes JT: The central nervous system in motor neurone disease. *J. Neurol. Neurosurg. Psychiatr.*, 1970; 33:338–357.

14. Lawyer T and Netsky MG: Amyotrophic lateral sclerosis. *Arch. Neurol.*, 1963; 8:117–127.

15. Hirano A, Kurland LT, and Sayre GP: Familial amyotrophic lateral sclerosis. *Arch. Neurol.*, 1967; 16:232–242.

16. Brody JA and Kurland LT: Amyotrophic lateral sclerosis and parkinsonism-dementia in Guam. In Spillane JD (ed): *Tropical Neurology*, 1973, pp 355–375.

17. Kurland LT and Mulder DW: Epidemiological investigations of amyotrophic lateral sclerosis. 1. Preliminary report on geographic distribution, with special reference to the Mariana Islands, including clinical and pathological observations. *Neurology*, 1954; 4:355-378, 438–448.

18. Reed DM and Brody JA: Amyotrophic lateral sclerosis and parkinsonism-dementia on Guam, 1945-1972. I. Descriptive epidemiology. *Am. J. Epidemiol.*, 1975; 101:287–301.

19. Garruto RM, Yanagihara R, and Gajdusek DC: Disappearance of high-incidence amyotrophic lateral sclerosis and parkinsonism-dementia on Guam. *Neurology*, 1985; 35:193–198.

20. Garruto RM, Gajdusek DC, and Chen KM, Amyotrophic lateral sclerosis among Chamorro migrants from Guam. *Ann. Neurol.*, 1980; 8:612–619.

21. Garruto RM, Gajdusek DC, and Chen K-M: Amyotrophic lateral sclerosis and parkinsonism-dementia among Filipino migrants to Guam. *Ann. Neurol.*, 1981; 10:341–350.

22. Brody JA, Edgar AH, and Gillespie MM: Amyotrophic lateral sclerosis. No increase among U.S. construction workers in Guam. *J. Am. Med. Assoc.*, 1978; 240:551–552.

23. Plato CC, Cruz MT, and Kurland LT: Amyotrophic lateral sclerosis/parkinsonism-dementia complex of Guam: further genetic investigations. *Am. J. Hum. Genet.*, 1969; 21:133–141.

24. Plato CC, Garruto RM, Fox KM, and Gajdusek DC: Amyotrophic lateral sclerosis and parkinsonism-dementia on Guam. A 25-year prospective case-control study. *Am. J. Epidemiol.*, 1986; 124:643–656.

25. Figlewicz DA, Garruto RM, Krizus A, et al: The Cu/Zn superoxide dismutase gene in ALS and Parkinsonism-dementia of Guam. *NeuroReport*, 1994; 5:557–560.

26. Gibbs CJ, Jr. and Gajdusek DC: An update on long-term in vivo and in vitro studies designed to identify a virus as the cause of amyotrophic lateral sclerosis, parkinsonism dementia, and Parkinson's disease. In Rowland LP (ed): *Human Motor Neuron Diseases*. New York, Raven Press, 1982, pp 343–353.

27. Garruto RM: Pacific paradigms of environmentally induced neurological disorders. Clinical, epidemiological and molecular perspectives. *NeuroToxicology*, 1991; 12:347–378.

28. Yase Y: The pathogenesis of amyotrophic lateral sclerosis. *Lancet*, 1972; 2:292–296.

29. Yase Y, Kumamotot T, Yoshimasu F, and Shinjo Y: Amyotrophic lateral sclerosis studies using neutron activation analysis studies. *Neurology (India)*, 1968; 16:46–59.

30. Yoshimasu F, Yasui M, Yase Y, et al: Studies on amyotrophic lateral sclerosis by neutron activation analysis. 2. Comparative study of analytical results on Guam PD, Japanese ALS and Alzheimer disease cases. *Folia Psychiatr. Neurol. Jpn.*, 1980; 34:75–82.

31. Garruto RM, Yanagihara R, Gajdusek DC, and Arion P: Concentrations of heavy metals and essential minerals in garden soil and drinking water in the western Pacific. In Chen KM and Yase Y (eds): *Amyotrophic Lateral Sclerosis in Asia and Oceania*. Taipei, Shyan-Fu Chou, National Taiwan University, 1983, pp 265–329.

32. Garruto RM: Search for the cause of amyotrophic lateral sclerosis and parkinsonism-dementia of Guam: deposition of heavy metals and essential minerals in the central nervous system. In Chen K-M and Yase Y (eds): *Amyotrophic Lateral Sclerosis in Asia and Oceania*. Taipei, National Taiwan University, 1984, pp 391–422.

33. Yase Y: The basic process of amyotrophic lateral sclerosis as reflected in the Kii Peninsula and Guam. In Bryun GW, Heijstee APJ (eds): *Proceedings of the Eleventh World Congress of Neurology*. Amsterdam, Excerpta Medica, 1988, pp 413–427.

34. Yanagihara R, Garruto RM, Gajdusek DC, et al: Calcium and vitamin D metabolism in Guamanian Chamorros with amyotrophic lateral sclerosis and parkinsonism-dementia. *Ann. Neurol.*, 1984; 15:42–48.

35. Yase Y: The basic process of amyotrophic lateral sclerosis as reflected in the Kii Peninsula and Guam. In den Hartog Jager WA, Bruyn GW, Heijstee APJ (eds): *Proceedings of the Eleventh World Congress of Neurology*. Amsterdam, Excerpta Medica, 1978, pp 413–427.

36. Ahlskog JE, Waring SC, Kurland LT, et al: Guamanian neurodegenerative disease: investigation of the calcium metabolism/heavy metal hypothesis. *Neurology*, 1995; 45:1340–1344.

37. Garruto RM, Plato CC, Yanagihara R, et al: Bone mass in Guamanian patients with amyotrophic lateral sclerosis and parkinsonism-dementia. *Am. J. Phys. Anthropol.*, 1989; 80:107–113.

38. Plato CC, Garruto RM, Yanigahara R, et al: Cortical bone loss and measurements of the second metacarpal bone. I. Comparisons between adult Guamanian Chamorros and American Caucasians. *Am. J. Phys. Anthropol.*, 1982; 59:461–465.

39. Yasui M, Yase Y, Kihira T, et al: Magnesium and calcium contents in CNS tissues of amyotrophic lateral sclerosis patients from the Kii Peninsula, Japan. *Eur. Neurol.*, 1992; 32:95–98.

40. Yasui M, Yase Y, Ota K, and Garruto RM: Aluminum deposition in the central nervous system of patients with amyotrophic lateral sclerosis from the Kii Peninsula of Japan. *NeuroToxicology*, 1991; 12:615–620.

41. Yoshimasu F, Yasui M, Yase Y, et al: Studies on amyotrophic lateral sclerosis by neutron activation analysis. 3. Systematic analysis of metals on Guamanian ALS and PD cases. *Folia Psychiatr. Neurol. Jpn.*, 1982; 36:173–180.

42. Hirano A: Progress in the pathology of motor neuron diseases. In Zimmerman HM (ed): *Progress in Neuropathology*, 2nd ed., New York, Grune & Stratton, 1973, pp 181–215.

43. Yoshida S: X-ray microanalytical studies on amyotrophic lateral sclerosis. I. Metal distribution compared with neuropathological findings in cervical spinal cord. *Clin. Neurol. (Japan)*, 1977; 17:299–309.

44. Traub RD, Rains TC, Garruto RM, et al: Brain destruction alone does not elevate aluminum. *Neurology*, 1981; 31:986–990.

45. Ikuta F and Makifuchi T: Distribution of aluminum in the spinal motor neurons of ALS, examined by wavelength-dispersive X-ray microanalysis. Annual Report of the research Committee of Motor Neuron Disease, Tokyo, The Ministry of Health and Welfare of Japan 1977; :66–69.

46. Garruto RM, Swyt C, Yanagihara R, et al: Intraneuronal co-localization of silicon with calcium and aluminum in amyotrophic lateral sclerosis and parkinsonism with dementia of Guam. *N. Engl. J. Med.*, 1986; 315:711–712.

47. Garruto RM, Swyt C, Fiori CE, et al: Intraneuronal deposition of calcium and aluminum in amyotrophic lateral sclerosis of Guam. *Lancet*, 1985; 2:1353.

48. Perl DP, Gajdusek DC, Garruto RM, et al: Intraneuronal aluminum accumulation in amyotrophic lateral sclerosis and parkinsonism-dementia of Guam. *Science*, 1982; 217:1053–1055.

49. Garruto RM, Fukatsu R, Yanagihara R, et al: Imaging of calcium and aluminum in neurofibrillary tangle-bearing neurons in parkinsonism-dementia of Guam. *Proc. Natl. Acad. Sci. U.S.A.*, 1984; 81:1875–1879.

50. Piccardo P, Yanagihara R, Garruto RM, et al: Histochemical and X-ray microanalytical localization of aluminum in amyotrophic lateral sclerosis and parkinsonism-dementia of Guam. *Acta Neuropathol.*, 1988; 77:1–4.

51. Linton R, Bryan SR, Cox XB, et al: Digital imaging studies of aluminum and calcium in neurofibrillary tangle-bearing neurons using SIMS (secondary ion mass spectrometry). *Trace Elem. Med.*, 1987; 4:99–104.

52. Yoshida S, Yase Y, Iwata S, et al: Comparative trace-elemental study on amyotrophic lateral sclerosis (ALS) and parkinsonism-dementia (PD) in the Kii Peninsula of Japan and Guam. *Wakayama Med. Rep.*, 1988; 30:41–53.

53. Yoshida S, Kihara T, Mitani K, et al: Intraneuronal localization of aluminum: possible interaction with nucleic acids and pathogenic role in amyotrophic lateral sclerosis. In Rose FC and Norris FH (eds): *ALS. New Advances in Toxicology and Epidemiology*. London, Smith-Gordon, 1990, pp 211–223.

54. Perl DP and Good PF: Microprobe analysis of aluminum accumulation in association with human central nervous system disease. *Environ. Geochem. Health*, 1990; 12:97–101.

55. Garruto RM, Shankar SK, Yanagihara R, et al: Low-calcium, high aluminum diet-induced motor neuron pathology in cynomolgus monkeys. *Acta Neuropathol.*, 1989; 78:210–219.

56. Yano I, Yoshida S, Uebayashi Y, et al: Degenerative changes in the central nervous system of Japanese monkeys induced by oral administration of aluminum salt. *Biomed. Res.*, 1987; 10:33–41.

57. Yasui M, Yase Y, Ota K, and Garruto RM: Evaluation of magnesium, calcium and aluminum metabolism in rats and monkeys maintained on calcium-deficient diets. *NeuroToxicology*, 1991; 12:603–614.

Chapter 12

Alzheimer's Disease and Aluminum

Walter J. Lukiw

CONTENTS

12.1 INTRODUCTION

In unraveling the complex etiopathogenesis of human brain disease, it becomes apparent that neurological disorders are never purely hereditary in nature, but more often than not there is a close interplay between the genetics of the host and environmental factors. Especially over the last 20 years, much attention has been focused on the environmental impact of aluminum on human mental health, partly because it has been strongly implicated in several devastating neurological disorders of the human mind, in particular in Alzheimer's disease (AD), and partly because of its ongoing mobilization, along with other metals, into our biosphere through acid precipitation. Importantly,

113

the liberation of this potent neurotoxin (at 8.3% w/w the most abundant metal in the earth's crust)[1]* into the biosphere will remain a persistent concern just as long as surface and ground waters continue to acidify. In this context, our aging population, the most rapidly growing segment of our society, is being exposed to this ubiquitous neurotoxic element at an increasing rate, and the contribution of aluminum to human neurological dysfunction may as yet have not reached its fullest potential.[2]

This chapter will review some of the effects and mechanisms of aluminum on the structure and function of human nervous tissue at the molecular level with specific emphasis on the connection in between aluminum and the etiopathology of AD, a fatal brain disease of uncertain etiology. Emphasis will be placed upon various aspects of the involvement of aluminum on the flow of genetic information in the human brain neocortex.

12.2 ALZHEIMER'S DISEASE (AD), ALUMINUM, AND SELECTIVE BRAIN CELL DEATH

12.2.1 AD and The Concept of Pathoclisis

More so than any other organ, the brain and central nervous system (CNS) are remarkably heterogeneous in cellular composition with a wide variety of cell types, each with their own distinct structures and physiological functions. *This cellular heterogeneity within the nervous system may also be the basis for the selective vulnerability of certain brain cell types to neurotoxic injury because of each cell's unique anatomical, neurobiochemical, or physiological signature.*

In a series of brilliantly conducted experiments with the aim of identifying *toxic effects upon brain cortical cells* and a goal of identifying cellular targets within the brain specific for classes of aniline, methylene, and toluidine dyes, the neuropathologist Franz Nissl (1860-1919) was first to demonstrate and acknowledge that the exposure of certain toxic compounds to the brain consistently resulted in a strictly defined and often characteristic interaction with only certain types of brain cells. His mentor, Emil Kraepelin (1856-1926), wrote in Nissl's obituary that *"(Nissl) was able to show through the methods of subacute maximal intoxication devised by him, that the nerve cells sometimes suffer completely distinctive changes through the different toxins which, without further consideration permit one to draw conclusions concerning the sort of pathological processes which are at work.*[3] This hypothesis concerning the selective destruction of specific brain cell components was later termed *pathoclisis*.[4] An important part of Nissl's work on cortical brain cell changes accompanying mental illness was carried out in collaboration with his colleague Alois Alzheimer (1864–1915), and both sought out the explanation of histopathologic findings in the cerebral cortex through the study of brain tissues obtained from senile humans, from animal experimentation, and from observations on the structural alterations in the brain following the administration of toxic substances.[3,5,6]

12.2.2 The Neocortical Substrate for AD and Aluminum

Since these original ideas were proposed by Franz Nissl, pathoclisis, or this neuronal selectivity towards neurotoxic insult, has been clearly and repeatedly demonstrated, for example, by the selective vulnerability of cerebellar granule cells to mercury (methylmercury) or by the destruction of the hippocampal mossy fiber system by lead. Regional and laminar analyses of atrophied nerve cells and patterns of cell death in AD-afflicted neocortex suggest that in this condition there is a particular vulnerability of a population of large pyramidal neurons, each with characteristic morphological and molecular profiles, in layers III and V of the association neocortex, to mechanisms associated with the AD process. These neurons, amongst the largest in the CNS, and noted for their complex synaptic relationships with other pyramidal neurons, are considered to be the major efferent

---

* On the basis of mass, after oxygen (46.1%, w/w) and silicon (28.5%, w/w), aluminum (8.3%, w/w) is the third most abundant element in the lithosphere of the earth, defined as the upper 14 km of the earth's crust; our moon's surface; and the most abundant metal (8.8%, w/w) in the biosphere of the earth.

elements of the association neocortex.[7–9] The selective destruction of brain cells within the internal and external pyramidal layers in AD-affected neocortex is thought to result in the disruption of the integrative neuronal network leading to a failure of the higher cognitive functions of the human brain, and this has led to the concept of AD as being a *phylogenetic disease*.[9] As discussed more fully below, there are many reports that one preferred target of aluminum deposition in the CNS is these same pyramidal cells so prominently affected by AD neuropathology.

12.3 ALUMINUM IS A POTENT NEUROTOXIN

Aluminum was recognized as a neurotoxin 100 years ago;[10] however it was not until 1921 that the specific effects of aluminum on the human CNS, such as memory loss and impaired coordination, were first documented in a medical journal.[11] Indeed, the cumulative knowledge of aluminum's cytotoxic and genotoxic effects within biological systems, and particularly neurobiological activities, are both remarkable and considerable, and have been the subject of numerous reports and reviews over the last 20 years.[2,12–24] For example, an extensive literature is currently available on the many toxic manifestations of aluminum in biology. Table 1 in Reference 2 lists some 125 biological effects published in over 190 scientific journals of the deleterious activities of aluminum on nuclear and cytoplasmic metabolism, effects on the cytoskeleton, membranes and membrane-bound enzymes, the synaptic apparatus and neurotransmission, effects on blood-borne elements, bone metabolism, and the central and peripheral nervous systems of mammals and nonhuman primates.

Table 12.1 summarizes data from some 36 reports of elevated amounts of aluminum in the central and peripheral nervous system associated with human neurodegenerative disease. These disorders include encephalopathy associated with aluminum workers (termed here as industrial encephalopathy), dialysis encephalopathy, Parkinson's disease, the amyotrophic lateral sclerosis-Parkinson's disease complex (ALS/PDC) peculiar to the island of Guam and the Kii Peninsula of Japan, and in particular, AD, or senile dementia of the Alzheimer type.

12.4 EVIDENCE FOR THE ASSOCIATION OF AD WITH ALUMINUM

Several independent lines of evidence have implicated aluminum's role in AD:

1. The cellular and biochemical changes occurring both in *in vitro* and *in vivo* test systems at aluminum concentrations equivalent to those found in various subcompartments of AD brain,
2. The slowing of the clinical progression of AD by desferrioxamine (DF; Ciba Geigy), a chelator drug which removes aluminum from the brain,
3. Epidemiologic evidence of increased incidence of AD in relation to the aluminum content of drinking water,
4. Neurotoxicologic studies and laboratory investigations, utilizing experimental aluminum encephalopathy (EAE), on the learning and memory performance in test animals, and
5. Similarities in the pathoclisis; the sites of toxic action of aluminum and the neocortical regions selectively affected by the AD process.

The many toxic effects of aluminum on cellular metabolism and the biochemical changes which are aluminum induced have been categorized (see Reference 2) and will not be discussed further here. Readers are encouraged to pursue investigating these many reports in order to make their own evaluation of the injurious effects of aluminum on neurobiological and neurogenetic systems. Similarly, the DF clinical trials and the epidemiological studies positively correlating the aluminum content of drinking water with an increased incidence of AD have been extensively criticized, discussed, and reviewed elsewhere.[22–33]

Table 12.1 Aluminum is Elevated in Human Neurological Disease

Neurological condition	Analytical method	Tissue	Ref.
Industrial encephalopathy[a]	NA	CNS	105
ALS/PDC[b]	NAA	CNS tissues	77
Alzheimer's disease	AA	Neocortex	13
Alzheimer's disease	AA	Neocortex	14
Alzheimer's disease	XMA	Senile plaques	106
Down's Syndrome with AD[c]	EAA	Neocortex	12
Dialysis encephalopathy	EAA	Neocortex	83
Alzheimer's disease	AA	Neocortex	107
Alzheimer's disease	AA	Neocortical nuclei	12–14
Alzheimer's disease	XMA	Hippocampal nuclei	84
ALS/PDC[b]	SEM/XS	Hippocampal nuclei	79
Alzheimer's disease	XMA	Tangles	85
Alzheimer's disease	AA	Tangle cores	108
Alzheimer's disease	INAA	Neocortex	70
Alzheimer's disease	AA	Ferritin	109
Alzheimer's disease	XMA	Senile plaques	110
Alzheimer's disease	LAMMA	Tangles	74
Alzheimer's disease	XMA	Senile plaques	110
Industrial encephalopathy[a]	NA	CNS	111
Alzheimer's disease	INAA	Neocortex	Corrigan, 1991*
Dialysis encephalopathy	NA	Blood plasma	113
Parkinson's disease	XMA	Substantia nigra	73
ALS/PDC[b]	NAA	Neocortex	78
Alzheimer's disease	EAA	Euchromatin	39
Parkinson's disease	LAMMA	Neuromelanin	75
Alzheimer's disease	EAA	Tangles	Fraser, 1992*
Alzheimer's disease	LAMMA	Tangles	75
Dialysis encephalopathy	NA	Blood plasma	114
Dialysis encephalopathy	EAA/SIMS	Neocortex	115
Alzheimer's disease	INAA/GFAAS/LAMMS	Neocortex nuclei	69
Alzheimer's disease	LAMMA	Tangles	81
ALS/PDC[b]	LAMMA	Tangles	81
Alzheimer's disease	SIMS	Senile plaques	115
Alzheimer's disease	Morin/GFAAS	Senile plaques	116
Alzheimer's disease	LAMMA	Tangles	82

Note: Analytical methods: AA = atomic absorption (flame), EAA = electrothermal atomic absorption (flameless), GFAAS = graphite furnace atomic absorption spectrometry, INAA = instrumental neutron activation analysis, LAMMA = laser microprobe mass analysis, LMMS = laser microprobe mass spectrometry, SEM/XS = scanning electron microscopy/X-ray spectrometry, SIMS = secondary ion mass spectrometry, XMA = X-ray microanalysis. NA = data not available; (*) refers to unpublished data or personal communication. "Senile plaques" and "tangles" refer to the insoluble amyloid deposits and neurofibrillary tangles, respectively, present in the cytoplasm of the diseased brain.

[a] Industrial exposure: aluminum workers showing cognitive deficits with encephalopathy.

[b] Guam and Kii Peninsula (Japan) amyotrophic lateral sclerosis and Parkinson dementia complex with neurofibrillary degeneration.

[c] Down's Syndrome with Alzheimer's disease.

The effects of aluminum on the flow of genetic information, the sequestration and compartmentalization effect of aluminum in AD-afflicted brain cell nuclei, the localization of aluminum in EAE, neurodegenerative disease, and AD, and some current observations on the effects of aluminum in tissue culture cells and in *in vitro* transcription systems are discussed more fully below.

12.5 THE EFFECTS OF ALUMINUM ON THE FLOW OF GENETIC INFORMATION

12.5.1 The Flow of Genetic Information in the Brain

The effects of aluminum on the flow of genetic information are, like the effects of aluminum on other biological processes, complex, multiple, and incompletely characterized. Generally speaking, eukaryotic gene expression is regulated through many fundamental genetic processes:

1. The control of chromatin structure,
2. The influence of epigenetic mechanisms such as DNA methylation and imprinting,
3. The initiation and rate of primary RNA transcript synthesis,
4. The processing of primary RNA transcripts into mature RNA message by capping, intron removal, and polyadenylation,
5. The shuttling of RNA message from the nucleus into the cytoplasm,
6. RNA message stability and longevity,
7. Intracellular RNA message targeting,
8. Translational control,
9. Posttranslational modification, and
10. The selective activation, inactivation, or compartmentalization of the gene products.[34]

In addition, all of these steps in the gene expression pathway can be modified, modulated, and regulated by other gene products, by signal transduction processes, by cell- and tissue-specific factors, and by many additional factors.[35] Indeed, there is substantial *in vivo* and *in vitro* evidence available in tissue culture cell systems, in susceptible animal species, in short post-mortem human brain, and in human brain biopsy tissue that each step in the gene expression pathway can be perturbed by aluminum at some exposure level.[2,14–24,34,36–39]

12.5.2 Genotoxic Mechanisms of Aluminum

Aluminum is particularly detrimental to normal nucleic acid function largely because of the extensive delocalized electron chemistry of the genetic material.[17–20,35–39] For example, when one considers the $\sim 5.6 \times 10^9$ base pairs, containing some $\sim 1.12 \times 10^{10}$ phosphates and $\sim 2.24 \times 10^{10}$ oxygen donor groups of DNA typical of a human cell and additional nascent RNA, primary and processed RNA message, and deoxy- and ribonucleotide mono-, di-, and triphosphates contained within a typical nuclear volume of about 1000 μm,[3,35,40] the nucleus of the cell contains the highest phosphate density, and hence potential aluminum binding capacity, of any cellular organelle. In addition, the dense positive charge of aluminum, with an unchanging valence of +3, an ionic radius of only 51 pm, and a charge-to-size ratio, $Z^2/r = 17.65$, may support the translocation of this small cation across endothelial, glial, or neuronal membrane barriers to ultimately anchor within the cell nucleus.[42–44] Moreover, as concentrations of inorganic phosphate approach 1 to 2 mM in blood plasma, 10 mM in the cytoplasm, and at least 60 mM within the nucleus, the high phosphate density within this latter organelle might provide a high-affinity, high-capacity sink to attract and trap aluminum within the nuclear matrix.[37–40,43–46] Also, nucleic acid polyphosphates of neuronal nuclei are relatively dispersed throughout the interphase nucleoplasm compared with cytoplasmic phosphate moieties which are relatively compartmentalized within the endoplasmic reticulum, mitochondria, and lysosomes.[35,45,47] The particularly large size of neuronal nuclei (up to 1200 μm^3 in a layer of 5 pyramidal nuclei in the neocortex), the extensive euchromatization of the genetic material within this organelle[40,48] as measured by the faster digestion kinetics when exposed to nuclease,[24,49–50] the qualitatively unique[51] and quantitatively high transcriptional output of RNA message from these kinds of nuclei,[52–54] and the extensively developed nuclear pore complex system[55] may make the respositories of genetic information particularly susceptible to the deleterious effects of this toxic metal.

There are numerous reports of preferred binding of aluminum to structures located within the eukaryotic nucleus[12–17,22,36–37,56–58] and particularly within brain cell nuclei.[13,14,39,42,44,58–63] Moreover, depending on the route, dose, and chemical species of the administered aluminum, the accumulation onto nuclear structures is a rapid process, measured typically in minutes to hours.[42,56] Elevated nuclear aluminum has also been correlated with an increase in neocortical H1 linker histone-DNA binding, which would be expected to have a compacting effect on chromatin,[37] a decreased rate of cell division and DNA synthesis,[60] an increase in the error rate of DNA replication, and an inhibition of both hormone-induced chromosome puffing and corticosterone receptor binding to DNA.[58,61] Aluminum-induced alterations in brain RNA message pool size *in vivo*[62] decreased RNA synthesis in aluminum-treated neuroblastoma cells[63] and caused a reduction of neurofilament light chain (NFL) RNA message in anterior horn cells in rabbits.[64,65] Each suggests an aluminum-induced impairment of normal nucleic acid metabolism and a deficit in both the readout and transmission of brain-specific genetic information.

12.5.3 Aluminum and Chromatin-Templated Transcription

While aluminum has a particularly high affinity for the dispersed chromatin normally associated with neuronal nuclei,[34] DNA strand separation during template-directed processes would transiently expose even more delocalized electron fields normally preoccupied with hydrogen bonding in native DNA structures. Euchromatic nuclear regions characteristic of actively transcribing brain cell DNA would be therefore in a transient and vulnerable *open* configuration susceptible to interaction with this neurotoxic cation.

In order to further evaluate the effects of different nano- to micromolar concentrations of ambient aluminum on the process of chromatin-directed transcription in an *in vitro* system, 0 to 3000 nM of aluminum chloride were incubated with preparations of mixed neuronal and glial temporal lobe brain cell nuclei obtained from ~1- to 12-h post-mortem human brains in the presence of [³H]UTP using a rapid nuclear runoff transcription assay.[66] In agreement with the results of Sarkander et al.[67] who used a rat brain nuclei assay reporter system: (1) the effect of aluminum on RNA polymerase II activity (in 100 mM Tris-HCl, pH 7.5, 3 mM Mn^{++}, 400 mM NH$_4$SO$_4$, pH 7.5) was more pronounced than on RNA polymerase I activity (in 100 mM Tris-HCl, pH 8.5, 6 mM Mg^{++}, α-amanitin; Table 12.2), and (2) the blocking of endogenous chromatin template with actinomycin D and the recovery of RNA synthesis by the addition of exogenous poly (dA-dT) template suggested that the inhibition of human brain cell RNA synthesis by aluminum chloride was caused by some impairment in the brain cell chromatin template rather than in a loss of activity in the RNA polymerase I or II enzymes.[24] Indeed, the effects of *the unique biophysical properties of aluminum* towards brain cell chromatin is manifest by the fact that in chromatin compaction/ultracentrifugation sedimentation experiments, out of 16 di- and trivalent metallic cations scrutinized, including Al^{+++}, Be^{++}, Ca^{++}, Cd^{++}, Co^{++}, Cu^{++}, Fe^{++}, Ga^{+++}, Hg^{++}, In^{+++}, Mg^{++}, Mn^{++}, Ni^{++}, Sc^{+++}, Sr^{++}, and Zn^{++}, only micromolar concentrations of Al^{+++} had the most profound effect on inducing brain cell neocortical chromatin to condense and precipitate; the effects on liver chromatins were much less apparent.[68]

12.5.4 Compartmentalization of Nuclear Aluminum in the AD Brain

Although many investigators have independently measured up to a twofold increase of aluminum in whole neocortex, in whole neocortical nuclei, in neuronal nuclei, and in both heterochromatin and euchromatin fractions when compared to control brains,[12,13–17,39,69] aluminum does not appear to be uniformly distributed within the nuclear compartment in AD-affected brains. Using nuclease digestion on nuclei preparations isolated in ultrapure aluminum-free reagents to generate a rapid release population of brain cell dinucleosomes, the basic dimeric repeat of chromatin,[34,35] the

Table 12.2 Aluminum and Human Brain Gene Transcription. Effect of Various Doses of Aluminum Chloride on *In Vitro* RNA Polymerase I (Ribosomal RNA Transcripts) and RNA Polymerase II (Messenger RNA Transcripts Coding for Protein) Activities in Isolated Human Cerebral Nuclei (Temporal Lobe Brodmann Area A22)

Ambient Al^{3+} (nm)	N[a]	RNA polymerase I	RNA polymerase II
0 (control)	4	100.0	100.0
50	3	98.7 +/- 4.3	105.2 +/- 5.5
100	3	94.4 +/- 3.5	23.5 +/- 2.0
500	2	57.6 +/- 2.6	21.0 +/- 3.1
1000	4	42.5 +/- 4.1	20.8 +/- 2.7
1500	2	39.2 +/- 3.6	18.8 +/- 4.1
3000	4	30.3 +/- 2.4	15.0 +/- 4.8

Note: Incorporation of [^{3}HUTP] (pmol × µg DNA^{-1}). RNA polymerase I and II assays were carried out in the presence of Mg^{2+} and α-amanitin under low ionic strength conditions (RNAP I) and in the presence of Mn^{2+} and 400 mM (NH$_4$)SO$_4$ (RNAP II), following the procedures described by Fei and Drake[66] and by Sarkander et al.[67]

[a] N = number of experiments.

aluminum content ratio, AD over control, increases on the dinucleosomes to 4.5 for all neocortical areas, to 4.7 for the frontal neocortex, a neocortical area only marginally affected in AD neuropathology, and to nearly 9 in temporal lobe neocortex, a region associated with marked AD neuropathological change.[7,58,69]

Moreover, in 15 normal, 21 AD, and 13 non-AD affected post-mortem brains, it was found that at the time of death there was an association of aluminum with human brain dinucleosomes isolated from AD brain frontal and temporal neocortex (x = 2669.8 +/- 806.7 µg/g DNA), an accumulation that was observed neither in normal controls (x = 885.4 +/- 371.6 SD µg/g DNA) nor in a non-AD dementia group (x = 603.3 +/- 306.9 SD µg/g DNA).[39]

Taken together, these data suggest that both the basic functional and structural levels of human brain chromatin organization are perturbed by the presence of aluminum salts, and this has important consequences for brain-specific DNA templates to be effectively transcribed. Importantly, both general and specific reductions in RNA message signals are a consistently reported feature of both AD[2,12,17,34,38,46,50,53,54] and EAE.[2,42,62–65,67,71]

12.6 EXPERIMENTAL ALUMINUM ENCEPHALOPATHY (EAE), NEURODEGENERATIVE DISEASE, AND AD

EAE is a model system whereby injection of an aluminum salt, typically aluminum as lactate or chloride, by intrathecal,[64] subcutaneous,[17,56] or intravenous[42] injection into an aluminum-susceptible animal leads to a rapid accumulation of aluminum onto the nuclear structures of the glia and large neurons.[12–14,42] A sequence of events resembling AD then ensues, including a progressive decline in higher cortical functions, an impairment of short-term memory, motor disturbances, and death within days or weeks depending on the aluminum dose, route of administration, and the particular species of aluminum salt used. Interestingly, using salts of the trivalent metal periodic table group IIIb (Sc, Y, La, Ac) and group IIIa (B, Al, Ga, In, Tl), only aluminum was capable of inducing a progressive encephalopathy in rabbits.[14,56] Morin histological staining of post-mortem EAE neural tissue often shows that the cytosol of astrocytes contributes strong initial binding sites for aluminum;[56,72] particularly high concentrations of aluminum are detectable in glial[22,72] or neuronal lysosomes[47] as well. *The compartmentalization of many neurobiochemical systems in the neuronal and glial cytosol might protect their biochemistry from the deleterious effects of this neurotoxin.* Specific aluminum foci within brain nuclei appear to be involved in subsequent aluminum binding.[24,39,42,56] For example, histochemical localization and subsequent subcellular fractionation studies[56] demonstrate that the increase in aluminum concentration occurs specifically on the largely euchromatic neocortical neuronal chromatins; the ultimate association of aluminum with neuronal interchromatin granules, heterochromatin, and in euchromatin again suggests that normal nuclear structures and functions are the main target for the deleterious effects of this element.[17,34,37,39,42] *Importantly, both the brains of humans with AD and experimental animals injected with aluminum salts showed remarkably similar loci for aluminum accumulation, that is, primarily within the genetic material of the largest nerve cells in the CNS.*

At concentrations found toxic to experimental animals in EAE, aluminum is found in a number of human neurological disorders including Parkinson's disease,[73] the neuromelanin of substantia nigra neurons,[74,75] amyotrophic lateral sclerosis (ALS; motor neuron disease[76–80]), the ALS-parkinsonian dementia complex (ALS-PDC) of Guam and the Kii Peninsula of Japan,[79–82] dialysis encephalopathy,[13,83] and in AD,[12–17,39,81,82,84,85] but there appears to be some variation in the particular cellular location of aluminum deposition in each neurological dysfunction. For example, two foci for aluminum deposition in ALS-PDC appear to be spinal tract lumbar motor neurons and the neurofibrillary tangle (NFT)-bearing hippocampal neurons.[79,82] Similarly in AD, while the major target for aluminum deposition is within neocortical neuronal chromatin, it has also been found to be associated with NFT, the amyloid cores of neuritic or senile plaques (SP), and with the iron binding and transport proteins ferritin and transferrin (see Table 12.1).

12.7 EAE AND CYTOSKELETAL GENE EXPRESSION

EAE, an informative model of the genetic response of brain cells to aluminum-induced neurotoxic injury, has suggested that certain gene classes, and in particular genes coding for several cytoskeletal elements, are highly sensitive to aluminum-induced neurotoxicity.[24,62,64,65] For example, in EAE in rabbits, intrathecal administration of 1% aluminum lactate results in decreased levels of the RNA messages for β–actin, α–tubulin, and NFL, a pattern analagous to the cytoskeletal RNA message reduction observed in AD.[24,46,50,53,54,86] Alterations in both cytoskeletal structure and function and in cellular morphology are a consistently reported feature of both EAE and AD.[64,65,86–90] These changes in gene expression following EAE appear to be a specific effect of aluminum intoxication on genetic processes rather than a response to axonal injury,[64,90] and accompanying the aluminum intoxication there is concomitant neuritic shrinkage, axonal atrophy, and a distortion in the geometry of neurons.[89–91] These changes have been proposed to be associated with the impairment in NFL RNA message generation *in vivo*, the determining control point in the expression of the single-copy neuron-specific NFL gene.[53,92] In addition, animals subjected to EAE are characterized by abnormal phosphorylation of neurofilaments in the perikarya of their neurons[89,90,] and exhibit neurofibrillary pathology in cell bodies and axons that are similar to AD changes.[89,93] For example, neuroblastoma cells treated with aluminum express epitopes associated with AD NFT[24,94] and the effects appear to be aluminum-specific since the neurotoxins iminodipropionitrile or acrylamide induce a different type of disruption of the neuronal cytoskeleton with no *in vivo* effects on cytoskeletal RNA message levels.[71,95] Moreover, there is a growing body of data on the ability of aluminum to act as a nucleating agent for both the tau[117] and β-amyloid[94,96] proteins which form the cores of the NFT and senile plaque, respectively. These aluminum-induced aggregations both microscopically and immunochemically resemble those NFT and SP deposits observed and described as being the hallmarks of AD neuropathology.[5,7,89,93–96]

12.8 IMPAIRMENT OF NERVE GROWTH FACTOR (NGF) INDUCTION OF CYTOSKELETAL GENES IN PHEOCHROMOCYTOMA (PC) CELLS EXPOSED TO ALUMINUM SALTS

Treatment of adrenal pheochromocytoma (PC) cells in tissue culture with nerve growth factor (NGF) arrests their mitosis and induces differentiation towards a neuronal phenotype with extensive neurite outgrowth.[24,98] Concurrently, there is a 10- to 15-fold increase in the rate of NFL RNA message generation when compared to the levels of β-actin. Using neurite outgrowth as an index of NGF-induced differentiation and Northern blotting and RT-PCR analysis of RNA message isolated from PC cells to monitor NFL levels, data from our laboratory have shown that *in vitro*, micromolar concentrations of aluminum lactate in the PC culture medium both inhibits neurite outgrowth and abolishes the phenomenon of NGF-induced NFL RNA message expression.[24,91,97] While PC cells preincubated with aluminum lactate were not able to respond normally to NGF-mediated induction of NFL gene expression and neurite extension, utilization of a technique involving *molecular shuttle chelaton*,[97] which uses aluminum chelators in tandem combination to remove nuclear-bound aluminum, was found to restore NGF-mediated gene induction and neurite extension *in vitro*.[24,97] *Therefore, both* in vitro *and* in vivo *data support the concept that one specific major neurotoxic target for aluminum in the cells of the CNS includes genetic components of the brain cell's cytoskeletal system.*[2,34,64,89]

12.9 THE SELECTIVITY OF THE ALUMINUM INTERACTION: NEUROLOGICAL DISEASE AND AD

Several related neurodegenerative disorders such as hypoxic encephalopathy, multiple infarct dementia, ALS, progressive supranuclear palsy,[39] or Creutzfeld-Jacob disease[99] do not elevate the concentration of aluminum in cells of the brain per se, suggesting that the ectopic accumulation of aluminum within neocortical pyramidal cell nuclei in AD is both pathoclisic and associated with AD etiopathogenesis. At present, however, it is still not clear what are the *exact primary pathogenic events* which lead to aluminum deposition and/or the development of AD. Genetic linkage studies suggest that AD is not a single homogeneous disorder;[100,101] while features of the hereditary or familial form of AD appear to be linked to genetic loci on at least four separate chromosomes, the molecular and genetic mechanisms associated with the more common idiopathic or sporadic form of AD remain largely unexplored. It is reasonable to speculate that both the familial and the sporadic forms of AD may include a genetic deficit in a gene product that is involved in the control of an age-related developmental program, or in gene products involved in the epithelial-blood-glial-neuronal-nuclear membrane barriers which affect aluminum uptake, transport, or excretion.

It is apparent that aluminum exerts neurotoxic effects via a *unique set of biophysical properties* which are detrimental towards a wide spectrum of neurobiological, and in particular, neurogenetic functions. The preferential interaction of aluminum with cellular and nuclear substrates which are phylogenetically specific to the highly evolved components of the human brain,[7,9,48] may explain, in part, this element's dramatic neurotoxicity towards elements and processes unique to the human CNS.[24,34,39] One attractive hypothesis is that a certain *critical mass of biological errors* must be induced by aluminum in the CNS to initiate the AD process.[20,21,24] Indeed, a lifetime of cumulative insults to brain cell metabolism would contribute to an increasing inability of CNS cells to maintain homeostasis, leading to progressive alterations in cellular barriers and escalatory mismetabolism reinforced by the process of positive feedback (see Reference 2, Figure 1).

12.10 ALZHEIMER'S DISEASE — THE SOCIAL AND ECONOMIC IMPACT

Finally, in order to give the reader some idea of the severity of the impact of AD on society and the economy, this irreversible neurological condition of the human brain is now the fourth leading cause of death in American adults, AD takes more than 100,000 lives annually, and represents the commonest form of severe intellectual impairment in the elderly, accounting for about 65% of all cases of dementia.[101–104] In the U.S., approximately 3 million persons are moderately or severely affected by this incapacitating disorder,[103] and according to the *Alzheimer Association*, AD is currently costing American society about $100 billion every year. With the elderly comprising the fastest growing segment of our population, unless a cure or a means of prevention is found more than 14 million Americans will be afflicted by the year 2050. In 1992, our federal government spent approximately $280 million for AD research, representing less than one-third of one cent for every dollar the disease now costs society.[104] In comparison, the federal investment in other diseases with similar economic or social impact such as cancer, heart disease, and AIDS is four to seven times higher. The *Alliance for Aging Research* has estimated that delaying the onset of AD by just 5 years would save an estimated $47 billion annually in our national health care costs. Importantly, in preliminary clinical trials, the aluminum chelator DF has shown promise in the treatment of AD, reducing neocortical aluminum concentrations to near control values,[32] slowing the clinical course of AD,[33] and improving the general quality of life for the AD patient.[32,33]

12.11 SUMMARY

Clearly, the involvement of aluminum in general CNS neurotoxicity, in brain genotoxicity, in EAE, and human neurological dysfunction including AD is substantiated by hundreds of reports. Aluminum has a wide array of adverse effects on the structural, biochemical, and genetic integrity of the CNS in mammals and humans and in many experimental systems, both *in vitro* and *in vivo*. However, the complex molecular mechanisms involved in aluminum toxicity towards neurobiological systems need to be more completely investigated. While features of the familial or hereditary form of AD appear to be genetically linked to several separate chromosomes, these together represent probably much less than a few percent of all AD cases, and the etiopathogenesis associated with the more common idiopathic or sporadic forms of AD strongly implicate epigenetic or environmental factors that include toxic elements such as aluminum. Members of our aging population at genetic risk for AD may be wise in choosing to limit their exposure to this ubiquitous neurotoxin since at this point in time the *prevention* of AD is much more easily attainable than the development of a *cure*.

REFERENCES

1. Mason BH and Moore CB: *Principles of Geochemistry,* 4th edition: John Wiley and Sons, New York, 1982, Table 3.5; Montgomery CW: *Environmental Geology,* 4th edition: W.C. Brown, Boston, 1995, Table 1.2; *CRC Handbook of Chemistry and Physics,* 70th edition: CRC Press, Boca Raton, 1989–1990, Table F-166 (Demayo).
2. Lukiw WJ and McLachlan DRC: Aluminum Neurotoxicity. In Chang L and Dyer R (eds), *Handbook of Neurotoxicology.* II: Effects and Mechanisms, Marcel Dekker, New York, 1995, 105–142.
3. Kraepelin E: Obituary to Franz Nissl. *J. Nerv. Ment. Dis.,* 1920; 51:207–214.
4. Vogt C and Vogt O: Erkrabkungen der grosshirnrinde in lichte der topistik, pathoklise und pathoarchitektonik. *J. Psychiatr. Neurol.,* 1922; 28:1–73. See also Lowndes HE et al., Keynote Introduction, pp 1–27 in Reference (2).
5. Alzheimer A: Uber eine eigenartige Erkrankung der Hirnrinde. Allgemeine eitschrift fur Psychiatrie und Psychisch-Gerichtlich Medizin 1907; 64:146–148.
6. Talbot JH: Franz Nissl (1860–1919); neuropathologist (editorial). *J. Am. Med. Assoc.,* 1968; 205:208–209.
7. Morrison JH, Hof P, Campbell M, et al: Cellular pathology in Alzheimer's disease; implications for corticocortical disconnection and differential vulnerability. In Rapoport S, Petit H, Leys D and Christen Y (eds): *Imaging, Cerebral Topography and Alzheimer's Disease,* Springer Verlag, New York, 1990, pp 19–40.
8. Katzman R and Saitoh R: Advances in Alzheimer's disease. *FASEB J.,* 1991; 5:278–285.
9. Rapoport SI: A phylogenetic hypothesis for Alzheimer's disease. In: Sinet P, Lamour Y, (eds): *Genetics and Alzheimer's Disease,* Springer Verlag, New York, 1988, pp 62–68.
10. Doellken P: Aluminum verursachten Lasionene im Zentralnervensystem. *Arch. Exp. Pathol. Pharmacol.,* 1897; 40:58–120.
11. Spofforth J: A case of aluminum poisoning. *Lancet,* 1921; i:1301.
12. Crapper DR: Functional consequences of neurofibrillary degeneration. In Gershon S and Terry RD (eds): *The Neurobiology of Aging,* Raven Press, New York, 1976, 405–432.
13. Crapper DR, Quittkat S and Krishnan SS: Intranuclear aluminum content in Alzheimer's disease, dialysis encephalopathy. *Acta Neuropath. (Berlin),* 1980; 50:19–24.
14. Crapper McLachlan DR and de Boni U: Aluminum in human brain disease. *Neurotoxicology,* 1980; 1: 3–16.
15. Haug A: Molecular aspects of aluminum toxicity. *CRC Crit. Rev. Plant Sci.,* 1984; 1:345–373.
16. Siegel N: Aluminum interaction with biomolecules: the molecular basis for aluminum toxicity. *Am. J. Kidney Dis.,* 1985; 6:353–357.
17. Crapper McLachlan DR: Aluminum and Alzheimer's disease. *Neurobiol. Aging,* 1986; 7:525–532.
18. Martin RB: Aluminum speciation in biology. *Alum. Biol. Med.,* 1992; 169:5–18.

19. Birchall JD and Chappell JS: The chemistry of aluminum and silicon in relation to Alzheimer's disease. *Clin. Chem.*, 1988; 34:265–267.

20. Joshi JG: Aluminum, a neurotoxin which affects diverse metabolic reactions. *Biofactors*, 1990; 2: 163–169.

21. Joshi JG: Neurochemical hypothesis: participation by aluminum in producing critical mass of colocalized errors in brain leads to neurological disease. *Comp. Biochem. Physiol.*, 1991; 100:103–105.

22. Crapper McLachlan DR, Kruck TPA, and Lukiw WJ: Would decreased aluminum ingestion reduce the incidence of Alzheimer's disease? *Can. Med. Assoc. J.*, 1991; 145:793–804.

23. McLachlan DR, Fraser PE, Jaikaran E, et al: Alzheimer's disease and other aluminum associated health conditions. In Chang L (ed): *Toxicology of Metals, General Toxicology, Carcinogenesis and Human Exposure*, Volume 1. Section IV: Clinical Aspects of Metal Toxicology: CRC Press, Boca Raton 1995, 1–24.

24. Lukiw WJ, Kruck TPA, Krishnan B, et al: Human brain neocortical aluminum in Alzheimer's disease (AD). In preparation, 1996.

25. Still CN and Kelly P: On the incidence of primary degenerative dementia vs. water fluoride content in South Carolina. *Neurotoxicology*, 1980; 4:125–131.

26. Vogt T: Water quality and health — study of a possible relationship between aluminum in drinking water and dementia (Sosiale og okonomiske studier 61, English abstract), Oslo: Central Bureau of Statistics of Norway, 1986.

27. Martyn CN, Osmond C, and Edwardson JA: Geographical relation between Alzheimer's disease and aluminum in drinking water. *Lancet*, 1989; 1:59–62.

28. Michel P, Commenges D, Dartigues JF, et al: Study of the relationship between Alzheimer's disease and aluminum in drinking water. Abstract 47. *Neurobiol. Aging*, 1990; 11:264.

29. Forbes WF, Hayward LM, and Agwani N: Dementia, aluminum and fluoride. *Lancet*, 1991; 338:1592–1593.

30. Neri LC and Hewitt D: Aluminum, Alzheimer's disease, and drinking water. *Lancet*, 1991; 338:390.

31. Flaten TP: Geographical associations between aluminum in drinking water and registered death rates with dementia (including Alzheimer's disease, Parkinson's disease and amyotrophic lateral sclerosis) in Norway. *Environ. Geochem. Health*, 12, 1992, 152–167.

32. McLachlan DRC, Dalton AJ, Kruck TPA, et al: Effect of Desferrioxamine on the clinical progress of Alzheimer disease. *Lancet*, 1991; 337:1304–1308.

33. Andrews DF, Crapper McLachlan DR, Dalton AJ, et al: Desferrioxamine for Alzheimer's disease. *Lancet*, 1991; 338:324–326.

34. Lukiw WJ, St. George Hyslop P, and McLachlan DRC: Chromatin structure, gene expression and nuclear aluminum in Alzheimer's disease. In Harrison PJ (ed): *Basic and Clinical Aspects of Neuroscience*, 6, Regulation of Gene Expression and Brain Function, Springer-Verlag, Berlin 1994, 6:31–45.

35. Hawkins JD: *Gene Structure and Expression*, 2nd Edition, Cambridge University Press, U.K.,1991.

36. Wedrychowski A, Schmidt W, and Hnilica L: The *in vivo* cross-linking of proteins and DNA by heavy metals. *J. Biol. Chem.*, 1986; 261:3370–3376.

37. Lukiw WJ, Kruck TPA, and McLachlan DRC: Alterations in human linker histone-DNA binding in the presence of aluminum salts *in vitro* and in Alzheimer's disease. *Neurotoxicology*, 1987; 8:291–302.

38. Lukiw WJ, Kruck TPA and Crapper McLachlan DR: Aluminum, intracellular liganding and the nucleus. *Lancet*, 1989; 1:781.

39. Lukiw WJ, Bergeron C, Wong L, et al: Nuclear compartmentalization of aluminum in Alzheimer's disease (AD). *Neurobiol. Aging*, 1992; 13:115–121.

40. Krstic RV: *Ultrastructure of the Mammalian Cell: An Atlas*. Springer Verlag, Berlin, 1979, 1–9.

41. Mortimer JA: Genetic and environmental risk factors for Alzheimer's disease: In Altman H (ed): Key questions and new approaches: *Alzheimer's Disease and Dementia: Problems, Prospects and Perspectives*. Plenum Press, New York, 1989, 85–100.

42. Wen G and Wisniewski HM: Histochemical localization of aluminum in the rabbit CNS. *Acta Neuropathol. (Berl.)*, 1985; 68:175–184.

43. Banks WA, Kastin AJ, and Fasold MB: Differential effect of aluminum on the blood-brain barrier, transport of peptides, technetium and albumin. *J. Pharmacol. Exp. Ther.*, 1988; 244:579–585.

44. Dobson CB and Itzhaki RF: Localization of aluminum in human neuroblastoma cells. *Life Chem. Rep.*, 1994; 11:189–193

45. Lukiw WJ, Kruck TPA, and McLachlan DRC: Linker histone-DNA complexes; enhanced stability in the presence of aluminum lactate and implications for Alzheimer's disease. *FEBS Lett.*, 1989; 253: 59–62.

46. Lukiw WJ and Crapper McLachlan DR: Chromatin structure and gene expression in Alzheimer's disease (AD). *Mol. Brain Res.*, 1990; 7:227–233.

47. Steckhoven J, Renkawek K, Otte-Holler I, et al: Exogenous aluminum accumulated in the lysosomes of cultured rat cortical neurons. *Neurosci. Lett.*, 1990; 119:71–74.

48. Kandel ER: Nerve cells and behavior. In Kandel E, Schwartz J, and Jessel T. (eds). *Principles of Neural Science*. Third Edition. Elsevier, New York, 1991; 2–32.

49. McLachlan DR, Lewis PN, Lukiw WJ, et al: Chromatin Structure in Dementia. *Ann. Neurol.*, 1984; 15:324–329.

50. McLachlan DRC, Lukiw WJ, Wong L, et al: Selective messenger RNA reduction in Alzheimer's disease. *Mol. Brain Res.*, 1988; 3:255–262.

51. Sutcliffe J, Milner R, Gottesfeld J, et al: Control of neuronal gene expression. *Science*, 1984; 255:1308–1315.
52. Thompson RJ: Studies on RNA synthesis in two populations of nuclei from the mammalian cerebral cortex. *J. Neurochem.*, 1973, 21;19–39.
53. Lukiw WJ, Wong L, and McLachlan DRC: Cytoskeletal messenger RNA stability in human neocortex: studies in normal aging and in Alzheimer's disease. *Int. J. Neurosci.*, 1990; 55:81–88.
54. Lukiw WJ, McLachlan DR, and Bazan NG: RNA message levels in normally aging and in Alzheimer's disease (AD) affected human temporal lobe neocortex. Abstract, XVth Washington International Spring Symposium, Neurodegenerative Diseases '95: Molecular and Cellular Mechanisms and Therapeutic Advances, 1995.
55. Dingwall C and Laskey R: The nuclear membrane. *Science*, 1992; 258:942–947.
56. De Boni U, Scott JW, and Crapper DR: Intracellular aluminum binding; a histochemical study. *Histochemistry*, 1974; 40:31–37.
57. Matsumoto H and Morimura S: Repressed template activity of chromatin of pea roots treated by aluminum. *Plant Cell Physiol.*, 1980; 21:951–959.
58. McLachlan DRC, Lukiw WJ, and Kruck TPA: New evidence for an active role of aluminum in Alzheimer's disease. *Can. J. Neurol. Sci.*, 1989; 16:490–497.
59. Shi B and Haug A: Aluminum uptake by neuroblastoma cells. *J. Neurochem.*, 1990; 55:551–558.
60. Berlyne G, Ben Ari J, Knoff E, et al: Aluminum toxicity in rats. *Lancet*, 1972; i:494–496.
61. Sanderson C, Crapper McLachlan DRC, and De Boni U: Inhibition of corticosterone binding *in vitro*, in rabbit hippocampus by chromatin bound aluminum. *Acta Neuropathol. (Berl.).*, 1982; 57:249–254.
62. Van Berkum MFA, Wong L, Lewis PN, et al: Total and poly(A) RNA yields during an aluminum encephalopathy in rabbit brains. *Neurochem. Res.*, 1986; 11:1347–1359.
63. Miller CA and Levine EM: Effects of aluminum salts on cultured neuroblastoma cells. *J. Neurochem.*, 1974; 22:751–758.
64. Muma NA, Troncoso JC, Hoffman P, et al: Aluminum neurotoxicity — altered expression of cytoskeletal genes. *Mol. Brain Res.*, 1988; 3:115–122.
65. Chambers CB, Singer SM, and Muma NA: Alterations in polyadenylated mRNA levels following administration of aluminum. Abstract 87.3, *Soc. Neurosci.*, 1994; 20:199.
66. Fei H and Drake TA: A rapid nuclear runoff transcription assay. *Biotechniques*, 1993; 15:838.
67. Sarkander H-I, Balb G, Schlosser H, et al: Blockade of neuronal brain RNA initiation sites by aluminum: a primary molecular mechanism of aluminum-induced neurofibrillary changes. In Cervos-Navarro J, Sarkander HI: *Brain Aging: Neuropathology and Neuropharmacology*. Raven Press, New York, 1983; 645–647.
68. Walker PR, LeBlanc J, and Sikorska M: Effects of aluminum and other cations on the structure of brain and liver chromatin. *Biochemistry*, 1989; 28:3911–3915.
69. Ehmann WD and Markesbery WR: A multi-technique approach to the study of aluminum in Alzheimer's disease brain. *Life Chem. Rep.*, 1994; 11:11–28.
70. Yoshimasu F, Yasui M, and Yoshida H: Aluminum in Alzheimer's disease in Japan and Parkinsonism-dementia in Guam. XII World Congress of Neurology, Hamburg Abstract 15–07–02, 1985.
71. Parhad IM, Krekoski CA, Mathew A, et al: Neuronal gene expression in aluminum myelopathy. *Cell. Mol. Neurobiol.*, 1989; 9:123–138.
72. Young JK: Alzheimer's disease and metal containing glia. *Med. Hypoth.*, 1992; 38:1–4.
73. Hirsch EC, Brandel JP, Galle P, et al: Iron and aluminum increase in the substantia nigra of patients with Parkinson's disease; an X-ray microanalysis. *J. Neurochem.*, 1991; 56:446–451
74. Good PF, Perl DP, Bierer LM, et al: Selective accumulation of aluminum and iron in the neurofibrillary tangles of Alzheimer's disease: a laser microprobe (LAMMA) study. *Ann. Neurol.*, 1992; 31:286–292.
75. Good, PF and Perl DP: A quantitative comparison of aluminum concentration in neurofibrillary tangles of Alzheimer's disease and Parkinsonian Dementia Complex of Guam by laser microprobe mass analysis. Abstract 115, *Neurobiol. Aging*, 1994; 15:S28.
76. Kobayashi K, Yumoto S, Nagai H, et al: ^{26}Al tracer experiment by accelerator mass spectrometry and its application to the studies for amyotrophic lateral sclerosis and Alzheimer's disease. I, *Proc. Jpn. Acad.*, 66B, 1990:189–192.
77. Yase Y: The pathogenesis of amyotrophic lateral sclerosis. *Lancet*, 1972; 2:292–296.
78. Yasui M, Yase Y, Ota K, et al: Aluminum deposition in the central nervous system of patients with amyotrophic lateral sclerosis from the Kii Peninsula of Japan. *Neurotoxicology*, 1991; 12:615–620.
79. Perl D, Gajdusek C, Garruto R, et al: Intraneuronal aluminum accumulation in amyotrophic lateral sclerosis and parkinsonism dementia of Guam. *Science*, 1982; 217:1053–1055.
80. Garruto RM: Pacific paradigms of environmentally induced neurological disorders: clinical, epidemiological and molecular perspectives. *Neurotoxicology*, 1991; 12:347–378.
81. Perl DP and Good PF: Practical approaches to microprobe analysis of neurofibrillary tangle-bearing neurons of Alzheimer's disease and related disorders. *Life Chem. Rep.*, 1994; 11:47–53.
82. Perl DP: Aluminum and Iron, potential pathogenic factors in Alzheimer's disease and related disorders. Unraveling Alzheimer's disease, *Clin. Res. Perspec.*, April 7, 1995, Austin, Texas, Abstract 1995.
83. Alfrey AC, LeGendre GR, and Kheany WD: The dialysis encephalopathy syndrome: possible aluminum intoxication. *N. Engl. J. Med.*, 1976; 294:184–188.

84. Perl DP and Brody AR: Alzheimer's disease: X-ray spectrometric evidence of aluminum accumulation in neurofibrillary tangle-bearing neurons. *Science*, 1980; 208:297–299.

85. Perl DP and Pendlebury WW: Aluminum accumulation in neurofibrillary tangle-bearing neurons of senile dementia, Alzheimer's type: Detection by intraneuronal X-ray spectrometry studies of unstained tissue sections. *J. Neuropathol. Exp. Neurol.*, 1984; 43:349–359.

86. Clark AW, Krekoski CA, and Parhad IM: Altered expression of genes for amyloid and cytoskeletal proteins in Alzheimer cortex. *Ann. Neurol.*, 1989; 25:331–339.

87. Uemura E and Ireland WP: Synaptic density in chronic animals with experimental neurofibrillary changes. *Exp. Neurol.*, 1984; 85:1–9.

88. Kosik KS, McCluskey AH, Walsh FX, et al: Axonal transport of cytoskeletal proteins in aluminum toxicity. *Neurochem. Pathol.*, 1985; 3:99–108.

89. Katsetos CD, Savory J, Herman MM, et al: Neuronal cytoskeletal lesions induced in the CNS by intraventribular and intravenous aluminum maltol in rabbits. *Neuropathol. Appl. Neurobiol.*, 1990; 16:511–518.

90. Troncoso J, March JL, Haner M, et al: Effect of aluminum and other multivalent cations on neurofilaments *in vitro*: an electron microscopic study. *J. Struct. Biol.*, 1990; 103:2–12.

91. Hoffman P, Cleveland D, Griffin J, et al: Neurofilament gene expression; a major determinant of axonal caliber. *Proc. Natl. Acad. Sci. U.S.A.*, 1987; 84:3472–3476.

92. Nakahira K, Ikenaka K, Wada K, et al: Structure of the 68-kDa neurofilament gene and regulation of its expression. *J. Biol. Chem.*, 1990; 265:19786–19791.

93. Bugiani O and Ghetti B: Progressing encephalomyelopathy with muscular atrophy, induced by aluminum powder. *Neurobiol. Aging*, 1982; 3:209–222.

94. Guy SP, Jones D, Mann DMA, et al: Human neuroblastoma cells treated with aluminum express an epitope associated with Alzheimer's disease neurofibrillary tangles. *Neurosci. Lett.*, 1991; 121:166–168.

95. Parhad I, Swedberg E, Hoar D, et al: Neurofilament gene expression following IDPN intoxication. *Mol. Brain Res.*, 1988; 4:293–301.

96. Mantyh PW, Ghilardi JR, Rogers S, et al: Aluminum, iron and zinc ions promote aggregation of physiological concentrations of beta-amyloid peptide. *J. Neurochem.*, 1993; 61:1171–1174.

97. Kruck TPA, Crapper McLachlan DR, Bergeron C, et al: Aluminum in neocortical nuclei — removal by shuttle chelation and relevance to Alzheimer's disease pharmacotherapy. In: *Alzheimer's Disease and Related Disorders: Selected Abstracts*, M. Nicolini, Editor, University of Padua. 1994.

98. Dickson G, Prentice H, Julien JP, et al: Nerve growth factor activates Thy-1 and neurofilament gene transcription in rat PC12 cells. *EMBO J*, 1986; 5:3449–3453.

99. Traub RD, Rains TC, Garruto RM, et al: Brain destruction alone does not elevate brain aluminum. *Neurology*, 1981; 31:986–990.

100. Nee LE, Elvidge R, and Sunduland T: Dementia of the Alzheimer type, Clinical and family study of 22 twin pairs. *Neurology*, 1987; 37:359–363.

101. McKhann G, Drachman D, and Folstein M: Clinical diagnosis of Alzheimer's disease, Report of the NINCDS-ADRDA Work Group under the auspices of Department of Health and Human Services Task Force on Alzheimer's Disease. *Neurology*, 1984; 34:939–944.

102. Hay J and Ernst R: The economic costs of Alzheimer's disease. *Am. J. Pub. Health*, 1987; 77:1169–1175.

103. Seventh Report of the Council on Alzheimer's Disease Progress in Research, U.S. Department of Health and Human Services, Washington, D.C., 1994.

104. American Pharmaceutical Manufacturer's Association Survey, 1993.

105. McLaughlin AIG, Kazantzis G, and King O; Pulmonary fibrosis and encephalopathy associated with the inhalation of aluminum dust. *Br. J. Ind. Med.*, 1962; 19:253–256.

106. Duckett F and Galle J: Mise en evidence de l'aluminum dans les plaques de la maladie d'Alzheimer: Etude a la microsonde de Castaing. *C. R. Acad. Sci. (Paris)*, 1976; 282:393–396.

107. Trapp GA: Studies of aluminum interaction with enzymes and proteins: the inhibition of hexokinase. *Neurotoxicology*, 1980; 1:89–100.

108. Masters CL, Simms G, Weinman NA, et al: Amyloid plaque core protein in Alzheimer's disease and Downs Syndrome. *P.N.A.S.*, 1985; 82:4245–4249.

109. Joshi JG, Fleming J, and Zimmerman A: Effects of long term intake of low levels of aluminum on some enzyme activities and ferritin. XIIIth World Neurology Congress, Hamburg. Supp., *J. Neurol.*, 1985; 232: 61.

110. Candy JM, Klinowski RH, Perry EK et al: Aluminosilicates and senile plaque formation in Alzheimer's disease. *Lancet*, 1986; 1:354–357.

111. Edwardson JA, Ferrier IN, McArthur FK, et al: Alzheimers disease and the aluminum hypothesis. In: *Aluminum in Chemistry, Biology and Medicine*, Nicolini, M., Zatta, P., and Corain, B., Editors. Raven Press, New York, 1991; 1:85–96.

112. Rifat SL, Eastwood MR, McLachlan DR, et al: Effect of exposure of miners to aluminum powder. *Lancet*, 1990, 336:1162–1165.

113. Moreno A, Dominguez P, and Dominguez C: High serum aluminum levels and acute reversible encephalopathy in a 4 year old boy with acute renal failure. *Eur. J. Pediatr.*, 1991; 150:513–514.

114. Bolla K, Briefel G, Spector D, et al: Neurocognitive effects of aluminum. *Arch. Neurol.*, 1992; 49:1021–1026.
115. Candy JM, Oakley AE, McArthur FK, et al: Microanalytical and molecular approaches to the study of aluminum in relation to Alzheimer's disease. *Life Chem. Rep.*, 1994; 11:55–69.
116. Favarato M, Zanoni S, Nicolini, M, et al: Histofluorimetric aluminum determination in the core of senile plaques from Alzheimer's disease. *Life Chem. Rep.*, 1994; 11:71–78.
117. Scott CW, Fieles A, Sygowski LA, et al: Aggregation of tau protein by aluminum. *Brain Res.*, 1993; 628:77–84.

Chapter 13

Dialysis Encephalopathy

Allen C. Alfrey

CONTENTS

13.1 INTRODUCTION

The neurotoxicity of aluminum has been well documented, being initially described by Siem in 1886.[1] Additional support for the neurotoxicity of aluminum was obtained by Dölken in 1897 who demonstrated anatomical alterations in the brains of animals who had received subcutaneous injections of aluminum.[1] More recent studies demonstrated that in certain species of animals, cats and rabbits, both local and systemic exposure to aluminum could cause histological alterations consisting of neurofibrillary changes in the brain.

However, because of its insoluble nature and the natural barriers for aluminum absorption imposed by the skin, lung, and gastrointestinal tract, the normal systemic burden of aluminum from usual environmental exposure has been extremely limited and humans have been largely protected from the toxicity of this element.[2] In fact only one, well-documented case of total body aluminum excess and presumed aluminum neurotoxicity was described in humans prior to 1972.[3]

In contrast to normal individuals, patients with chronic renal failure are at an increased risk of aluminum loading. This is a consequence of multiple factors including (1) defective barriers to aluminum absorption associated with the uremic state,[4] (2) the inability to eliminate any absorbed aluminum because of compromised renal function,[5] and (3) increased exposure to aluminum sources which can either circumvent or overwhelm the usual barriers to aluminum absorption. In regards to the oral ingestion of aluminum, it has been shown in animals with experimentally induced uremia that aluminum absorption is markedly enhanced.[4] The factors responsible for the enhanced aluminum absorption in the uremic state have not been defined. However, since aluminum is normally absorbed passively through the paracellular pathways it seems possible that the integrity of the tight junctions may be altered in uremia.[6]

The kidneys have been shown to be the only effective avenue by which any systemically administered aluminum can be eliminated.[5] Normally, the small amount of aluminum absorbed is rapidly eliminated by the kidney, with the body burden of this element maintained at extremely low levels throughout life.[2] As a consequence of the enhanced absorption and inability to eliminate aluminum, increased tissue stores of aluminum have been found in 85% of nondialyzed uremic patients, with no known history of excess environmental aluminum exposure.[2] In addition to the above disturbances in aluminum metabolism, the uremic patient can also have an increased oral and parenteral exposure to aluminum that can further enhance the body burden of this element.

It was first shown in 1941 that orally administered aluminum compounds could decrease the gastrointestinal absorption of phosphorus, normalize serum phosphorus levels, increase serum calcium, and cause symptomatic improvement of renal bone disease in uremic children.[7] By 1970, orally administered aluminum compounds were routinely given to uremic patients to control serum phosphorus levels with the purpose of preventing metastatic calcification and secondary hyperparathyroidism. The use of these agents increased the usual oral intake of aluminum from dietary sources of only 5 to 10 mg/day to gram quantities.[8] An additional risk of systemic loading of aluminum in dialyzed uremic patients is through aluminum-contaminated dialysate.[5] Aluminum is naturally present in very low concentrations in most freshwater sources; however, municipal water treatment facilities commonly add aluminum as a flocculant to clarify the water. This in turn leaves an aluminum residue in the water. If the residual aluminum is not removed from the water used to prepare dialysate, this results in a corresponding enrichment of the dialysate. Any aluminum present as a contaminant in dialysate is readily transferred to the uremic patient throughout the dialysis procedure.[5] This is because any transported aluminum is bound to transferrin in plasma, thereby maintaining a continuous gradient for transport from dialysate to plasma.[9] Because of the plasma binding of aluminum it is not removed by dialysis even if the dialysate is aluminum free.[9]

Uremic patients who have taken large amounts of orally administered aluminum containing phosphate binding gels or been exposed to aluminum-contaminated dialysate have been found to have greatly increased total body aluminum.[2] Although the tissues most affected are liver, spleen, bone, and brain, aluminum levels are frequently increased in virtually all tissues.[2]

13.2　EVIDENCE FOR ALUMINUM NEUROTOXICITY IN UREMIC PATIENTS

In 1972 a new and distinct neurological disease was noted to occur in a number of uremic patients receiving chronic dialytic therapy in Denver.[10] This disease was subsequently recognized to occur worldwide. It soon became apparent that it occurred in epidemic proportions in some dialysis centers, while in dialysis centers in other geographical areas it was rarely if ever seen.[11-14] This was the first evidence suggesting that an environmental toxin was responsible for dialysis encephalopathy. Based on strong biochemical, epidemiological, and clinical data it was shown that the toxin responsible for this disease was aluminum. In 1976 it was reported that one trace element, aluminum, was consistently higher in brain gray matter of patients dying of dialysis encephalopathy than in nonuremic controls and dialysis patients dying of other causes (Figure 13.1).[11] This finding was subsequently confirmed by a number of other investigative groups (Table 13.1).[11-13] Additional evidence incriminating aluminum in the pathogenesis of dialysis encephalopathy came from large epidemiological studies carried out in the United Kingdom and France.[12-14] These studies showed that dialysis centers with a high incidence of dialysis encephalopathy had large amounts of aluminum in the water used to prepare the dialysate. In contrast, centers with a low incidence of dialysis encephalopathy used water for dialysate preparation which was largely free of aluminum contamination. It was further found that centers with a large number of cases of dialysis encephalopathy could markedly decrease the incidence of this disease in their dialysis population by removing aluminum from the water used to prepare the dialysate.

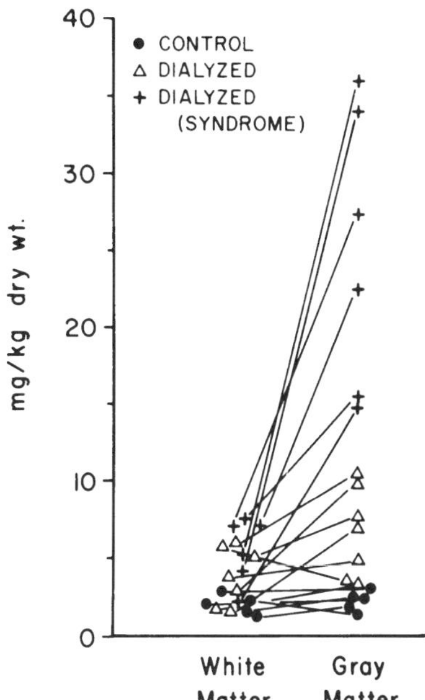

FIGURE 13.1 Aluminum levels in brain gray and white matter of control subjects and uremic patients with and without the encephalopathy syndrome.

Table 13.1 Brain Aluminum Values

Normal	Dialysis patients	Dialysis encephalopathy	Ref.
2.2 ± 1.3	8.5 ± 3.3 (21)	24.5 ± 9.9 (34)	2
1.7 ± 0.6	4.7 ± 1.8 (11)	21.0 ± 18.5 (6)	13
—	—	33.0 ± 9.6 (8)	12

Note: Al mg/kg dry weight gray matter. Data given as mean ±1 SD. Numbers in parentheses are number of patients studied.

Further evidence supporting aluminum as the toxin responsible for dialysis encephalopathy was obtained when the number of cases of dialysis encephalopathy decreased from 0.4% (229 of 61,450 patients) in 1980 to less than 0.1% (129 of 140,555 patients) of the dialysis population in the U.S. in 1990. During this interval, the number of dialysis units adequately treating water to remove aluminum increased from 26% in 1980 to greater than 99% in 1990. The final evidence for aluminum neurotoxicity comes from more recent reports showing that aluminum toxicity can be reversed and death prevented by chelating and removing the aluminum present in patients with dialysis encephalopathy.[15]

13.3 CLINICAL FEATURES

13.3.1 Chronic Aluminum Neurotoxicity

The dialyzed uremic patients with the neurological disease first described in 1972 had clinical features[10] which strongly resembled the description in 1962 of a fatal dementing neurological disease in an aluminum processing plant worker that was attributed to aluminum toxicity.[3] The classical type of aluminum neurotoxicity, dialysis encephalopathy or dialysis dementia, occurs after years of parenteral aluminum exposure through aluminum-contaminated dialysate.[11] It may rarely occur after prolonged administration of oral aluminum compounds. The onset of symptoms is insidious, with the initial manifestation being an intermittent speech disturbance characterized by a stuttering and stammering quality. The speech disturbance is subsequently accompanied by personality changes, parietal lobe alterations (such as directional disorientation), seizures, and visual and auditory hallucinations. Later in the course asterixis and severe myoclonic jerks develop. From 7 to 9 months after the initial onset of symptoms the patient becomes totally mute and unable to perform any purposeful movements. Death rapidly ensues.[10,11] The patient with classical aluminum neurotoxicity frequently has additional manifestations of aluminum toxicity such as fracturing osteomalacia, proximal myopathy and a microcytic hypochromic anemia.

13.3.2 Acute Aluminum Neurotoxicity

More recently it has been recognized that there are two distinct types of aluminum neurotoxicity (Table 13.2). The second type represents a much more acute and fulminate form of neurotoxicity. It occurs under three different conditions, all associated with markedly elevated serum aluminum levels. They include the oral administration of citrate compounds in association with aluminum compounds,[16] the performance of dialysis with dialysate highly contaminated with aluminum, and extreme elevation of plasma aluminum levels in association with deferoxamine chelation therapy.[17] Symptoms of acute aluminum intoxication may occur after months of ingestion of the combination of citrate and aluminum compounds. However, it is not uncommon for symptoms to develop in a matter of weeks following the combined ingestion of these two compounds.[16] Acute aluminum neurotoxicity which occurs as a consequence of high aluminum levels in the dialysate typically develops during the course of the dialysis procedure and is manifested as grand mal seizures. Symptoms related to deferoxamine therapy commonly occur within days of beginning deferoxamine treatment.[17]

TABLE 13.2 Aluminum Neurotoxicity

	Acute	Chronic
Etiology	Extremely high dialysate Al, deferoxamine therapy, concomitant oral administration of citrate and aluminum compounds	Chronic oral or parenteral Al exposure
Clinical Features		
Onset	Sudden, abrupt	Gradual, insidious
Neurological	Agitation	Speech disturbance
	Confusion	Personality changes
	Seizures	Hallucinations
	Myoclonic jerks	Seizures
	Coma	Myoclonus
Other	None	Osteomalacia
		Microcytic anemia
Laboratory	Plasma Al >500 µg/l	Plasma Al 100–200 µg/l
	EEG changes	EEG changes

It would appear that virtually all nondialyzed uremic adults and children who have developed aluminum neurotoxicity have done so from the ingestion of the combination of aluminum compounds with citrate.[16,18] Citrate, it is felt, opens the paracellular pathways between cells in the small intestine and increases the solubility of aluminum compounds in the intestinal fluids, both of which markedly enhance aluminum absorption.[6] This results in extremely high plasma aluminum levels. These high plasma aluminum levels as well as possible aluminum-citrate complexes may facilitate aluminum entry into the central nervous system. Although the patient may have mild neurological findings of aluminum neurotoxicity such as speech disturbance, more commonly there is an acute explosive onset of symptoms. The major findings in adults include confusion, myoclonic jerks, agitation, grand mal seizures, obtundation, coma, and death. The symptoms are usually unrelenting, with death occurring in a matter of days to weeks.[16]

The neurotoxicity in children that is caused by the concomitant ingestion of aluminum and citrate compounds is somewhat different from that seen in adults. The disease is more insidious in onset and characterized by intellectual impairment, seizures, and regression of verbal and motor skills.[18] This is exemplified by the finding that children who have just started to walk and talk may lose these skills. The frequency of seizures in children is equal to or even greater than that found in adults with acute aluminum intoxication. With the acute form of aluminum neurotoxicity evidence of aluminum toxicity in other systems, i.e., hematopoietic and skeletal systems, is characteristically absent.

13.4 DIAGNOSTIC FEATURES

The diagnosis of aluminum neurotoxicity is largely based on a history of either oral ingestion of aluminum compounds, usually in association with citrate, or parenteral exposure to aluminum in a uremic patient, the demonstration of the classical clinical features, and exclusion of other neurological conditions by appropriate studies. As stated above, the disease is more difficult to diagnose in children since the features are not as classic or consistent. Additional help in the diagnosis of classical aluminum neurotoxicity is the finding of other evidence of aluminum intoxication such as aluminum-associated osteomalacia and a microcytic hypochromic anemia not associated with iron deficiency.

Measurement of aluminum levels, although not diagnostic, is helpful in identifying patients who are either intoxicated or in danger of becoming intoxicated. Plasma aluminum levels of greater than 100 µg/l (normal <3 µg/l) have been found in patients undergoing excess aluminum loading; such levels are usually present in patients with dialysis encephalopathy.However, a plasma aluminum level of less than 100 µg/l does not exclude the diagnosis. Conversely, since plasma is a transport system for aluminum and reflects current loading, levels of greater than 100 µg/l do not necessarily indicate that total body aluminum is markedly increased. Although bone aluminum content does not correlate well with brain aluminum levels, it does reflect total body aluminum.[2,19] Therefore, a quantitative bone aluminum assay is a simple way of documenting excess total body aluminum. In addition there are histochemical stains, aluminon and solochrome azurine,[19] which are also useful in documenting aluminum overload in bone in patients with toxicity.

More recently, a deferoxamine (DFO) infusion test has been used to estimate total body aluminum.[20] In this regard, the amount of rise of plasma aluminum 24 to 48 h following DFO infusion (usually 40 mg/kg) has been shown to correlate reasonably well with bone aluminum content.

Diagnosis of acute aluminum intoxication is dependent on finding extremely increased plasma aluminum levels, usually in excess of 500 µg/l (18.5 µmol/l).[16]

In the classical form of aluminum neurotoxicity there is no increase in aluminum levels or other abnormalities of the cerebrospinal fluid. However, there is some evidence in acute aluminum neurotoxicity, at least when induced by deferoxamine, that cerebrospinal fluid aluminum levels may be increased.[21] The CT scan may be normal or only show mild cortical atrophy commonly found in uremic patients.

The most useful laboratory test supporting the diagnosis of aluminum neurotoxicity is the electroencephalogram (EEG)[10] (Figure 13.2). Unlike most metabolic encephalopathies, including uremic encephalopathy, which exhibit a generalized slowing of the EEG, in aluminum intoxication the background rhythm is relatively normal and there are multifocal bursts of slow or delta wave activity frequently accompanied by spikes. In the classical or chronic form of intoxication the EEG changes may precede the development of any clinical symptoms by 4 to 6 months.

13.5 HISTOLOGICAL ALTERATIONS WITH ALUMINUM NEUROTOXICITY

Extensive histological studies were carried out on the brains obtained from 12 patients who had died of dialysis encephalopathy.[22] Changes noted included cytoplasmic and nuclear shrinkage in

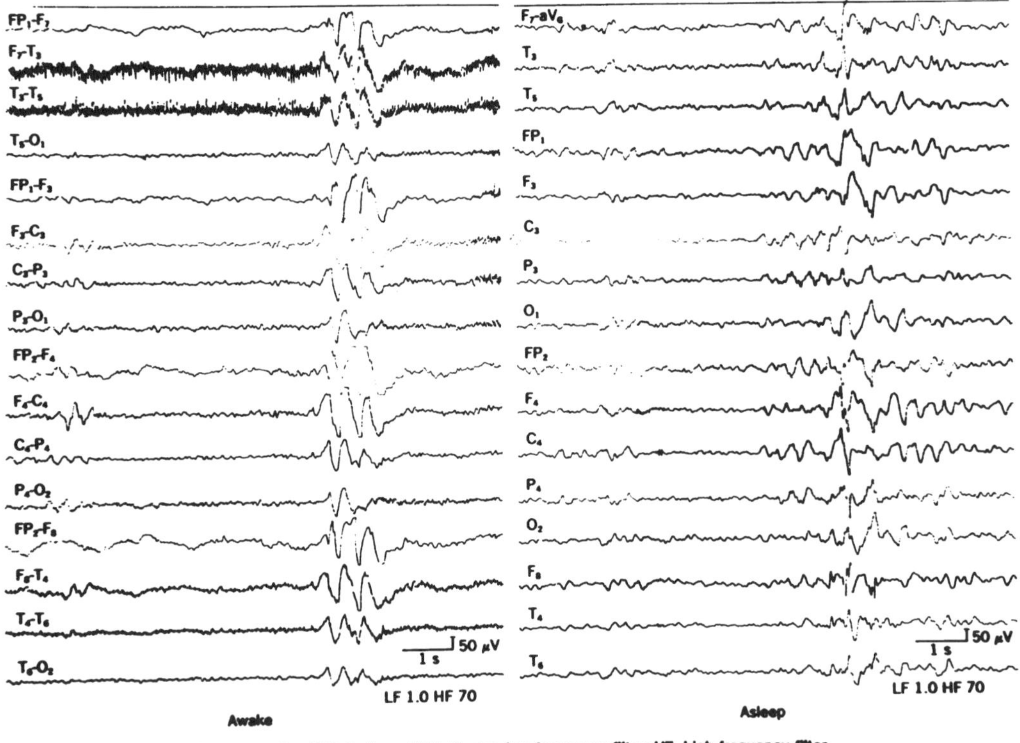

FIGURE 13.2 Typical electroencephalogram showing multifocal bursts of delta activity with relatively normal background activity.

many neurons, irregularity of the cytoplasmic outline, increased amounts of lipofuscin pigment, loss of Nissl substance, and nuclear hyperchromicity. Mild spongiform changes were also noted in the cerebral cortex. Significant negative findings included no apparent neuronal loss, senile plaques, or neurofibrillary tangles. The findings noted were felt to be nonspecific since similar alterations were found in the brains of dialysis patients dying of other causes who had no clinical findings of dialysis encephalopathy.[22]

More recently, histological studies were carried out on brains obtained from 15 dialysis patients who were exposed chronically to aluminum, mainly in the form of oral compounds, and had high blood aluminum levels.[23] However, none of these patients had dialysis encephalopathy. Mean brain aluminum was 6.3 mg/kg in the dialysis patients vs. 2.5 mg/kg in the control, with considerable overlap in whole-brain aluminum between these two groups. Using dynamic secondary ion mass spectrometry, focal accumulation of aluminum was found to be present in the cortical pyramidal neurons only in the uremic group. Of 15 dialysis patients, 8 had insoluble amyloid beta protein but there was no association between its presence and the accumulation of aluminum. Alzheimer's disease-like changes in the processing of tau protein were also described. However, neurofibrillary tangles were not observed and only two dialysis patients had protease-resistant paired helical filaments.[23]

Aluminum is primarily deposited in the gray matter of the brain with aluminum levels in the white matter being normal or only slightly elevated.[11] Galle et al.,[24] using energy-dispersive X-ray microanalysis, reported the localization of aluminum in lysosomes of cortical neurons in the gray matter of brain obtained from two patients who died of dialysis encephalopathy. Using laser microprobe analysis, Reusche and Seydel[25] found that aluminum was present in inclusions in the cytoplasm of choroidal epithelium, glia, and neurons in the brains of patients with dialysis encephalopathy.

13.6 MECHANISM OF ALUMINUM NEUROTOXICITY

Experimentally, aluminum has been shown to cause a number of histological and biochemical alterations in neural tissue. Numerous *in vivo* studies have shown that aluminum can produce proliferation of disorganized tangles of 10-nm-diameter filaments of a nonhelix and nonpaired nature in rabbits. In fish chronic aluminum exposure results in damage to the nervous system, including the olfactory bulb. Histological changes noted include chromatolysis, eosinophilic inclusions, neuronal loss, neuritic thickening, and plaque-like structures that are immunoreactive to antibodies against tau. *In vitro* studies dialyzing tau against aluminum citrate caused the formation of paired helix-like aggregates.[26] It has been suggested the aluminum-induced aggregation could involve noncovalent electrostatic cross-linking. It has been further implied that aluminum may contribute to the assembly of tau into paired helical filaments following hyperphosphorylation. The possibility that aluminum may stimulate kinase activity, causing anomalous phosphorylation, was also suggested.

However, neither animal models exposed to aluminum nor humans with dialysis encephalopathy develop paired helical filaments. Therefore, if aluminum does play such a role *in vivo*, additional mechanisms must also be involved. Aluminum has also been shown to inhibit the uptake of choline, glutamate, norepinephrine, and serotonin into rat synaptosomes as well as inhibit choline transport in rat brain and human erythrocytes and reduce neural choline acetyltransferase activity in the rabbit brain.[27] Aluminum has also been shown to induce osmotic fragility and acanthocyte formation and cause a marked decrease in membrane fluidity in suspended erythrocytes.[28] Based on these findings it has been suggested that aluminum may affect the fluidity of neuronal membranes influencing receptor function, especially in regards to serotonin. Aluminum has also been shown to bind to calmodulin, inducing configurational changes blocking the activity of the calcium-calmodulin-dependent phosphodiesterase.[29] Several recent studies have shown that aluminum increases the permeability of the blood-brain barrier to a variety of agents which could allow neurotoxins to reach the central nervous system.[30]

The speciation of aluminum may also be an important determinant of neurotoxicity.[9] Acute neurotoxicity of aluminum appears to result from the smaller molecular species, i.e., citrate and deferoxamine complex, suggesting that they more easily cross the blood-brain barrier. It has also been suggested that aluminum may enter the central nervous system as aluminum-transferrin complex via the transferrin receptor on the brain microvasculature.

Finally, it is possible that the uremic state predisposes the individual to brain aluminum uptake and toxicity. This is suggested by the fact that aluminum neurotoxicity has not been noted in nonuremic patients receiving comparable parenteral aluminum loading through total parenteral nutrition solutions that the uremic patient receives from contaminated dialysate.

13.7 PREVENTION AND TREATMENT

13.7.1 Prevention

It is easier to prevent than treat aluminum intoxication. This can be done in uremic patients by insuring that dialysis is performed with aluminum-free dialysate and restricting the oral ingestion of aluminum compounds. In 1981 recommended permissible levels of aluminum (<10 µg/l) as a contaminant in the water used for dialysate preparation were published by AAMI and accepted by the nephrology community. Comparable permissible levels for aluminum contamination in dialysate were also developed by Canada and the European Common Community. With compliance of the recommended standards for permissible levels of aluminum in the dialysate, epidemics of aluminum neurotoxicity have virtually disappeared in dialysis populations.

Aluminum loading and toxicity can also result from orally administered aluminum containing phosphate binding gels. Approximately 20% of dialyzed uremic patients maintained with aluminum-free dialysate but receiving aluminum-containing phosphate binding gels manifested aluminum skeletal toxicity. Similarly, uremic children, dialyzed as well as nondialyzed, receiving these aluminum gels are also at risk of developing aluminum skeletal and neurological toxicity.[18] It was not until 1985 that it was appreciated that orally administered citrate compounds played a major role in the causation of aluminum neurotoxicity in patients receiving orally administered aluminum compounds.[16] Retrospectively, it seems likely that virtually all nondialyzed uremic children who developed aluminum toxicity did so through the concomitant administration of aluminum gels, given for phosphate control, and sodium citrate used for the correction of uremic acidosis.[18] Aluminum compounds are no longer recommended to be prescribed to children. Calcium carbonate and calcium acetate are increasingly being used as phosphate binders, replacing aluminum binders. Although the calcium salts are not as effective as aluminum in binding phosphate they are usually adequate for controlling serum phosphate. At the present time it would appear that aluminum phosphate binders have been largely replaced in uremic populations by calcium compounds.

13.7.2 Treatment

Diazepam administration can result in dramatic clinical improvement of dialysis encephalopathy. It is one of the most effective means of treating or preventing seizures.[31] Since dialysis frequently exacerbates symptoms, especially seizures, diazepam given before dialysis may be especially useful. However, the improvement is only temporary in aluminum-intoxicated patients and, in fact, the progression of the disease remains unchanged.

Aluminum neurotoxicity was initially thought to be uniformly fatal, even if aluminum exposure was eliminated or kidney function was returned to normal via transplantation. However, the recent use of deferoxamine (DFO) to chelate the aluminum has effected a cure in a number of patients with the classical, or chronic, form of aluminum neurotoxicity. Since the chances for recovery are much better if the disease is recognized and treated early, prompt diagnosis of the disease is imperative. To date, all reported adult patients with acute aluminum neurotoxicity have died irrespective of chelation therapy.[16] Although disease progression may be arrested in children by chelation and renal transplantation, there is usually no reversal or improvement of symptoms present at the time of chelation. The child is typically left with severe motor and learning disabilities.[18]

Successful treatment for aluminum intoxication has only been reported for uremic patients. In this population deferoxamine chelation has been successful in both decreasing the body burden of aluminum and reversing neurological[15] as well as skeletal and hematological toxicity. Only one study has been reported on chelating aluminum in a nonuremic population who had a presumed isolated increase in aluminum in the central nervous system.[32] However, in this study it was not documented that aluminum was actually chelated or removed from the body.

Treatment of aluminum intoxication initially involves the removal of all parenteral and oral exposures to aluminum. Following removal of parenteral exposure through the dialysate, a number of patients with osteomalacia and far fewer with dialysis encephalopathy have recovered from aluminum intoxication without additional treatment. Chelation therapy for aluminum intoxication has been widely used and adequately evaluated only in patients with advanced renal failure usually requiring dialysis. A number of different approaches to chelation have been recommended. Unlike iron, the amount of aluminum chelated appears to be determined by the size of the chelatable aluminum pool and not the dose of deferoxamine administered.[33] It has been found that 1 or 2 g of deferoxamine is equally as effective as 4 g in chelation of aluminum as determined by the rise in plasma aluminum following the administration of deferoxamine. The most effective spacing between dosages of deferoxamine has not been defined so it is unclear whether 1 to 2 g of deferoxamine should be given once, or three times weekly intravenously during the last hour of dialysis. At a consensus conference held in Paris in June 1992, it was suggested that deferoxamine be given once weekly in an even lower dose of 5 to 10 mg/kg/bw during the last 30 min of a dialysis session. Deferoxamine is equally effective when given intramuscularly or intraperitoneally as when given intravenously.[34]

Treatment of aluminum-associated bone disease with deferoxamine usually results in symptomatic improvement in regards to bone pain and proximal muscle weakness in 2 to 3 months. However, treatment should be continued for around 6 months and discontinued when there is only a minimal rise in plasma aluminum levels following deferoxamine administration, demonstrating depletion of the chelatable aluminum pool. Symptomatic improvement has been found to occur as rapidly in patients given a weekly dose of 0.5 g as in patients given 6 g of deferoxamine. Aluminum-induced anemia responds in a similar time period as symptomatic bone disease. In contrast, aluminum-associated neurological toxicity requires a more extended duration of treatment with deferoxamine. Although progression of the neurological disease may stabilize, improvement frequently requires 10 to 12 months of continuous treatment with deferoxamine. With adequate treatment EEG abnormalities may resolve, making EEG measurement important in determining the duration of therapy. Following discontinuation of therapy, especially if the duration was too short, recurrence of aluminum neurotoxicity is not uncommon.[15]

Therapy with deferoxamine is not without some risk. As described above, patients with marked aluminum overload may develop acute neurological toxicity following deferoxamine therapy. Acute infections with Yersinia and mucormycosis have also been reported in patients receiving deferoxamine. It is felt that ferroxamine is an iron supplier for these nonsiderophore-producing organisms, promoting their growth. Therefore, prior to treatment, diagnosis of aluminum intoxication should be reasonably well established and evidence of aluminum overload demonstrated either by a rise of plasma aluminum greater than 150 to 200 µg/l following deferoxamine infusion or increased bone aluminum levels in association with classical histological findings of aluminum-associated bone disease.[19,20]

REFERENCES

1. Siem (1886):Quoted in: Dölken: Uber die wirkung des aluminum, mit besonderer berucksichtigung der durch das aluminum verursachten lasionen im centralnervensystem. *Naunym-Schmiedebergs Arch. Exp. Pathol. Pharmakol.*, 1897; 40:98–120.
2. Alfrey AC: Aluminum metabolism in uremia. *Neurotoxicology*, 1980; 1:43–53.
3. McLaughlin AIG, Kazantzis G, King E, Teare D, et al: Pulmonary fibrosis and encephalopathy associated with the inhalation of aluminum dust. *Br. J. Ind. Med.*, 1962; 19:253–263.
4. Ittel TH, Buddington B, Miller NL, and Alfrey AC: Enhanced gastrointestinal absorption of aluminum in uremic rats. *Kidney Int.*, 1987; 32:821–826.
5. Kovalchik MT, Kaehny WD, Jackson T, and Alfrey AC: Aluminum kinetics during hemodialysis. *J. Lab. Clin. Med.*, 1987; 92:712–716.
6. Froment DH, Molitoris BA, Buddington B, Miller NL, et al: Site and mechanism of enhanced gastrointestinal absorption of aluminum by citrate. *Kidney Int.*, 1989; 36:978–984.
7. Freeman S and Freeman WMC: Phosphorus retention in children with chronic renal insufficiency. The effect of diet and of the ingestion of aluminum hydroxide. *Am. J. Dis. Child.*, 1941; 61:981–1002.
8. Kaehny WD, Hegg AP, and Alfrey AC: Gastrointestinal absorption of aluminum from aluminum-containing antacids. *N. Engl. J. Med.*, 1977; 296:1389–1390.
9. Martin RB: The chemistry of aluminum as related to biology and medicine. *Clin. Chem.*, 1986; 32:1797–1806.
10. Alfrey AC, Mishell MM, Burks J, Contiguglia SR, et al: Syndrome of dyspraxia and multifocal seizures associated with chronic hemodialysis. *Trans. Am. Soc. Artif. Intern. Organs*, 1972; 18:257–261.
11. Alfrey AC, LeGendre GR, and Kaehny WD: The dialysis encephalopathy syndrome. Possible aluminum intoxication. *N. Engl. J. Med.*, 1976; 294:184–188.
12. Cartier F, Allain P, Garv J, Chatel M, et al: Encephalopathia myclonique progressive des dialyses. Role de l'eau utilisee pour l'hemodialyse. *Nouv. Presse Med.*, 1978; 7:97–102.
13. McDermott JR, Smith AI, Ward MK, Fawcett RWP, et al: Brain-aluminum concentration in dialysis encephalopathy. *Lancet*, 1978; 1:901–903.
14. Platts MM, Goode GC, and Hislop, JS: Composition of the domestic water supply and the incidence of fractures and encephalopathy in patients on home dialysis. *Br. Med. J.*, 1977; 2:657–660.
15. Ackrill P and Day PP: The use of desferrioxamine in dialysis-associated aluminum disease. In Bourke E, Mallick NP, and Pollak VE (eds): *Moving Points in Nephrology*. Contrib. Nephrol., Basel, Karger, 1993, pp 125–134.
16. Bakir AA, Hryhorczuk DO, Berman E, and Dunea G: Acute fatal hyperaluminemic encephalopathy in undialyzed and recently dialyzed uremic patients. *Trans. Am. Soc. Artif. Intern. Organs*, 1986; 32:171–176.

17. Sherrard DJ, Walker JV, and Boykin JL: Precipitation of dialysis dementia by deferoxamine treatment of aluminum related bone disease. *Am. J. Kidney Dis.*, 1988; 12:126–130.

18. Sedman AB, Wilkening GN, Warady BA, Lum GM, et al: Encephalopathy in childhood secondary to aluminum toxicity. *J. Pediatr.*, 1984; 105:836–838.

19. Maloney NA, Ott S, Alfrey AC, Coburn JW, et al: Histologic quantitation of aluminum in iliac bone from patients with renal failure. *J. Lab. Clin. Med.*, 1982; 99:206–216.

20. Milliner DS, Nebeker HG, Ott SM, Andress DL, et al: Use of the deferoxamine infusion test in the diagnosis of aluminum-related osteodystrophy. *Ann. Intern. Med.*, 1984; 101:775–780.

21. Ellenberg R, King AL, Sica DA, Posner M, et al: Cerebrospinal fluid aluminum levels following deferoxamine. *Am. J. Kidney Dis.*, 1990;16:157–159.

22. Burks JS, Alfrey AC, Huddlestone J, Norenberg MD, et al: A fatal encephalopathy in chronic hemodialysis patients. *Lancet*, 1976; 1:764–768.

23. Harrington CR, Wischik CM, McArthur FK, Taylor GA, et al: Alzheimer's disease-like changes in tau protein processing: association with aluminum accumulation in brains of renal dialysis patients. *Lancet*, 1994; 343:993–997.

24. Galle P, Chatel M, Berry JP, and Menault F: Encephalopathie myoclonique progressive des dialyses. Presence d'aluminum et forte concentration dans les Ivsosomes des cellules cerebrates. *Nouv. Presse Med.*, 1979; 8:4091–4094.

25. Reusche E and Seydel U: Dialysis-associated encephalopathy. Light and electron microscopic morphology and topography with evidence of aluminum by laser microprobe mass analysis. *Acta Neuropathol.*, 1993; 86:249–258.

26. Letterier JF, Langui D, Probst A, and Ulrich J: A molecular mechanism for the induction of neurofilament bundling by aluminum ions. *J. Neurochem.*, 1992; 58:2060–2070.

27. Lai JCK, Lim L, and Davison AN: Effects of Ca^{2+}, Mn^{2+} and Al^{3+} on rat brain synaptosomal uptake of noradrenaline and serotonin. *J. Inorg. Biochem.*, 1982; 17:215–225.

28. Zatta P, Perazzolo M, and Corain B: Tris acetylacetonate aluminum (III) induces osmotic fragility and acanthocyte formation in suspended erythrocytes. *Toxicol. Lett.*, 1989; 45:15–21.

29. Siegel N and Haug A: Aluminum interaction with calmodulin — evidence for altered structure and function from optical and enzymatic studies. *Biochim. Biophys. Acta*, 1983; 744:36–45.

30. Banks WA and Kastin AJ: Aluminum increases permeability of the blood brain barrier to labelled DSIP and β-endorphin: possible implications for senile and dialysis dementia. *Lancet*, 1983; ii:1227–1229.

31. Nadel AM and Wilson WP: Dialysis encephalopathy: A possible seizure disorder. *Neurology*, 1976; 26:1130–1134.

32. Crapper McLachlan DR, Dalton AJ, Kruck TPA, Bell MY, et al: Intramuscular desferrioxamine in patients with Alzheimer's disease. *Lancet*, 1991; 337:1304–1308.

33. Ciancioni C, Poignet JL, Mauras Y, Planthier G, et al: Plasma aluminum and iron kinetics in hemodialyzed patients after IV infusion of desferrioxamine. *Trans. Am. Soc. Artif. Intern. Organs*, 1984; 30:385–479.

34. Molitoris BA, Alfrey PS, Miller NL, Hasbargen JA, et al: Efficacy of intramuscular and intraperitoneal deferoxamine for aluminum chelation. *Kidney Int.*, 1987; 31:986–991.

SECTION 5

ZINC

Chapter 14

Role of Zinc in the Central Nervous System

Masayuki Yasui, Kiichiro Ota, and Vincent A. Murphy

CONTENTS

14.1 INTRODUCTION

Zinc (Zn) is an essential trace metal in humans and animals, especially during pregnancy. Zn is absorbed from the small intestine and its absorption is inhibited by fiber, phytates, calcium (Ca), and copper (Cu) and is enhanced by amino acids, peptides, and chelating agents. Elimination of Zn occurs primarily through pancreatic and intestinal secretion. Zn is loosely bound to albumin and other proteins in the blood and 99% of the metal in the body is inside cells. Zn deficiency results in defects of the central nervous system (CNS) and in peripheral neuropathy.[1] Enzymatic Zn, which is firmly incorporated into the structure of Zn-metalloenzymes, plays no unique role in neural tissue, whereas ionic Zn (Zn^{2+}) is the fraction of tissue Zn that appears to be associated with secretory-signaling functions of certain neurons. Zn plays an important role in synaptic function in the brain, as it is present in synaptic vesicles, is released with neuronal activity, and provides modulation of different neurotransmitters. The clinical symptoms of Zn deficiency other than those in the CNS are described elsewhere.[2] We focus herein on the roles of Zn in CNS function and describe CNS disorders related to Zn deficiency.

14.2 ROLES OF ZINC IN THE CNS

Zn can influence synaptic transmission by modulating the functions of various channels within the CNS. Presynaptic uptake of Ca through voltage-gated channels and postsynaptic uptake through glutamate-activated channels is reduced in synaptosomes taken from Zn-deficient animals.[1] Addition of Zn to these synaptosomes reduced uptake of Ca. Therefore, a certain level of Zn is necessary for Ca-channel function and at higher levels Zn can modulate channel function by blocking Ca entry. Zn as well as magnesium (Mg) are implicated in blockade of the N-methyl-D-aspartate (NMDA) receptor channel. In the suprachiasmatic nucleus, Zn enhances a voltage-dependent potassium current, presumably by modifying a potassium channel.[3] Zn can enhance the flow of chloride through γ-aminobutyric acid (GABA) chloride channels by modifying the release of GABA[4] or can noncompetitively block the channel thereby reducing chloride flow.[5,6]

Zn^{2+} may act as an endogenous ligand for various receptors.[7] The σ receptor in the rat hippocampus is an example. Sigma receptors are a class of membrane proteins that recognize a wide variety of psychoactive drugs, including pentazocine, dextromethorphan, phencyclidine, and haloperidol. Focal electrical stimulation of Zn-containing mossy fibers in the hilar region of the hippocampus reduced specific binding of ligands to the σ receptor, and this reduced binding was blocked using the Zn chelator, metallothionein peptide 1. Several experimental findings indicate that Zn is released presynaptically and is available to act postsynaptically as an endogenous ligand. Excitation of hippocampal slices causes a release of Zn^{2+} into the extracellular compartment. Electrical stimulation of mossy fibers causes preferential uptake and release of Zn^{2+}. Kainic acid treatment causes a loss of Zn from presynaptic neurons in limbic and cerebrocortical regions and leads to an accumulation of excessive concentrations of Zn^{2+} in postsynaptic regions.[8] Zn^{2+} acting as an endogenous ligand has been postulated for the NMDA, GABA, and opiate receptors as well as the sigma receptor.[7]

Modulation of drug binding to the dopamine uptake complex is another role for Zn in the CNS.[9] In the rat caudate putamen, Zn decreases dopamine uptake and enhances the binding of dopamine uptake inhibitors, such as cocaine. This enhanced binding may be related to prevention of oxidation of a sulfhydryl group within the complex. Other cations (Cu^{2+}, Hg^{2+}, and Cd^{2+}) and N-methylmaleimide, which oxidize the sulfhydryl group, reduce binding of dopamine uptake inhibitors.

Zn appears to be involved in myelination both during development and in disease states. Maternal Zn deficiency was imposed during oligodendroglial cell proliferation and myelination in monkeys and changes in myelin protein profiles occurred in the offspring.[10] These alterations in the myelin protein profiles provide evidence that Zn is necessary for proper brain myelination. Early loss of myelin basic protein (MBP) and the accumulation of hydrolytic enzymes are hallmarks of active plaque in multiple sclerosis. Zn appears to decrease endogenous proteolysis of myelin proteins, especially MBP, caused by increases in alkalinity and alkaline earth metals (K^+, Na^+, Ca^{2+}, and Mg^{2+}). This proteolysis involves a release of extrinsic proteins from myelin, including proteinase activity, which is abolished by Zn. The immobilization of MBP and prevention of endogenous breakdown caused by Zn and its possible relationship to myelin diseases requires further study.[6]

Zn deprivation in rats during prenatal or early postnatal life leads to altered learning ability and impaired cognitive and emotional behavior. The hippocampus is involved in the processing of memory and integration of emotion. Normally, the hippocampus contains the highest Zn concentration of the major brain areas, and these high Zn levels are located in the intrahippocampal mossy fiber pathway. Electrophysiological studies have revealed impaired hippocampal function in Zn deficiency. Successive low-frequency stimulation of hippocampal mossy fiber axons in Zn-deficient rats produced synaptic responses of declining amplitude. This response is specific to mossy fiber axons and supports the notion that Zn is essential for hippocampal function and, therefore, normal memory and emotional behavior.

The synaptic terminals of the mossy fibers are uniquely rich in Zn. These fibers are the prototypical Zn-containing neurons where Zn is sequestered in vesicles, taken up by high-affinity mechanisms, and released by Ca and voltage-dependent exocytosis. In particular, it has been observed that a reduction in hippocampal Zn levels, obtained with chelating agents, can induce spatial memory deficits in adult animals. During aging, reduced bioavailability of Zn, evidenced by decreased serum Zn levels, can potentially alter CNS functions by altering the hippocampal pool of Zn^{2+}. Since aged rats develop spatial memory impairment to varying degrees, the relationship between memory performance and hippocampal Zn levels has received attention.[11] In aged rats, the memory deficit is associated with specific localized changes in hippocampal Zn levels, supporting the role of Zn^{2+} in synaptic function. The different degrees of mossy fiber Zn depletion observed in impaired and nonimpaired animals may reflect differences in the decline of serum Zn levels during aging, as only a chronic depression of serum Zn results in a significant reduction of mossy fiber Zn. This is true with dietary restriction as well, for even severe deficiency which rapidly reduces serum Zn levels does not decrease hippocampal Zn concentration. Differences in response could also result from a difference in the permeability of the blood-brain barrier to Zn and in the function of the molecular mechanisms regulating Zn turnover at the synapse.[11]

Recent evidence that the Zn^{2+} ion is a potent modulator of excitatory amino acid (EAA) receptors and a possible factor in seizures and excitative toxicity has spurred interest in CNS neurons that contain Zn in presynaptic regions.[12] *In vitro*, Zn inhibits the binding of glutamate and aspartate to hippocampal membranes, but nutritional Zn deficiency does not alter binding. Zn is a noncompetitive antagonist of NMDA receptors and also antagonizes inhibitory GABA transmission. Zn attenuates NMDA receptor-mediated neurotoxicity and reduces the postsynaptic excitatory response from this receptor.[13] All of the Zn-containing fiber systems are systems that release an EAA as a transmitter. Efferent projections from CA3 hippocampal pyramidal cells are glutamergic, and EAA generally serve as neurotransmitters in the mossy fiber pathway. In general, it appears Zn modulates the neuronal response, particularly in the hippocampus, by exerting an inhibitory effect on EAA receptors.[8] Zn-containing neurons comprise a substantial part of the synaptic input to limbic and cerebrocortical neurons. It is therefore plausible that one physiologic role of Zn is to coregulate transmission at certain limbic and cerebrocortical EAA synapses. Synaptic Zn may participate in the modification of synaptic strength that underlies development and/or experiential learning.

14.3 ZINC-IMPLICATED CNS DISORDERS

The replication, repair, and transcription of DNA involves many enzymes such as nucleases, nucleotidases, nucleotidyl transferase, DNA polymerase, etc., all of which contain Zn. Postnatal Zn deficiency in rats results in profound DNA dysfunction: decrease of the total amount of DNA, impairment of thymidine incorporation in DNA, abnormal DNA-RNA protein synthesis, poor learning capacity, and alterations of behavioral performance. The implications of Zn metabolism in brain development and function are apparent in rats at all stages of ontogenesis, with Zn deficiency imposed in the early prenatal period resulting in severe neural dysmorphogenesis, including hydrocephalus and spina bifida. Late prenatal and early postnatal Zn deprivation is reflected mainly in impaired brain function and associated behavioral abnormalities which occur more slowly and are less persistent than the teratological outcome associated with an early prenatal deficit.[14] Zn deficiency is associated with impaired long-term memory and some reduction in short-term recall and learning ability, coupled with diminished activity, lack of emotional control, and increased responsiveness to stress. These functional brain deficits have a single etiology, but probably arise from several overlapping neurological disturbances. Overall, the focus in evaluating zinc in the brain should probably lie less with acute prenatal zinc deficiency, and more with the effects of chronic suboptimal zinc status on mental alertness and performance at all stages of the life cycle in humans.[8]

The teratogenicity of Zn deficiency is well established. Typical abnormalities with severe Zn deficiency in the rat include cleft lip, brain and eye malformations, and numerous abnormalities of the heart, lung, and urogenital systems. In humans, maternal Zn deficiency (low serum Zn levels) can produce a wide range of effects including infertility, prolonged labor, growth retardation, birth defects, and embryonic and fetal death. The postnatal complications of prenatal Zn deficiency can also induce impaired immunocompetence and abnormal behavioral patterns. There is considerable evidence from epidemiological and intervention studies that marginal to severe maternal Zn deficiency may be a complication of some human pregnancies. For example, disease-induced alterations in maternal Zn metabolism, such as with diabetes and alcoholism, have been postulated to increase the risk for a poor pregnancy outcome. Maternal Zn deficiency results from poor dietary intake, tissue injury, and/or maternal stress.[15]

Neurofibrillary tangles (NFT) in human encephalopathies of various etiologies may result from a common pathogenic mechanism, a functional Zn decrease leading to a deficiency of the DNA-metabolizing Zn enzymes and giving rise to abnormal neuronal DNA and synthesis of pathological proteins. In Guamanian patients with amyotrophic lateral sclerosis (ALS) and parkinsonism-dementia (PD), the Zn content of the frontal gray matter in ALS and PD patients was significantly decreased compared with that of Guamanian age-matched controls; however, no significant differences between these groups were observed in Zn levels in white matter from the frontal region or in gray or white matter from the occipital region.[16] Zn content in gray matter from both frontal and occipital regions was less in ALS and PD than controls (Figure 14.1). Fe content in gray and white matter from both frontal and occipital regions was greater in PD than in controls for a reference. As demonstrated in Figure 14.2, the Zn content in CNS white matter in patients with multiple sclerosis was markedly low compared with that in patients with ALS, spinocerebellar degeneration, and controls. This result indicates that the pathogenesis of multiple sclerosis involves the action of Zn.[17] The gross pathology in Zn deficiency includes a disturbed growth rate, parakeratosis and skin lesions, impaired reproduction and disease resistance, and abnormal posture and locomotor function. The latter signs direct interest to the peripheral nervous system. The painful neuropathy and decreased nerve conduction velocity (NCV) associated with diabetes could be associated with the alteration in Zn metabolism seen in this disease.[8]

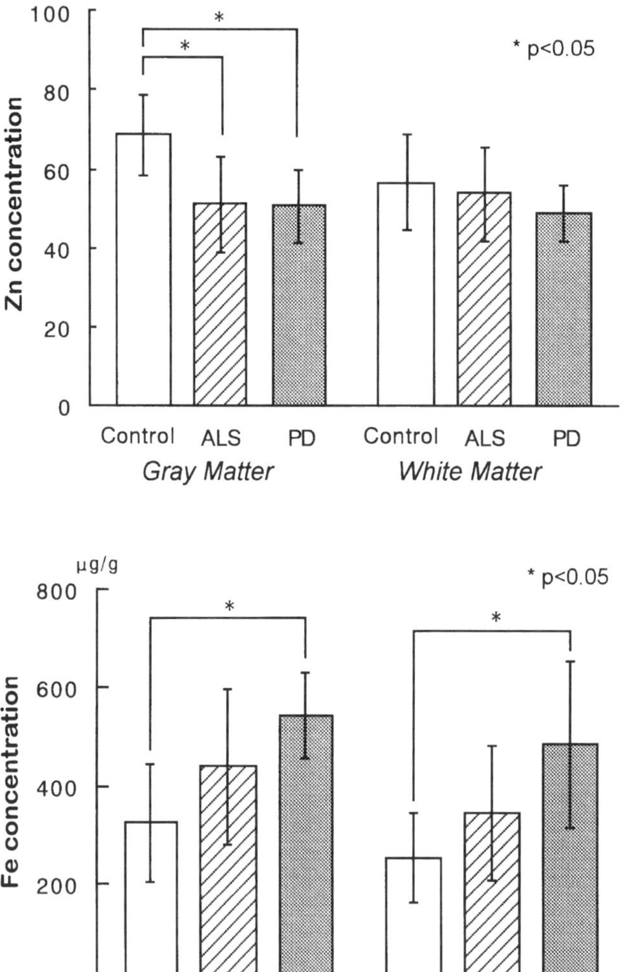

FIGURE 14.1 Zinc and iron concentrations in gray and white matter of frontal and occipital lobes of eight patients with Guamanian amyotrophic lateral sclerosis (ALS), four with Guamanian parkinsonism-dementia (PD), and four neurologically normal Guamanian controls. Values are expressed as mean ± SD.

Together with our previous data, these data indicate that Zn deficiency plays an important role in the pathogenesis of ALS and PD. Ca and Mg deficiency promotes the entry into the brain of toxic metals that may displace Zn. The blood-brain barrier (BBB) is altered and abnormal metals

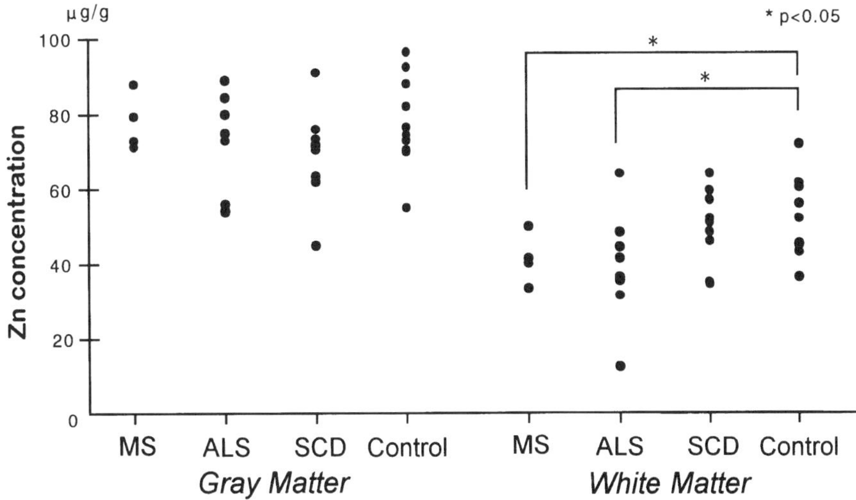

FIGURE 14.2 Zinc concentration (μg/g dry weight) in gray matter and white matter of a patient with multiple sclerosis (MS), five patients with Japanese amyotrophic lateral sclerosis (ALS), five with spinocerebellar degeneration (SCD), and five controls (μg/g dry weight, mean ± SE).

may reach the brain in Down's syndrome and Alzheimer's disease. These metals (Fe and aluminum) become encrusted in the amyloid and their levels increase in the brain, whereas levels of Zn decrease, especially in the hippocampus. A deficiency involved in the Zn enzymes of neuronal detoxification, glutamate catabolism, and some neurotransmitter metabolism may also contribute to the neuronal dysfunction in these encephalopathies. A nontoxic Zn compound crossing the BBB may be useful for the treatment of these encephalopathies, especially for Alzheimer's disease.[18]

Zn deficiency may be the common factor for all NFT encephalopathies with or without amyloidosis (AM). In human lead encephalopathy, NFT may be formed by displacement of Zn by lead and consequent induction of functional Zn deficiency. Lead exposure decreases the concentration of Zn in the brain, with accumulation of lead in place of Zn in the hippocampic mossy fiber system. Lead is fixed in certain Zn enzymes and inhibits their activity. In Alzheimer's disease with NFT and senile plaques (SP), the Zn level decreases in the cerebral cortex, especially in the hippocampus. Genetically determined amyloidosis (AM) induces 21 chromosome codes near the AM, the zinc enzyme superoxide dismutase (SOD). The plasma level of SOD is higher in Down's syndrome and Alzheimer's disease than in controls.[19] In Alzheimer's disease Zn is encrusted in AM-SP.

Functional Zn deficiency may produce NFT and other neuronal disturbances by causing a deficiency in Zn enzymes. A decrease in Zn-dependent DNA enzymes results in decreased DNA, impaired thymidine incorporation into DNA, and abnormal DNA-RNA protein synthesis that can lead to poor learning capacity and decreased performance on mnemonic tests. Zn deficiency could result in decreased levels of glutamate (GLU) dehydrogenase, which catabolizes GLU, resulting in an increase in neurotoxic GLU, especially in the hippocampal mossy fiber system where GLU is a major neurotransmitter. Experimental excess of GLU produces NFT. In addition to enzymes involved in DNA synthesis, Zn-dependent enzymes involved in neuronal detoxification, such as SOD, may be reduced during Zn deficiency. Hippocampal areas normally rich in Zn-dependent detoxification enzymes would then be at risk for increased neurotoxicity. Other Zn-dependent enzymes are involved in the metabolism of neurotransmitter, including neuropeptides (carboxypeptidase, enkephalinase), histamine (especially in the hippocampus), GABA, and acetylcholine (nerve growth factor (NGF)-activated acetylcholine synthesis).[20]

CuZn SOD transcripts are present in great abundance in the pyramidal neurons of the CA1-CA4 fields, and in granule cells of the dentate gyrus. This cellular distribution is similar to that obtained with the anti-CuZn SOD antiserum. The biochemical pathways leading to the generation of superoxide radicals are especially active in these neurons, requiring an active transcription of the CuZn-SOD gene. The hippocampal formation is particularly vulnerable to the pathological mechanisms that operate in Alzheimer's disease, adult Down's syndrome and, to a lesser extent, in normal human aging. All the associated neurodegenerative events, including cell death, SP, NFT, and granulovacuolar degeneration are expressed disproportionately in the hippocampus during the disease process.

Altered activity of cellular antioxidant systems has been implicated in the neuronal cell loss associated with aging and with degenerative diseases of the CNS like Down's syndrome and Alzheimer's disease. Part of the system of antioxidant defense is CuZn SOD, which catalyzes the dismutation of superoxide anions (O_2^-) into oxygen and hydrogen peroxide (H_2O_2) and presumably protects oxidizable compounds like catecholamines and subcellular zones such as membranes against the toxic effects of O_2^-. O_2^- are highly reactive radicals formed by the one-electron reduction of oxygen and by autooxidizing catecholamines, and may be particularly important in the genesis of cell damage. Moreover, the harm may arise from the augmenting of activation of H_2O_2 (Fenton reactions) with the generation of highly reactive species: hydroxy radicals ($OH\cdot$) or single molecular oxygen (O_2) leading to powerful oxidations of cell components.

The CuZn SOD protein is a ubiquitous enzyme in the distribution of the enzyme activity in the human brain, thus requiring a high CuZn SOD content to facilitate the removal of toxic radicals. The gene coding for CuZn SOD protein in the human is located on chromosome 21, and in Down's patients increased Cu SOD activity in various tissues including the brain reflects a gene dosage effect. It is noteworthy that patients with Down's syndrome develop accelerated ageing, and the brain histopathological changes are reminiscent of those in Alzheimer's disease. Moreover, CuZn SOD activity was significantly elevated in fibroblasts in familial Alzheimer's disease. Recent publications further support this hypothesis. Mouse cells expressing the transfected human CuZn SOD gene show an induction of glutathione peroxidase activity, an enzyme which catalyzes the reduction of H_2O_2. In addition, PC12 cells overexpressing the human CuSOD gene have impaired neurotransmitter uptake resulting from modifications of the membrane properties of chromaffin granules, likely secondary to lipid peroxidation. The abundance of CuZn SOD mRNA and protein in the pyramidal neurons of the human hippocampus suggests that these neurons represent a subset with distinct characteristics. This may account for their disproportionate vulnerability in ageing or in neurodegenerative disease such as Alzheimer's disease.[18,19]

14.4 CONCLUSION

It is clear from the discussion above that Zn plays numerous important roles in the CNS. Enzymes important for DNA synthesis, oxygen radical detoxification, and neurotransmitter synthesis have Zn as a cofactor. Zn also is important for modulation of binding to neuroreceptors, particularly the excitatory amino acid systems. Lastly, Zn^{2+} can act as a neurotransmitter, being stored and released from synapses and acting on pre- and postsynaptic sites. The hippocampus, which is rich in Zn, is particularly vulnerable to deficiency of this metal and clinical manifestations of Zn deficiency such as memory deficits, learning disorders and alterations in emotional behavior are indicative of hippocampal malfunction. Deficiency of Zn in the brain appears to be associated with neurodegenerative diseases such as AD, PD, and ALS. The association of these diseases with excitatory amino acid systems and metal imbalances provides further evidence that Zn may play a role in these diseases. Although much is known about the role of Zn in the CNS, the pathogenesis of CNS disorders related to Zn deficiency is not yet understood and further investigation is needed.

REFERENCES

1. Browing JD and O'Dell BL: Low zinc status in guinea pigs impairs calcium uptake by brain synaptosomes. *J. Nutr.*, 1994; 124:436–443.

2. Falchuk KH: Disturbances in trace element metabolism. In Wilson JD, Braunwald E, Isselbacher KJ, et al. (eds): *Harrison's Principles of Internal Medicine*, 12th edition. New York, McGraw-Hill, 1992, pp 443–445.

3. Huang RC, Peng YW, and Yau KW: Zinc modulation of a transient potassium current and histochemical localization of the metal in neurons of the suprachiasmatic nucleus. *Proc. Natl. Acad. Sci. U.S.A.*, 1993; 90:11806–10.

4. Xie X and Smart TG: Properties of GABA-mediated synaptic potentials induced by zinc in adult rat hippocampal pyramidal neurons. *J. Physiol.*, 1993; 460:503–23.

5. Davies MF, Maguire PA, and Loew GH: Zinc selectively inhibits flux through benzodiazepene-insensitive gamma-aminobutyric acid chloride channels in cortical and cerebellar microsacs. *Mol. Pharmacol.*, 1993; 44:876–881.

6. Berlet HH, Ilzhofer H, and Nohe M: Metal cations and binding of extrinsic proteins of bovine myeline. *Biol. Chem. Hoppe-Seyler*, 1987; 368:1246.

7. Connor MA and Chavkin C: Ionic zinc may function as an endogenous ligand for the halperiod-sensitive σ2 receptor in rat brain. *Mol. Pharmacol.*, 1992; 42:471–479.

8. O'Dell BL: Roles of zinc and copper in nervous system. *Prog. Clin. Biol. Res.*, 1993; 380:147–162.

9. Richfield EK: Zinc modulation of drug binding, cocaine affinity states, and dopamine uptake on the dopamine uptake complex. *Mol. Pharmacol.*, 1993; 43:100–108.

10. Liu H, Oteiza PI, Gershwin ME, et al: Effects of maternal marginal zinc deficiency on myelin protein profiles in the suckling rat and infant rheusus monkey. *Biol. Trace Elem. Res.*, 1992; 34:55–66.

11. Guidolin D, Polato P, Venturin G, et al: Correlation between zinc level in hippocampal mossy fibers and spatial memory in aged rats. *Ann. N.Y. Acad. Sci.*, 1992; 673:187–193.

12. Frederickson CJ, Rampy BA, Reamy-Rampy S, et al: Distribution of histochemical reactive zinc in the forebrain of the rat. *J. Chem. Neuroanat.*, 1992; 5:521–530.

13. Legendre P and Westbrook GL: Noncompetitive inhibition of γ-aminobutyric acid channels by Zn. *Mol. Pharmacol.*, 1990; 39:267–274.

14. Oosterwegel M, Timmerman J, Leiden J, et al: Expression of GATA-3 during lymphocyte differentiation and mouse embryogenesis. *Dev. Immunol.*, 1992; 3:1–11.

15. Keen CL, Taaubeneck MW, Daston GP, et al.: Primary and secondary zinc deficiency as factors underlying abnormal CNS development. *Ann. N.Y. Acad. Sci.*, 1993; 678:37–47.

16. Yasui M and Ota K: Experimental and clinical studies on dysregulation of magnesium metabolism and the aetiopathogenesis of multiple sclerosis. *Magnesium Res.*, 1992; 4:295–302.

17. Yasui M, Ota K, and Garruto RM: Concentration of zinc and iron in the brains of Guamanian patients with amyotrophic lateral sclerosis and parkinsonism-dementia. *NeuroToxicology*, 1993; 14:445–450.

18. Ceballos I, Javoy-Agid F, Delacourte A, et al.: Neuronal localization of copper-zinc superoxide dismutase protein and mRNA within the human hippocampus from control and Alzheimer's disease brain. *Free. Rad. Res. Commun.*, 1991 (Pt 2); 12–13:571–580.

19. Laboyrie PM, Roos RA, and Verspaget HW: Copper-zinc superoxide dismutase in cerebrospinal fluid: implications for oxygen-radical stress in the central nervous system. *Clin. Neurol. Neurosurg.*, 1992; 94:S106–107.

20. Constantinidis J: The hypothesis of zinc deficiency in the pathogenesis of neurofibrillary tangles. *Med. Hypothesis*, 1991; 35:319–323.

Chapter 15

Zinc-Toxicity in Alzheimer's Disease: An Emerging Hypothesis

*Jean Constantinidis**

CONTENTS

---

* Note added in proof: Dr. Constantinidis passed away prior to the submission of revisions to this manuscript. At the discretion of the editors and out of respect for the many years of work contributed by Dr. Constantinidis, we have chosen to publish this submission in spite of the controversial nature of the arguements presented. Final editorial and factual modifications were made at the discretion of the editors.

15.1 INTRODUCTION

In Alzheimer's disease (**AD**), we have postulated that the aggregation to cortical extracellular β-amyloid (**βA₄**) deposits of the (very probably) genetically determined and diffusely formed pre-AM seems to be activated by free zinc, preceded by the pericytes-microglia on the walls of cortical microvessels, thus disturbing the blood-brain barrier (**BBB**). Therefore, toxic metals may enter the brain (their levels increase) and displace the zinc in enzymes, which become dysfunctional, and in synaptic vesicles, leading to (1) production of the neurofibrillary pathology: the neuritic paired helical fillaments (**PHF**) which with the βA₄ core form the senile plaques (**SP**) and in the neuronal bodies the neurofibrillary tangles (**NFT**) (constituted by PHF); (2) neuronal suffering by deficient detoxification; and (3) disturbances in synaptic neurotransmitter function. The βA₄ deposits alone are asymptomatic, the clinical symptoms of dementia are related to the PHF-NFT. The *timing window* βA₄-NFT is of about 1.7 to 10 years. The neuroleptics which activate the entrance of metals in the brain and decrease the brain zinc, reduce the incidence and delay the production of βA₄ deposits in elderly schizophrenics (**SCHIZ**) chronically treated with NLPT; but when the βA₄ deposits are formed, the chronic NLPT treatment accentuates and accelerates the βA₄-induced neurofibrillary pathology.

A similar mechanism is postulated for the NFT production in some encephalopathies without βA₄: entrance into the brain of toxic metals (heavy metal intoxication, viral and polytraumatic BBB disturbances), displacing and decreasing the zinc.

We suggest that during the early asymptomatic AD, when the βA₄ deposits are formed, periodically repeated treatment with a metal chelating agent capable of crossing the BBB to decrease the brain zinc may inhibit βA₄ formation. During advanced AD, we propose the periodically repeated chelating treatment to decrease in the brain the high levels of toxic metals, followed by periodically repeated treatment with a zinc compound capable of crossing the BBB to increase the brain zinc, promote its restitution in enzymes and synaptic vesicles, restore their functions, stop or delay worsening of the neurofibrillary pathology, improve neuronal detoxification and the neurotransmitter's synaptic activity, and therefore, delay or stop the evolution of clinical deterioration and even improve dementia. Without chelating pretreatment, zinc aspartate administered i.v. and *per os* leads to promising improvement. Research is in progress that is scheduled to evaluate the potential therapeutic efficacy of various chelating agents and zinc compounds which may become drugs to use against the Alzheimer's brain pathology.

This chapter results from the synthesis of:

1. Findings from the author's research during the last 33 years, published in more than 200 papers, the first in 1962[1] and summarized in two recent monographs [1993],[2,3] concerning the various aspects of AD: brain lesions, clinical symptoms, anatomoclinical correlations, disturbances of neurotransmitters, etiopathogenesis of βA₄, pathogenesis of NFT, role of toxic metals and zinc, effects of the neuroleptics, encephalopathies with NFT but without βA₄, and preventive and therapeutic approaches;
2. Findings published by other researchers during the last 60 years, concerning the above aspects of AD and particularly concerning the role of toxic metals and zinc.

The complete list of publications reported and discussed in this chapter may be found in the author's monographs [1993],[2,3] and only the 30 more-representative ones are in the list of references in this chapter. The synthesis of these findings allows us to delimit the profile concerning the problems exposed in this chapter, as follows below.

Cortical and subcortical neuronal NFT are observed in some human encephalopathies of various etiologies:

- In heavy metal intoxication.
- In alimentation particularities facilitating penetration into the brain of toxic metals, as in parkinsonism-dementia complex of Guam.

- In viral encephalitides and in the polytraumatic (boxer's) *encephalopathia pugilistica*, with BBB disturbances and possible penetration into the brain of toxic metals.
- In AD, with the (very probably) genetically determined β-amyloidosis of cortical microvessels, producing BBB disturbances and allowing penetration into the brain of toxic metals.

We postulate that despite the variability of the etiologies, the pathogenetic mechanism of the NFT production is similar in all the above encephalopathies: penetration in the brain of toxic metals and then displacing and decreasing the zinc in enzymes, which become dysfunctional, leading to the synthesis of the abnormal protein tau constituting the PHF-NFT.

15.2 βA$_4$ DEPOSITS PRECEDE NEUROFIBRILLARY PATHOLOGY RELATED TO CLINICAL SYMPTOMS[4]

When the βA$_4$ deposits are formed in the cortex there are no immediately neighboring NRTC alterations. The presence of βA$_4$ deposits produces an NRTC reaction in the parenchyma, i.e., the formation of PHF in the neighboring neurites constituting the SP. This neuritic component should be related to the first clinical symptom (amnesia) of AD, while the βA$_4$ deposits alone are asymptomatic. Later on, production of NFT in the neuronal bodies of the hippocampus is related to the worsening of amnesia and NFT in the neocortex is related to the instrumental disorders: aphasia, agnosia, and apraxia.

The *timing-window* between the formation of βA$_4$ deposits and the production of the first NFT (in the hippocampal allocortex) may be evaluated as of about **1.7 y**. An average *timing-window* of 2.1 y should be necessary for the βA$_4$-induced production of NFT in the hippocampal isocortex.[2,3]

The *timing-window* between the senile plaque formation and the onset of amnesia and the production of the first NFT (in the hippocampal allocortex) may be estimated to be about **14 months**; this *timing-window* is of about 37 months for the production of NFT in the hippocampal isocortex and of about 47 to 67 months for the production of NFT in the neocortex, related to their progressively increased densities.[2,3]

15.3 BRAIN ZINC IN EARLY AND ADVANCED AD

15.3.1 Formation of βA$_4$ Deposits and Free Extracellular Brain Zinc in Early AD

In AD, the *primum movens* is the diffusely pre-AM production; at a second stage the βA$_4$ deposits are formed within the cerebral cortex, followed later by the neurofibrillary pathology.[2] The hippocampus, where the senile plaque and the PHF-NFT are focused at the onset of the disease is the area of the brain with the highest zinc content [Euler, 1962; Haug, 1973]. Free extracellular zinc seems to activate Alzheimer's βA$_4$ deposit formations from pre-AM material.[7] The free extracellular brain zinc may be increased in early, symptomless AD when βA$_4$ deposits are in formation and not yet involved in the neurofibrillary pathology, but zinc estimations in very early cases have not been performed.

15.3.2 In Advanced AD the Intraneuronal Brain Zinc Is Decreased

In advanced AD with neurofibrillary pathology, the zinc decreases, especially in the hippocampus as has been estimated in 1977 by atomic absorption[8] and in 1987 and 1991 by neutron activation,[9,10] while in non-AD-aged brains (40 to 99 years) the brain zinc level remains relatively constant. The decrease of the brain zinc in advanced AD seems to be intraneuronal as may be suggested from the decreased staining of the hippocampal mossy fibers by the Timm's method.[8,11,12]

The possibility that in advanced AD the intraneuronal brain zinc decreases while the extracellular zinc level remains high may not be excluded, but is difficult to investigate. Probably both intraneuronal and extracellular zinc decrease in advanced AD, as low zinc levels have also been found in the CSF.

15.3.3 Probable Pathogenetic Mechanism of the Brain Zinc Decrease in Advanced AD

βA_4 deposition within the walls of the capillaries may be associated with disturbances of the BBB.[5,14]

Because of the βA_4-induced BBB disturbances, toxic metals reach the cerebral cortex. They are encrusted in the β-AM core of the capillary SP, where they can be observed by Gomori's staining for iron[1] and by the Timm's method of staining for heavy metals;[11,12] the SP may also be encrusted with calcium, aluminum, and silicon.

Many toxic metals are increased in the AD brain: iron,[15] aluminum,[16] manganese,[17] calcium, silicon,[9] and mercury[18] — This increase of toxic metals in the AD brain may be related to the decrease of zinc by metal to metal displacement,[12] especially in zinc enzymes which become dysfunctional and induce the production of NFT.

The importance of the microvascular βA_4 (and, therefore, BBB disturbances) in the pathogenesis of the NFT is illustrated by the fact that in older primates, with nonmicrovascular βA_4, NFT are not produced, while in the small lemurine primate *Microcebus murinus*, with microvascular βA_4 when aged 4 years, NFT are produced when aged 8 years.[19]

15.4 ZINC-INDUCED ENZYMATIC DYSFUNCTION

Postnatal zinc deficiency in rats results in a decrease of the total amount of DNA in the brain, impairment of thymidine incorporation in DNA, and abnormal DNA, RNA protein synthesis.

15.4.1 Deficiency of Glutamate (GLU) Dehydrogenase (GDH)

GDH deficiency results in a neurotoxic increase of GLU. This may relate to the production of PHF-NFT:[11] as may be concluded by the following:

- Zinc deficiency (and therefore the GDH activity) produces high concentrations of GLU in animals.
- At the onset of AD there is in the brain an excess of GLU; in mildly to moderately demented AD patients the concentration of the CSF free GLU is significantly increased.
- GLU produces PHF in cultured human neurons, and intrathecally injected kainic acid, a potent GLU agonist, in the rat lumbar region, produced aggregates of phosphorylated neurofilaments.
- The hippocampal areas where the NFT are formed at the onset of the disease, and which present the highest density of these neuronal lesions, are the areas receiving important GLU pathways: CA1, Sommer's sector, and area entorhinalis, while the areas which do not receive GLU pathways: the gyrus dentatus and presubiculum, are spared from NFT.[2]
- In advanced AD, the cortical GLU neurons are also destroyed by the NFT, resulting in a decrease of cortical GLU and an increase of cortical and subcortical GLU receptors, i.e., GLU hypoactivity.[2]

15.4.2 Deficiency of the Enzymes of Neuronal Detoxification

Deficiency of superoxide dismutase (SOD), carbonic anhydrase, and lactate dehydrogenase leads to neuronal toxicity.[12] The hippocampal areas CA1 and Sommer's sector, whose neurons are the most affected by the NFT, normally present the highest level of SOD, related to their intense activity and metabolic turnover producing toxic substances; probably zinc decrease and the corresponding functional deficiency of this detoxicant zinc enzyme are particularly important for these neurons. A functional deficiency of SOD in AD (because of the displacement of zinc by other metals) should lead to a compensatory quantitative increase of SOD, dysfunctional but immunoreactive.

Using SOD monoclonal antibodies, we have observed an increased SOD immunoreactivity in the neurons of AD hippocampus, compared to controls (unpublished observations). Ceballos et al. [1992] did not find such an increase of SOD in the hippocampus of AD, but they observed that both the normal and degenerating neurons were labeled by the anti-SOD antiserum. The enzymatic activity of the abundant SOD that we observed in the hippocampal neurons of AD must be evaluated.

15.4.3 Dysfunction of Synaptic Neurotransmitters

Zinc is localized in synaptic vesicles and intervenes as a modulator in the synaptic activity of some neurotransmitters. Neuronal dysfunction in AD may also be related to the decrease of zinc in synaptic vesicles and the resultant disturbances in synaptic neurotransmission.

15.5 TOXIC METALS IN ENCEPHALOPATHIES WITH NFT BUT WITHOUT βA_4 DEPOSITS

Lead — In the human encephalopathy caused by lead intoxication: *encephalopathia saturnica*, NFT without βA_4 or senile plaques has been observed in the hippocampus.[20] Using the methodology of Niklowitz,[21] we produced PHF, characteristic of AD, in the neurons of the hippocampal CA 1 and Sommer's areas after i.p. administration of tetraethyl lead to the rabbit (unpublished observations). The administration of lead decreases the concentration of zinc in the brain and accumulates in the place of zinc in the hippocampal mossy fiber system, is fixed in zinc enzymes inhibiting their activity, and induces reversal learning deficit by hippocampal dysfunction. In experimental and human lead encephalopathies, the PHF-NFT could be formed by the displacement of zinc by lead in some enzymes, resulting in zinc enzymatic deficiency.

15.6 POTENTIAL OF INTERVENTION

15.6.1 Formation of the βA_4 Deposits and the βA_4-Induced Production of NFT Influenced by the Administration of Drugs: The Neuroleptics

We tried to verify this hypothesis for the neuroleptics (NLPT), chronically administered to schizophrenics since 1953, which seem to activate the βA_4-induced abnormal entry into the brain of toxic metals and the zinc deficiency, as may be concluded by the following discussion:

In the hippocampus of schizophrenics treated with NLPT, zinc measurements by atomic absorption gave 50% subnormal values compared to controls;[23] and NLPT treatment (haloperidol) induces an increase of the iron transport across the BBB, as demonstrated by a scan, after application of (Fe-52) citrate to the rhesus monkey.[24] The resulting high iron content in the brain displaces the zinc, which decreases.

To estimate the eventual relationships between the NLPT treatment with the AD brain pathology, we investigated all the brains, preserved in formalin, of elderly schizophrenics who died at more than 70 years of age and came to autopsy in the University Psychiatric Hospital of Geneva:[3]

- 41 not treated with NLPT (NLPT–) who died from 1932 through 1952 at an average age of 79 years; and
- 61 treated with NLPT (NLPT+) who died from 1953 through 1990, age matched (79 years) to the NLPT.

From these patients, samples of the hippocampus and the temporal, occipital, and frontal cortex have been studied by silver staining for the AM deposits with their NRTC component and the NFT (Table 15.1). Cases of NLPT+ have been studied morphometrically (in immunohistologically stained sections with anti-tau antibodies) for the densities of NRTC plaques and of NFT for their stages of maturation and for the neuronal loss (Table 15.2).

Table 15.1 Amyloid-Senile Plaques (AM-SP) and Neurofibrillary Tangles (NFT) in Brains of Schizophrenics Dead at More Than 70 Years of Age in the University Psychiatric Hospital of Geneva

		NLPT–		NLPT+	
Incidence (%) of βA_4 deposits		60	>	56	
Incidence (%) of NFT (among those with β-AM)		36	<	74	$p <.05$
Age at death (years) total of cases		79 ± 7	=	79 ± 6	
Age at death (years) when βA_4–	(16 cases)	77 ± 5	=	77 ± 6	(26 cases)
Age at death (years) when βA_4+	(25 cases)	80 ± 3	=	80 ± 3	(35 cases)
Age at death (years) when βA_4+ NFT–	(16 cases)	76 ± 5	<	79 ± 9	(9 cases)
Age at death (years) when βA_4+ NFT+	(9 cases)	86 ± 9	>	81 ± 3	(26 cases)
Timing window (years) from βA_4 to NFT		10	>	2	

Note: Nontreated with neuroleptics, death at 1932–1952 (NLPT–): 41 cases. Treated with neuroleptics, death at 1953–1990 (NLPT+): 61 cases.

From Constantinidis, J., *Editions Médecine et Hygiene*, Geneva, 1993. With permission.

Table 15.2 Comparative Immunohistological Morphometry of the Neurofibrillary Pathology: of the NRTC.SP, of the NFT, of Their Evolutive Stages, and of the Neuronal Loss

Density of	**NLPT–**		**NLPT+**	**p<**
NRTC.SP		<		.01
NFT		<		
NFT early stages (pretangles)		>		.01
NFT late stages (ghost tangles)		<		.01
Remaining intact		>		.01
Neuronal bodies				

Note: 9 Brains of elderly schizophrenics (average age: 86) nontreated with neuroleptics (NLPT–). 7 Brains of age-matched schizophrenics (average age 86) treated (NLPT+).

From Constantinidis, J., *Editions Médecine et Hygiene*, Geneva, 1993. With permission.

From our findings in Tables 15.1 and 15.2, it may be concluded that:

- The chronic NLPT treatment slightly decreases the incidence of βA_4 deposits (56% vs. 60%) and slightly delays their formation (79 years vs. 76 years, if we take into account the cases NLPT+ and NLPT– who died after the formation of βA_4 deposits but before the production of NFT). As zinc activates the formation of βA_4 deposits,[7] the NLPT-induced decrease of the brain zinc[22] may explain the relative inhibition of their formation.
- When the βA_4 deposits have been formed, the NLPT treatment enhances significantly the incidence of NFT (which is two times higher in NLPT+: 74% vs. 36%, $p <0.05$), increases their density (and the density of NRTC.SP), causes earlier production (81 years vs. 86 years) which is accelerated (2 years vs. 10 years, with more late and less early stages of NFT maturation), and enhances the neuronal loss.[3]

15.6.2 Contrary to Neuroleptics, Other Drugs Administered During the *Timing-Window* βA_4-NFT May Inhibit the βA_4-Induced Neurofibrillary Pathology: Metal Chelating Agents Followed by Zinc Compounds

We postulated[11,12] that during the *timing-window* βA_4-NFT, an excess of zinc will, by normalizing the activity of the zinc enzymes, stop or delay the PHF-NFT production and therefore worsening of the clinical symptoms of dementia. These symptoms, if yet present, may also be improved with zinc treatment by normalization of the neuronal detoxification and of the neurotransmitter synaptic function.

In support of our theory, the βA_4-induced high levels of toxic metals in the brain are responsible for the decrease of zinc and for its displacement in the enzymes. The toxic metals clearance of the brain by a periodically repeated administration of a metal chelating agent capable of crossing the BBB will be beneficial before each periodically repeated administration of zinc compounds.

15.7 ZINC COMPOUNDS PROPOSED FOR THE TREATMENT OF AD

A trial with zinc sulfate per os that we published in 1980[11] was not conclusive. This inorganic zinc salt is probably not well absorbed in the gut (produces gastrointestinal troubles) and probably does not cross the BBB. In 1981, Burnet[25] proposed to administer zinc in AD. Unsuccessful trials with zinc sulfate in AD were also performed in 1990 by Rhijn et al.[26]

Important zinc compounds are protein-peptidic-amino acid-zinc complexes because proteins in food intake influence zinc absorption. The zinc is transported in the plasma linked on albumin and globulins. In AD we found a correlation of serum zinc with albumin but not with globulins, and a low zinc content in β-globulins.[12] From these findings we may suppose that in AD there is a deficiency of zinc linked on globulins. The zinc amino acid complex for the treatment of AD would be (after entering the blood) preferentially linked to globulins and so transported into the brain.

A valuable profile are zinc complexes with essential amino acids which, when in the blood, will be linked on the plasma proteins. Zinc salts with nonessential amino acids may also be considered appropriate preparations in AD. Such compounds are zinc aspartate, zinc orotate, and zinc pidolate. Other appropriate zinc compounds that cross the BBB easily are zinc ethanol aminophosphate (Zn.EAP) and zinc with insulin (insulin potentiation).

In 1989 we started preliminary trials with zinc aspartate administered per os and i.v. in AD patients and we reported our first results in 1990 and 1991.[12,27] A group of 10 patients (8 female and 2 male), 5 of them presenile (aged 56, 62, 64, 67, and 67) and 5 senile (aged 73, 74, 76, 79, and 82) were selected according to the criteria for AD of the NICDS-ADRDA group. The 10 patients received zinc aspartate during periods from 3 months to 1 year.

- All of them received per os 3 pills every day; each pill contains 50 mg of zinc bis-(DL-hydrogenaspartate) = 9.22 mg Zn.
- Three of them also received i.v. 1 ampule every 2 days; each ampule contains 30 mg zinc bis-(DL-hydrogenaspartate) = 0.091 mM Zn^{2+} = 5.96 mg Zn^{2+}.
- The daily requirement for zinc obtained by normal nutrition is 5.25 mg. Every day these patients received a zinc supplement higher than the maximum daily need. No other nootropic or neuroleptic treatment was given during the trial period.
- The clinical state was evaluated before and after the treatment by psychometric tests and by reports of patients' relatives.

For two patients, aged 62 and 67, given only per os treatment, no improvement was noted. For eight patients, improvement of memory, understanding, communication, and social contact was evident. For one patient, aged 79, who was relatively less demented, and who received zinc aspartate both per os and i.v., the improvement of memory was qualified as "unbelievable" both by the medical staff and patients' relatives. The discontinuation of the treatment for some period, because of drug supply problems, resulted in a decrease and even disappearance of the improvement in all improved patients.

Funfgeld who, following our suggestion performed similar trials with zinc aspartate, reported[28] the disability score of Hoehn and Yahr dropped from category 3 to category 1, and the computerized EEG showed a marked acceleration.

Nachev et al. who, following our suggestion performed similar trials with another (not specified) zinc compound,[29] reported improvement of various degrees in memory, orientation, speech, and visuospatial skills, which disappeared 1 month after stopping the treatment.

15.8 FUTURE RESEARCH DIRECTIONS

15.8.1 Animal Models of Alzheimer Brain Pathology

Natural animal model of Alzheimer brain pathology (AM and NFT): the Lemurine primate *Microcebus murinus* — With AM deposits when aged 4 years and also with NFT in cortical neurons when aged 8 years (see Section 15.3 in this chapter).

Experimental animal model of Alzheimer-type βA_4 deposits: the transgenic mouse — Transgenic mice at 6 to 9 months of age begin to exhibit deposits of βA_4 in the hippocampus, corpus callosum, and cerebral cortex; by 8 months many deposits (30 to 200 μm) are seen; at ≥ 9 months the density of the AM plaques increase until the βA_4 staining pattern resembles that of AD. Many plaques are closely associated with distorted neurites; synaptic and dendritic densities are reduced in the molecular layer of the hippocampal dentate gyrus, but microvascular βA_4 and NFT are not produced.[30]

Experimental animal model of Alzheimer type neuronal PHF: the lead intoxicated rabbit — Alzheimer type PHF are produced in the hippocampal neurons of the rabbit by i.p. administration of lead compounds (see Section 15.5 in this chapter).

15.8.2 Research Concerning the Alzheimer Brain Pathology on Animal Models

- Levels of zinc and other metals in the brain of *Microcebus murinus*, aged 4 years and 8 years.
- Immunoreactivity and enzymatic activity of the zinc enzymes, SOD and GDH:
  - In cortical neurons of *Microcebus murinus*, aged 4 years and 8 years.
  - In the hippocampal PHF-containing neurons of the lead-intoxicated rabbit.
- Eventual DNA changes in the hippocampal neurons containing PHF, of the lead intoxicated rabbit.
- Eventual enhancement by repetitive NLPT administration:
  - Of the NFT production in cortical neurons of *Microcebus murinus* when aged 8 years, following chronic administration of NLPT after the formation of AM deposits when aged 4 years.
  - Of the lead-induced PHF formation in the hippocampal neurons of the lead-intoxicated rabbit.

Research on metal chelating agents capable of crossing the BBB should investigate the efficacy of various chelating agents in inhibiting *in vitro* the zinc-induced aggregation of AM (see Section 15.3 in this chapter) and to resolubilize the already formed zinc-AM aggregates.

- Estimate by neutron activation analysis the decrease of zinc and other metals in the rat brain after administration of chelating agents.

- Evaluate the eventual inhibition of βA_4 deposit formations.

Research on zinc compounds should evaluate by neutron activation analysis the increase of brain zinc in rats after administration of various zinc compounds (see Section 15.7 in this chapter) with always the same per dose quantity of metal zinc.

15.8.3 Creation of New Drugs for Preventive and Therapeutic Treatment Against the Production of Alzheimer Brain Lesions and Dysfunctions

More effective nontoxic chelating agents are needed:

- To inhibit the *in vitro* AM aggregations.
- To decrease zinc and other metals in the rat brain.
- To protect against the formation of β-AM deposits in the cerebral cortex of transgenic mice and of *Microcebus murinus*.

Nontoxic zinc compounds are needed which for the same quantity of administered zinc metal are more effective:

- To increase the zinc in the rat brain.
- To protect against the formation of PHF-NFT in the hippocampal neurons of the lead-intoxicated rabbit and in the cortical neurons of *Microcebus murinus*.
- After the regulatory investigations, may be tried in AD patients and produced commercially for broad therapeutic use.

REFERENCES

1. Constantinidis J: Le fer sérique et les dépôts ferreux cérébraux chez les vieillards déments. *Rev. Fr. Géront.*, (*Paris*), 1962, 8:267–284.
2. Constantinidis J: Maladie d'Alzheimer, de l'amyloïde à la lésion neurofibrillaire: effets des neuroleptiques et des composés à zinc. Monographie, *Editions Médecine et Hygiène*, Genève, 1993.
3. Constantinidis J: Alzheimer's amyloid-induced neurofibrillary pathology and chronic treatment with neuroleptics, monography, *Editions Médecine et Hygiène*, Geneva, 1993.
4. Constantinidis J and Richard J: Alzheimer's Disease. In: Vinken PJ, Bruyn GW, and Klawans HL: *Handbook of Clinical Neurology*, a revised series, vol. 46, "Neurobehavioural Disorders", Frederiks JAM (Ed.), Amsterdam, Elsevier Science, 1985, pp. 247–282.
5. Constantinidis J and Ajuriaguerra J de: Les altérations cérébrales au cours du vieillissement. *Acta Neurol. Psychiatr. Hellen.* (*Thessaloniki*), 1968, 7:79–115.
6. Constantinidis J: Is Alzheimer's disease a major form of senile dementia? Clinical, anatomical and genetic data. In: Katzman R, Terry RD, and Bick KL (Eds.): Alzheimer's disease, senile dementia and related disorders, *Aging*. New York. Raven Press, 1978, 7:15-25.
7. Bush AI, Pettingell WH, and Multhaup G: Rapid induction of Alzheimer's beta Amyloid formation by zinc. *Science*, 1994, 265:1464–1467.
8. Constantinidis J, Richard J, and Tissot R: Démences dégénératives de Pick et d'Alzheimer et métabolisme du zinc. *Méd. Hyg.* (*Genève*), 1977, 35:3706–3710.
9. Ward NI and Mason JA: Neutron activation analysis techniques for identifying elemental status in Alzheimer's disease. *J. Radioanal. Nucl. Chem. Art*, 1987, 113:515–526.
10. Corrigan FM, Reynolds GB, and Ward NI: Reduction of zinc and selenium in brain in Alzheimer's disease. *Trace Elem. Med.*, 1991, 8:1–5.
11. Constantinidis J: Maladie de Pick, démences à plaques séniles et lésions neurofibrillaires d'Alzheimer, métabolisme du zinc et acides aminés neurotransmetteurs. In: Tissot R (Ed.): *Etats Déficitaires Cérébraux Liés à l'Age*, Genève, Georg, 1980, 201–228.
12. Constantinidis J: Démence d'Alzheimer et zinc. *Schweiz. Arch. Neurol. Psychiatr.*, 1990, 141: 523–556.

13. Constantinidis J, Wisniewski HM, and Wegiel J: Pericytes and microglia: the amyloid-producing cells in Alzheimer's Disease. *Clin. Neuropathol.*, 1992, 11:22–23.

14. Wisniewski HM and Kozlowski PB: Evidence of blood brain barrier changes in senile dementia of the Alzheimer type (SDAT). *Ann. N.Y. Acad. Sci.*, 1982, 396:119–129.

15. Braunmuhl A, von: Ueber den Eisenstoffwechsel bei seniler involution und seniler Entartung. In: Lubarsch O, Henke F, und Rössle R (Eds.): *Handbuch der Speziellen Pathologische Anatomie und Histologie*. Berlin, Springer-Verlag, 1957, 13:391–393.

16. Crapper DR, Krishnann SS, and Dalton AJ: Brain aluminum distribution in Alzheimer's disease and experimental neurofibrillary degeneration. *Science*, 1973, 180:511–513.

17. Banta RG, Clark DB, and Markesbery WR: Elevated manganese levels associated with Alzheimer's disease and extrapyramidal signs. *Neurology (Minneap.)*, 1975, 25:356.

18. Thompson CM, Markesbery WR, and Ehmann WD: Regional brain trace-element studies in Alzheimer's Disease. *Neurotoxicology*, 1988, 9:1–8.

19. Bons M, Mestre N, and Petter A: Senile plaques and neurofibrillary changes in the brain of aged lemurine primate, *Microcebus murinus*. *Neurobiol. Aging*, 1991, 13:99–105.

20. Niklowitz WJ and Mandybur TI: Neurofibrillary changes following childhood lead encephalopathy. Case report. *J. Neuropathol. Exp. Neurol.*, 1975, 34:445–455.

21. Niklowitz WJ: Neurofibrillary changes after acute experimental lead poisoning. *Neurology (Minneap.)*, 1975, 25:927–934.

22. Yasui M, Ota K, and Garruto RM: Concentrations of zinc and iron in the brains of Guamanian patients with amyotrophic lateral sclerosis and Parkinsonism-dementia. *Neurotoxicology*, 1993, 14:445–450.

23. McLardy T: Hippocampal zinc and structural deficit in brains from chronic alcoholics and some schizophrenics. *Orthomol. Psychiatr.*, 1975, 4:32–36.

24. Leenders K, Antonini A, Grunther I, et al: Blood to brain iron transport: Influence of Haloperidol. *J. Neurol.*, 1992, 239 (Suppl. 2):31–32.

25. Burnet FM: A possible role of zinc in the pathology of dementia. *Lancet*, 1981, 1/8213: 188–196.

26. Rhijn AG, Prior CA, and Corrigan M: Dietary supplementation with zinc sulfate, sodium selenite and fatty acids in early dementia of Alzheimer's type. *J. Nutr. Med.*, 1990, 1:259–266.

27. Constantinidis J, Hatziantoniou JS, and Mentenopoulos G: Alzheimer's Disease: cerebral zinc deficiency and preliminary therapeutic trials by zinc aspartate. 5th Congress Intern. Psychogeriatr. Assoc., Rome, Abstracts, 1991, p. 69.

28. Funfgeld EW: Alzheimer's Disease and SDAT: A further multitherapeutical approach — the zinc-aspartate treatment. 4th Annual Meeting, Am. Psychogeriatr. Assoc. FL Abstracts, Febr. 1991.

29. Nachev CK, Bonchev P, Kirov GK, and Kissiova K: Zinc deficiency and Alzheimer's disease: a new approach in: *Psychiatry, A World Perspective*, Stefanis C. et al. (Eds), New York, Elsevier Science, 1990, Vol 2:496–504.

30. Games D, Adams D, Alessandrini R, et al: Alzheimer-type neuropathology in transgenic mice. *Nature*, 1995, 373:523–527.

Chapter 16

Pathogenesis of Methyl Mercury Neurotoxicity

M. Anthony Verity

CONTENTS

16.1 INTRODUCTION

While human exposure to mercury occurs via inorganic and organic forms, clinically relevant studies of mercury neurotoxicity follow exposure to the alkyl (methyl) mercury derivatives (MeHg). The following chapter by Professor L. Chang discusses the clinical features of human and experimental neuropathology associated with mercury exposure. In this discussion, experimental studies elucidating the pathogenesis of organic mercurial neurotoxicity will be summarized. Despite numerous studies devoted to the identification of a single locus for the toxicity, it is now believed that a multifactorial pathogenesis accounts for the end-stage neuronal injury. Although many toxic effects of MeHg have been identified in neuron systems, the role of astrocytes in mediating and/or modulating neuronal toxicity has assumed importance. Recent reviews have examined differing aspects of methyl mercury toxicity.[1–4]

16.2 PHARMACOKINETICS

Methyl mercury toxicity commonly results following ingestion. The bulk of gastrointestinal MeHg undergoes fecal excretion with a minor component undergoing demethylation or absorption. Once absorbed, MeHg may enter red cells and be covalently bound to glutathione and protein cysteine groups.[5] The preferential sensitivity and toxicity of MeHg for dorsal root ganglion cells is related to the lack of an efficient blood-nerve barrier, in marked contrast to the role played by blood-brain barrier mechanisms. Mechanisms underlying the uptake or efflux of MeHg in brain appear dependent upon binding to high-affinity −SH groups on specific translocator molecules. Brain uptake follows binding of MeHg to L-cysteine and translocation at the neutral amino acid carrier site.[6] Similarly, *in vitro* studies using astrocyte cultures have provided evidence for a similar role of −SH-containing molecules in the efflux of MeHg which also appears specific for the MeHg-cysteine conjugate.[4] Glutathione plays a major role in the modulation of MeHg toxicity: firstly, as a major cellular antioxidant mitigating the lipoperoxidative injury induced by MeHg; secondly, in participating in the stress protein response following low-dose MeHg exposure, and thirdly, conjugation with MeHg providing an accelerated efflux from glia and renal epithelium[7] (Fujiyama et al., personal communication).

16.3 MECHANISMS OF METHYL MERCURY NEUROTOXICITY

The molecular basis for neurotoxicity is multifactorial (see Table 16.1) and the ultimate morphological abnormality or pathology is the result of overlapping toxic mechanisms acting directly or indirectly on neuronal-astrocyte interactions. Early ultrastructural observations (reviewed in Chapter 17) identified disorganization of rough endoplasmic reticulum and ribosome dissociation, suggesting a defect in protein synthesis. Subsequent biochemical studies characterized a major and early inhibition in protein and macromolecular synthesis.[8–10]

Table 16.1 Major Pathways Involved in the Pathogenesis of Methyl Mercury Neurotoxicity

1.	Inhibition of protein and macromolecular synthesis
2.	Mitochondrial dysfunction
3.	Defective Ca and ion flux: abnormal neurotransmitter homeostasis
4.	Initiation of oxidative (stress) injury
5.	Microtubule disaggregation
6.	Posttranslation phosphorylation

16.3.1 Protein Translation Inhibition

Following the original studies defining the inhibition of protein synthesis *in vivo* and *in vitro* by MeHg,[2] Cheung and Verity[9] presented evidence that the methyl mercury-induced lesion was in the pH 5 enzyme component of the postmicrosomal supernatant. Further studies[10] demonstrated that the mercurial did not directly induce disaggregation of brain polyribosomes nor change the proportion of 80 S monoribosomes detected on sucrose density gradients. If anything, initiation complex formation was stimulated by methyl mercury. Methyl mercury inhibited polyuridylic acid-directed incorporation of [3H]-phenylalanine, identifying a principal defect in peptide elongation. No inhibition of ribosomal peptidyltransferase activity was found, nor inhibition of translocation mediated by elongation factor-2. However, selected inhibition of certain aminoacyl-tRNA synthetases was observed, identifying a pivotal (primary) defect of *in vivo* and *in vitro* brain translation. Confirmation of these observations was observed when [3H]-phenylalanyl-tRNA was substituted for [3H]-phenylalanine in the incorporation assay. Under these circumstances the radiolabeled precursor tRNA should bypass the requirement for aminoacylation and no MeHg-induced inhibition of incorporation was observed, also suggesting that the distal elongation steps in translation were not a target for MeHg *in vivo*.

Sarafian and Verity,[11] using isolated cerebellar granule cells, demonstrated coincident inhibition of RNA and protein synthesis; 50% inhibition was found at ~8 μM MeHg. Studies on the apparent mechanism of transcription inhibition in cerebellar neurons pointed to a direct inhibitory effect on phosphouridine kinase enzymes.[12] Reduced intracellular levels of UTP and UDP were observed, pointing to the necessity for consideration of a defect in ADP/ATP ratio to account for the abnormality in uridine phosphorylation (*vide infra*).

16.3.2 Mitochondrial Dysfunction: Role of Defective Energy Charge

Verity et al.,[13] using a reproducible model of subacute MeHg intoxication, demonstrated an inhibition of synaptosomal respiration following *in vivo* or *in vitro* MeHg. In particular, *in vitro* studies at 5 to 15 μM MeHg revealed stimulation of State 4 respiration, loss of respiratory control, and inhibition of State 3 respiration at [MeHg] >15 μM. Inhibition of electron transfer occurred at MeHg >20 μM but isosmotic replacement of KCl in the incubation medium by mannitol reversed the MeHg stimulation of State 4 respiration, coupling State 4 stimulation to induction of K^+-induced translocation, as has been confirmed by Sone et al.[14]

The interaction between mercurial-induced mitochondrial dysfunction and inhibition of protein synthesis was examined in synaptosomes.[15] Incubation of synaptosomes in the presence of MeHg resulted in a coincident, dose-dependent decline in ATP and protein synthesis analogous to the effect of valinomycin, a known K^+ ionophore on synaptosome protein synthesis.[16] A similar coordinate decline in energy charge and protein synthesis was observed in cerebellar perikarya in suspension.[17] Further lines of evidence for methyl mercury modification of cellular energy charge have been summarized.[3] Inhibition of protein synthesis is poorly correlated with absolute [ATP], but the activity of initiation and elongation are strongly dependent upon the ratio ADP/ATP and the accumulation of ADP and AMP.[18] ATP levels may be maintained in the presence of defective oxidative phosphorylation by phosphoryl transfer from creatine phosphate. However, phosphocreatine kinase is extremely sensitive to MeHg[3] thereby providing a major block in maintaining energy potential (see Section 16.3.3).

16.3.3 Defective Ca²⁺, Ion Homeostasis, and Neurotransmitter Release/Uptake

Methyl mercury influences ion, water, and nonelectrolyte transport. Walum[19] analyzed the efflux of [³H]-2-deoxy-D-glucose-6-phosphate from neuroblastoma cells following uptake of labeled 2-deoxy-D-glucose and found an accelerated efflux at 1 μM MeHg, which is significantly more sensitive than inorganic mercury. Astrocyte swelling was observed at 10 μM MeHg accompanied by inhibition of rubidium (K^+) and glutamate uptake.[20] Such presumed changes in K^+ influx/efflux would be expected to rapidly depolarize the membrane potential and reduce transmembrane gradients coupled and needed for glutamate uptake.

Methyl mercury stimulates the spontaneous release of acetylcholine at the neuromuscular junction[21] and from synaptosomes.[22,23] The increased spontaneous release is dependent partly upon external Ca^{2+} but occurs in the absence of external Ca^{2+}.[23] Intraterminal mitochondria contain Ca^{2+}, cycling of which is dependent upon the membrane potential, oxidative phosphorylation, ATP, and function of mitochondrial Ca^{2+} transporters.[24] Methyl mercury induces Ca^{2+} efflux from isolated mitochondria and such efflux is likely coupled to the mercurial-induced inhibition of membrane potential and oxidative phosphorylation (see above). Hence, due to either accelerated Ca^{2+} influx at the synaptosome membrane, and/or mercurial-induced Ca^{2+} release from intraterminal mitochondrial stores, the appropriate stimulus for increased transmitter release is available. Ruthenium red inhibits Ca^{2+} influx/efflux and it would be expected that the mercurial-stimulated release of acetylcholine would be inhibited in a Ca^{2+}-free medium in the presence of ruthenium red.[23]

Perturbation of intracellular Ca^{2+} homeostasis may lead to irreversible cellular degeneration (Figure 16.1) and evidence has been gathered to demonstrate MeHg disturbance of Ca^{2+} homeostasis.[3] The prior description of abnormal mitochondrial Ca^{2+} sequestration is supported by the demonstration of increased synaptosomal Ca^{2+} uptake and content,[25] Ca^{2+}-dependent activation of intracellular neuronal phospholipase A_2, and MeHg-induced increase of A23187 releasable Ca^{2+} in cerebellar granule culture.[27]

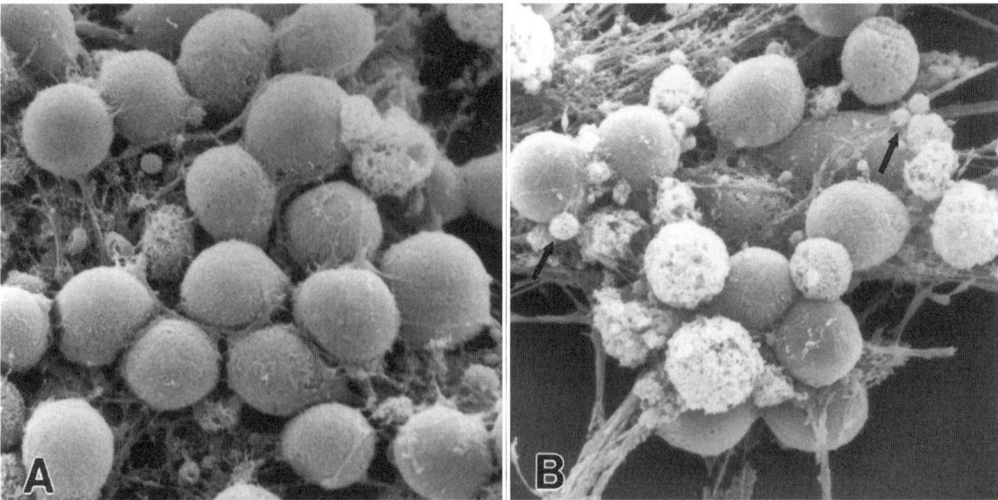

FIGURE 16.1 Scanning electron micrograph of a 14-day *in vitro* cerebellar granule neuron culture incubated with methyl mercury. (A) Aggregation of cerebellar neurons revealing intact plasma membrane and numerous neuritic processes. (B) Granule culture 24 h following incubation with 2 μM methyl mercury. Many neurons reveal multiple plasmalemmal lytic lesions while occasional membrane blebs may be seen. It is hypothesized that blebs may rupture to produce lytic lesion although localized membrane hydrolysis may result in direct membrane lysis. (× 4000.)

16.3.4 Role of Glutathione and the Initiation of Oxidative (Stress) Injury

Methyl mercury initiates membrane lipoperoxidation and free radical-induced injury in biological tissues including brain.[28,29] The brain is sensitive to oxidative injury and vitamin E/selenium has provided some protection against MeHg neurotoxicity *in vivo*.[30] Lipoperoxidation induced by MeHg in cerebellar granule cell culture was coupled to GSH reduction and decreased viability in a dose-dependent manner. However, the mercurial-induced lipoperoxidation was not the direct cause of neuron cytotoxicity since malonaldehyde production could be blocked by α-tocopherol or EDTA without significant protection.[31]

Numerous pathways exist for the induction of oxidative injury by methyl mercury. Abnormalities in signal transduction (see below) and neurotransmitter function may lead to phospholipase A_2 activation,[26] increased arachidonate production, and superoxide radical formation. Also, mercurial-GSH binding promotes depletion and failure of appropriate substrate for GSH/GSSG cycling and H_2O_2 metabolism. Thirdly, the previously discussed effect on mitochondria and electron transport allows for accelerated superoxide ion production, hydrogen peroxide formation, and ultimate accelerated lipid peroxidation.[32]

Glutathione exerts major protective effects against MeHg cytotoxicity. GSH is low in neurons compared to astrocytes and glial processes,[33] identifying a possible mechanism for neuronal sensitivity to the toxin compared to glia. We have examined this question further. Depletion of GSH followed treatment of cerebellar granule culture or glial culture with buthionine sulfoximine (BSO), an irreversible inhibitor of α-glutamyl cysteine synthetase. Although the depletion of glutathione was similar in both cultures, the neuronal culture was markedly sensitized to cytotoxicity by MeHg.[31] Low-dose MeHg incubation for 24 h induced GSH and such induction was inhibited by preincubation with BSO in the presence of MeHg. It is likely that the increase of intracellular GSH is mediated through α-glutamyl cysteine synthetase activation. Woods et al.[34] demonstrated that the increase in GSH content is due to transcriptional activation of the gene for α-glutamyl cysteine synthetase, a proximate enzyme in the pathway for GSH synthesis.

Two further studies highlight the importance of GSH in providing cellular resistance to MeHg. Miura and Clarkson[35] isolated PC12 cells resistant to the toxin. Analysis of the resistant cells revealed fourfold higher levels of intracellular GSH compared to control cells and an increased rate of MeHg efflux, presumed due to the known efflux mechanism whereby MeHg is covalently linked to GSH.[4,7] In other experiments, our laboratory selectively transfected a neuronal cell line with the anti-apoptotic gene, BCL-2. Neuronal expression of BCL-2 provided significant protection against MeHg-induced free radical production and cytotoxicity. Increased intracellular levels of GSH were observed to be two- to fourfold higher in BCL-2 cells compared to control.

While we have suggested that glutathione protects cells against MeHg-induced oxidative damage, *in vitro* culture studies do not reveal a strict correlation between GSH content and cytoprotection. It is possible that the capacity of neurons or astrocytes to synthesize glutathione in the presence of challenge may be of greater significance than the glutathione level per se.

16.3.5   Microtubules and Cytoskeletal Disaggregation

Pathological data obtained following methyl mercury poisoning of children *in utero* revealed a major effect on neuroblast genesis, dysplasia, and migration, confirmed in experimental systems.[37] Radioautographic studies suggested a disturbance in migration of proliferating ventricular neuroblasts, a decrease in the mitotic index, and scattered single-cell degeneration in the proliferating neuroblasts. An observed loss of microtubules confirmed previous studies in other cell systems in which an early effect on microtubule disaggregation was noted.[38,39] Moreover, major effects on cell cycle kinetics were observed, explained in terms of disrupted mitotic activity and modification of G_2/M phases of the cell cycle.[40]

The hypothesis identifying microtubule disorganization as a proximate event in MeHg neurotoxicity is attractive. Microtubules form the structural basis of the mitotic spindle and control cell growth, neuritogenesis, axon elongation, and cell shape. Defects in this cellular system have been associated with mercurial neurotoxicity (Figure 16.2). Studies with colchicine in neuro-derived culture systems have revealed an analogous sequence of events. For instance, inhibition of neuroblastoma differentiation is followed by increased cell sloughing without a decrease in cell viability.[38] The regulation of microtubule assembly and stability depends upon numerous factors (Figure 16.3), including an adequate supply of tubulin subunits and/or microtubule-associated proteins (MAP); adequate control of microtubule assembly/disassembly under the influence of GTP, glutathione, posttranslational phosphorylation, and available –SH groups; recognition that different isoforms of tubulin may confer specificity in function; and intracellular compartmentation and further coordinate assembly with other selected receptor molecules, e.g., plasma membrane binding sites.

Possible sites of methyl mercury/microtubule interaction are summarized in Figure 16.3. The major protein components (tubulin, MAPs, and tau) are synthesized at different rates depending upon age, neurite differentiation, and degree of tubulin polymerization. Both MeHg and colchicine are known to depolymerize microtubules and depress the rate of tubulin synthesis.[41] Moreover, tubulin mRNA levels decline in response to MeHg, identifying a selective inhibition of transcription. The process of assembly/disassembly of microtubules is strictly modulated by GTP, –SH accessibility, cellular glutathione (likely reflected in the GSH/GSSG potential), and Zn^{2+}. Methyl mercury influences all these modulators: GTP decline is secondary to reduced ATP synthesis and is also important in initiation/elongation steps of translation; methyl mercury covalently binds glutathione and other available –SH groups even on microtubules. The modification of microtubule assembly in brain extracts by Zn^{2+} was associated with alterations in –SH group content, and assembly is critically dependent upon oxidative stress.[42]

16.3.6   Phosphorylation and Posttranslational Protein Modification

Many neuronal functions are influenced by the phosphorylation/dephosphorylation state of selective proteins. Cultured cerebellar granule neurons exposed to 0.5 to 3 μM MeHg for 24 h reveal stimulation of protein phosphorylation which appears to be neuron specific.[43] Two-dimensional polyacrylamide gel electrophoresis revealed protein selectivity, with hyperphosphorylation involving

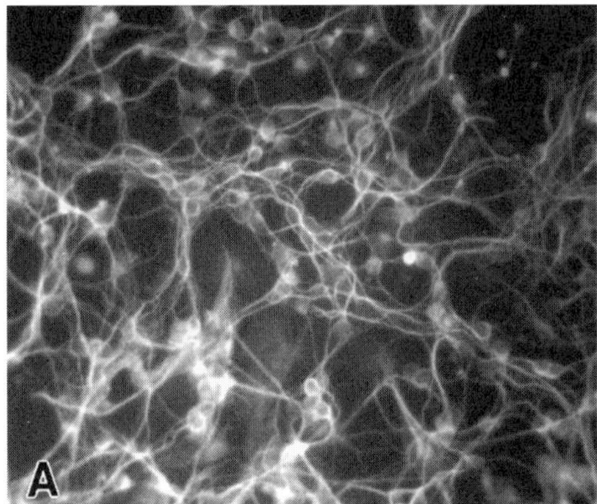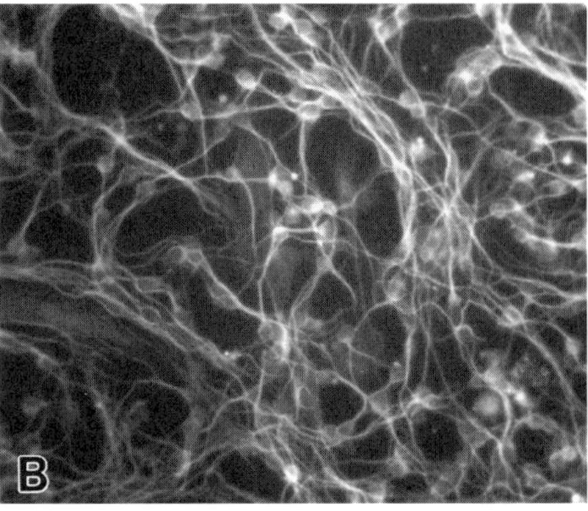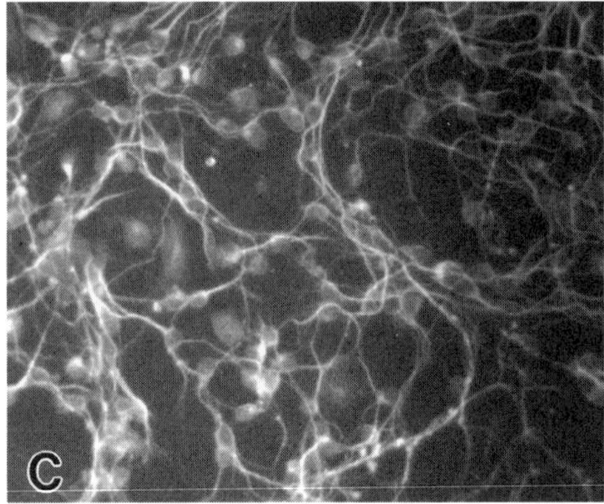

FIGURE 16.2 Effect of methyl mercury on β-tubulin immunofluorescence of a 1-day *in vitro* cerebellar granular cell culture. (A) Control revealing nonaggregated cerebellar neurons with diffuse, interconnecting single and fasciculated neurite extensions; (B) 24 h incubation with 0.2 μM methyl mercury; (C) 24 h postincubation with 0.5 μM methyl mercury. No considerable reduction in embedded-tubulin fluorescence with irregular and truncated neurite processes. At this time, no increased cytotoxicity was evident. (× 140.)

protein species 58, 68, and 75 kDa which are closely allied to tau and β-tubulin. Although unproven, these observations suggest a modification of tubulin and MAPs phosphorylation, known to powerfully influence microtubule assembly/disassembly.

Enhanced protein phosphorylation may reflect stimulation of Ca^{2+}-dependent protein kinases. We consider this likely as an increase in cytoplasmic $[Ca^{2+}]$ has been observed (see previous discussion). While *in vitro* studies did not reveal a selective protein kinase stimulation,[45] an increase in inositol phosphate was found, implying an activation of protein kinase C associated with the coproduction of diacylglycerol (DAG), known to activate PKC. The interaction of these major second messenger pathways (Ca^{2+}, calmodulin, inositol phosphate, DAG) will have profound effects on cellular integrity and differentiation, and most certainly will influence microtubule pathophysiology.

16.4 CONCLUSION

This chapter reviews some of the major mechanistic considerations accounting for mercury neurotoxicity. Certainly, no single pathway is exclusive and it is suggested that interactions and synergy between the proposed mechanisms exist. Major questions remain. For instance, why is the organic

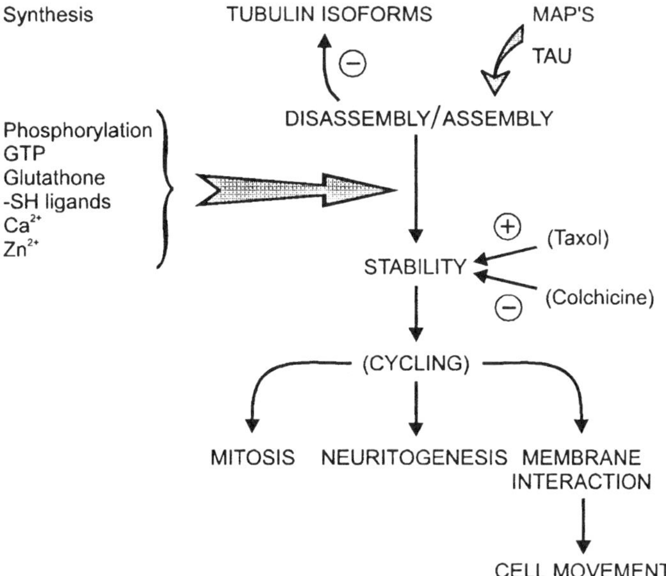

FIGURE 16.3 Schematic summary of principal steps in microtubule assembly and sites of methyl mercury interaction. Changes in synthesis of tubulin isoforms and accessory factors provide available substrate for assembly/disassembly. Disassembled tubulin exerts negative feedback on further tubulin biosynthesis. The rate of assembly and ultimate stability is modulated by GTP, intracellular glutathione, and phosphorylation of accessory proteins (which may promote inhibition of assembly). Ultimate stability may be increased by taxol or decreased by colchicine. Ultimate function of microtubules known in mitosis, neurite differentiation, membrane interaction, and cell movement. Effects in actin/microtubule-membrane interaction may promote blebbing and subsequent membrane lysis.

mercurial relatively toxic for the nervous system? Moreover, given such relative selectivity, why do certain topographical areas (calcarine cortex, cerebellar granule cells, dorsal root ganglion) manifest selective sensitivity compared to other zones. While *in utero*/neonatal neurotoxicity reveals broad sensitivity, selective vulnerability only appears once neurons have undergone significant differentiation. In the later case the development of specialized receptors, e.g., NMDA or microtubule arrays, are needed for expression of cytotoxicity. In conclusion, extracellular and intercellular events must be critically evaluated. Studies identifying glia-neuronal interaction, or failure thereof, in conjunction with the need for appropriate cell-matrix signaling are likely to provide clues for an understanding of the ultimate pathogenesis of methyl mercury neurotoxicity.

REFERENCES

1. Atchison WD and Hare MF: Mechanisms of methyl mercury neurotoxicity. *FASEB J.*, 8:622–629, 1994.
2. Chang LW and Verity MA: Mercury neurotoxicity. Effects and mechanisms. In: Chang LW and Dyer RS (Eds). *Handbook of Neurotoxicology*. Marcel Dekker, New York, pp 31–59, 1995.
3. Verity MA: Nervous system: In *Organ-Specific Metal Toxicology*. Goyer RA, Waalkes MP, and Klassen CD (Eds). Academic Press, pp 199–235.
4. Aschner M: Methyl mercury in astrocytes — What possible significance? *Neurotoxicology*, (In Press, 1996).
5. Lorscheider FL, Vimy MJ, and Summers AO: Mercury exposure from "silver" tooth fillings. Emerging evidence questions a traditional dental paradigm. *FASEB J.*, 9:504–508, 1995.
6. Aschner M and Clarkson TW: Uptake of methyl mercury in the rat brain. Effects of amino acids. *Brain Res.*, 462:31–39, 1988.
7. Naganuma A, Oda-Urano N, Tanaka T, et al: Possible role of hepatic glutathione in transport of methyl mercury into mouse kidney. *Biochem. Pharmacol.*, 37:291–296, 1988.
8. Yoshino Y, Mozai T, and Nakao K: Biochemical changes in the brain in rats poisoned with an alkyl mercuric compound, with special reference to the inhibition of protein synthesis in brain, cortex slice. *J. Neurochem.*, 13:1223–1230, 1966.
9. Cheung MK and Verity MA: Experimental methyl mercury neurotoxicity. Similar *in vivo* and *in vitro* perturbation of brain cell-free protein synthesis. *Exp. Mol. Pathol.*, 38:230–242, 1983.

10. Cheung MK and Verity MA: Experimental methyl mercury neurotoxicity. Locus of mercurial inhibition of protein synthesis *in vivo* and *in vitro*. *J. Neurochem.*, 44:1799–1803, 1985.

11. Sarafian T and Verity MA: Inhibition of RNA and protein synthesis in isolated cerebellar cells by *in vitro* and *in vivo* methyl mercury. *Neurochem. Pathol.*, 3:27–39, 1985.

12. Sarafian T and Verity MA: Mechanism of apparent transcription inhibition by methyl mercury in cerebellar neurons. *J. Neurochem.*, 47:625–631, 1986.

13. Verity MA, Brown WJ, and Cheung M: Organic mercurial encephalopathy: *in vivo* and *in vitro* effects of methyl mercury on synaptosomal respiration. *J. Neurochem.*, 25:759–766, 1975.

14. Sone N, Larsstuvold MK, and Kagawa Y: Effect of methyl mercury on phosphorylation, transport and oxidation in mammalian mitochondria. *J. Biochem.*, 82:859–868, 1977.

15. Cheung MK and Verity MA: Methyl mercury inhibition of synaptosome protein synthesis. Role of mitochondrial dysfunction. *Environ. Res.*, 24:286–298, 1981.

16. Verity MA, Cheung MK, and Brown WJ: Studies on valinomycin inhibition of synaptosome fraction protein synthesis. *Biochem. J.*, 196:25–32, 1981.

17. Sarafian TA, Cheung MK, and Verity MA: *In vitro* methyl mercury inhibition of protein synthesis in neonatal cerebellar perikarya. *Neuropathol. Appl. Neurobiol.*, 10:85–100, 1984.

18. Hucul JA, Henshaw EC, and Young DA: Nucleoside diphosphate regulation of overall rates of protein biosynthesis acting at the level of initiation. *J. Biol. Chem.*, 260:15585–15591, 1985.

19. Walum E: Membrane lesions in cultured mouse neuroblastoma cells exposed to methyl compounds. *Toxicology*, 25:67-74, 1982.

20. Aschner M, Eberle MB, Miller K, et al: Interactions of methyl mercury with rat primary astrocyte cultures. Inhibition of rubidium and glutamate uptake and induction of swelling. *Brain Res.*, 530:245–250, 1990.

21. Atchison WD: Effects of activation of Na^+ and Ca^{2+} entry on spontaneous release of acetylcholine induced by methyl mercury. *J. Pharmacol. Exp. Ther.*, 241:131–139, 1987.

22. Minnema DJ, Cooper GP, and Greenland RD: Effects of methyl mercury on neurotransmitter release from rat brain synaptosomes. *Toxicol. Appl. Pharmacol.*, 99:510–521, 1989.

23. Levesque PC, Hare MF, and Atchison WD: Inhibition of mitochondrial Ca^{2+} release diminishes the effectiveness of methyl mercury to release acetylcholine from synaptosomes. *Toxicol. Appl. Pharmacol.*, 115:11–20, 1992.

24. Nicholls DG and Akerman KEO: Biochemical approaches to the study of cytosolic calcium regulation in nerve endings. *Philos. Trans. R. Soc. London*, 296:115–122, 1981.

25. Kauppinen RA, Komulainen H, and Taipale H: Cellular mechanisms underlying the increase in cytosolic free-calcium concentrated induced by methyl mercury in cerebral cortical synaptosomes from guinea pig. *J. Pharmacol. Exp. Ther.*, 248:1248-1254, 1989.

26. Verity MA, Sarafian T, Pacifici EHK, et al: Phospholipase A_2 stimulation by methyl mercury in neuron culture. *J. Neurochem.*, 62:705–714, 1994.

27. Sarafian TA: Methyl mercury increases intracellular Ca^{2+} and inositol phosphate levels in cultured cerebellar granule neurons. *J. Neurochem.*, 61:648–657, 1993.

28. Sarafian T and Verity MA: Oxidative mechanisms underlying methyl mercury neurotoxicity. *Int. J. Dev. Neurosci.*, 9:147–153, 1991.

29. Ali SF, LeBel CP, and Bondy SC: Reactive oxygen species formation as biomarkers of methyl mercury and trimethyltin neurotoxicity. *Neurotoxicology*, 13:637–648, 1992.

30. Chang LW, Gilbert M, and Sprecher J: Modification of methyl mercury neurotoxicity by vitamin E. *Environ. Res.*, 17:356–366, 1978.

31. Verity MA and Sarafian T: Role of oxidative injury in the pathogenesis of methyl mercury neurotoxicity. In: *Advances in Mercury Toxicology*, Suzuki T, Imura N, and Clarkson TW (Eds.), Plenum Press, New York, pp 209–222, 1991.

32. Zoccarato F, Deana R, Cavallini L, et al: Generation of hydrogen peroxide by cerebral-cortex synaptosomes. Stimulation by ionomycin and plasma-membrane depolarization. *Eur. J. Biochem.*, 180:473–478, 1989.

33. Beiswanger CM, Diegmann MH, Novak RF, et al: Developmental changes in the cellular distribution of glutathione and glutathione S-transferases in the murine nervous system. *Neurotoxicology*, 16:425–440, 1995.

34. Woods JS, Davis HA, and Baer RP: Enhancement of γ-glutamyl cysteine synthetase mRNA in rat kidney by methyl mercury. *Arch. Biochem. Biophys.*, 296:350–353, 1992.

35. Miura K and Clarkson TW: Reduced methyl mercury accumulation in a methyl mercury-resistant rat pheochromocytoma PC12 cell line. *Toxicol. Appl. Pharmacol.*, 118:39–45, 1993.

36. Sarafian TA, Vartavarian L, Kane DJ, et al. Bcl-2 expression decreases methyl mercury induced tree-radical generation and cell killing in a neural cell line. *Toxicol. Lett.*, 74:149–155, 1994.

37. Choi BH: Effects of methyl mercury on neuroepithelial germinal cells in the developing telencephalic vesicles of mice. *Acta Neuropathol.*, 81:359–365, 1991.

38. Koerker RL: The cytotoxicity of methyl mercuric hydroxide and colchicine in cultured mouse neuroblastoma cells. *Toxicol. Appl. Pharmacol.*, 53:458–469, 1980.

39. Miura K, Inokawa M, and Imura N: Effects of methyl mercury and some metal ions on microtubule networks in mouse glioma cells and *in vitro* tubulin polymerization. *Toxicol. Appl. Pharmacol.*, 73:218–231, 1984.

40. Zucker RM, Elstein KH, Easterling RE, et al: Flow cytometric analysis of the mechanism of methyl mercury cytotoxicity. *Am. J. Pathol.*, 137:1187–1198, 1990.

41. Miura K and Imura N: Microtubules: A susceptible target of methyl mercury cytotoxicity. In: *Advances in Mercury Toxicology*, Suzuki T, Imura N, and Clarkson TW (Eds.), Plenum Press, New York, pp 241–253, 1991.

42. Orrenius S, McConkey DJ, and Nicotera P: Mechanisms of oxidant-induced cell damage. In: *Oxy-Radicals in Molecular Biology and Pathology*. Alan R. Liss, New York, pp 327–339, 1988.

43. Sarafian TA and Verity MA: Changes in protein phosphorylation in cultured neurons after exposure to methyl mercury. *Ann. N.Y. Acad. Sci.*, 679:65–77, 1992.

44. Hoshi M, Akiyama T, Schinohara Y, et al: Protein kinase C-catalyzed phosphorylation of the microtubule-binding domain of microtubule-associated protein 2 inhibits its ability to induce tubulin polymerization. *Eur. J. Biochem.*, 174:225–230, 1988.

45. Sarafian TA: Methyl mercury increases intracellular Ca^{2+} and inositol phosphate levels in cultured cerebellar granule cells. *J. Neurochem.*, 61:648–657, 1993.

Chapter 17

Mercury-Related Neurological Syndromes and Disorders

Louis W. Chang

CONTENTS

17.1 INTRODUCTION

Mercury, among all metals, has been widely used by humans since the dawn of civilization. Even at the present time mercury is still being used in medications, industry, and agriculture. On the other hand, mercury is also one of the most insidious toxic chemicals known.

Mercury (Hg) has an atomic weight 200.6, atomic number 80, melting point -38.9°C, boiling point 356.6°C, and a density of 13.6. In the metallic state, mercury exists as a liquid at room temperature. The silverish metallic liquid of mercury is sometimes referred to as "quicksilver".

Mercury exists as elemental (metallic) metal, inorganic mercury salts (mercurous or mercuric), and organic mercury compounds (aryl- or alkylmercury). While all forms of mercury have different toxicological properties, vapor from elemental mercury (mercury vapor) and alkylmercury compounds are most neurotoxic.

The various aspects of mercury neurotoxicity are covered in other chapters in this volume (see Chapters 2, 16, 18, and 19). The present chapter is intended to supplement and complement these chapters without undue overlaps. The prime objective of this chapter is to compare and contrast the biological impacts of various species of mercury and to present the different clinical syndromes that may be induced by different species of mercury compounds.

The general metabolism of different species of mercury and their compounds have been extensively reviewed by Berlin.[1] The essence of these excellent conceptions will be presented in this chapter. Extensive and detailed information on mercury neurotoxicity has also been reviewed recently by this author.[2,33] Readers are encouraged to seek more detailed information on this subject from these publications.

17.2 ELEMENTAL MERCURY AND MERCURY VAPOR

Metallic mercury is very volatile and vaporizes readily even at room temperature. A saturated atmosphere of mercury vapor contains approximately 20 mg/m^3 Hg at 25°C.

Once inhaled, mercury vapor is absorbed rapidly through the alveolar membrane.[3] Mercury vapor is also rapidly oxidized to mercuric mercury in the blood. The oxidation of mercury vapor (Hg^0) to mercuric ion (Hg^{2+}) can be reduced by alcohol and aminotriazole[4] and the elimination of mercury via exhalation is thus enhanced.

It has been demonstrated both in rodents and in primates that ten times more mercury is accumulated in the brain after a single exposure to mercury vapor, than after intravenous injection of a similar dose or concentration of mercuric mercury.[3] It is suggested that aside from oxidized mercuric ion (Hg^{2+}) transportation in the blood, mercury vapor (Hg^0) is also dissolved in the blood and may be released by the blood to the brain directly. Mercury vapor (Hg^0) crosses the blood-brain barrier readily and thus poses a much more neurotoxic hazard than the mercuric mercury (Hg^{2+}).

The average whole-body biological half-time of inhaled mercury is approximately 60 days.[5] The half-life of mercury in the brain is probably even longer. Berlin and coworkers[6] demonstrated that, in squirrel monkeys, inhaled mercury is accumulated in the cerebral cortex, especially the occipital and parietal cortices, cerebellar cortex, and various brainstem nuclei.

In chronic exposures, neurological changes are prominent.[2] Mercury vapor intoxication is characterized by an asthenic-vegetative syndrome of unspecific symptoms and signs such as generalized weakness, fatigue, anorexia, weight loss, gastrointestinal disturbances, and erethism (insomnia, shyness, loss of memory, personality changes, hyperexcitabililty, and depression). This syndrome has also been referred to as "micromercurialism" by Trachtenberg.[7]

In more severe cases, patients display a characteristic mercurial tremor which appears as fine muscular tremors interrupted by coarse shaking movements. It usually begins in the fingers, lips, and eyelids. Such tremor is an intentional tremor which disappears during sleep. Aside from mercurial erethism, delerium and hallucination may also occur in severe cases. Gingivitis and ptyalism are frequently associated disturbances in these patients also. ALS-like symptoms or those resembling motoneuron disease have also been reported.[8] However, this phenomenon is still controversial. Future investigation will be needed.

The allowable time-weighted average (TWA) threshold limit value (TLV) for an 8-h day and a 40-h week on mercury vapor is 0.01 mg/m^3. Occupational exposure to mercury concentrations in air above this TLV would increase health risk. Micromercurialism has not been reported at exposure concentrations below 0.01 mg/m^3.[1]

Although, on a group basis, elevated levels of mercury in blood and urine may be associated with prolonged exposure to mercury vapor, this association is not strong on an individual basis. Caution must be exercised on individual evaluation.

17.3 INORGANIC MERCURY SALTS

Mercury from inorganic mercury salts is not well absorbed in the gastrointestinal tract. It is estimated that less than 10% of ingested mercuric chloride is absorbed.[1] Skin absorption of mercury in humans has not been conclusively studied. Animal data, however, show that in 5 h time about 8% dermal absorption is expected.[1]

Mercuric ions in the blood are bound to the erythrocytes and plasma, with a slightly larger amount probably in the plasma.

17.3.1 Mercurous Mercury

Inorganic mercury salts may be divided into two main groups: the monovalent mercurous mercury and divalent mercuric mercury.

Inorganic mercury salts, especially the mercurous salts, when applied to the skin induce erythema and severe exfoliative dermatitis.[1] An interesting human episode of mercurous mercury poisoning, a specific form of hypersensitivity to mercury, has been reported in children between 4 months and 4 years of age. These children used a mercurous mercury-containing compound (calomel) as teething powder in the early 20th century.[9,10] This syndrome, referred to as "pink disease", is characterized by a general rash over the body. This condition is frequently accompanied by painful extremities (acrodynia) which is believed to be the result of stimulation of the sympathetic nervous system by mercury.[11] Other symptoms are swelling of the hands, chills, irritations on the hands, feet, and face accompanied by desquamation, hair loss, hyperkeratosis, and ulceration. Increased irritability, photophobia, insomnia, anorexia, and perspiration are also frequent complaints.[9,10]

Two cases of adult poisoning involving accidental overdose from a laxative containing mercurous chloride have also been reported.[12] At autopsy, cerebral atrophy and cerebellar granule cell loss were found.

17.3.2 Mercuric Mercury

In acute poisoning, the critical organs are the kidney and gastrointestinal tract. Gastric pain and vomiting may ensue, with abdominal pain and bloody diarrhea. This may lead to circulatory collapse and death. If the patient survives the intestinal trauma, the clinical concern will be on the kidneys. Extensive renal tubular necrosis would occur leading to renal failure, anuria, and uremia.[13]

Chronic exposure to mercuric mercury is usually a consequence of occupational exposure where mercury, e.g., mercuric nitrate or oxide, enters the body either via inhalation (dusts of mercuric salt) or via dermal contact. The best-known example in this situation is the "Mad Hatter" syndrome, with characteristic neurological disturbances similar to those seen in mercury vapor poisoning: micromercurialism, erethism, tremor, personality and neuropsychological changes, and incoordination.[13,14]

17.4 ORGANIC MERCURY COMPOUNDS

There are two main classifications of organic mercury compounds: (1) those which are unstable in the mammalian biologic system, such as arylmercury and alkoxyalkylmercury, and (2) those which are stable in the mammalian system, such as the short-chain alkylmercuric compounds.

17.4.1 Arylmercury and Alkoxyalkylmercury

This category of organic mercury can be best exemplified by phenylmercury (arylmercury) and by methoxyethylmercury (alkoxyalkylmercury). Both phenylmercury and methoxyethylmercury are absorbed more efficiently than inorganic mercuric salts in the gastrointestinal tract or via inhalation. Dermal absorption can also take place. However, little data are available for quantitative conclusions.

After absorption, phenylmercury and methoxyethylmercury compounds are rapidly decomposed, mainly in the liver, to inorganic mercuric ions (Hg^{2+}).[15] The rate of biodegradation is most rapid for methoxyethylmercury (within 24 h). For phenylmercury, biodegradation requires approximately 96 h. In contrast to inorganic mercury, 90% of phenylmercury is bound to red blood cells. Although phenylmercury and methoxymethylmercury penetrate cell membrane more readily than inorganic mercury, neither of these mercurys crosses the blood-brain barrier to a greater extent than the inorganic mercury ion (Hg^{2+}).[1,16]

Phenylmercury is primarily excreted via feces (bile in liver); the inorganic mercury ions generated from biodegradation are mainly excreted in the urine via the kidney.

Because of the rapid biotransformation of the arylmercury and alkoxyalkylmercury compounds in the biological system, the general distribution and toxicity of these two categories of organic mercury compounds are similar to those of inorganic mercury.[16]

Aside from the general encephalopathy similar to that seen in inorganic mercury poisoning, there are also some claims that phenylmercury can induce degeneration of the spinal motoneurons resembling those in ALS and motoneuron diseases.[8,17–19] However, a distributional study by Gage and Swain[20] failed to demonstrate significant mercury in the central nervous system. Other studies, both in animals and in humans, also failed to show any specific neuronal changes in the CNS.[13,14,21–24] It is interesting to note that a recent study by Arvidson,[25] using modern tracing techniques, demonstrated an accumulation of mercury (Hg^{2+}) in spinal cord and brainstem motoneurons following intramuscular injection of inorganic mercury. This observation certainly renewed the interest in possible relationship between mercury and motoneuron degeneration. Further investigations are needed to elucidate this phenomenon.

17.4.2 Alkylmercury

The best-known neurotoxic examples of alkylmercury are methylmercury and ethylmercury; both of these are short-chain organomercuric compounds. The association of methylmercury with the massive outbreak of poisonings in Japan in the 1950s and 1960s ("Minamata disease")[26] alerted the world that methylmercury is a potent environmental toxicant which, through food-chain contamination, poses a serious health hazard to humans. In the 1970s, another significant outbreak of methylmercury poisoning was reported in Iraq where the residents were exposed to mercury-contaminated grains.[27]

Aside from pollutions by humans, methylmercury may also be produced via methylation of inorganic mercury by microbiological actions.[16] In the environment, methylmercury may enter into the food chain (e.g., fish) and be consumed by humans. Once consumed, methylmercury is absorbed rapidly by the gastrointestinal tract.[1,16,28] Methylmercury is also readily absorbed via inhalation and dermal contact in industrial or occupational situations.[6,16] Most of the methylmercury in the blood is bound and carried by red blood cells. In contrast to arylmercury and inorganic mercury, alkylmercury is relatively stable in the mammalian tissues. The whole body biological half-time for methylmercury is approximately 76 days.[28]

Methylmercury is also found to incorporate in the hair follicle and growing hair. The amount of hair-mercury is proportional to blood-mercury concentration at time of mercury incorporation and hair formation. The blood:hair concentration of mercury is calculated to be 250:1.[29] Thus, hair-mercury may be used as a "marker" in the assessment of mercury exposure in a given span of time.[30,31]

Methylmercury crosses the blood brain barrier readily and brain tissues have a high affinity towards methylmercury. Although methylmercury accumulates in the brain at a slower rate than in other organs (e.g., liver and kidney), methylmercury in the brain tissues is relatively stable and resists degradation to inorganic form. With time, the level of alkylmercury in the brain would eventually accumulate at least 3 to 6 times higher than that in the blood[16] to reach the threshold for toxicity. This slow accumulation in the brain may explain the delayed signs and symptoms of intoxication several weeks after exposure. Data from the Minamata episode suggest that overt neurological signs and symptoms of alkylmercury poisoning correlate with the brain mercury level of 10 ppm or more.[32]

The clinical symptoms in alkylmercury poisoning may vary slightly in accordance to the severity of involvement and the age and sex of the patient. The primary clinical signs and symptoms in Minamata disease are visual disturbance (constriction of visual field), sensory disturbance, and cerebellar ataxia.[26] The overall clinical signs and symptoms of Minamata disease is presented in Table 17.1.

An increasing constriction of the visual field (tunnel vision), which may eventually lead to total blindness, represents a characteristic clinical symptom of methylmercury poisoning. Sensory disturbance usually starts as paresthesia and tingling sensations in the fingers and around the mouth, followed by total numbness of the extremities. Cerebellar ataxia (drunken gait) develops in almost all patients, together with tremor and general weakness of the extremities.[26]

Table 17.1 Frequency of Clinical Signs and Symptoms
in Minamata Disease

Symptom or sign	Frequency (%)
Constriction of visual fields	100
Sensory disturbance	100
Ataxia	100
Impairment of speech	88
Impairment of hearing	85
Impairment of gain	82
Tremor	76
Mental disturbance	71
Exaggerated tendon reflexes	38
Hypersalivation	24
Hyperhydrosis	24
Muscular rigidity	21
Ballism	15
Chorea	15
Pathologic reflexes	12
Athetosis	9
Contractures	9

(From Takeuchi, T. In: M. Kutsuma (ed.), *Minamata
Disease*, Study Group of Minamata Disease, Kumamoto
University, Japan, 1968.)

The topographical sites of neuropathology correlate well with neurological symptoms and mercury distribution: cerebellar cortex (ataxia), calcarine cortex (constriction of visual field), and dorsal root spinal ganglia (sensory disturbance).[21,26,33]

Primary sensory neuropathy is probably one of the most sensitive parameters in methylmercury poisoning. Chang and co-workers first demonstrated the extensive damage of the dorsal root ganglion neurons and fibers in rats after exposure to methylmercury.[2,33] These observations were later confirmed by other investigators.[34] In the dorsal root ganglia, degeneration of both dorsal root fibers and neurons occur. The earliest lesion in the fibers seems to begin at the node of Ranvier with accumulation of cellular debris and organelles. Axoplasmic and myelin degradation usually follow.[2,33]

The neuronal changes in the dorsal root ganglia begin with degranulation and disintegration of the rough endoplasmic reticulum. This change corresponds with the biochemical data, which indicate a reduction in neuronal RNA,[35] a breakdown of polysomal structure,[36] and a decrease in RNA and in protein synthesis[37–39] in these neurons after exposure to alkylmercury.

Histopathological changes in the cerebellum may serve as a characteristic diagnostic criterion for methylmercury poisoning. Cerebellar granule cell loss acquires a characteristic pattern: early intoxication involves severe cell losses at the depth of the sulci, with proliferation of Bergmann's glial fibers.[2,33] Widespread destruction of the granule cells throughout the cerebellum eventually occurs in prolonged intoxication. Most of the Purkinje neurons are spared in methylmercury intoxication.

17.5 CONCLUDING REMARKS

Mercury can exist in different forms. Although all species of mercury (inorganic, organic, etc.) are neurotoxic to some extent, the action and extent of toxicity are different. The various neurological dysfunction and pathology induced by the various species of mercury are summarized in Table 17.2. Therefore, it is inaccurate and dangerous to discuss mercury neurotoxicity without identifying the specific speciation(s) of mercury in the discussion. Intermixing information from one species of mercury to another in a discussion should be conducted with caution and authority. Risk assessment and guidelines set for mercury, therefore, should also be done with caution. Overgeneralization will simply produce more confusion for this issue.

Table 17.2 Neurotoxic Effects of Mercury

Mercury species	Primary neurological effects	Primary pathological lesions
Elemental mercury vapor	Mad Hatter's Syndrome Asthenic-vegetative syndromes Erethism and micromercurialism Intentional tremor	Cerebral gray, cerebellum, brainstem nuclei
Inorganic mercury		
Mercurous salts	Pink's Disease Acrodynia	Cerebral gray, cerebellum
Mercuric salts	May resemble those of mercury vapor (rare)	Kidneys are primary targets; Neuro-lesions, if any, may resemble those induced by mercury vapor
Organic mercury		
Aryl- and alkoxyalkylmercury	ALS-like and motor neuron disease-like syndromes (rare and unconfirmed)	Kidneys are the primary organ affected Some claimed lesions in the anterior horns of the spinal cord and motor cortex
Alkylmercury	Minamata disease with sensory disturbance, constriction of visual fields, and cerebellar ataxia	Dorsal root ganglia, calcarine cortex, and cerebellum

From Chang, L.W. and Verity, M.A., *Handbook of Neurotoxicology*, Chang, L.W. and Dyer, R.S., Eds., Marcel Dekker, New York, 1995. With permission.

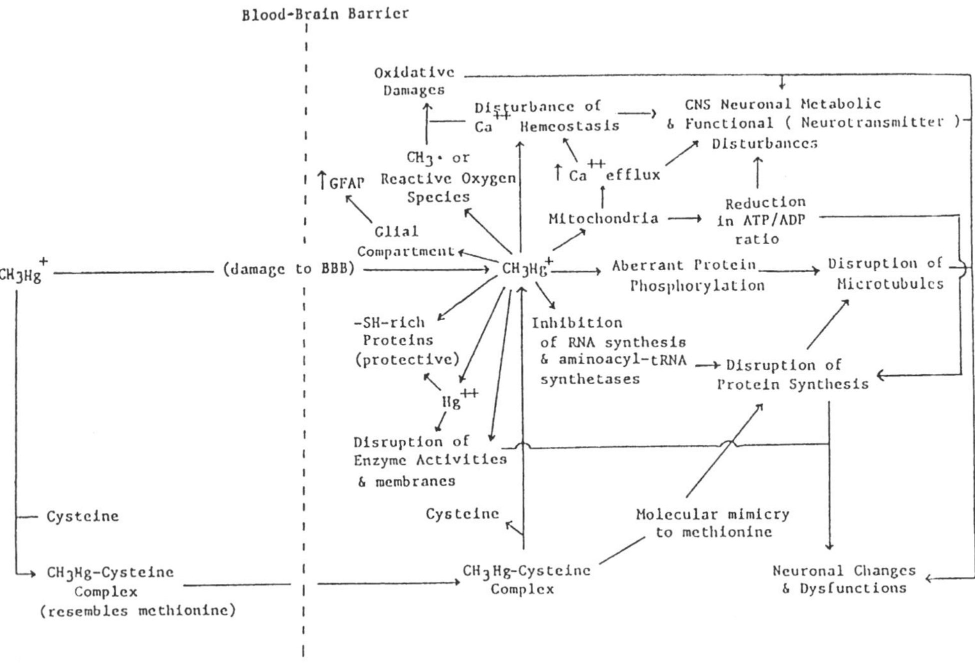

FIGURE 17.1 Neurotoxic mechanism of actions of methylmercury.

The biomolecular mechanisms of mercury neurotoxicity are perhaps one of the most important, yet complex, subjects in metal toxicology. Since a detailed discussion on this aspect has recently been published by this author[2] and Chapter 16 in this volume is also devoted to this subject, it will not be elaborated on in this chapter; however, a brief summary on the various mechanisms of action of mercury may be outlined as follows:

1. Disruption of blood-brain barrier function
2. Induction of oxidative stress and damage
3. Disturbance of calcium hemostasis and ion channels of the cells
4. Aberrant protein phosphorylation and neuronal cytoskeletal system
5. Inhibition of aminoacyl-tRNA synthetases
6. Disruption of RNA and protein syntheses
7. Disruption of cell membrane and enzyme systems
8. Interference with normal mitochondrial functions and the energy production system

These mechanisms of action are not mutually exclusive. In fact, they may occur simultaneously and have a close interrelationship to each other, as illustrated in Figure 17.1.[2]

REFERENCES

1. Berlin, M: Mercury. In: Friberg, L., Nordberg, G.F., and Vouk, V.B. (eds.), *Handbook on the Toxicology of Metals*, Vol. 2, 2nd edition, Amsterdam, Elsevier, 1968, pp. 387–445.
2. Chang LW and Verity MA: Mercury neurotoxicity: effects and mechanisms. In: Chang, L.W. and Dyer, R.S. (eds.), *Handbook of Neurotoxicology*, Marcel Dekker, New York, 1995, pp. 31–59.
3. Berlin M, Nordberg G, and Serenius F: On the site and mechanism of mercury vapor resorption in the lung. *Arch. Environ. Health*, 1969; 18:42–50.
4. Magos L, Sugata Y, and Clarkson TW: Effect of 3-amino-1,2,3-thiazole on mercury uptake by in vivo human blood samples and by whole rats. *Toxicol. Appl. Pharmacol.*, 1974; 28:267–373.
5. Cherian MG, Hursh JB, Clarkson TW, and Allen J: Radioactive mercury distribution in biological fluids and excretion in human subjects after inhalation of mercury vapor. *Arch. Environ. Health*, 1978; 33:109–114.
6. Berlin M, Carlsen J, and Norseth T: Dose-dependence of methylmercury metabolism. A study of distribution, biotransformation, and excretion in the squirrel monkey. *Arch. Environ. Health*, 1975; 30:307-313.
7. Trachtenberg, IM: The chronic action of mercury on the organism, current aspects of the problem on micromercurialism and its prophylaxis. *Zdorv'ja Kiev*, 1969; 21:7–10 (in Russian).
8. Vroom FQ and Greer M: Mercury vapor intoxication. *Brain*, 1972; 95:305–318.
9. Swift H: Erythroedema. Australasian Medical Congress, Transactions of 10th Session, 1914, Auckland, New Zealand, pp. 36–42.
10. Warkany J and Hubbard DM: Acrodynia and mercury. *J. Pediatr.*, 1953; 42:365–369.
11. Cheek D: Acrodynia. In: *Brennemann's Practice of Pediatrics*, Harper & Row, Hagerstown, N.Y., 1980, pp. 110–124.
12. Davis LE, Wands JR, Weiss SA, Price DL, and Girling EF: Central nervous system intoxication from mercurous chloride laxatives. *Arch. Neurol.*, 1974; 30:428–431.
13. WHO: Environmental Health Criteria: Inorganic Mercury. Geneva: World Health Organization, 1981.
14. Stopford, W: Industrial exposure to mercury. In: J.O. Nriagn (ed.), *The Biogeochemistry of Mercury in the Environment*, Elsevier/North Holland, Amsterdam, 1979, Chapter 15, pp. 367-397.
15. Gage JC: Mechanisms for the biodegradation of organic mercury compounds: the action of ascorbate and of soluble proteins. *Toxicol. Appl. Pharmacol.*, 1975; 32:225-238.
16. WHO: Environmental Health Criteria 101. Methylmercury. Geneva: World Health Organization. 1990.
17. Brown I: Chronic mercurialism: a cause of the clinical syndrome of amyotrophic lateral sclerosis. *A.M.A. Arch. Neurol. Psychiatr.*, 1954; 674–681.
18. Kantarjian A: A syndrome clinically resembling amyotrophic lateral sclerosis following chronic mercurialism. *Neurology*, 1964; 639–644.
19. Adams CR, Ziegler DK, and Lin JT: Mercury intoxication simulating amyotrophic lateral sclerosis. *J. Am. Med. Assoc.*, 1983; 250:642–643.
20. Gage JC and Swain AAB: The toxicity of alkyl and aryl mercury salts. *Biochem. Pharmacol.*, 1961; 8:77 (Abst. No. 250).
21. Currier RD and Haerer AF: Amyotrophic lateral sclerosis and metallic toxins. *Arch. Environ. Health*, 1968; 17:712–719.
22. Conradi S, Ronnevi D, and Norris F: Motor neuron disease and toxic metals. In: Lewis P. Roland (ed.), *Human Motor Neuron Diseases*, Raven Press, New York, 1982, pp. 201–231.
23. Roberts MC, Seawright AA, and Ng JC: Chronic phenylmercuric acetate toxicity in a horse. *Vet. Hum. Toxicol.*, 1979; 25:321–327.
24. Yanagihara R: Heavy metals and essential minerals in motor neuron disease. In: L.P. Rowland (ed.), *Human Motor Neuron Diseases*, Raven Press, New York, 1982, pp. 233–247.
25. Arvidson B: Inorganic mercury is transported from muscular nerve terminals to spinal and brainstem motor neurons. *Muscle Nerve*, 1992; 15:1089–1094.

26. Takeuchi T: Pathology of Minamata disease. In: M. Kutsuma (ed.), *Minamata Disease*, Study Group of Minamata Disease, Kumamoto University, Japan, 1968, pp. 141–228.

27. Bakir J, Damluji SF, Amin-Zaki L, and Murtadha M, et al: Methylmercury poisoning in Iraq. *Science*, 1973; 181:230–241.

28. Miettinen, JK: Absorption and elimination of dietary mercury (Hg^{++}) and methylmercury in man. In: M.W. Miller and T.W Clarkson (eds.), *Mercurials and Mercaptans. Mercury*, Charles C Thomas, Springfield, IL, 1973, pp. 233–243.

29. Skerfving S: Methylmercury exposure mercury levels in blood and hair and health status in Swedes consuming contaminated fish. *Toxicology*, 1974; 2:3–23.

30. Subramanian R: Metals in hair as an indicator for metal burden of the body. In: H.K. Dillon and M.H. Ho (eds.), *Biological Monitoring of Exposure to Chemicals — Metals*, John Wiley & Sons, New York, 1991, Chapter 20, pp. 255–261.

31. Katz SA and Katz RB: Use of hair analysis for evaluating mercury intoxication of the human body: a review. *J. Appl. Toxicol.*, 1992; 12:79-84.

32. Berlin M: Dose-response relations and diagnostic indices of mercury concentrations in critical organs upon exposure to mercury and mercurials. In: G.F. Nordberg (ed.), *Effects and Dose Response Relationships of Toxic Metals*, 1976, pp. 235–245.

33. Chang LW: Toxico-neurology and neuropathology induced by metals. In: L.W. Chang (ed.), *Toxicology of Metals*, CRC Press/Lewis Publisher, FL, 1996 (in press).

34. Herman SP, Klein R, Talley FA, and Krigman MR: An ultrastructural study of methylmercury-induced primary sensory neuropathy in rats. *Lab. Invest.*, 1973; 28:104–118.

35. Chang LW, Desnoyers PA, and Hartmann HA: Quantitative cytochemical studies of RNA in experimental mercury poisoning. I. Changes in RNA content. *J. Neuropathol. Exp. Neurol.*, 1972; 32:489–501.

36. Sugano H, Omata S, and Tsubaki H: Methylmercury inhibition of protein synthesis in brain tissue. I. Effects of methylmercury and heavy metals on cell-free protein synthesis in rat brain and liver. In: Studies on the Health Effects of Alkylmercury in Japan. Environmental Agency, Tokyo, Japan, 1975, pp. 129–136.

37. Chang LW, Martin AH, and Hartmann HA: Quantitative autoradiographic study on the RNA synthesis in the neurons after mercury intoxication. *Exp. Neurol.*, 1972; 37:62–67.

38. Yoshino Y, Mozai T, and Nakao K: Biochemical changes in the brain in rats poisoned with an alkyl mercuric compound, with special reference to the inhibition of protein synthesis in brain cortex slice. *J. Neurochem.*, 1966; 13:1223–1230.

39. Cavanagh JB and Chen FCK: Amino acid incorporation in protein during the "silent phase" before organo-mercury and p-bromophenylacetylurea neuropathy in the rat. *Acta Neuropathol. (Berlin)*, 1971; 19:216–224.

Chapter 18

Neurotoxicity of Methylmercury; Minamata and the Amazon

Masazumi Harada

CONTENTS

18.1 INTRODUCTION

The incident of Minamata, which had been subjected to the world's largest mercury poisoning, and the Amazon, where poisoning is still in progress although no patients have currently been reported, are discussed.

Although much has been known since ancient times about the toxicity of inorganic mercury (Hg), it was in the beginning of the 19th century that the clinical picture of inorganic Hg poisoning was elucidated.[1] On the other hand, the history of organic Hg poisoning dates from relatively recent years. In the 1860s, a report was published on the toxic effects of organic Hg on the central nervous system. In 1910, its efficacy for germination and sterilization of plants became known and the knowledge triggered its large-scale production in Europe as an agricultural chemical. With the increase of production, the incidence of organic Hg poisoning increased.[2] A case of poisoning reported by Hunter and colleagues in 1940 is quite well known (Hunter-Russell syndrome).[3,4] They described in detail occupational poisoning by methylmercury (MeHg) which occurred in a plant for producing MeHg as an agrochemical. And yet it was only after the occurrence of Minamata Disease (M. D.) that neurotoxicity of MeHg came to be known to the world.[5] Research on MeHg poisoning, technical measures for prevention, and the legal regulations have made remarkable progress even though there still remain many facts that need to be clarified.

18.2 MECHANISM OF OCCURRENCE AND BACKGROUND OF MINAMATA DISEASE

M. D. was first discovered in 1956 in Minamata, Kumamoto Prefecture, Japan followed by the cases discovered in 1965 along the Agano River in Niigata Prefecture (second Minamata Disease).[5,6] M. D. is an indirect poisoning (through the food chain) that occurred in humans who ingested fishes and shellfish contaminated by methylmercury and is distinguished from methylmercury poisoning which had developed as a result of direct exposure.[7] The former is a disease occurring as a result of exposure to methylmercury through foods for an extended period of time while the latter is an acute incident which develops as a result of massive exposure during a relatively short period of time.

In Canada[8] and Brazil[9] there are currently problems of methylation of inorganic Hg which takes place in a natural environment. In the case of M. D., inorganic Hg which was used as a catalyst in the industrial production of acetaldehyde turned into MeHg and contaminated the environment. It had been known in certain circles that MeHg was formed during acetaldehyde production as evidenced by the papers of Professor Zangger of Zurich University (1930)[10] and Professor Koelish of Munich University (1937).[11]

Based on the estimated Hg intake and Hg concentrations in fishes and shellfish ingested by M. D. patients in Kumamoto, Niigata, and Iraq, the amount of accumulation in the human body which would trigger manifestation of typical symptoms of MeHg poisoning is calculated to be 100 mg in terms of MeHg and the amount to trigger the minimum symptoms (sensory disturbance) is 25 mg.[12]

As the biological half life of MeHg is said to be 39 to 70 days, the provisional tolerable weekly intake (PTWI) of MeHg for an adult (body weight, 50 kg) is estimated to be 0.17 mg.[13] However, these figures are calculated from the data involving typical cases of acute poisoning and should be regarded as a mere model. A theory postulates that the biological half life is 180 to 288 days, and these figures are inappropriate for considering chronic M. D. or long-term trace amount exposure.[14]

Nevertheless, 200 mg/l of mercury in blood and 50 mg/g in hair are the provisionally established standards for judgment, and anyone with higher concentrations is considered to be at a risk of poisoning.[13]

Based on the above, the allowable level for Hg in fish and shellfish was set at 0.5 ppm, but the figure was revised to 0.4 ppm (0.3 ppm in terms of MeHg) in 1973 in view of the more than average ingestion of fish by the Japanese. PTWI of MeHg for an adult is set at 200 mg/kg) (WHO).[13]

The reality is not so simple. In Minamata, mercury discharge was started in 1932 and lasted for a long time. At the peak of its contamination (1956), a 4-kg cat developed the disease in 30 to 60 days and died. Even after the discharge was completely stopped in 1968, contamination continued until quite recently.[5] More than 100 ppm of mercury was detected in mud samples taken extensively from Minamata Bay in 1974, and there is a ban on catching some fishes because they contain mercury beyond the allowable level. This means that the inhabitants have been exposed to varying degrees of contamination for an extended period of time. Along the coast of the Shiranui Sea, 200,000 people lived in the area where cats had died frenzied death in 1960 (Figure 18.1). Most of them were engaged in fishery and fishery-related industries and their fish and shellfish intake was far greater than that of the average Japanese (108.9 g/day maximum); about 400 g/day for a household head, 165 g/day for a housewife, and 221 g for a child.[15]

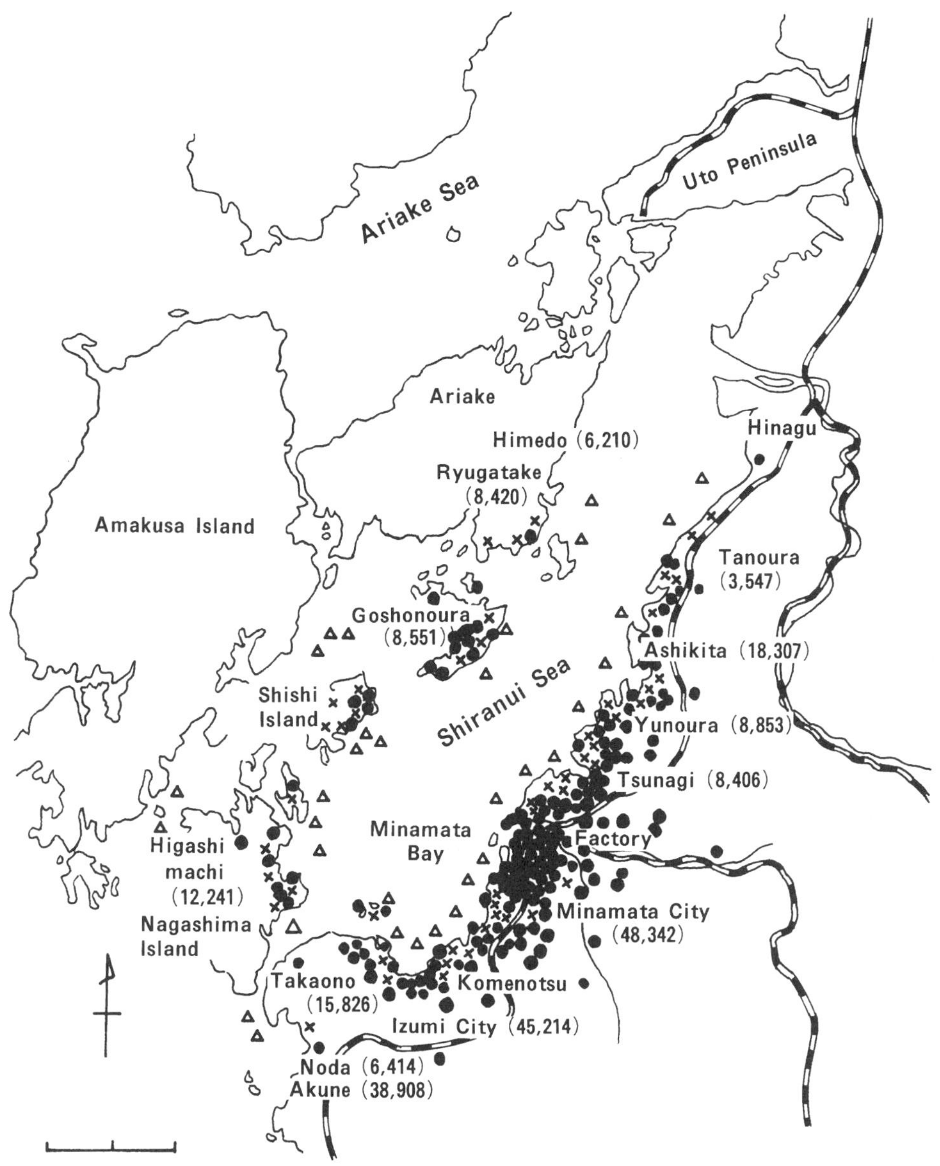

FIGURE 18.1 Map of the Shiranui Sea. (From Harada M: *Crit. Rev. Toxicol.*, 1995; 25: 1–24. With permission.)

The whole picture of M.D. is still not sufficiently clear because the cause had not been known in early years and because of gross negligence on the part of the administration in later years to perform strict health screenings of the inhabitants and environmental surveys.

18.3 CLINICAL SYMPTOMS OF MINAMATA DISEASE

18.3.1 Typical Minamata Disease

MeHg mainly damages the central nervous system. A typical case of acute poisoning develops symptoms called Hunter-Russell syndrome, consisting of sensory disturbances, ataxia, dysarthria, auditory disturbance, and constriction of the visual field. Tokuomi reported that 100% of M. D. patients in early years manifested superficial sensory disturbances, 100% constriction of the visual field, 93.5% ataxia, 88.2% dysarthria, 85.3% auditory disturbance, and 75.8% tremor. That these patients were seriously ill is clear from the fact that as many as 82.4% developed disturbance of gait (1960).[5,12]

Early signs and symptoms of the Second Minamata Disease (in Niigata) were similar to those described above, but frequently differed; 92% for superficial sensory disturbances, 33% for constriction of the visual field, 41% for ataxia, 35% for dysarthria, 69% for auditory disturbance, 26% for tremor, and 31% for disturbance of gait (1965).[6,12]

Differences in the frequencies of M. D. symptoms are not related to quality but to the method employed in picking the patients. In Minamata where the etiology was unknown, only severely ill patients with the typical symptoms of M. D. were picked. In Niigata, those who had manifested such symptoms were picked up in the health screening in the communities considered to have been contaminated with organic Hg, and those whose hair Hg level exceeded 50 ppm were regarded as M. D. patients.[6,12]

18.3.2 Subjective Symptoms

Subjective symptoms were frequently seen in those who were mildly and slightly ill, and more so in those in the chronic stage of illness than in the acute stage.[16] These symptoms were varied, diverse, and protracted. Despite their diversity, however, the symptoms could be classified into several types: those related to sensory disturbances, those related to ataxia, and those related to vision, visual field, and hearing. There are many other subjective symptoms related to autonomic disturbances, vesticular disorders, and circulatory disorders which are seen commonly in the patients in Minamata and Niigata.[6,16] They include headache, pains in the joints, muscle pains, forgetfulness, vertigo and dizziness, trembling, convulsion, fainting, muscular contracture, sense of fatigue, and susceptibility to cold.

18.3.3 Sensory Disturbances

Sensory disturbances occur around the mouth, and the glove and stocking type occurs at the limb extremities as in polyneuropathy. Hemilateral sensory disturbances are sometimes observed, but they rarely occur as isolated symptoms.[5,7,15] These types of sensory disturbances are considered as peripheral nervous system damages. In fact, the sensory fibers of the peripheral nerves are selectively damaged,[14] but the author suspects these damages to be those of CNS for the following reasons: tendon reflex is accelerated in many cases, the symptoms easily change, there is difficulty in discerning the distance between two points, the ability to discern shape, texture, etc. of a thing is impaired, and it is difficult to demonstrate the delay in neurotransmission. Moreover, these types of sensory disturbances are considered to be the primary symptoms observed in an adult. Similar types of sensory disturbances are observed in the inhabitants of Indian reservations in Canada[8] and in a district in China[17] contaminated by MeHg. Disturbances in taste and smell are also observed.

18.3.4 Other Manifestations

Ataxia is a typical disorder of the cerebellum, and so is dysarthria. A typical case is confirmed by ataxic gait, adiadochokinesis, the results of the finger-nose test and the knee-heel test, and intentional tremor, although these signs in chronic or mild cases are often perceived as awkwardness and slowness in motion. Where typical ataxia cannot be observed, tests should be performed to detect abnormalities in electroophthalmography (EOG) and optokinetic nystagmus pattern (OKP).[7,15]

Auditory disturbances are frequently observed and they are characterized by neurological hypacusia, particularly posterior labyrinthine (cortical) deafness. A typical complaint is "I can hear the voice but cannot understand what is being said if spoken rapidly."[7]

Constriction of the visual field is manifested in such a late stage of the disease that its presence is relied on to establish the diagnosis.[5,7,12,15] In other words, the visual constriction is low in incidence and is likely to change easily. Depression of the visual field which precedes constriction may be useful for establishing the diagnosis. Also seen are muscular atrophy, hemilateral symptoms, and spinal symptoms in some patients.[6,7]

18.4 PATHOLOGICAL FINDINGS

18.4.1 Acute and Subacute Minamata Disease

Pathological findings were also marked with common features:[14,18] damage to the central nervous system, particularly the cerebral cortex and cerebellar cortex, was notable. In addition, the disease was further characterized by notable damage to the calcarine region of the occipital lobe (visual center), the pre- and postcentral cortex (motor and sensory centers), and temporal cortex (auditory center). In relation to the cerebellar cortex, deciduation of granular cells was remarkable while Purkinje cells were relatively well retained, showing a granular cell-type cerebellar atrophy. As for the peripheral nerve system, destruction and demyelination of corsal roots or sensory nerve fibers were characteristically notable.

18.4.2 Congenital Minamata Disease

The pathological findings of congenital M. D. are general atrophy and hypoplasia of the brain cortex and abnormality of the cytoarchitecture, remaining matrix cells, hypoplasia of the corpus callosum, intramedullary preservation of the nerve cells, and dysmyelination of the pyramidal tract. In the cerebellum, hypoplasia of the granular cell layer and other layers as well as degeneration of granular cells (characteristic of Minamata Disease) were also observed.[18,19,22]

18.4.3 Chronic Minamata Disease

While degrees vary greatly, the symptoms of chronic M. D. are essentially the same as those for acute and typical M. D. The pathological findings of chronic cases of M. D. are characterized by lesions (1) in the cerebral cortex, (2) cerebellar cortex, and (3) changes in peripheral nerve fibers.[14,15,20] Moreover, there are extreme variations in the degree of brain damage suffered, consequently resulting in a great variety of clinical symptoms of chronic M. D.[15]

There were inhabitants of the contaminated areas and family members of the patients who manifested one or more of the lesions, but not all three. For instance, some suffered only from disorders of peripheral nerve fibers, while others had no lesions in the cerebral cortex. These findings are not pathologically determined as M. D., but should be regarded as important in considering the total effect of environmental contamination on human bodies.

18.5 CONGENITAL MINAMATA DISEASE

18.5.1 Clinical Symptoms

A major toxic characteristic of organic Hg is that the substance easily passes through the placenta and damages the brain of the fetus.[5] It became clear in or around 1960 that many children developed signs of infantile cerebral palsy in the highly affected areas. These signs were congenital, and yet no child had ingested contaminated fish. From early days the relation with M. D. was suspected, but 3 to 6 years had passed since their birth and no Hg analysis of the hair of mothers or infants at the time of childbirth had been performed. Yet, the incidence was abnormally high. Between 1955 and 1958, 14 children, or 7.5%, were found with cerebral palsy among 188 babies born in the three districts with high incidences of M. D. When three more children with an incomplete form of the disease are added to this group, the ratio becomes 9.0%. The incidence of cerebral palsy among the general population in Japan is 0.02 to 0.23%, and is 0.59% at the maximum.[5,21–23] According to the author's investigation, to date there have been as many as 64 children born with the disease.[15] In the early stage of illness, their signs and symptoms were identical, indicating that they had the identical disease. Mental retardation (100%), primitive reflexes (100%), strabismus (77%), cerebellar ataxia (100%), dysarthria (100%), chorea and athetosis (95%), deformed limbs (100%), hypersalivation (95%), epileptic attacks (82%), and growth disorders (100%) were observed at high incidences, indicating that the disease is due to grave and extensive damage of the CNS.[19,22]

18.5.2 Epidemiology

Patients with congenital disorders were born between 1952 and 1963 in the fishing villages along the coast of the Shiranui Sea. Thus, there was concurrence in the districts and timing of occurrences of M. D. and congenital cerebral palsy.

In addition, 64% of their families had acute M. D. and 100% had chronic M. D. Apparently, asymptomatic mothers of the patients were confirmed to be manifesting mild symptoms of M. D. upon detailed examinations. All the mothers had sensory disturbances of the extremities of the limbs, 76% slight ataxia, 48% visual constriction, 76% auditory disturbances, and 43% dysarthria. And yet, the mothers' symptoms were milder than those of their children or any other members of the family.[15,22]

As mentioned before, the Hg content in their hair at birth is not known. Fortunately, there is an old custom in Japan to preserve the umbilical cord of the newborn. These cords were analyzed for MeHg and showed high values in those born between 1953 and 1965, showing that acetaldehyde production in the Chisso Plant and MeHg levels in the umbilical cords of the newborn were clearly in parallel. The level was higher than 1.0 ppm for those diagnosed as congenital M. D. patients.[22,23]

18.5.3 Experiments and Etiology

Experiments also demonstrated that MeHg passed through the placenta into the brain of the fetus. Out of 64 congenital M. D. patients, 13 severely ill patients who had been diagnosed as such have died. The pathological examinations of the brains of six autopsied patients included the characteristic findings of MeHg poisoning and those suggesting damage inflicted in the latter half of the fetal period. These were diagnosed to be from MeHg poisoning which had occurred in the latter half of the fetal period in pathological terms.[18,19]

These patients were so severely ill pathologically and clinically that chances are many mild or atypical cases have been overlooked. In 1970, we observed 29.1% of children with mental retardation accompanying inferior movements in the heavily contaminated area between 1955 and 1958, 17.5% with mental disorders, 21% with sensory disturbances, 12% with vagueness in speech, and 9% with inferior movements such as the inability to stand on one foot in a primary school in the contaminated district which points out the actual conditions of mild cases of poisoning caused by exposure to MeHg during the fetal stage.[15,22]

18.6 CHRONIC MINAMATA DISEASE

18.6.1 Background

As about 200,000 persons have been exposed to varying degrees of MeHg (see Figure 18.1), the health effects are also considered to differ from those of the severe vegetable state and those with the Hunter-Russell syndrome to atypical or moderate cases from the beginning. In early years, however, atypical or moderate cases were excluded from diagnosis of M. D. Up to 1970, only 121 persons were officially recognized as M. D. patients (23 were congenital M. D. patients).

As environmental contamination in the Shiranui Sea reached a peak around 1960 and ingestion of fishes and shellfish had also reached a peak before 1960, typical and severe cases occurred before 1960. Although contamination continued, fisheries recovered gradually from around 1965. After all the processes requiring use of Hg were ceased in 1968, contamination gradually decreased. And yet in 1978, higher than 100 ppm (250 to 1000 ppm) total Hg was detected from sludges in the bay, while in May 1971 Hg in fish and shellfish was also elevated beyond the tolerable level, leading to a ban on fishing in the bay.[25]

Against such a background, M. D. progressed gradually in a large number of patients. There are currently more than 2200 persons who have been officially recognized as M. D. patients (1995); but more than 10,000 are suspected of the disease. The author distinguishes these patients from those of the initial stage by calling them chronic M. D. patients.[7,15]

18.6.2 Characteristics

Chronic M. D. is first characterized by a very gradual progress (as discussed above) of 5 to 10 years. The progress is also varied, and may be classified into "gradually progressive type", "delayed onset type", or "escalatingly progressive type". Secondly, the disease is characterized by diverse and varied symptoms. As one symptom is particularly pronounced while others are latent, the patient may appear to suffer from an entirely different disease. Thirdly, patients develop new symptoms or the disease progresses for a certain period of time even if exposure to MeHg is ceased or decreased.

The author has performed clinical tests on several thousand people with health disorders who used to ingest large quantities of contaminated fish from the Shiranui Sea, or whose family members had M. D. Most of these people applied for official recognition of their M. D. with the Committee for Official Recognition of M. D. Some were recognized officially and others had their applications rejected. The committee's criteria for diagnosing M. D. include presence of the sensory disturbances and either cerebellar ataxia or visual constriction. This is why many of the patients are not officially recognized. Apart from the official recognition, the inhabitants of the contaminated areas do show varying degrees of symptoms.

18.6.3 Clinical Symptoms

We have sorted out the data of 2383 persons who show various symptoms. Those with the sensory disturbances of the glove and stocking type account for an overwhelming number of people (64.6%) (Figure 18.2), followed by auditory disturbance (49.2%), incoordination (33.6%), muscular weakness (35.3%), tremor (27.7%), dysarthria (22.1%), and constriction of the visual field (16.7%).[15]

Of 296 workers at Chisso, those with sensory disturbances of the glove and stocking type account for 27.0% (Figure 18.2), those with auditory disturbance for 53.3%, incoordination for 6.4%, and dysarthria for 3.7%.

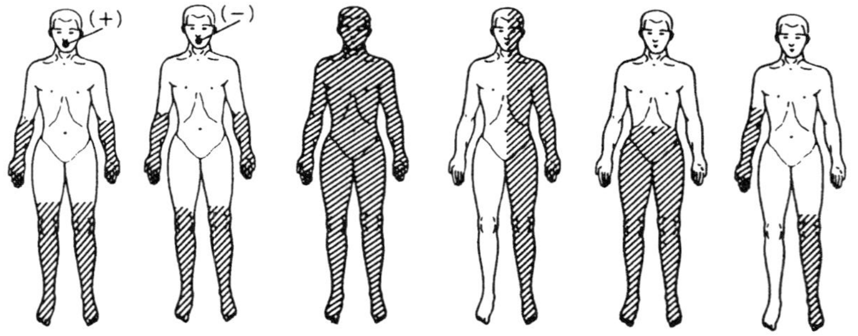

Types / Objects	Numbers of case	Glove & stocking +Perioral	Generalized	Hemiplegic	Vertebral	Irregular & uncertain	Reference
Mothers of Cong. M. d.	28	6 — 22, 28 (100)	0	1 (3.5)	0	1 (3.5)	Harada. M. (1974)
Family members of acute M. d.	145	44 — 65, 109 (75.1)	0	11 (7.5)	4 (2.7)	4 (2.7)	Harada. M. (1972)
Residents of Minamata areas	928	70 — 145, 215 (23.1)	14 (1.5)	51 (5.4)	21 (2.2)	16 (1.7)	Tatetsu. S. (1973)
Harada's material	2383	147 — 1393, 1540 (64.6)	184 (7.7)	199 (8.3)	84 (3.5)	196 (8.2)	Harada. M. (1992)
Laborers of Chisso	296	9 — 71, 80 (27.0)	2 (0.6)	7 (2.3)	8 (2.7)	31 (10.4)	Harada. M. (1990)
Canada Indian	89	2 — 13, 15 (16.8)	4 (4.4)	3 (3.3)	0	16 (17.9)	Harada. M. (1977)

※Some cases overlapped each other types.
( ) are %.

FIGURE 18.2 Types of Sensory Disturbances.

The issue of what is the minimum M. D. symptom is replaced with that of to what degree of patients should compensation be paid. The issue of the least minimum symptom of M. D. is bound to keep its importance in the future issues of Hg poisoning.

18.7 MERCURY POLLUTION IN THE AMAZON

18.7.1 Mercury Pollution in the Amazon Basin

The area upstream of the Amazon River in Brazil has long been known for its gold mining. As a large quantity of Hg (metal) is used in gold mining, environmental contamination by Hg is posing a grave problem.

It is said that 2 to 10 g of mercury is needed to obtain 1 g of gold. There is a report that 1800 to 2000 tons of Hg were used during a recent 8-yr period in the entire Amazon region, although the precise amount of Hg used is not known because the amount of gold mined is also not known. In February 1989 the Brazilian Government enforced the law banning use of Hg, but we have failed to see the actual effects.[9,26]

There are five stages from environmental contamination to the onset of M. D. In the first stage, there occurs inorganic Hg poisoning by direct exposure to Hg used (either via the respiratory tract as vapor or through the skin by contact). In the second stage, Hg discharged into the environment contaminates the air, soil, and water and becomes organic by methylation in the natural world. In the third stage, organic Hg is taken in by fishes and shellfish and is concentrated through the food chain. In the fourth stage, people accumulate mercury in their body as they eat fish and shellfish contaminated by Hg. In the fifth stage, M. D. develops.

18.7.2 Mercury Content in Hair

The author confirmed by on-site survey that many of the gold miners have developed severe cases of inorganic Hg poisoning. The Hg level in the workers' hair is high; maximum 113.2 ppm and mean 22.2 ppm in one place, and 1.6 to 82.6 ppm and mean 12.3 ppm in another; 80% of the Hg was inorganic Hg.[9]

The process of methylation in the second stage is not fully clear. There are reports, however, that Hg is detected in the river water or the mud of the river bed. It is therefore clear that there is localized Hg contamination in the waters of Amazon as demonstrated by the steady elevation of Hg levels in fishes and shellfish. According to a report, the maximum level is 5.9 mg/g in the Tocatinse River, 2.57 mg/g in the Tapajos River, and 1.47 to 2.7 mg/g in the Madiras River. It has been confirmed further that 69.1 to 121.6% of Hg in fish is MeHg.[9]

The Hg content in the fishermen's hair in the fishing district downstream of the mine is rising. Our investigation revealed maximum levels of 31.1 ppm (mean 15.8 ± 8.9) in people's hair in Rainha Village, 42.6 ppm (mean 14.1 ± 10.3) in Brasilia Legal Village, and 12.6 ppm (mean 6.1 ± 12.6) in Pont de Pedras Village in the Tapajos Basin. In July (dry season) 1992, we observed the highest levels at Brasilia Legal Village with the maximum total Hg level of 151.2 ppm (mean 35.9 ± 36.8). These values are definitely high. And more than 80% (84.9 to 94.1%) of Hg was MeHg (Figure 18.3). None of the people from whom these hair samples had been taken showed symptoms which indicated M. D. In other words, the area has already reached the fourth stage. When it enters the fifth stage, there will arise the issue "What are the minimum symptoms of M. D."? Continued observation is warranted.[9,26]

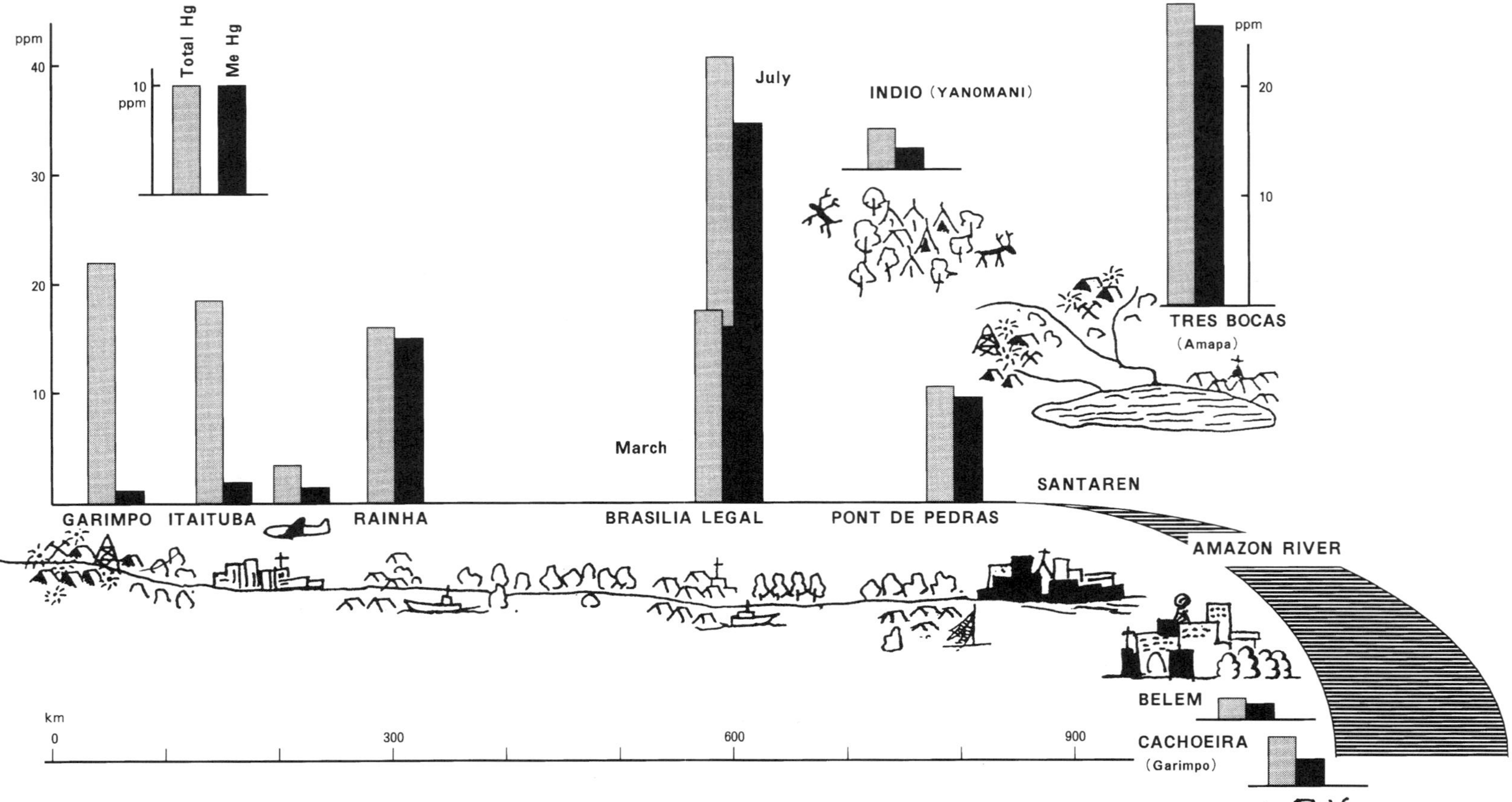

FIGURE 18.3 Total Hg and MeHg Concentration in Hair — Tapajos River Basin.

18.8 SAFETY OF MERCURY

18.8.1 Safety Level in Pregnancy

IPCS proposed the following challenge in 1990: "The current provisional safety level of 50 ppm for Hg in hair should be reviewed in the light of safety to fetuses as well."[13] This level of 50 ppm was based on the adult patients in relatively acute onset of the disease. It is necessary to examine if this value is applicable to fetuses as well as to chronic or mild M. D., since the effects on fetuses and the disease with the minimum signs and symptoms will be crucial problems in future. The cases in Canada,[8] China,[17] and Brazil[9,26] are the forwarning signs.

IPCS raised the question of fetotoxicity because of three reports, all of which discussed the relation between the Hg levels in the hair of pregnant mothers and the brain development of their children by clinical epidemiological surveys. The minimum Hg levels in mothers whose children showed the effect of Hg were 14 to 18 ppm according to Marsh et al.,[27] 13.0 to 23.9 ppm according to McKeown-Eyssen,[28] and 13 to 15 ppm and a maximum of 25 ppm according to Kjellström.[29] The MeHg level in the preserved umbilical cords of congenital M. D. children was approximately 1.0 ppm or higher.[22,23] According to Dalgaard et al., the relation between the MeHg level in the umbilical cord and the Hg level in maternal hair can be expressed by the following equation:[30]

$$Hg(ppm) \text{ in maternal hair} = 19.5 \times Hg(ppm/dry\ weight) \text{ in umbilical cord} + 17.9 \qquad (1)$$

When this equation is used to calculate the levels for the aforementioned patients, the minimum Hg level in the hair of a mother who gave birth to a congenital M. D. child would be 21.5 ppm.

As suggested by IPCS, the safety standard for pregnant women should perhaps be lower than that for the general public.

18.8.2 Consideration of Safety

One theory to explain the mechanism of onset of poisoning, including Hg poisoning, is accumulation. That is, there is a prescribed threshold value in a body and unless the level of a toxic substance exceeds this threshold, no poisoning is manifested. According to this theory, safety can be assured by setting the daily intake allowance to maintain the accumulation below the minimum amount to manifest the symptoms considering the biological half life. However, the theory cannot explain chronic M. D. such as delayed M. D. and effects on health by long-term trace contamination. Assuming that there is a prescribed threshold value for onset of the symptoms, individual cells have no such threshold values. It is easier to explain the mechanism of the onset of chronic M. D. by saying that a certain number of cells should be damaged before any clinical symptoms become manifest, since even a trace amount of Hg does affect the body.[7] At any rate, the current safety standard should be fully reviewed of the safety of fetuses if a relatively small degree of contamination continues for an extended period of time. This should carefully be pursued in the case of Brazil.[26]

REFERENCES

1. Cohen MM and Flora GC: Cerebral intoxication. In Baker AB and Baker LH (eds): *Clinical Neurology.* London, Harper and Row, 1971, Vol.2, Chap. 20, pp 1–66.
2. Lundgren KD and Swensson A: Occupational poisoning by alkyl mercury compounds. *J. Ind. Hyg. Toxicol.*, 1949; 31:190–200.
3. Hunter D, Bomford RR, and Russell DS: Poisoning by methyl mercury compounds. *Q. Med. J.*, 1940; 9:193–213.
4. Hunter D and Russell DS: Focal cerebral and cerebellar atrophy in a human subject due to organic mercury compounds. *J. Neurol. Neurosurg. Psychiatr.*, 1954; 17: 235–241.
5. Minamata Disease Research Group: Minamata Disease. Medical School of Kumamoto University, Kumamoto, Japan, 1968.

6. Tsubaki T and Irukayama K: Minamata Disease: Methylmercury poisoning in Minamata and Niigata, Japan. Amsterdam, Elsevier, 1977.

7. Harada M: Minamata disease: Organic mercury poisoning caused by ingestion of contaminated fish. In Patrice Jellife EF and Jellife DB (eds): *Adverse Effects of Food*. New York, Plenum Press, 1982, pp 135–148.

8. Harada M, Fujino T, Akagi T, and Nishigaki S: Mercury contamination in human hair at Indian reserves in Canada. *Kumamoto Med. J.*, 1977; 30:57–64.

9. Harada M: Preliminary field survey in Tapajos River basin, Amazon. In Proc. Int. Symp. "Assessment of Environmental Pollution and Health Effects from Methylmercury", National Institute for Minamata Disease, WHO, Kumamoto (Japan), 1993, pp 33–40.

10. Zangger Hr: Erfahrungen über Quecksilbervergiftungen. *Arch. Gewerbepathol. Gewerbehyg*, 1980; 1:52–538.

11. Koelsch F: Gesundheitsschädigen durch oranische Quecksilberver-bindungen. *Arch. Gewerepathol. Gewerbehyg.*, 1987; 8:113–116.

12. An Expert Group of National Institute of Public Health: Methyl-mercury in fish, A toxicologic-epidemiologic evaluation of risks. Stockholm, Uno S. Andersons Tryckeri, 1971.

13. WHO: IPCS, Methylmercury, Environmental health criteria 101. Geneva, WHO, 1990.

14. Takeuchi T: Human effects of methylmercury as an environmental neurotoxicant. In Blum K and Manzo L (eds): *Neurotoxicology*. New York, Marcel Dekker, 1985, pp 345–367.

15. Harada M: Minamata Disease: Methylmercury poisoning in Japan caused by environmental pollution. *Crit. Rev. Toxicol.*, 1995; 25:1–24.

16. Fujino T: Clinical and epidemiological studies on chronic Minamata Disease. Part 1. Study on Katsurajima island. *Kumamoto Med. J.*, 1994; 44:139–155.

17. Soong TR: Epidemiological research on the health effect of residents along the Sonhua River polluted by methylmercury. In Proc. Int. Symp. "Assessment of environmental pollution and health effects from methylmercury", National Institute for Minamata Disease, WHO, Kumamoto, 1993, pp 165–169.

18. Takeuchi T, Eto K, and Eto N: Neuropathology of childhood cases of methylmercury poisoning with prolonged symptoms, with particular reference to the decortication syndrome. *Neurotoxicology*, 1979; 1:1–20.

19. Takeuchi T: Pathology of fetal Minamata disease. The effect of methylmercury on intrauterine life of human being. *Pediatrician*, 1977; 6:69–87.

20. Eto K and Takeuchi T: A pathological study of prolonged cases of Minamata disease with particular reference of 83 autopsy cases. *Acta Pathol. Jpn.*, 1978; 28:565–584.

21. Harada M: Neuropsychiatric disturbances due to organic mercury poisoning during the prenatal period. *Psychiatr. Neurol. Jpn.* 1964; 66:429–468 (in Japanese).

22. Harada M: Congenital Minamata disease. Intrauterine methyl-mercury poisoning. In John L Sever: *Teratogen Update, Environmental Birth Defects Risks*. New York, Alan R. Liss, 1986, pp 123–126.

23. Harada M: Intrauterine poisoning: Clinical and epidemiological studies and significance of the problem. *Bull. Inst. Const. Med. Kumamoto Univ.*, 1976; Suppl. 24:1–60.

24. Harada M: Minamata disease as a social and medical problem. *Jpn. Q.*, 1978; 25:20–34.

25. Harada M: Environmental contamination and human rights, Case of Minamata Disease. *Ind. Environ. Crisis Q.*, 1994; 8:141–154.

26. Akagi H, Kinjo Y, Branches F, et al: Methylmercury pollution in Tapajos River basin, Amazon. *Environ. Sci.*, 1994; 3:25–32.

27. Marsh DO, Clarkson TW, Cox C, et al: Fetal methylmercury poisoning. Relation between concentration in single strands of maternal hair and child effects. *Arch Neurol.*, 1987; 44:1017–1022.

28. McKeown-Eyssen GE, Ruedy J. and Neims A: Methylmercury exposure in northern Quebec. II. Neurological findings in children. *Am. J. Epidemiol.*, 1983; 18:470–479.

29. Kjellström T, Friberg L, Lind B, et al: Physiological and mental development of children with prenatal exposure to mercury from fish, Stage 2, Interviews and psychological test at age 6. National Swedish Environmental Protection Board, 1988, pp 1–63 (Report No.3642).

30. Dalgaard C, Grandjean P, Joergensen PJ, and Weihe P: Mercury in the umbilical cord: Implications for risk assessment for Minamata Disease. *Environ. Health Perspect.*, 1994; 102:548–550.

Chapter 19

Minamata Disease in Niigata: Epidemiology and Legal-Social Issues

Kiyotaro Kondo

CONTENTS

19.1 INTRODUCTION

Minamata disease (MD) is a specific neurological affliction among the local residents who ingested fish or shellfish contaminated with alkylmercurial compounds originated from industrial wastage.

Two well-documented outbreaks of MD are known; the first occurring in Kumamoto and the second in Niigata Prefectures, Japan as presented in the foregoing chapter and the present chapter. They are abbreviated as MD-1 and MD-2, respectively. Various aspects of MD-2 are already reported in numerous reports and two books.[1,2] While they are minimally quoted, this chapter deals mainly with unpublished data on overall incidence patterns of MD-2 based on recent epidemiological studies and discusses some social and legal issues. Sections 19.2 and 19.3 summarize reported studies, Sections 19.4 and 19.5 respectively, describe the mechanisms of official diagnosis, report patterns of incidence based on the official decisions, and discuss some social and legal implications.

19.2 DISCOVERY AND THE CENSUS SURVEY

The first patient, I.K. aged 31, with Hunter-Russell syndrome, was suspected as MD by the late Professor Tsubaki, Department of Neurology, Niigata University, on 28 January 1965. High total mercury (Hg) value, 116 ppm, was established in the patient's hair. Two male cases, aged 28 and 55, were also diagnosed. A male aged 63 who died of "unexplained psychiatric disease" in the fall of 1964 was diagnosed as MD based on clinical records.

All four patients lived along the Agano River and a preliminary survey disclosed that they ate large amounts of the river fish. We worried whether the cases were due to the mechanism of MD. There were a few known mechanisms of alkylmercury poisoning, especially methylmercury (MeHg): accidents in factory, misuses of Hg-containing fungicides and pesticides, and MD. We decided to discover more cases, to give prompt treatments, and to collect adequate data to evaluate mechanism(s) of the outbreak, especially to establish whether the epidemic was the second MD. Few new cases were discovered and we reported the incidence to the Prefectural Government on 31 May.

Reports on MD-1 showed us ways to avoid mistakes and pitfalls in clinical care and research. Social upheavals caused by MD-1 motivated us to quickly introduce a census in the area that was possibly polluted, which was not done in MD-1. It was decided on 4 June and was carried out from 14 June on a total of 22,741 residents in 4,464 households in a total of 65 communities in 2 cities, 2 towns, and a village directly located along the river. It was later extended upstream to 50 km from the sea to a town where a chemical plant existed which was regarded as one of the possible sources of Hg. Collected information for each household included number of members, economical code, recent deaths, dead/missing pet animals, source of drinking water, etc. For each member of a household, inquiry was made about sex, date of birth, ingestion of river fish, recent pregnancy/delivery, clinical signs/symptoms of organic mercury toxicity, use of agrochemicals and fungicides, etc. Hair samples were collected. Whereas studies on MD-2 were thus based on the census of the population at risk, studies on MD-1 was clinical based on full-blown cases that were admitted to hospitals.

Reports based on this large-scale survey disclosed highlights such as follows, which gave fundamental information for the control of MD-2 and which was finally attributed to the chemical plant that dumped Hg catalysts in the river in late 1964:

1.   Histories of exposure absolutely excluded all causative possibilities except MD.
2.   Total Hg in hair showed association with the amount in the river fish ingested, especially a carp, *Hemibarbus barbus* (Table 19.1).
3.   Clinical study resulted in the detection of 26 cases including 7 incipient cases whose only manifestation was sensory impairment but characterized by high hair Hg by the end of 1965.
4.   A total of 9 clinically normal residents with elevated hair Hg (>200 ppm) were identified, admitted, and subjected to Hg excretion remedies, and an additional 50 to 60 normal residents with levels of 50 to 199 ppm were given free medication (asymptomatic Hg carriers).
5.   Approximately 80 each of pregnant and breast-feeding women were identified, and special maternal-child health care was made available for those with high hair Hg (>50 ppm), in an attempt to prevent the "fetal" type of MD among their children.

19.3 CLINICAL MANIFESTATIONS

Descriptions in the foregoing chapter by Harada are valid also in MD-2. Onset is usually acute or subacute. There was one case of the fulminant type and one case of the fetal type in Niigata, scarcity of the latter compared with MD-1 owing to preventive measures.

Table 19.1 Ingestion of the Agano River Fish and Total Mercury in Hair

Hair Hg (ppm)	Male						Female					
	0	1	2	3	4	Total	0	1	2	3	4	Total
0.0–4.9	0	0	0	0	0	0	2	1	0	0	0	3
5.0–9.9	1	5	1	0	0	7	4	2	0	0	0	6
10.0–14.9	4	3	1	0	0	8	4	4	1	0	0	9
15.0–19.9	2	4	1	0	0	7	3	1	0	0	0	4
20.0–24.9	1	1	2	0	0	4	2	1	1	1	1	6
25.0–29.9	0	3	2	0	0	5	4	2	2	0	1	9
30.0–34.9	0	0	0	0	0	0	0	2	1	0	0	3
35.0–39.9	2	2	1	0	0	5	0	1	2	0	0	3
40.0–44.9	0	0	0	0	0	0	1	0	2	0	0	3
45.0–49.9	1	0	1	0	1	3	1	0	1	0	0	2
50.0–54.9	2	0	1	0	0	3	0	0	1	0	1	2
55.0–59.9	0	1	0	0	0	1	0	0	1	0	0	1
60.0–64.9	1	0	1	1	0	3	0	0	2	1	0	3
65.0–69.9	0	1	1	0	1	3	0	0	1	0	0	1
70.0–74.9	0	0	0	0	1	1	0	0	0	0	0	0
75.0–79.9	0	0	0	0	0	0	0	0	1	0	1	2
80.0–84.9	0	0	1	0	0	1	0	0	0	0	0	0
85.0–89.9	0	1	0	0	0	1	0	0	0	0	0	0
90.0–94.9	0	1	0	0	0	1	0	0	1	0	0	1
95.0–99.9	0	1	0	0	0	1	0	0	0	0	0	0
100.0–109.9	0	0	1	0	1	2	0	0	0	0	0	0
110.0–119.9	0	0	0	0	1	1	0	1	0	0	0	1
120.0–129.9	0	0	0	0	0	0	0	0	0	0	0	0
130.0–139.9	0	0	0	0	0	0	0	0	0	0	0	0
140.0–149.9	0	0	0	0	0	0	0	0	1	0	0	1
150.0–199.9	0	1	1	0	0	2	0	0	0	2	0	2
200.0–249.9	0	1	0	0	0	1	0	0	0	0	0	0
250.0–	0	0	3	2	6	11	0	0	1	1	0	2
Hair Samples	14	25	18	3	11	71	21	15	19	5	4	64
Residents without hair samples	347	171	32	2	8	560	413	142	25	0	6	586
Total	361	196	50	5	19	631	434	157	44	5	10	650
%	57.2	31.1	7.9	0.8	3.0	100.0	66.8	24.2	6.8	0.8	1.5	100.1
Hair sample collection (%)	3.9	12.8	36.0	60.0	57.9	11.3	4.8	9.6	43.2	100.0	40.0	9.8
Average Hg (ppm)	28.1	43.0	104.6	248.7	201.9	89.0	18.4	24.7	67.3	140.2	44.8	45.5

Note: This table shows the results in six settlements where the first seven cases lived. More hair samples were taken in 56 additional settlements, but they are not shown in the table. River fish codes 0–4 corresponded to ingestion of the Agano River fish around June 1965 in 0–4, 5–9, 10–14, 15–19, or 20 or more meals per week.

So-called "delayed-onset type" with prolonged latent periods is debatable. Besides a follow-up program for 26 cases, asymptomatic Hg carriers were also monitored. While seven cases with sensory symptoms later developed incomplete Hunter-Russell syndrome, some Hg carriers manifested the syndrome in varying degrees within a few years after exposure. Such cases reportedly never ate river fish since ingestion was immediately prohibited by the Prefectural Government after the end of the census. Whether they ate no fish may be questionable, but Hg contents of the fish rapidly returned to safe levels in those periods. The authors therefore referred to these cases as the "delayed-onset type", a concept strongly criticized by experimental toxicologists and foreign neurologists who did not see similar cases.

Mechanism(s) of the delay is unknown. Incubation time of an epidemic due to a point exposure represents a log-normal distribution. Exposure in MD-2 is approximately single occurring in the end of 1964, as shown by Hg contents in cut segments of very long hair samples from several female patients. It may be that acute cases merely represent a tail of the distribution of the log-normal incubation time, and "delayed onset type" is by no means a type, but the main body of MD.

19.4 OFFICIAL DIAGNOSTIC DECISIONS

This section shows how medicine and administration collaborated to deal with official diagnosis of MD in favor of the victims.

19.4.1 Diagnostic Criteria

As a basis for aiding the victims of MD-2, the Niigata Prefectural Government asked Dr. Tsubaki to develop diagnostic criteria. Table 19.2 represents the final revision of his criteria, which was the basis of the current voluminous and meticulous *Conditions for Judgments* officiated by the Japan Environment Agency in 1977.

Table 19.2 Diagnostic Criteria of Minamata Disease

1. Ingestion of large amounts of fish (from the Agano River)
2. Elevated Hg contents in hair or other specimens from a patient
3. Syndrome at least partly comprising the following
      Sensory symptoms
      Concentric visual field defects
      Auditory impairments
      Cerebellar signs
4. Successful differentiation from similar diseases

Note: This is the revised version by Tsubaki of the criteria he originally proposed for the Niigata Prefectural Government in late 1965. This revised version is the basis of the current "Conditions" (see text); (1) and (2) identify the exposure, (3) and (4) assure that very mild, incomplete, or atypical cases be recognized insofar as other diseases are reasonably excluded.

19.4.2 Administrative Decisions

As MD has been a social-political issue, diagnosis was inevitably a prerequisite for administrative aid and compensation by the polluting company already convicted. The Conditions applied were to make decisions since 1971 in the following three phases:

1. Phase A. Application. Any residents in the Agano River area may apply for the official diagnosis. A certifying letter of a local physician is required to receive the application.
2. Phase B. Ranking. An expert medical committee chaired by Tsubaki classified the evidence on how a given applicant satisfies the Conditions. The following "Ranks" were used; Ranks 1 to 3 for definite, probable, and possible MD, Ranks 4 and 5 for unclassifiable cases and non-MD diseases.
3. Phase C. Administrative decisions. Ranks 1 to 3 were accepted as MD, while Ranks 4 and 5 were rejected.

Such a system generated serious criticisms not only by patients' organizations but by academic circles and the general public against the extremely rigorous criteria employed in the early stages in MD-1, when only patients with the typical full-blown syndrome were accepted as MD. Dr. Ohishi, Minister of Environment, declared in a congressional committee in 1972 that his agency shall develop a scheme with which patients are officially recognized when they are likely to have MD with a probability "in excess of 50%" (first political solution).

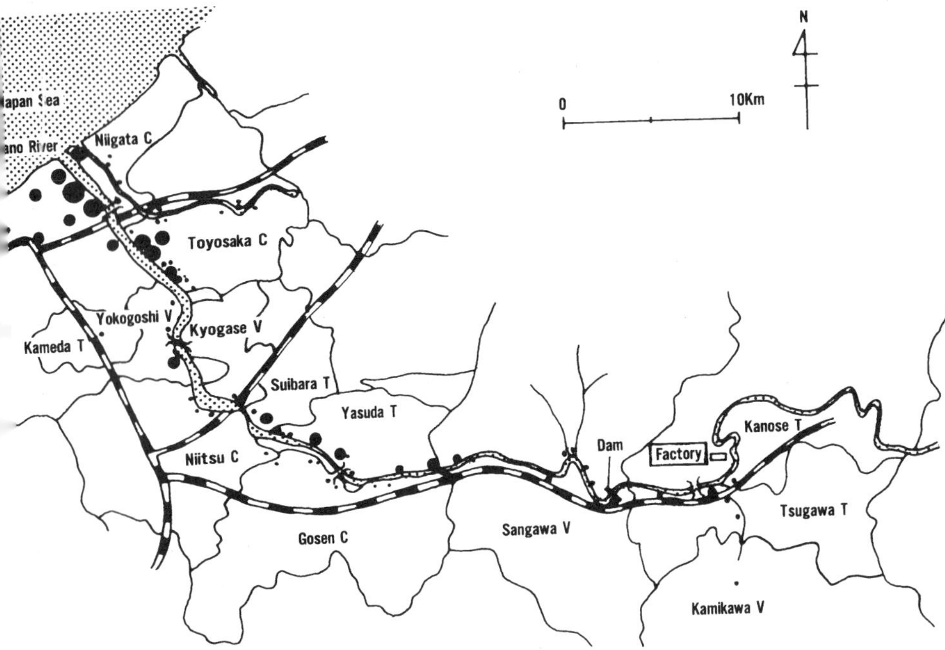

.1 Agano River tributary and Minamata disease. Black spots indicate settlements where officially recognized inamata disease live. Size of dots are roughly parallel to the number of the cases which totaled 690. Numbers 324 in Niigata city, 171 in Toyosaka city, 80 in Yasuda town, 23 in Suibara town, 1 in Kyogase village, 18 village, 10 in Gosen city, 6 in Niitsu city, 3 in Kameda town, 23 in Sangawa village, 3 in Kanose town, 3 in llage, and 25 in Tsugawa town. The source of mercury is a factory in Kanose town about 50 km upstream

9.3 Incidence Rates of Officially Recognized Cases and the Rejected Applicants for ata Disease, Niigata, Both Sexes Pooled

		Fish eaters					Total
ish code[a]	0	1	2	3	4	Total	
ts (N)	13,398	4,164	494	73	124	4,855	18,253
	73.4	22.8	2.7	0.4	0.7	—	100.0
nts n ($n^0 + n^1$)	265	247	83	14	31	375	640
tion rate (n/N%)	2.0	5.9	16.8	19.2	25.0	7.7	3.5
ition rate (n^1/N%)	29.8	42.1	61.4	64.3	61.3	48.8	40.9
ized cases (n^1)	79	104	51	9	19	183	262
ce rate (n^1/N%)	0.6	2.5	10.3	12.3	15.3	3.8	1.4
d cases (n^0)	186	143	32	5	12	192	378
ce rate (n^0/N%)	1.4	3.4	6.5	6.8	9.7	4.0	2.1

ish codes: see Table 19.1.

ncidence rates are not annual but cumulated cases through the process of the Official Recognition by the number of residents, N. Note that application is not exhaustive but voluntary. Dose response in gnized as well as the rejected cases eliminating Code-0 were significant ($p < 0.000$, df = 2).

Correlation of Hair Hg and Ingestion of Contaminated Fish

7 hair samples were collected in June 1965, disclosing the stated correlation (Table 19.1). few residents with the elevated hair Hg (>20 ppm), who denied ingestion of the fish, ed in the Code-0. Some were attributed to the misuse of Hg-containing agrochemicals ut about 9% of those who actually ate fish were estimated to have said "No" in the e-negative answers).

The stated *Conditions for Judgments* were thus developed wh
above phases. This system is extremely biased to be generous to the
practically included extremely mild cases with sensory symptoms
well be due to causes other than MeHg. The expert medical cor
Classification of Ranks, upon which the Environment Agency adm
nosis" of each applicant. Accepted applicants are therefore by no mea
point of view, but administratively they are dealt with as "patients'

Cases ranked 4 showed unexplained sensory symptoms only,
distribution. They do not satisfy the Conditions, but may repre:
pyramid", at least for the protesting groups' and the applicants' poi
was 100% (about 20 million yen per capita) when accepted, but ze
in Rank 4 was strongly objected to.

The cut-off point was very subtle, however, because it practic
interpretation of questionable ataxia which was often obscured by

Overdiagnosis appeared likely in the accepted cases, where;
insisted that rejected cases are in fact true victims of the pollutio
The threshold for paresthesias was 25 mg, while it was set at 35
Iraqi cases of MeHg-containing agrochemical poisoning.[3] This is c
that there are numerous monosymptomatic sensory cases of in fa

19.4.3 Results of the Official Diagnosis

A total of 690 applicants were accepted as MD (Figure 19.1). By
prised 16.5, 16.5, and 67.0% of the cases, respectively. Reliak
accurate in the upper two Ranks. The preponderance of Rank 3 is
for this extremely miserable disease, but the diagnosis is highly do
were rejected. Rank 4 predominated among the rejected cases ow
physicians to exclude other diseases (Rank 5), in Phase A whic
entire process of Official Diagnosis.

In 1979, some rejected applicants accused the nation and
reasons including their disobedience on the Official Diagnosis, (
MD showing sensory symptoms only, a concept never accepted t
experts.

19.5 INCIDENCE PATTERNS

The census data were not utilized in the Official Diagnosis. T
analyze epidemiological patterns of MD-2 and evaluate the vali

19.5.1 Materials

Questionnaires of the census for 8,914 male and 9,345 female
contained 262 and 379 accepted and rejected applicants. Table
river fish code (Code) as defined in the footnote. n^0/N and n^1/N a
of the rejection and acceptance, with full awareness that n^0 and
tions, not on exhaustively identified applicable individuals. $n^0/$
the Code, or a dose-response effect which is strong evidence
The clear-cut dose-response in the accepted cases indicates va
same in the rejected cases creates far more public concerns and
at least some rejected cases are also victims of the pollutio
plaintiffs.

FIGURE 1
cases with 1
of cases we
in Yokogost
Kamikawa
from the se;

Table
Mina;

River

Resid
%
Applic
Applic
Recog
Recog
Incide
Rejec;
Incide;

[a] River

Note:
divide
the rec

19.5.2

A total of 3
There wer
thus classi
then legal,
census (fal

19.5.3 Diagnostic Accuracy of Acceptance

Ingestion of river fish is the cause in true cases, whereas this dietary behavior appears in the false cases exactly at the same proportion as the local residents. Therefore;

$$100 \times y = x(1 - p) + (100 - x)a \tag{1}$$

where:

> y = % fish ingestion among the accepted cases
> p = false-positive answer rate about fish ingestion in the census (0.09)
> a = % fish ingestion among the residents

Thus, x = 100 (y − a)/[1 + (p + a)], which can be calculated if y, p, and a are available, y was (262 − 79)/262 = 0.698 (Table 19.3); p was 0.090. With these figures, x was 67.1%, which is obviously excessive because application is recommended by local physicians and the ingestion can be one reason to recommend it, while a = 0.507 among the rejected cases who have reasons for applying but who did not satisfy the Conditions clinically: x was 47.4% in this case, indicating that about half of the accepted cases are in fact patients with other diseases resembling MD.

It is impressive that the official diagnosis was so generous as it included about 50% of unapplicable cases resembling MD.

19.5.4 Diagnostic Accuracy of the Rejection

The attributable risk percent (ARP) is $100 \times (R^1 - R^0) / R^1$, where R^1 and R^0 are incidence rates in the exposed and unexposed groups, respectively. When l^1 and l^0 ($l^0 + l^1 = 1$) are proportions of both groups in the total population, the population attributable risk percent (PAR) is

$$\mathrm{PAR} = \frac{(R^1 l^1 + R^0 l^0) - R^0}{R^1 l^1 + R^0 l^0} \times 100$$

which shows how much is due to the exposure among the total cases in the total population.

The population relative risk (PRR) defined as

$$\mathrm{PRR} = \frac{R^1 l^1 + R^0 l^0}{R^0 (l^1 + l^0)}$$

indicates how the incidence rate increased due to the exposure relative to the value when there was no exposure.

R^0 calculated from Table 19.3 is an underestimate, however, because those who in fact ate no fish had no reason to apply because local people knew that MD-2 occurs only when they ate polluted river fish. An interval estimation of APR was made therefore.

Underestimates — Incidence rate in Code 1 was used instead of R^0 assuming that this grade of ingestion is not harmful which is in fact slightly harmful. R^0 was 3.4 and R^1(Codes 1 to 4) was 4.0, ARP was 13.2%, and since Codes 1 to 4 correspond to 50.8% of the total, 6.6% of the rejected cases are influenced by Hg. PRR was 3.9%.

Overestimates — Overlooking the underestimation of R^0, the effect of MeHg is overestimated. Assuming that those with the symptom classified as Rank 4 who ate no fish have 1/2 of the motives to apply than those who ate the fish, ARP was 29.8%, 14.9% of the rejected cases were influenced by MeHg, and PRR was 10.1%. Some 7 to 15% of the rejected cases, in fact, suffered more or less from the harmful effects of MeHg in the pathogenesis of their sensory symptoms, which is less than the acceptable level in the Conditions. Due to the pollution, there was an increase of sensory symptoms in the local population classified as Rank 4 by 5 to 11%. Despite diagnoses as generous as accepting the same number of false cases, there were still few unrecognizable victims whose only symptom was sensory.

Conclusions — This analysis revealed that

1. Official recognition was as generous as to include 50% non-MD cases to rescue mild cases difficult to diagnose, precisely according to the principle established by the first political solution.
2. Even so, rejected applicants contained about 10% of cases influenced by MeHg, who displayed sensory symptoms only but were indistinguishable clinically from cases due to other causes.

Of course, both the census data and the material of the official diagnosis may not be totally free from errors. The former, however, is the only existing defined population at risk for MD, as given in the latter data. These results, if not without biases, are the only available quantitative clues in the present social conflicts on MD.

19.5.5 Social-Legal Issues

MD still represents unsolved medical, legal, and social problems nearly 40 years after its discovery. In current suits still in the district court level, major issues are twofold: (1) responsibility of the Government, and (2) validity of the diagnosis. Details are too complex to explain here, but MD still gives serious lessons to our society.

19.5.6 Chronic Safety Levels

Based on experiences including acute cases in MD-2, FAO/WHO (1972) proposed 0.3 mg (0.2 mg MeHg) as the Allowable Weekly Intake,[4] and the Japan Ministry of Health and Welfare revised this value to 0.25 mg (0.17 mg) for use in Japan, taking average body weights into account.

Hg intake in Code 1 corresponded to about 1.5 mg Hg per week based on the average fish intake multiplied by the average Hg contents of the Agano River fish: about 5 ppm in June 1965. Few applicable cases including the delayed-onset type emerged from this Code, indicating that this level of intake is slightly beyond the upper allowable limit; 0.25 mg is 1/6 of this level, and may be acceptable as safe even for chronic effects.

19.6 CONCLUSIONS

"Minamata" has been a hallmark of one category of human disasters in the twentieth century. In MD-1, it is believed that about 10,000 victims who live throughout the polluted bay area are still unrecognized and uncompensated. Various problems have been solved that allowed the pollution to occur, and Japan now has rigorous legislation and technologies in fighting health hazards and environmental destruction due to economical/industrial activities.

Lessons from MD are valuable. That minimal manifestation of MD is subjective and nonspecific is the source of endless conflicts which make its final social-legal solution based on science extremely difficult. The controversies now affect tripartite national rights: legislation, justice, and administration.

Of prime importance is to assess possible dangers to health and environment, and to provide funds within the budgets which are required to prevent adverse consequences before introduction of any new developments. Precious sacrifice experienced in MD and other public nuisances enforced this doctrine, which did not exist when MD occurred.

Coalition parties proposed a scheme for political solution of MD on 21 June 1995, accepting the plaintiffs' claims with apologies that the rejected cases in the current suits are "patients" of MD. This appeared inevitable in view of the aging of the plaintiffs, being already in their 70s or 80s who cannot wait until the lengthy debates end in the Supreme Court. Our current Government insists that they are "gentle to people", and accepted this proposal. We now have three concepts of MDs, therefore: (1) medically acceptable MD showing at least an incomplete syndrome, (2) administrative MD which includes 50% of other diseases not caused by Hg, and (3) newly created "political" MD which includes 90% of other diseases, according to the present analysis. There are uncertainties how and who compensate these political patients, however (second political solution).

This is ridiculous from a purely scientific point of view, but to the present author, who witnessed MD for the last 30 years, this "solution" is inevitable if not desirable, because the company/administration complex first ignored MD-1 and then afterwards dealt with the victims in such an unsatisfactory way. An end must be officially put to the saga on MD in favor of the victims while they are alive, and this is the only way to do so after more than 30 years since the discovery of MD.

ADDENDUM

The Government has been the co-plaintiff with the polluting companies, and repeatedly declared that they will never yield unless ordered to by the Prime Minister or are defeated in the Supreme Court. In June 1995, however, the Prime Minister officially expressed desire for a reconciliation, and the Government agreed to "regret" the occurrence of MD, and recognized that the rejected applicants are "not necessarily completely concluded that they are not influenced by mercury, and there are reasons that they seek for relief."

On 29 October 1995, patients' organizations in MD-1 officially accepted these clauses, and the Government is to pay 2.6 million Yen per capita, and apologizes for MD, putting the matter to an end 40 years after its discovery. On 27 February 1996, all plaintiffs with MD-2 also accepted the same conditions to reconcile with the accused.

ACKNOWLEDGMENTS

This study was supported by grants from the Japan Environment Agency. Cordial thanks are due to Ms. Mayumi Yamamoto for her excellent technical assistance.

REFERENCES

1. Tsubaki T and Irukayama K (eds): Minamata Disease; *Methylmercury Poisoning in Minamata and Niigata, Japan.* Amsterdam, Elsevier Scientific, 1977.
2. Tsubaki T and Takahashi H (eds): Recent Advances in Minamata Disease Studies; *Methylmercury Poisoning in Minamata and Niigata, Japan.* Tokyo, Kodansha, 1986.
3. Bakir F, Damluji, SF, Amin-Zaki L, et al: Methylmercury poisoning in Iraq. *Science*, 1973; 181: 230–241.
4. FAO/WHO Expert Committee on Food Additives: Evaluation of Certain Food Additives and the Contaminants Mercury, Lead and Cadmium. Geneva, 1972. World Health Organization.

SECTION 7
MAGNESIUM

Chapter 20

Mechanisms of Action on the Nervous System in Magnesium Deficiency and Dementia

Jean Durlach and Pierre Bac

CONTENTS

20.1 INTRODUCTION

Whatever the age, nervous forms of magnesium deficit represent the most commonly seen form in clinical practice.[1] First of all, it seems very important to discriminate between the two types of magnesium deficit: magnesium deficiency and magnesium depletion. In the case of **magnesium deficiency**, the disorder corresponds to an insufficient magnesium intake: it merely requires oral physiological magnesium supplementation. In the case of **magnesium depletion** the disorder which induces magnesium deficit is related to a dysregulation of the control mechanisms of magnesium metabolism, either failure of the mechanisms which insure magnesium homeostasis or intervention of endogenous or iatrogenic perturbating factors of the magnesium status. **Magnesium depletion** requires more or less specific correction of its causal dysregulation.[1] It should not be permitted today to extrapolate from physiological data observed in overt acute magnesium deficiency to physiological consequences of chronic magnesium deficiency. Although acute and chronic magnesium deficiencies are specifically reversible through oral magnesium supplementation with physiological doses, the experimental and clinical symptoms may differ. The typical pattern of chronic magnesium deficiency is latent whereas overt signs are observed in acute magnesium deficiency.

The discrepancy between the patent and latent nervous forms of magnesium deficiency suggests that in the latent form there are compensatory factors which antagonize the nervous hyperexcitability observed in the overt form.

The aim of this review is to study:

1. The mechanisms of action of magnesium deficiency on the nervous system as demonstrated by
    *   Electrophysiological data testifying to diffuse nervous hyperexcitability (NHE)
    *   Biochemical data showing the mechanisms which may induce overt and latent NHE
    *   Neuromuscular and neurotic psychiatric data
2. The links between some types of magnesium depletions and dementias.
3. Therapeutical implications.

20.2 MECHANISMS OF ACTION OF MAGNESIUM DEFICIENCY ON THE NERVOUS SYSTEM

We will analyze the mechanisms of action of magnesium deficiency on the nervous system successively through electrophysiological, biochemical, and clinical data.

20.2.1 Electrophysiological Data

Experimental and clinical electrophysiological procedures allow us to neurophysiologically examine the effects of magnesium deficiency on the cortical, subcortical, and peripheral levels of the nervous system.

Standard electroencephalography (EEG) in humans exhibits "diffuse irritative tracings" without focal lesions or paroxysmal discharges. The recordings contain spikes, a pointed appearance of alpha and/or theta waves, which is facilitated more often by hyperventilation than by intermittent photic stimulation. The polygraphic study of afternoon sleep may complete the data of standard EEG: brevity of the time required to fall asleep, superficial character of the sleep, frequency of awakening, hypnoagnosia.[1] The EEG in the magnesium-deficient rat (electrocorticography) allows us to make observations similar to those found in humans. Mg deficiency induces electrocorticographic alterations in the rat analogous with those seen in Mg deficiency in humans. Sleep quality analysis shows particularly similar alterations of the hypnograms.[1,2]

Electronystagmography shows functional impairment at the level of the second vestibulo-ocular motoneuron which becomes apparent mainly as the variable association of "paroxysmal ocular states", i.e., a sequence of repetitive ocular movements with vertical predominance over a period averaging 20 s, irregularity of evoked nystagmic responses, excessive bilateral labyrinthine reflex activity, oblique or rotary nystagmus, prolonged latency during pendular tests, and differences between pendular nystagmic responses with the eyes open and closed. Alterations in optokinetic tests confirm the importance of subcortical functional disturbances.[1]

Moreover, behavioral changes in the Mg-deficient rats are due to hyperexcitability which first arises in deeper structures — in the whole limbic system and in the hippocampus, particularly — and which secondarily becomes generalized by projecting on the neocortices.[3]

Electromyography (EMG) shows peripheral NHE. A series of autorhythmic events is observed (singlets, multiplets, complex tonicoclonic activity) beating for more than 2 min during one (or several) of the three facilitations tests: tourniquet-induced ischemia lasting 10 min, a postischemic phase measured 10 min after removal of the tourniquet, and finally hyperventilation for a maximum of 5 min. A repetitive EMG, whatever the intensity, constitutes the principal neurophysiological mark of NHE due to Mg deficiency.[1]

Skin conductance reflex might appear attractive as an investigation tool of diffuse NHE due to Mg deficiency,[4] but this procedure still requires further validation.

These neurophysiological procedures, mainly EEG, ENG, and EMG, allow us to examine the effects of Mg deficiency on the nervous system: Mg deficiency causes diffuse neuromuscular hyperexcitability operating from the center to the periphery. NHE affects the nervous system as a whole, but comparing the impairments in its various sectors is less important than determining the intra- or extracellular origin of NHE.[1,5]

The long duration of experimental Mg deficiency required to produce manifestations of NHE has led to the hypothesis of concurrent reduction of intracellular Mg which represents "the bulk" of the Mg pool. In Mg deficiency induced in young rats, there is a direct correlation between the severity of hyperexcitability and the decrease in the brain Mg level. Clinical evidence includes patients with Mg deficiency but without abnormalities of extracellular Mg, hypo- or normo-magnesemia in the same patient at different times, and a reduction of the mean erythrocyte or lymphocyte Mg contents, two forms of intracellular Mg. This demonstrates the possible role of intracellular Mg in NHE due to Mg deficiency.

This NHE is not always accompanied by low levels of cerebral Mg: these are only observed in severe Mg deficiency in young rats. Usually, no changes have been found in brain Mg concentrations during the course of Mg deficiency in adult rats.[5–7] It should be stressed that in Mg deficiency there exist complex mechanisms to maintain normal, and even increased, Mg concentration in the tissue which are of vital importance in brain, liver, and brown fat.[5–8] With reduced plasma Mg levels, an extracellular origin of NHE due to Mg deficiency may intervene, but plasma normo-magnesemia has also been observed with some Mg deficiencies.

An analysis of the relationship between extra- and intracellular concentrations, clinical findings, and electronystagmographic tracings have shown that more symptoms of NHE occur when Mg deficiency predominates in one of the intra- or extracellular compartments rather than when it affects both equally.[1,5] This electroclinical observation highlights the importance of Mg distribution disturbances in the physiopathological consequences of Mg deficiency on the nervous system.

20.2.2 Biochemical Data

During Mg deficiency, a complex neuroendocrinometabolic and renal regulation may intervene for compensating its systemic effects. The role of the blood-brain barrier which attenuates the effect of the systemic regulating factors of Mg status and of the humoral consequences of decompensated Mg deficiency in the nervous system must not be overlooked. The biochemical factors capable of increasing nervous excitability are essentially local.[1,5]

20.2.2.1 Factors Inducing NHE

Extrapolating from data observed *in vitro*, *in situ*, or in other pharmacological manipulations to the physiological basis of an Mg deficiency simply due to insufficient intake[1] remains a methodological error.

Pharmacological Mg excess causes some systemic reactions which are not the opposite of physiological effects of Mg. For example **pharmacological** load of Mg **increases** release of calcitonin and nitric oxide (NO).[9,10] In contrast, **physiological** Mg supplementation, far from acting similarly, **reduces** high levels of calcitonin[1] (as well as of calcitonin gene-related peptide[11]) and of NO[12] released in the case of Mg deficiency.

The great stability of brain Mg during Mg deficiency particularly disagrees with the very notion of extrapolating from *in vitro* or *in situ*, extra- or intracellular Mg data[1,5] to *in vivo* physiological data. This leads to suggesting an updated scheme of the factors which cause NHE[5]: Mg deficiency would induce a diffuse NHE through a neuronal depolarization which derives from the sum of its direct cellular effects in the neural cells and from several mediated reactions.[5]

20.2.2.1.1 *Direct Cellular Effects*

Mg deficiency results in three basic effects: disturbances in cellular Ca distribution, decreased second messenger nucleotidic ratio,[5] and increased susceptibility to peroxydation.[12–14] Through membranous and postmembranous alterations, Mg deficiency brings about a cellular Ca load with subcellular distribution modifications.[5] Mg deficiency reduces 3′,5′-cyclic adenosine monophosphate (cAMP) concentration and increases 3′,5′-cyclic guanosine monophosphate (cGMP) concentration, perhaps through inhibition of adenylate cyclase and activation of guanylate cyclase.[5] Mg-deficient animals show an increased susceptibility to *in vivo* oxydative stress and the tissues of these animals are more susceptible to *in vitro* peroxydation, affecting lipid particularly.[12–14] Protein oxydation in Mg-deficient rat brains occurs early. A significant increase of protein carbonyls is observed within 2 to 3 weeks of a Mg-deficient diet. These changes take place prior to any detectable tissue damage, dysfunction, or changes in cellular glutathione.[15]

Mg deficiency may increase formation of free radicals directly, but also indirectly through free-radical-triggered mechanisms.[12–15]

20.2.2.1.2 *Mediated Local Effects*

NHE due to Mg deficit is also linked to modifications in the turnover of various types of neurotransmitters: monoamines, amino acids, but also nitric oxide, neuropeptides, and cytokines.

Neurotransmitters — NHE due to Mg deficiency mainly depends on modifications in the turnover of several neuromediators and neuromodulators. They associate an increased turnover of the monoamines: serotonin (5HT), acetylcholine, catecholamines (dopamine and noradrenaline, mainly), and of excitatory amino acids (aspartic and glutamic acids, mainly) with a decreased turnover of inhibitory amino acids (γ-amino butyric acid and taurine, mainly).[5] As a great number of *in vitro* studies on Mg and NMDA[16] receptors have suggested that the latter had predominant and almost exclusive importance, their *in vivo* role has often been overestimated.[2] If hyper NMDA receptivity enters into the mechanisms of NHE due to Mg deficiency as confirmed *in vivo*,[17] a hyperreceptivity concerning other non-NMDA receptors of excitatory amino acids may also intervene.[18] The genuine complexity of biology must not be disregarded just because the present trend is towards focusing on NMDA receptors at the expense of many other receptors.

Nitric oxide, peptides, and cytokines — An increased production of nitric oxide and of various inflammatory peptides — such as substance P, CGRP, and VIP — is observed in Mg-deficient rats. All these substances might directly intervene as neurotransmitters in the physiopathology of NHE due to Mg deficiency; but NO could also mediate an increase in cGMP whereas inflammatory neuropeptides might stimulate production of inflammatory cytokines and of free radicals.[12–14] During the progression of Mg deficiency in a rodent model, dramatic increases of inflammatory cytokines were observed: interleukins 1 and 6 (IL1, IL6) and tumor necrosis factor α (TNFα). Increase of these various cytokines was neither concomitant nor constant, according to species and strains.[12–14,19] So far, their importance in the physiopathology of NHE has not been clearly defined but we have observed with P. Maurois that audiogenic seizures in Mg-deficient mice might be correlated with possible TNFα release. According to the strains of mice, there is a parallelism between production of TNFα induced by bacterial lipopolysaccharides and NHE. However, coexistence does not mean causality and further research focusing on the effects of specific TNFα antibody on the production of audiogenic seizures will be necessary. Opioid peptide activity could be reduced since, in the complex mechanisms of opioid action, Mg at the physiological level may be most often an agonist of δ, μ, and κ opioid receptors.[5,20]

The sum of these direct and mediated local factors may bring about overt NHE due to Mg deficiency. Frequency of latent forms of this NHE postulates the existence of local compensatory factors which may control NHE.[5]

20.2.2.2   NHE Compensatory Factors

The local compensatory factors instrumental in the latency of NHE due to Mg deficiency may also be direct and mediated.

20.2.2.2.1   Direct Compensatory Factors

Since Mg can more or less be replaced by a natural polyamine in many biochemical reactions, an "Mg-substitutive" increase in polyamines might decrease the direct cellular effects of Mg deficiency. This "Mg-vicariant" increase in polyamines may be a factor regulating the alterations of protein synthesis and particularly that of Ca^{2+} and Mg^{2+} binding proteins.[5]

Increased formation of free radicals may be antagonized by the cell antioxidant system: enzymes such as superoxide dismutase and glutathione peroxidase, antioxidant vitamins such as E, A, and C, selenium, and sulfur compounds such as glutathione and taurine.[8,14–18]

20.2.2.2.2   Mediated Compensatory Factors

Some types of adenylate cyclase-receptors may contribute to a compensatory increase in the cAMP/cGMP ratio.[5] The main compensatory factors are mediated by the increase of several physiological neuroprotective agents: inhibitory aminoacids[1,5,10,21–23] and perhaps melatonin.[24] The particular efficiency of N-acetyl-amino compounds such as Mg N-acetyl-amino taurinate and melatonin might depend on a decreased activity of the Mg-dependent N-acetyl-amino transferase in the nervous system (EC 2.3.1.5.).[10,21–24] The main mediated compensatory factor is taurine (TA) with the help of its peptidic congener: γ-L-glutamyl taurine (GTA).[5]

When these direct and mediated compensatory factors are effective, NHE remains latent. It is patent when compensatory factors are insufficient. Their failure may depend on several reasons:

1. Sufficient amounts of amino acids — precursors of neuromediators and neuromodulators — must be available in the nervous system. It is important to have qualitatively and quantitatively a sufficient protein intake and sufficient amino acid transport across the blood-brain barrier, eventually facilitated by a homeostatic reactive hyperinsulinism. Both may be lacking.
2. If a compensatory high release of cAMP counteracts the decrease of cAMP/cGMP ratio, it inhibits the TA biosynthesis through the reduction of cystine dioxygenase activity.
3. Efficient brain metabolism and, particularly enzymatic activity, is necessary. However, Mg deficiency frequently alters protein biosynthesis and induces enzymatic hypoactivity.
4. Finally, the number of excitatory factors appears larger than that of compensatory factors during Mg deficiency in the nervous system.[5]

However, one should not overlook the schematic nature of this general pattern which covers both a homogeneous explanation of the diffuse character of the symptomatic NHE due to Mg deficiency and the possibility of the Mg deficient latent form.

Because of the heterogenicity of NHE due to Mg deficiency, a special study of each parameter of this scheme in each brain area is necessary to obtain a better understanding of these complex phenomena.[5]

20.2.3   Neuromuscular and Psychiatric Data

NHE due to Mg deficiency results in a nonspecific clinical pattern which associates peripheral and autonomic neuromuscular signs and central or rather psychiatric symptoms. Neuromuscular disturbances include acroparesthesias, muscle fasciculations, cramps, and myalgias occurring more frequently than tetanoid or tetanic attacks, and various autonomic functional complaints such as cardiac palpitations, precordial pain, extrasystolae, Raynaud's syndrome, hepatobiliary dyskinesia, gastrointestinal cramps and spasms, and asthma-like dyspnea.

Psychiatric symptoms consist of anxiety, hyperemotionality, asthenia, headache, insomnia, dizziness, nervous fits, lipothymias, and sensations of a "lump in the throat" and of "blocked breathing". On encountering this nonspecific pattern, the signs of neuromuscular hyperexcitability are of much greater importance. Chvostek's sign must be sought systematically. The sign is positive in 85% of cases examined. Trousseau's sign, less sensitive than Chvostek's sign, is observed only in cases of obvious hyperexcitability. The hyperventilation test can complete the search for Chvostek's sign and may give greater sensitivity to the Trousseau's sign (Von Bonsdorff's test).

Neurophysiological tracings and routine Mg assessment (at least plasma and red blood cell Mg; if possible evaluation of Mg intake and daily magnesuria, calcemia, and calciuria) may complete the clinical examination. However, the diagnosis of Mg deficiency mainly requires an Mg oral loading test. The dose of Mg to be administered is 5 mg/kg/day for at least 1 month. At this physiological dose level, oral magnesium supplement is totally devoid of the pharmacodynamic effects of parenteral magnesium. Correction of symptoms by this oral Mg load constitutes the best proof that they were due to Mg deficiency and may represent the beginning of its treatment.[1,4,10]

Psychiatric forms of Mg deficiency have been well identified. Personality disorders are of the neurotic type. For example the Minnesota Multiphasic Personality Inventory (MMPI) finds a direct correlation between the "neurotic triad" (hypochondria, depression, and hysteria) and the EMG marks of NHE due to Mg deficiency.[1,4] With all the psychometric evaluations, and with the DSM III R interview particularly, the clinical pattern induced through Mg deficiency was always neurotic (for example: generalized anxiety, panic attack disorders, and depression) but never psychotic. Mg deficiency never induces dementia.[1,4,25,26] Although a neurosis pattern due to Mg deficiency is frequently observed and simply cured through oral physiological supplementation, neuroses are preeminently conditioning factors for stress. Neuroses may therefore very frequently produce secondary Mg depletion. They require their own specific antineurotic treatment and not mere oral Mg physiological supplementation, but both genuine forms of neurosis due to primary neural Mg deficiency and Mg depletion secondary to a neurosis may exist. These two conditions may be concomitant and reinforce each other. In these stressful patients it may be difficult to establish the primacy of one or the other. In practice, physiological oral Mg supplements may be added to psychiatric treatments, at least at the start.[1,10]

It is imperative to emphasize that the nervous consequences of Mg deficiency remain functional with anatomical integrity for a long time. They are completely reversible since they can be restored to normal with simple oral physiological Mg supplementation,[1,5,10] but it should also be pointed out that a prolongation of untreated chronic Mg deficiency can produce irreversible lesions[1,5] with histological changes: morphological changes in the rat hippocampus, degeneration of the Purkinje cells, glial aberration of positive Gomori cells, and neurovasculitis. When Mg deficiency secondary to alcoholism was corrected, no alcoholic encephalopathy was observed within a period of 5 years.[1,5]

These processes could account for the clinical and paraclinical data that persist after treatment of NHE due to Mg deficiency.[1,5] They are especially of concern during the early development of the nervous system. The constitutional characteristics of the nervous forms of primary chronic Mg deficiency could arise from undetected maternal Mg deficiency.[1,5] An early maternal Mg deficiency could be the fountainhead of more severe impairments: sudden infant death syndrome,[8,28] some forms of infantile convulsions or psychiatric disturbances,[1,5] and even in adults, cardiovascular diseases and noninsulin-dependent diabetes mellitus.[28] The protocol of the multicenter trials of maternal Mg physiological supplementation should be followed not only on the mother, the fetus, and the neonate, but also on the child throughout life from infancy to older age.[28]

20.3 MAGNESIUM DEPLETION AND DEMENTIAS

Although Mg deficiency might not result in dementia, some types of Mg depletion can play a role in the physiopathology of several types of dementia.

20.3.1 Experimental Models of Mg Depletions

Various types of more or less severe Mg depletion are used: genetic models (in rats and mice) and acquired models: either secondary to an irreversible (or partially reversible) cause (such as traumatic brain injury) or reversible. In the latter case, the models associate a low Mg intake with diverse types of Mg stress.[10,22,23] Two types of experimental Mg depletions will be highlighted because of their possible link with dementia.

20.3.1.1 Neurological Degeneration Due to Al Load and Low Mg Intake

Garden soil and drinking water in some Western Pacific areas with high incidence of amyotrophic lateral sclerosis and parkinsonism-dementia (ALS-PD) contain high concentrations of polluting metals such as Al, Fe, and Mn, and low concentrations of common metals such as Mg and Ca. Decreased exposure to traditional sources of foodstuffs and drinking water resulted in a dramatic decline in ALS-PD.

These data as well as the links between aluminum load, magnesium status, and dialysis encephalopathy — more hypothetically, Alzheimer's disease — highlight the interest of corresponding experimental studies. With a high Al diet alone, Al content in the nervous system in rats showed no difference with a control group although serum Al was high. No degenerative process was observed. However, with an insufficient intake of Mg the same Al load induced an increase in Al and Ca concentrations in the nervous system and neurodegeneration with precipitation of insoluble hydroxyapatites.[28]

20.3.1.2 Mg Depletion Due to Kainic Acid Plus Mg Deficiency

Hippocampal injury of ageing may originate from an increased calcium influx in pyramidal neurones resulting from the deleterious effects of increased release of excitatory aminoacids associated with a decrease of neuroprotective factors. Kainic acid acting through its specific receptors generates toxicity in the hippocampus, whereas Mg deficiency — a model of accelerated ageing — decreases Mg neuroprotection.[23] In this model physiological Mg supplementation and pharmacological doses of Na acetyltaurinate were ineffective. On the other hand, Mg acetyltaurinate at pharmacological doses had preventive and curative effects in both the short and long terms.[23]

These two types of experimental Mg depletion models may be useful for screening various treatments of psychiatric disturbances possibly linked with Al load and ageing insults.[21,23,29]

20.3.2 Possible Links Between Dementias and Mg Depletion

Established links between some types of dementias and Mg depletion are presently scarce, but the experimental models of two types of Mg depletion, perhaps related to the physiopathology of some dementias, constitute promising tests for screening potentially efficient drugs both in these Mg depletions and in the related types of dementias.

20.4 THERAPEUTIC IMPLICATIONS

It seems obvious to contrast the specific, easy, and efficient treatment of the nervous form of Mg deficiency with the difficult problems set by the treatment of certain types of Mg depletion playing a possible role in the physiopathology of some dementias.

20.4.1 Treatment of Magnesium Deficiency

Physiological oral Mg supplementation (5 mg/kg/day) is simple and can be carried out in the diet or with Mg salts. To correct *in vivo* experimental or clinical Mg deficiency all Mg salts have a comparable bioavailability, but evidently their anions have their own importance. This treatment is totally atoxic since it palliates Mg deficiency by simply normalizing the Mg intake.[1,10] It is able to cure all the functional symptoms of Mg deficiency: signs of neuromuscular hyperexcitability and psychiatric symptoms which frequently mimic a neurotic pattern. It prevents irreversible stigmata of NHE due to primary or secondary magnesium deficiency, alcoholic encephalopathy in chronic alcoholism particularly.[1,5] It is necessary to highlight the curative and preventive importance of oral physiological maternal Mg supplementation, not only during pregnancy but also in the child throughout life from infancy to older age, to possibly prevent the so-called constitutional factor of neurolability, some cases of sudden infant death syndrome, infantile convulsions, or psychiatric diseases, and even in adult cardiovascular diseases and noninsulin-dependent diabetes mellitus.[1,5,8,10,28]

20.4.2 Magnesium Therapy and Dementias

With perhaps the one exception of the treatment of autism through very high pharmacological doses of vitamin B_6 and high doses of Mg[1,5,10,25,30] we cannot presently control the dysregulations of Mg status in Alzheimer's disease, dialysis encephalopathy, and ALS-PD. We can only advise some prophylactic measures. However, if an insufficient Mg intake (i.e., Mg deficiency) would add up to depletion cases, then an Mg deficit would be observed associating deficiency and depletion. The correction of Mg deficiency through simple oral supplementation therefore constitutes an adjuvant treatment of this possible component of Mg deficit.[1,10]

In some cases, the interest of pharmacological Mg therapy may be discussed. However, pharmacological Mg therapy may induce toxicity since it creates Mg overload. High oral doses of Mg (10 mg/kg/day) are advisable for chronic indications and the parenteral route is suitable for acute indications. Mg infusions can only be envisaged in intensive care units with careful monitoring of pulse, blood pressure, deep tendon reflexes, hourly diuresis, and electrocardiogram and respiratory recordings. It is presently difficult to evaluate the chronic toxicity of long-term high oral Mg doses: they may bring latent complications which may reduce life span.[1,10] Neuromuscular hypoexcitability due to hypermagnesemia only occurs when plasma Mg is more than twice normal levels. The blood-brain barrier gives priority to the peripheral action of Mg overload.[1,10] *In vivo*, this neuroprotection seems essentially indirect through the beneficial effects on antithrombotic platelet and endothelial functions and on vasospasm, mainly by acting as a calcium antagonist.[1,9,10] Except for a pathological disruption of the blood-brain barrier, direct neuroprotection is observed with massively increased plasma Mg which cannot be practically carried out in human beings.[10] In experimental models, the best protective effects with pharmacological doses of Mg were obtained with Mg acetyltaurinate, an Mg salt of the N-acetylamino-derived compound from taurine, the most neuroprotective inhibitory aminoacid.[21–23] Further study should evaluate its clinical therapeutic effects.

20.5 CONCLUSION

Two different types of links between Mg deficit and the nervous system should be emphasized.

1. NHE due to Mg deficiency with neuromuscular and psychiatric symptoms is well recognized nowadays. Induced by insufficient Mg intake, the primary or secondary acute or chronic nervous forms of Mg deficiency remain reversible over a long period by simply normalizing the Mg intake. Untreated chronic forms may however bring about irreversible organic disorders. The psychiatric forms of Mg deficiency may fit into a neurotic pattern, but never result in dementia.

2. In contrast, the relationships between some types of Mg depletion due to various dysregulations of Mg status and some dementias have not been clearly defined yet. However, some epidemiological, experimental and clinical data are promising and new paths remain open in this very interesting field of neuroscience research.

REFERENCES

1. Durlach J: Magnesium in clinical practice. London-Paris, John Libbey Eurotext, 1988, pp.360.
2. Depoortere H., Francon D., and Llopis J.: Effects of a magnesium deficient diet on sleep in rats. *Neuropsychobiology*, 1993; 27:237–245.
3. Goto Y, Nakamura M, Abe S, et al: Physiological correlates of abnormal behaviors in magnesium-deficient rats. *Epilepsy Res.*, 1993; 15:81–89.
4. Sandrini G, Ruiz G, Alfonsi E, et al: Effects of Mg salt administration on skin conductance response in neuronal hyperexcitability syndrome. *Magnesium Res.*, 1989; 1/2:122–123.
5. Durlach J, Poenaru S, Rouhani S, et al: The control of central neural hyperexcitability in magnesium deficiency. In: *Nutrients and Brain Function*, Ed., W.B. Essmann, Basel, Karger, 1987, pp 49–71.
6. Poenaru S, Aymard P, Durlach J, et al: Regional distribution of magnesium in normal and Mg deficient rats (abst.). *Magnesium Res.*, 1991; 4:246.
7. Lerma A, Planells E, Aranda P, et al: Evolution of Mg deficiency in rats. *Ann. Nutr. Metabol.*, 1993; 37:210–217.
8. Durlach J, Durlach V, Rayssiguier Y, et al: Magnesium and thermoregulation. I. Newborn and infant. Is sudden infant death syndrome a Mg-dependent disease of the transition from chemical to physical thermoregulation? *Magnesium Res.*, 1991; 4:137–152.
9. Kemp PA, Gardiner SM, March JE, et al: Effects of N^6-nitro-l arginine methyl ester on regional haemodynamic responses to $MgSO_4$ in conscious rats. *J. Pharmacol.*, 1994; 111:325–331.
10. Durlach J, Durlach V, Bac P, et al: Magnesium and therapeutics. *Magnesium Res.*, 1994; 7:313–328.
11. Weglicki WB, Tong Mak I, and Phillips TM: Blockade of cardiac inflammation in Mg^{2+} deficiency by substance P receptors inhibition. *Circ. Res.*, 1994; 74 1009–1013.
12. Rayssiguier Y, Mazur A, Gueux E, et al: Mg deficiency affects lipid metabolism and atherosclerosis processes by a mechanism involving inflammation and oxidative stress (abst.). *Magnesium Res.*, 1994; 7 (Supp. l): 46–47.
13. Rayssiguier Y, Gueux E, Bussière L, et al: Dietary Mg affects susceptibility of lipoproteins and tissue to peroxidation in rats. *J. Am. Coll. Nutr.*, 1993; 12:133–137.
14. Rayssiguier Y, Durlach J, Gueux E, et al: Mg and ageing. I. Experimental data: importance of oxidative damage. *Magnesium Res.*, 1993; 6:369–378.
15. Stafford RE, Mak IT, Kramer JH, et al: Protein oxidation in Mg deficient rat brains and kidneys. *Biochem. Biophys. Res. Commun.*, 1993; 196:596–600.
16. Smalheiser NR and Swanson DR,: Assessing a gap in the biomedical literature: Mg deficiency and neurologic disease. *Neurosci. Res. Commun.*, 1994; 15:1–9.
17. Nakamura M, Abe S, Goto Y, et al: In vivo assessment of prevention of white-noise-induced seizure in Mg-deficient rats by NMDA receptor blockers. *Epilepsy Res.*, 1994; 17: 249–256.
18. Nakamura M, Abe S, and Akazawa K: AMPA — but not NMDA — receptor blocker NBQX prevents seizure induction in Mg-deficient rats. *Magnesium Res.*, 1995; 8:55–56.
19. Rayssiguier Y, Malpuech C, Nowacki W, et al: Evaluation of the inflammatory state during Mg deficiency in the rat (abst.). *Magnesium Res.*, 1994; 7 (Supp. l): 51.
20. Benyhe S, Szucs M, Varga E, et al: Cation and guanine nucleotide effects on ligand bindings properties of mu and delta receptors in rat brain membranes. *Acta Biochim. Biophys. Hung.*, 1989; 24:69–81.
21. Durlach J, Durlach V, Bac P, et al: Mg and ageing. II. Clinical data: aetiological mechanisms and physiopathological consequences of Mg deficit in the elderly. *Magnesium Res.*, 1993; 6 374–394.
22. Bac P, Herrenknecht C, Binet P, et al: Audiogenic seizures in Mg-deficient mice: effects of Mg pyrrolidone-2-carboxylate, Mg acetyltaurinate, Mg chloride and vitamin B6. *Magnesium Res.*, 1993; 6:11–19.
23. Bac P, Herrenknecht C, Binet P, et al: Effects of various Mg salts on the action of systemic kainic acid in Mg deficient rats: a new model of accelerated hippocampal ageing-like injury? (abst.). *Magnesium Res.*, 1993; 6:300–301.
24. Velloso A: Magnesium, free radicals and longevity. *Magnesium Res.*, 1994; 7 (Supp. l): 48.
25. Galland L: Magnesium, stress and neuropsychiatric disorders. *Magnesium Trace Elem.*, 1991–1992; 10:287–301.

26. Kirov GK, Birch NJ, Steadman P, et al: Plasma Mg levels in a population of psychiatric patients: correlation with symptoms. *Neuropsychobiology*, 1994; 30:73–78.
27. Durlach J: Death from infancy to older age and marginal Mg deficiency. How long should follow-up of the consequences of undernutrition in pregnancy be continued? *Magnesium Res.*, 1993; 6:297–298.
28. Mitani K: Relationship between neurological diseases due to Al load, especially amyotrophic lateral sclerosis and magnesium status. *Magnesium Res.*, 1992; 5:203–213.
29. Durlach J: Magnesium depletion and pathogenesis of Alzheimer's disease. *Magnesium Res.*, 1990; 3:217–218.
30. Tolbert L, Haigler T, Waits M, et al: Brief report: lack of response in an autistic population to a low dose clinical trial of pyridoxine plus magnesium. *J. Autism Dev. Dis.*, 1993; 23:193–199.

Chapter 21

Magnesium: Regulation of Cell Minerals and Its Therapeutic Use

Michel Bara

CONTENTS

21.1 INTRODUCTION

Magnesium ions play an important role in a large number of cellular processes (membranes, cytoplasm, nucleus, subcellular organites, etc.). Magnesium ions regulate various biochemical reactions (cofactor for hundreds of enzymes and a role in protein synthesis), act as a cofactor in transmembrane ionic movements, and interact with numerous minerals.

21.2 GENERAL CONSIDERATIONS

21.2.1 Importance of Magnesium in the Cell and in the Nervous System

In the cell, magnesium ions possess powerful membrane stabilizing properties due to their structural and enzymatic roles. Magnesium is necessary for the insertion of protein into membranes and for the formation of phospholipids with parallel electrostatic polarizing effects. Magnesium has profound effects on solute and water transport in various cells and intracellular Mg^{2+} is known to interact with several channels[1]. Magnesium is an essential cofactor for the activity of nerve growth factor *in vitro*, and it may also promote neuronal adhesion to substances and contribute to extradural neuronal membrane stability. Moreover, excess of Mg^{2+} ions in the extracellular fluid depresses synaptic transmission and causes rectification of the ionic channels coupled to the *N*-methyl-D-aspartate receptor.

21.2.2 Screening-Binding Concept

To understand the cellular effects of Mg^{2+}, it is important to clarify the screening-binding concept. The membranous and cellular effects of Mg^{2+} are principally due to the electrostatic interactions between Mg^{2+} and external negatively charged surface groups. There are two categories of electrostatic interactions:

1. In the screening process, the cations remain mobile, most likely fully hydrated and unbound, being held loosely by hydrogen bonds in a diffuse layer close to the surface. The term "screening" is used to show the existence of ions at a distance of a few angstroms from the membrane surface in the aqueous diffuse layer and to distinguish this effect from the phenomenon of specific adsorption. In this case, the membrane polar groups (positive and negative), whatever their location and distribution, are masked by the hydration shell around the cations and this reduces considerably the repulsive or attractive forces between the Mg^{2+} ions and the surface charges. The screening process induces an increase in membrane stability and in membrane resistance.
2. In the binding processes, Mg^{2+} ions have lost their hydration shells, are not mobile, and the interactions between cations and surface polar groups are direct and possible. A new theory[2] indicates that the binding process induces either a decrease or an increase in membrane resistance and in membrane stability. The induction of these two processes is generally due to the concentration of magnesium salts. In a binding process, the increase or decrease in membrane stability depends on the distribution and the location of the external sites. The magnesium salts influence the packing of the phospholipids in membrane and can lead to membrane alterations, such as morphological changes and/or site distribution changes and ionic transfer modifications.

The specific screening-binding property of Mg^{2+} ions elicits a relationship between internal and external magnesium and other cell minerals. For example, in nerve, it has been suggested that magnesium and calcium may decrease the magnitude of the potential at the outer surface (due to Na^+ and K^+ concentrations) by screening existing negative charges on the membrane as well as by binding to specific sites.[3] This property of Mg^{2+} ions may be carried out either on common or on toxic minerals, regulating their cellular actions.

21.3 MAGNESIUM: REGULATION OF CELL MINERALS

21.3.1 Interactions Between Intra- and Extracellular Magnesium and Common Metals

In this chapter, the interactions are limited to ionic channels but Mg interacts with other systems: Na/K, Ca/Mg-ATPases, and Na/Ca exchanger, for examples.

21.3.1.1 Sodium

In cells, Mg^{2+} ions interfere essentially with two Na^+ exchange pathways.[3]

Paracellular exchange — Extracellular Mg (e-Mg^{2+}) blocks the passive movement of Na$^+$ across the heart membrane cells by a screening effect on fixed charges. In a human model (amniotic membrane), Mg^{2+} has a biphasic effect: at lower oral concentrations, the transfer is decreased by a screening effect; at higher parenteral concentrations, it is increased because Mg^{2+} ions release the surface charges which become accessible to Na$^+$ ions.

Na$^+$ channels — Generally, intracellular Mg (i-Mg^{2+}) decreases the Na$^+$ channel conductance. This effect is due to an i-Mg block; i-Mg acts as a fast blocker rather than gradually decreasing current by the screening of surface charges.[3] In cerebellar granule cells,[4] a model is proposed in which Na$^+$ and Mg^{2+} occupy the sodium channels competitively. Na$^+$ and Mg^{2+} have to bind to the sodium channels during their passage through the membrane. The blockage of Na$^+$ currents caused by Mg^{2+} ions occupying the Na$^+$ channels, may be explained by either (1) i-Mg entering into the sodium channels and binding to an inside site during the membrane depolarization, or (2) by the strong binding force between i-Mg and the binding site retarding the release of Mg^{2+} from the channel. Generally, e-Mg^{2+} blocks the Na$^+$ channels in tight membranes and opens it in leaky membranes.[3]

21.3.1.2   Potassium

Interactions between Mg^{2+} and K$^+$ have been reported in various cells.[5] Inhibitory effects of i-Mg^{2+} have been reported for outward currents through several K$^+$ channels.[3] As a first example, depletion of i-Mg increases outward muscarinic K$^+$ currents, but decreases inward currents, thereby reducing the inwardly rectifying property of the channels. i-Mg acts as a fast ionic blocker by plugging the open channel pore from the inside of the cell and preventing outward K$^+$ passage. As a second example, the outward single-channel current through the inwardly rectifying K$^+$ channel is blocked by i-Mg^{2+}. This channel fluctuates between sublevels which are blocked by i-Mg^{2+} in a voltage-dependent manner.

Activation effects of e-Mg^{2+} on K$^+$ ion transfer have been reported on artificial bilayer membranes and on the human amniotic membrane model.[5]

21.3.1.3   Calcium

In the cell, the interactions between intra- and extracellular Mg^{2+} and Ca^{2+} are numerous. There is often an inverse relationship between Mg^{2+} and Ca^{2+}, and this may be reflected in their activities at different sites.[1] Mg^{2+} ions regulate the Ca^{2+} concentration and the reverse also occurs. Indeed, in synaptosomes, an elevation of the external calcium concentration (e-Ca^{2+}) significantly reduces i-Mg^{2+} concentration and this result indicates that the intrasynaptosomal Mg^{2+} activity is partially regulated by an Na/Mg exchange mechanism controlled by e-Ca^{2+}.[6] The concentration relationship is also observed in rat cortical slices, where excitatory amino acids inhibit carbachol-stimulated phosphoinositide breakdown by lowering i-Ca^{2+} through a mechanism dependent on e-Mg^{2+}.[7]

In other cells, particularly rat aortic smooth muscle, e-Mg^{2+} regulates the level of i-Ca^{2+}. The regulatory mechanism is coordinated by an Mg^{2+} effect on the Ca^{2+} release, which is reduced in skinned skeletal muscle fibers of frog.[8] These data introduce the effect of Mg^{2+} on the calcium channels. Generally, Mg^{2+} is a potent inhibitor of Ca^{2+} currents in various cells. e-Mg^{2+} blocks a single Ca channel current in guinea-pig ventricular cells and competitively blocks I-Ca by increasing the apparent dissociation constant for Ca^{2+} in guinea-pig smooth muscle cells.[9] In myocytes, an increase of e-Mg^{2+} reduces the magnitude of the voltage-dependent inward Ca^{2+} current I-Ca.

This effect is typical of both T- and L-type Ca^{2+} currents.[10] The blockage effect of Mg^{2+} implicates an inhibition of the development of the slow regenerative potential resulting from inward current through the Ca^{2+} channels. This inhibitory effect is also revealed by the fact that e-Mg^{2+} can not be substituted for Ca^{2+} as the charge carrier in inward current channels in guinea-pig neocortical neurones and in cholinergic synapse between two neurones.[11]

Elevation of i-Mg^{2+} in cardiac myocytes caused a reduction (16%) in peak I-Ca amplitude.[12] Inactivation of Ca channels can occur by both calcium-dependent and purely voltage-dependent mechanisms, and a component of voltage-dependent inactivation can be modulated by changes in cytoplasmic Mg^{2+}. In cerebellar granule neurones, the N-type Ca^{2+} channels are preferentially inhibited by elevation of i-Mg^{2+}.[13] Also, different Ca^{2+}-type channels (T, L, N) are inhibited either by e-Mg^{2+} or by i-Mg^{2+}. In this case, Mg^{2+} ions lodge within the pore and create an open-channel block.

Another phenomenon of Mg^{2+} inhibition has been described. In isolated guinea-pig ventricular myocytes, elevations of i-Mg^{2+} shorten the action potential and suppress the Ca^{2+} current. These data suggest that cytosolic Mg^{2+} regulates Ca^{2+} channel activity. In the same way, in Lymnaea neurones, i-Mg^{2+} is one important regulator of the voltage-independent calcium channels. i-Mg^{2+} rapidly and reversibly inhibits the activity of this calcium channel, the primary effect of Mg^{2+} being to promote long closings of the channel. In this case, the mechanism of i-Mg^{2+} inhibition is distinct from open-channel block, a phenomenon seen previously in a variety of Ca^{2+} channels.[14] Moreover, in the mouse neuromuscular junction a noncompetitive antagonism of Ca^{2+} by Mg^{2+} has been shown. e-Mg^{2+} and i-Mg^{2+} ions regulate common mineral concentrations and transfer in the cells. The Mg^{2+} effect is generally an inhibition of the transfer through Na^+, K^+, and Ca^{2+} channels.

21.3.2 Interactions Between Magnesium and Other Metals

A number of metal ions have mutagenic and/or carcinogenic activities. These metal ions have been classified as (1) well-established carcinogenic metals or metalloids (As, Be, Cr, Ni); (2) cocarcinogenic metals (Cd, Co, Pb); and (3) weak cocarcinogenic metals (Al, Cu, Fe, Zn). The interactions between Mg^{2+} and these metals and with another metal (Hg) have been observed.[15]

21.3.2.1 Well-Established Carcinogenic Metals

Mg^{2+} ions have an antagonistic action towards the carcinogenesis induced by Be and Ni.[1] However, there are few studies on the relationship between Mg and As, though the additional effects of As and Pb reduce the brain Mg. There is a similar effect between Mg deficiency and chromium hypersensitivity: an increase of the glutathione level and a deficiency of protein metabolism.[1] Mg^{2+} and Ni^{2+} often have a similar cellular action. For example, e-Mg^{2+} and e-Ni^{2+} block Ca^{2+} currents and K^+ currents in frog skeletal muscle. In rat skeletal muscle, there is an apparent competitive inhibition between i-Mg^{2+} and i-K^+, and Ni^{2+} was found to have similar concentration- and voltage-dependent blocking effects, but not identical to those of i-Mg^{2+}.[16] Moreover, e-Mg^{2+} and e-Ni^{2+} decrease the membrane fluidity of rat cerebral neurone cortex.[17] In the human amniotic membrane model, no antagonism between Mg-As, Mg-Be, and Mg-Cr is observed, but there is a noncompetitive inhibition between Mg and Ni (Mg and Ni are fixed on different sites).[15]

21.3.2.2 Cocarcinogenic Metals

The interactions between magnesium and cocarcinogenic metals generally have two forms: opposite or similar effects.

Opposite effects — As a first example, Co^{2+} ions block Mg^{2+} channels on the outer surface of the apical membrane of the distal tubule cells. Mg^{2+} has an activating effect on alkaline phosphatase activity of intestinal microvilli, while Co^{2+} is inactive. Moreover, K^+ currents of dissociated neurones from adult guinea-pig hippocampus are depressed by substituting Co^{2+} or Cd^{2+} for e-Ca^{2+}, but a similar effect is not obtained by substituting Mg^{2+} for Ca^{2+}.[18] As a second example, the presence of Mg^{2+} diminishes the toxic effect of Cd and Pb on the intestinal transport of D-glucose and L-tyrosine in rat.[19] Mg^{2+} interferes with the entry of Cd^{2+} into frog motor nerve endings and there is a competitive inhibition between Mg-Cd on the external fixed groups of the human amnion model.[15] As a third example, the inhibitory effect of Pb^{2+} on calcium-dependent K^+ channels from human red cell membrane vesicles is not observed when Mg^{2+} and Ca^{2+} are added before Pb^{2+}.[20] Moreover, a competitive inhibition between Mg-Pb is observed on the human amnion model[15] (Figure 21.1).

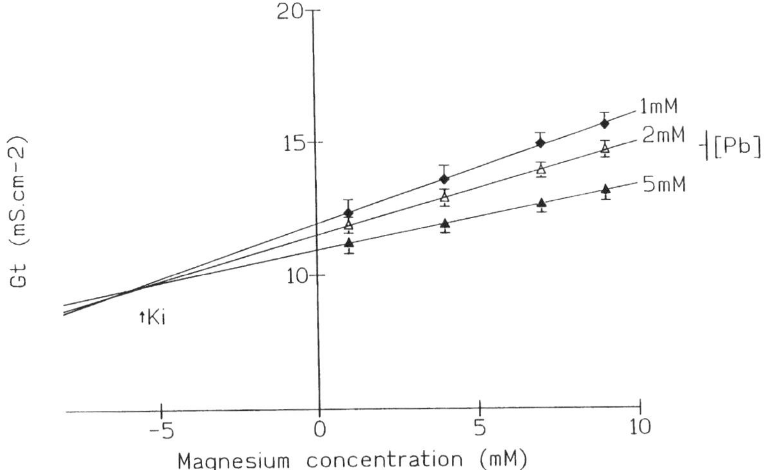

FIGURE 21.1 Schematic representation of the Dixon's curves showing the relationship between the external magnesium concentration and the total conductance Gt across the human amniotic membrane model when the Pb concentration is modified. These curves indicate a competitive inhibition between Mg and Pb (Ki = inhibition constant).

Similar effects — In guinea-pig ventromedial hypothalamic neurones, Ca^{2+} conductance blockage with Co^{2+} or replacement of Ca^{2+} by Mg^{2+} abolishes the low-threshold response. Mg^{2+} and Co^{2+} protect against anoxic damage of the hippocampus in rat brain.[21] Furthermore, Mg^{2+} and Cd^{2+} have a common blocking action on single Ca^{2+} currents and on inward Ca^{2+}-dependent currents in the second-order neurones of cockroach ocelli and in the presynaptic terminal arbors of barnacle phosphoreceptors.[22]

21.3.2.3   Weak Cocarcinogenic Metals

In general, magnesium ions and weak cocarcinogenic metals have opposite effects in the cell. For example, Al impairs transport of Mg into brain tissues and acts at multiple sites to displace Mg^{2+} in synaptosomes of rat brain.[23] There are opposite effects of Cu^{2+} and Mg^{2+} on N-acetyl transferase activity in the pineal gland: Cu^{2+} is a partial noncompetitive inhibitor, Mg^{2+} is an activator. Mg^{2+} reduces the immediate effect of Cu^{2+} exposure on an electroolfactogram in the olfactory epithelium of Atlantic salmon.[24] Moreover, Mg^{2+} inhibits Ni-induced carcinogenesis in the rat kidney, while iron accelerates this effect.[25] Furthermore, Mg^{2+} ions stimulate the (^3H)MK-801 binding to N-methyl-D-aspartate receptor channel in rodent brain and Zn^{2+} ions inhibit this binding.[26] Ca^{2+} and Mg^{2+} remove Zn^{2+}, which is the most potent inhibitor of NMDA receptor-induced Ca^{2+} uptake. Just the opposite, Mg^{2+} and Zn^{2+} activate dolichol kinase in mammalian brain and depress NMDA response in rat pyriform cortex neurones.[26]

21.3.2.4   Mercury

The interactions between Mg^{2+} and Hg^{2+} are of two types: similar effects such as activation of three types of membrane currents in identified molluscan pacemaker neurones; and antagonistic effects in the case where Hg^{2+} inhibits the active transport of D-glucose in the rat small intestine, whereas Mg^{2+} diminishes the toxic effect of Hg^{2+}, rendering it less effective. On the other hand, Hg^{2+} causes a competitive inhibition of Mg^{2+}-ATPase and the inhibition is Mg^{2+} concentration dependent in rat brain microsomes.[27] In the human amniotic model, there is a noncompetitive inhibition between Mg^{2+} and Hg^{2+} on the external charged groups (Figure 21.2).

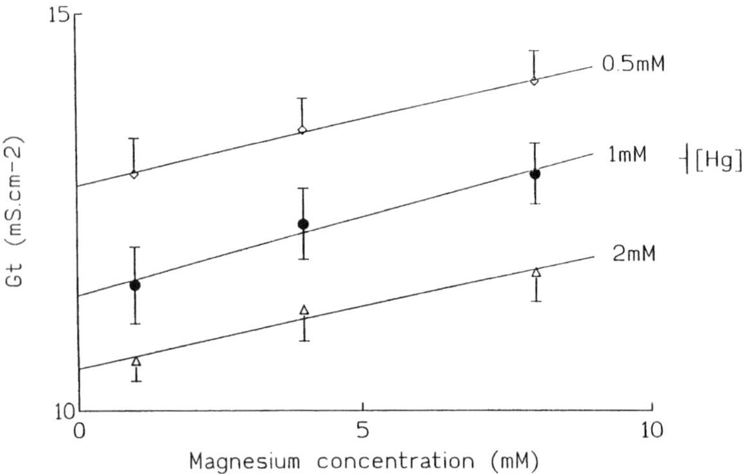

FIGURE 21.2 Schematic representation of the Dixon's curves showing the relationship between the external magnesium concentration and the total conductance Gt across the human amniotic membrane model when the Hg concentration is modified. These curves indicate a noncompetitive inhibition between Mg and Hg.

21.4 THERAPEUTIC USE

Two different types of therapy with magnesium may be used: physiological oral magnesium supplement which is totally atoxic since it palliates magnesium deficiencies by simply normalizing the magnesium intake, and pharmacological magnesium therapy which may induce toxicity since it creates iatrogenic magnesium overload.[28] The cellular properties of Mg^{2+} towards minerals, such as blockage of channels and competitive inhibition against pollutants, may be exploited in therapeutic use. For example, the fact that e-Mg^{2+} or i-Mg^{2+} ions block the Ca^{2+} channels may be put to use. Mg^{2+} ions protect neurones against anoxic damage in the rat hippocampal slices and significantly improve recovery of both dentate granule cells and CA_1 pyramidal cells after anoxia. Indeed, Ca^{2+} has been suggested to be the final common mediator of cell damage. Mg^{2+}, a nonspecific antagonist of Ca^{2+}, inhibits Ca^{2+} uptake process, may reduce cell damage, and is able to counteract the cell damage induced by entry of Ca^{2+}. These data indicate a neuroprotective effect of Mg^{2+}. Any potential clinical benefit of Mg^{2+} would have to be tested in an *in vivo* model, but serious problems, such as the limited permeability of the blood-brain barrier to Mg^{2+} would have to be overcome.[21,29]

In the same way, studies[30] *in vitro* and *in vivo* show that phencyclidine hydrochloride (PCP) can induce cerebral arterial and arteriolar spasms, in psychotomimetic concentrations, by acting on a specific PCP receptor, which is followed by rupture of cerebral and postcapillary venules. Mg^{2+}, which has the ability to block Ca^{2+} channels, neurotransmitter release, intracellular Ca^{2+} release, and the NMDA-glutamate receptor channel, might block the PCP receptor which subserves cerebral contractile events and thereby prevent rupture of microvessels.

In rat, Mg aspartate HCl, administered intravenously, attenuates cerebrospasms induced by PCP. Since Mg^{2+} prevents rupture of the cerebral microvessels, it may prove clinically useful in prevention and treatment of PCP-intoxicated victims. On the other hand, the property of inhibition of T- and L-type Ca^{2+} currents by Mg^{2+} may still be used. Indeed, T-current may participate in the late diastolic slope of subsidiary pacemakers, and inhibition of T-type Ca^{2+} currents by Mg^{2+} may contribute to the antiarrhythmic effects of Mg^{2+} on atrial muscle.[9] The competitive and noncompetitive inhibition of Mg^{2+} towards some pollutants, which may remove these metals from the receptor sites, may be exploited. For example, pharmacological magnesium therapy may be used in the treatment for acute intoxication by aluminum phosphide, as well as for intoxication by the major pollutants Pb, Cd, and perhaps Hg, ultimately alternating with chelation treatment.[28]

21.5 CONCLUSION

Extra- and intracellular magnesium regulate various minerals in the cell. Generally, Mg^{2+} ions block the transfer of common minerals (Na^+, K^+, Ca^{2+}). The relationships between Mg^{2+} and pollutant minerals are complicated: either their effects are similar or there are opposite and competitive actions. The properties of magnesium towards minerals may be used in therapeutics, particularly the channel blockage and the competition on fixed sites properties.

ACKNOWLEDGMENT

We thank M. M. Bouyer for excellent technical assistance in documentation research.

REFERENCES

1. Durlach J: *Magnesium in Clinical Practice*. John Libbey, London, 1988.
2. Bara M, Guiet-Bara A, and Durlach J: A qualitative theory of the screening-binding effects of magnesium salts on epithelial cell membranes. *Magnesium Res.*, 1989; 2 243–247.
3. Bara M, Guiet-Bara A, and Durlach J: Regulations of sodium and potassium pathways by magnesium in cell membranes. *Magnesium Res.*, 1993; 6:167–177.
4. Lin F, Conti F, and Moran O: Competitive blockage of sodium channel by intracellular magnesium ions in central mammalian neurones. *Biophys. J.*, 1991; 19:109–118.
5. Bara M and Guiet-Bara A: Potassium, magnesium and membranes. *Magnesium*, 1984; 3:212–225.
6. Heinonen E and Akerman KED: Intracellular free magnesium in synaptosomes measured with entrapped eriochrome blue. *Biochim. Biophys. Acta*, 1987; 898:331–337.
7. Lee HM and Fain JN: Magnesium-dependent inhibition of agonist-stimulated phosphoinositide breakdown in rat cortical slices by excitatory amino acids. *J. Neurochem.*, 1992; 59:953–962.
8. Lamb GD and Stephenson DG: Effect of magnesium ion on the control of calcium ion release in skeletal muscle fibers of the toad. *J. Physiol. (London)*, 1991; 434:507–528.
9. Smirnov SV, Ganitkevich VY, and Shuba MF: Mechanism of bivalent cation action on the calcium conductance of a single smooth muscle cell membrane. *Biol. Membr.*, 1986; 3:704–715.
10. Wu J and Lipsius SL: Effects of extracellular magnesium on T- and L-type calcium currents in single atrial myocytes. *Am. J. Physiol.*, 1990; 259:H1842–H1850.
11. Franz P, Galvan M, and Constanti A: Calcium-dependent action potentials and associated inward currents in guinea-pig neocortical neurons in vitro. *Brain Res.*, 1986; 366:262–271.
12. Hartzell HC and White RE: Effects of magnesium on inactivation of the voltage-gated calcium current in cardiac myocytes. *J. Gen. Physiol.*, 1989; 94:745–768.
13. Pearson HA, Sutton KG, Scott RH, et al: Ca^{2+} currents in cerebellar granule neurones: role of internal Mg^{2+} in altering characteristics and antagonist effects. *Neuropharmacology*, 1993; 32:1171–1183.
14. Strong JA and Scott SA: Divalent-sensitive voltage-independent calcium channels in Lymnaea neurons: permeation properties and inhibition by intracellular magnesium. *J. Neurosci.*, 1992; 12 2993–3003.
15. Durlach J, Bara M, Guiet-Bara A, et al: Relationship between magnesium, cancer and carcinogenic or anticancer metals. *Anticancer Res.*, 1986; 6:1353–1362.
16. Ferguson WB: Competitive magnesium block of a large-conductance, calcium-activated potassium channel in rat skeletal muscle: calcium, strontium and nickel also block. *J. Gen. Physiol.*, 1991; 98:163–182.
17. Ohba S, Hiramatsu M, Edamatsu R, et al: Metal ions affect neuronal membrane fluidity of rat cerebral cortex. *Neurochem Res.*, 1994; 19:237–241.
18. Sah P, Gibb AJ, and Gage PW: Potassium current activated by depolarization of dissociated neurons from adult guinea-pig hippocampus. *J. Gen. Physiol.*, 1988; 92:263–278.
19. Iturri SJ and Pena A: Heavy metal-induced inhibition of active transport in the rat small intestine in vitro: interaction with other ions. *Comp. Biochem. Physiol.*, 1986; 84:363–368.
20. Alvarez J, Garcia-Sancho J, and Herreros B: Inhibition of calcium-dependent potassium channels by lead in one-step inside-out vesicles from human red cell membranes. *Biochim. Biophys. Acta*, 1986; 857:291–294.

21. Kass IS, Cottrell JE, and Chambers G: Magnesium and cobalt, not nimodipine, protect neurons against anoxic damage in the rat hippocampal slice. *Anesthesiology*, 1988; 69:710–715.

22. Hayashi JH and Stuart AE: Currents in the presynaptic terminal arbors of barnacle photoreceptors. *Vision Neurosci.*, 1993; 10:261–270.

23. Mundy WR, Kodavanti PRS, Dulchinos VF, et al: Aluminum alters calcium transport in plasma membrane and endoplasmic reticulum from rat brain. *J. Biochem. Toxicol.*, 1994; 9:17–23.

24. Bjerselius R, Winberg S, Winberg Y, et al: Calcium protects olfactory receptor function against acute copper (II) toxicity in Atlantic salmon. *Aquat. Toxicol.*, 1993; 25:125–137.

25. Kasprzak KS, Diwan BA, and Rice JM: Iron accelerates while magnesium inhibits nickel-induced carcinogenesis in the rat kidney. *Toxicology*, 1994; 90:129–140.

26. Sharif NA, Nunes JL, and Whiting RL: Pharmacological characterization of the N-methyl-D-aspartate receptor-channel in rodent and dog brain and rat spinal cord using tritiated MK-801 binding. *Neurochem. Res.*, 1991; 16:563–570.

27. Chetty CS, McBride V, Sands S, et al: Effects in vitro of mercury on rat brain magnesium-ATPase. *Arch. Int. Physiol. Biochem.*, 1990; 98:261–268.

28. Durlach J, Durlach V, Bac P, et al: Magnesium and therapeutics. *Magnesium Res.*, 1994; 7:313–328.

29. Tsuda T, Kogure K, Nishioka K, et al: Magnesium administered up to 24 h following reperfusion prevents ischemic damage of the CA1 neurons in the rat hippocampus. *Neuroscience*, 1991; 44:335–342.

30. Huang QF, Gebrewold A, Altura BT, et al: Magnesium protects against PCP-induced cerebrovasospasms and vascular damage in rat brain. *Magnesium Trace Elem.*, 1990; 9:44–46.

Chapter 22

Magnesium-Related Neurological Disorders

Masayuki Yasui, Kiichiro Ota, and Vincent A. Murphy

CONTENTS

22.1 INTRODUCTION

Magnesium (Mg) is an essential cofactor for many enzymatic reactions, especially those involved in energy metabolism. Mg toxicity occurs primarily after intravenous exposure or during renal failure and can be associated with neurological symptoms resulting from peripheral neuromuscular blockade. More relevant clinically than Mg toxicity, Mg deficiency occurs due to inadequate intake, malabsorption, and renal loss of Mg. Prolonged deficiency of Mg in both animals and humans results in neurological disturbances, namely, hyperexcitability, convulsions, memory loss, and various psychiatric symptoms ranging from apathy to psychosis. These neurological symptoms are reversible with restoration of Mg concentration in the body either by supplementation or correction of the underlying disease causing Mg imbalance.

Although the role of Mg in the normal function of the central nervous system (CNS) is not completely understood, the metal appears to be important for modulation of excitatory actions in the CNS. In animal studies, increased epileptiform activity and decreased threshold to seizures is associated with decreased levels of Mg in the cerebrospinal fluid (CSF) or brain. Clinically, high-dose Mg therapy is effective in treating the convulsions resulting from eclampsia.[1] The excitatory actions of the amino acid neurotransmitters, glutamate and aspartate, via the receptors for *N*-methyl-D-aspartate (NMDA), are blocked by physiological concentrations of Mg.[2]

Further studies are necessary to increase our understanding of the neurological effects of Mg deficiency on the CNS. We present here a brief review of Mg regulation and describe the role of Mg in some neurological disorders.

22.2 MAGNESIUM REGULATION AND DISTRIBUTION

Figure 22.1 depicts the factors that regulate Mg concentrations in the body. The Mg concentration in the body is tightly regulated in order to maintain the physiological functions and activities of various types of cells, especially those in the nervous system.[3] Mg primarily enters the body via

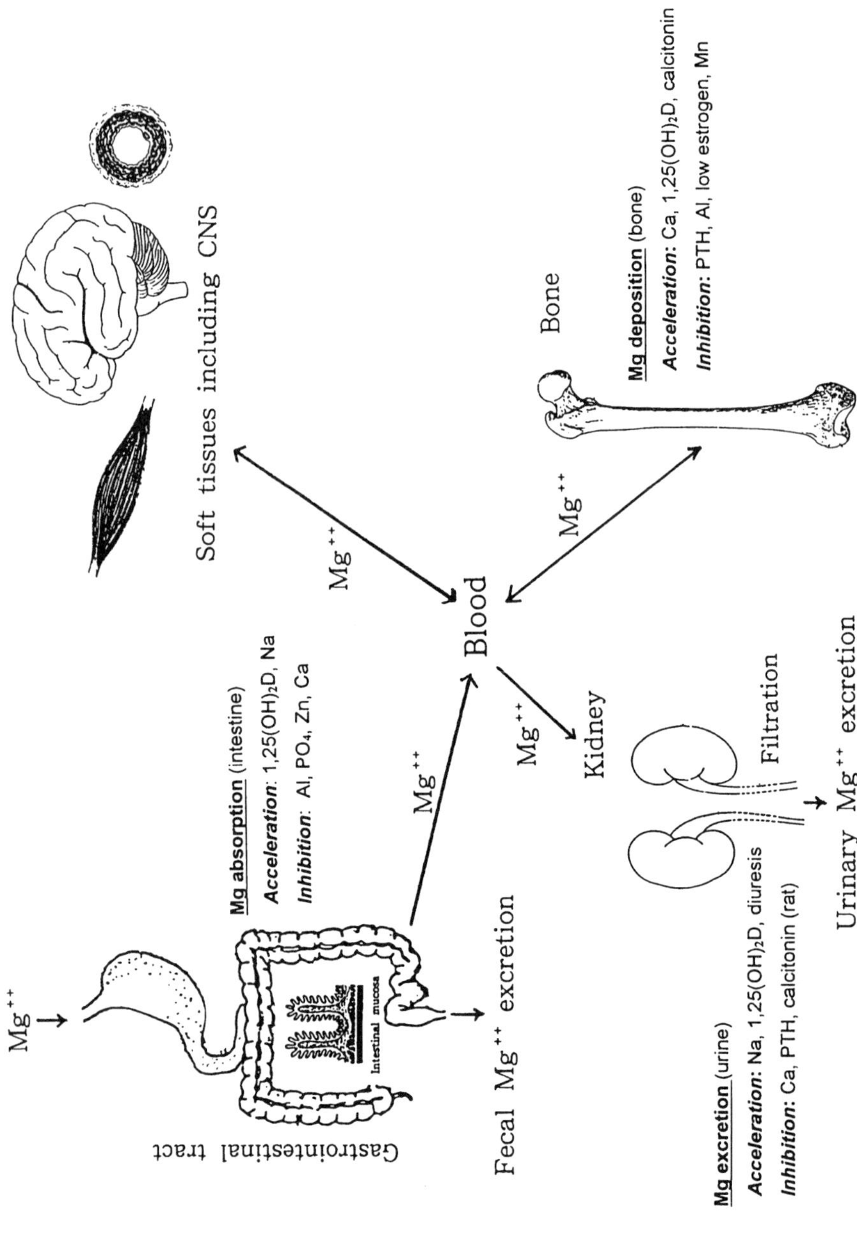

FIGURE 22.1 Factors that regulate magnesium concentration in the body.

absorption from the intestine. Vitamin D enhances absorption of Mg whereas aluminum (Al), calcium (Ca), phosphate, and zinc (Zn) decrease absorption. Once absorbed, Mg concentrations in the serum are regulated by renal excretion and bone deposition. During Mg deficiency, intestinal absorption is increased and Mg renal elimination is decreased.[4,5] Bone deposition is regulated by factors that influence bone Ca deposition. Vitamin D and calcitonin enhance bone deposition whereas parathyroid hormone (PTH) and deficiency of vitamin D or estrogen decrease deposition. Metals can compete with Mg, thereby decreasing bone deposition as is the case for manganese (Mn) and Al, but can also stimulate deposition by increasing levels of hormones that enhance deposition as is the case with Ca. Renal elimination of Mg is regulated at the ascending limb of the loop of Henle. Hypocalcemia decreases Mg renal elimination. Conversely, elevated levels of serum Ca or Mg increase renal elimination. Parathyroid hormone directly decreases Mg renal excretion, but its effect is complicated by increases in serum Ca which increase Mg renal excretion. Enhanced Mg elimination occurs during conditions of diuresis, such as diabetes and hyperaldosteronism. These regulatory sites can be overcome by prolonged changes in serum Mg resulting in changes in tissue Mg concentration.

Tissue regulation results from cell systems that regulate the concentration of Mg in the cell such as membrane ion pumps and intracellular binding sites. Some of the membrane ion pumps are Mg-Na exchange centers in the cell membrane and Mg-H exchange and Mg single transport centers in mitochondria, and the major intracellular binding sites are RNA and ATP.[6] Adrenergic hormones via β-receptors and prostaglandin E1 inhibit Mg influx, whereas phorbol esters increase influx. Intracellular concentration is about four times extracellular concentration. The higher intracellular Mg compared to extracellular Mg explains the higher total levels of Mg in brain regions containing more cellular elements (cortices and hippocampus) compared to those with fewer cellular elements (medulla and spinal cord).

The blood-brain barrier and choroid plexus have an essential role in the regulation of CSF and brain Mg. Transport systems that regulate Mg concentration in the CSF and brain extracellular fluid are located at the choroid plexus and cerebral capillaries, respectively.[7,8] These systems are acutely effective in preventing large fluctuations in Mg concentrations in the brain and CSF during rapid changes in serum Mg. In conjunction with cellular regulatory systems, the CNS levels of Mg are well maintained. However, during extended periods of Mg deficiency or during multiple ion deficiencies (low Ca-Mg diet), CSF and brain levels of Mg decrease.[3] Even acutely, sufficient Mg can enter the brain and CSF in rats after intraperitoneal injections of Mg sulfate to decrease the electrical seizure threshold.[9]

It appears that δ-opioid receptors may be involved in the regulation of endogenous levels and movements of ions in the CNS of the rat, including Mg;[10] sixty min after administration of a single dose of 0.2 µg of encephalon (DPDPE) intraventricularly, transient decreases of the endogenous Ca^{++} and Mg^{++} contents in the parietal cortex, hippocampus, and striatum are found with no change in the levels of monovalent ions or Mn^{++}. A time-dependent downregulation in Ca^{++} and Mg^{++} content was also demonstrated. The action of DPDPE was also found to be dose dependent.

22.3 Mg-RELATED CNS DISORDERS

Mg deficiency increases reaction to noise, excitement, and bodily contact. Decreased Mg levels occur during delirium tremens and trauma, with recovery related to the return of Mg levels to normal. Lower CNS Mg levels are found in amyotrophic lateral sclerosis (ALS) and parkinsonism-dementia (PD). Mg is involved as a cofactor in many vital enzymatic reactions. It is also important in the maintenance of membrane electric potential. Diagnosis of disturbances depending on Mg must often be based on clinical judgment. Hypomagnesemia is frequently associated with hypokalemia and hypocalcemia. Hypermagnesemia presents as neuromuscular, CNS, and cardiac abnormalities. Inadequate dietary intake or absorption of Mg occurs in alcoholism, catabolic states, and gastrointestinal disease.

22.3.1 Excitatory Disorders

Mg deficiency has long been associated with increased CNS excitation. Prolonged Mg deficiency as evidenced by excessive skeletal depletion of the metal can result in severe neurological symptoms, including seizures, coma and, death.[4] Neurological disturbances (hyperexcitability, convulsions, and psychiatric symptoms ranging from apathy to psychosis) occur with Mg deficiency and can be corrected with Mg replacement.[2] Lowering of Mg in the CSF or brain results in increased epileptic activity and decreased seizure thresholds in animals. Mg blocks excitatory NMDA receptors, and NMDA antagonists can reduce the epileptiform activity of Mg deficiency. Alcohol-induced hypomagnesemia is known to be a causative factor in epileptic seizures.[11]

Delirium tremens is described as being linked with chronic depression of the CNS by alcohol and compensatory hyperactivity of neurotransmitters. The sudden discontinuance of alcohol intake induces excessive levels of adrenergic neurotransmitters. A state of noradrenergic hyperactivity in patients with delirium tremens may be responsible for a reduction of Mg in blood and tissues. The Mg blood level may drop to less than 1 mM/l, precipitating seizures that can be treated with Mg sulfate intravenously. Noradrenergic hyperactivity which is responsible for reduced Mg can be prevented by intravenous administration of alcohol, or if this fails, clonidine can be used to block noradrenaline action on the CNS.[12]

The effects of parenteral Mg sulfate ($MgSO_4$) administration on electroencephalographic seizures induced by hyperbolic oxygen (HBO) in awake rats has been reported;[13] sixteen rats chronically implanted with electrocorticographic electrodes were preinjected intraperitoneally with either vehicle or 3 mM/kg $MgSO_4$. The time until development of an electric ictal seizure was measured, and the median times found after the Mg treatment are almost double that in vehicle administration. Mg action seems to be via activation of inhibitory γ-amino butyric acid (GABA)-nergic and glycinergic pathways as well as by blocking of excitatory glutamatergic pathways induced by the HBO action in the CNS, resulting in effective delay of CNS O_2 toxicity.[13]

A major role has been assigned to calcium-regulated intraneuronal enzymatic processes in the alterations of neuronal excitability and production of seizure activity. Calcium entry into neurons is regulated by specific excitability amino acid (EAA) receptor-linked channels. One such type of a receptor-channel complex, characterized by binding of the EAA analogue NMDA, has its channel blocked by Mg^{++}.[14] Other blockers of the NMDA receptor-channel have been shown to be very potent anticonvulsants in seizures induced by maximum electroshock and by HBO.

Under certain experimental conditions, the activation of NMDA/low affinity quisqualate receptors located presynaptically on excitatory fiber terminals and/or postsynaptically on GABAnergic interneurons is essential in hippocampal GABA release.[15] A functional interaction is proposed between glycine and extracellular Mg^{++} in the modulation of the quisqualate NMDA-gated ionophores involved in this process. The balance and plasticity of the above sophisticated system is considered to be essential for several important physiological functions, for example, for generation of long-term potentiation, memory formation, learning, arousal, and emotions.[16]

Mg appears to play a role in the etiopathogenic conditions involved in the onset of migraine.[17] Patients suffering from migraine with and without aura and tension-type headache have lower levels of serum, erythrocyte, mononuclear blood cell, and salivary Mg concentrations during interictal periods. During migraine attacks (ictal periods) serum levels of Mg can decrease further. These lower Mg concentrations in the periphery are a reflection of decreased brain intracellular concentrations as determined by magnetoencephalography and ^{31}P magnetic resonance spectroscopy.[18] Central neuronal hyperexcitability in migraine patients is likely the result of decreased Mg.

22.3.2 Trauma

Nuclear magnetic resonance (NMR) studies of CNS trauma have shown that intracellular free Mg^{++} concentration declines after brain injury.[19] This fall in free Mg concentration was associated with a decrease in brain total tissue Mg concentration. Declines in both free and total tissue Mg concentration can be reduced and neurological outcome improved by removal of brain injury factors, such as excitatory amino acids and opioid peptides. The tissue Mg concentration is significantly reduced by trauma and recovers more slowly in ethanol-treated rats. Furthermore, neurological pathological change is compensated by adjustment of the Mg content in the CNS. Accordingly, it has been suggested that the Mg, and in particular the cytosolic Mg^{++} concentration, is important in determining the degree of neurological pathological symptoms after traumatic injury in the CNS. The Mg concentration in CNS tissues is known to be important in the development of irreversible tissue damage in patients with traumatic brain injury[20] and treatment of neurotrauma can be directed toward restoration of cellular Mg homeostasis.[21]

Mg deficiency, especially in the brain, leads to alteration of cell membrane permeability and decreases ATP transfer reactions necessary for energy-dependent pumps, such as Na-K ATPase. Inhibition of presynaptic Na-K ATPase destroys the sodium gradient which drives the uptake of acidic amino acids. This gradient also assists in the maintenance of cell volume and homeostasis of intracellular Ca^{++}. This breakdown of the sodium gradient results in the release of excitatory amino acids, cell swelling, and increases in intracellular Ca^{++}. Increases in excitatory amino acids result in stimulation of NMDA receptors which can lead to massive depolarization leading to seizures and neuronal death. Stimulation of NMDA receptors also contributes to increasing intracellular Ca^{++}. Elevated intracellular Ca^{++} leads to the release of catecholamines, opioid peptides, and increased synthesis of prostaglandins. Catecholamines and prostaglandins decrease cerebral blood flow and thereby decrease the delivery of energy substrates to the cells. Opioid peptides and catecholamines can further exacerbate the Mg deficiency. All of these phenomena serve to exhaust the cellular energy supply and can contribute to cell death.[1,22]

Direct and indirect evidence indicates that Na-K ATPase activity is reduced or insufficient to maintain ionic balance during and immediately after traumatic events such as episodes of ischemia, hypoglycemia, epilepsy, or after administration of exotoxins (glutamate agonists). Inhibition of this enzyme results in neuronal death and it has been hypothesized that inhibition or attenuation of this enzyme contributes to the production of central neuropathy. Pharmacological manipulation of the various phenomena that contribute to inhibition of Na-K ATPase, Mg deficiency, and neuronal death (catecholamines, excitatory amino acids, Ca^{++}, lipid metabolism, opioid peptides, and decreased energy metabolism) may improve functional neurological outcome after brain injury.[23]

Mg is associated with alcohol- and phencyclidine (PCP)-induced vascular damage. Studies in rats suggest that high alcohol ingestion can result in severe vasospasm, ischemia, and rupture of blood vessels, as a consequence of decreased intracellular Mg^{++}.[24] Administration of high doses of alcohol in rats produces hemorrhagic stroke and is associated with deficits in brain energy metabolism which leads to intracellular Ca^{++} overload and decreases in intracellular pH. Prior to these events is a rapid drop in intracellular Mg^{++}; pretreatment of rats with Mg prevents the intracellular Mg^{++} decrease and prevents the other events. PCP can induce cerebral arterial spasms by acting on specific vascular receptors, which lead to rupture of cerebral venules.[25] Mg prevents cerebrovasospasms and shifts the concentration-effect curve of PCP to higher concentrations, suggesting Mg may modify the binding of PCP to vascular receptors. Protective effects of Mg could also result from direct effects on cerebral vasculature. Image-slitting television microscope studies in the rat have shown that Mg has local vasodilator effects on brain microvessels at physiologic concentrations and is antispasmotic at lower doses.[26] These findings indicate that Mg may be useful for treatment of cerebral vascular effects due to alcohol and PCP intoxication.

Alcohol also exacerbates the behavioral and neurochemical effects of spinal cord trauma in the rat. Alcohol enhances phospholipid hydrolysis with free fatty acid and thromboxane accumulation, increases release of excitatory amino acids, and decreases tissue Mg levels. These changes act to exacerbate secondary tissue damage and diminish neurological recovery after spinal cord injury associated with acute alcohol intoxication.[27]

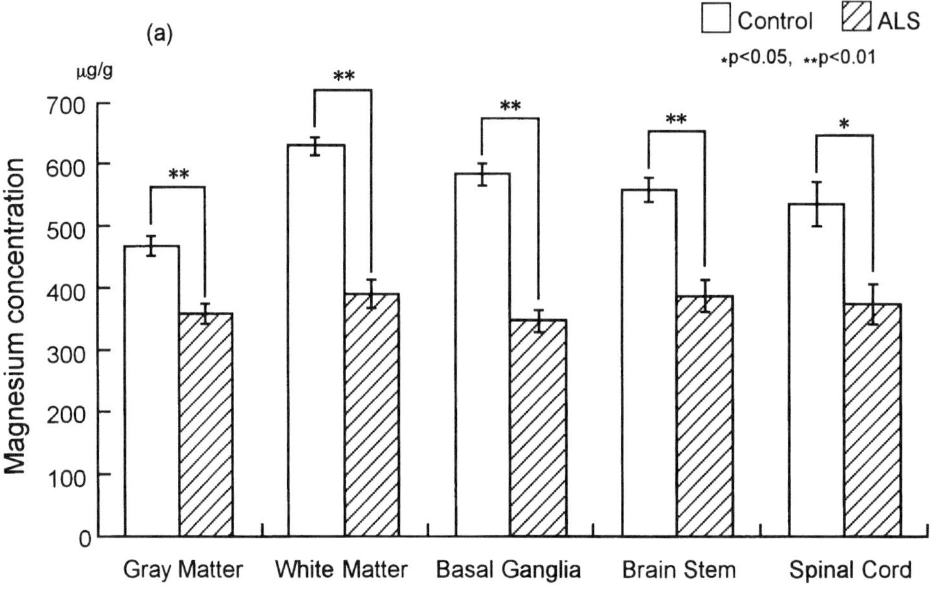

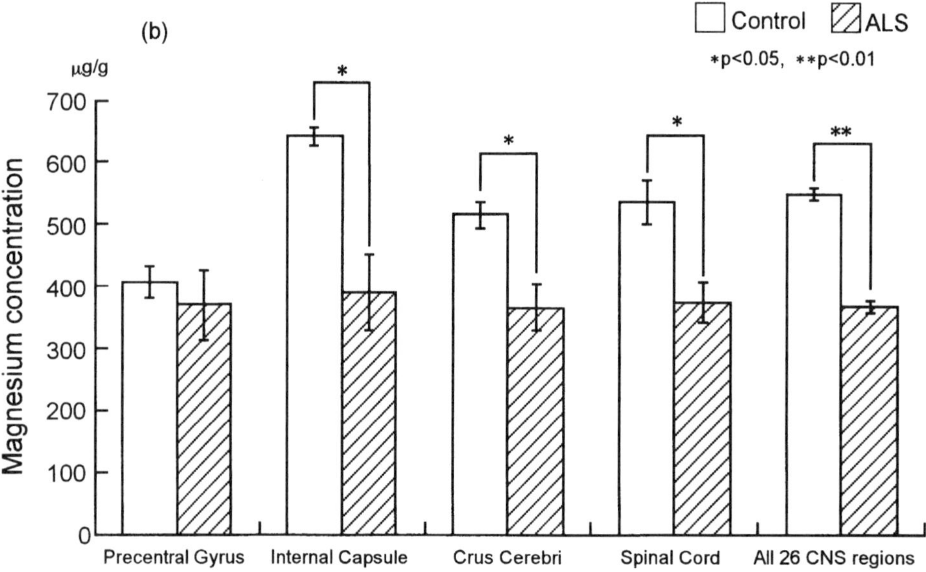

FIGURE 22.2 (a): Magnesium concentration (µg/g dry weight, mean ± SE) in cortical gray matter, white matter, basal ganglia, brain stem, and spinal cord of six cases with amyotrophic lateral sclerosis (ALS) and five neurologically normal controls (** p <.01: significant differences from controls). (b): Magnesium concentration (µg/g dry weight, mean ± SE) in areas with ALS-related neuropathology and all 26 CNS regions of 6 cases with ALS and 5 neurologically normal controls.

22.3.3   Neurodegenerative Diseases

Abnormal interactions of Mg with other metals such as Ca and Al have been noted in disturbances of neurotransmitter or neuroregulator release in CNS tissues.[28] Inhibition of vital enzymatic activities in the CNS due to Mg deficiency was suggested as a causative factor in CNS degeneration. Our studies.[29–31] have consistently demonstrated depressed Mg in ALS, multiple sclerosis, and Parkinson's disease. The Mg content of patients with ALS is shown in Figure 22.2. Patients with

ALS had significantly lower levels of Mg in the CNS than controls. These disorders have been associated with a shortage of Mg in either drinking water or the diet and in dysregulation of Mg *in situ*. An epidemiological study of soil and drinking water in areas where there is a high frequency of CNS degenerative diseases such as ALS and PD found low levels of Mg in soil and water.[32] Mg deficiency has been associated with Alzheimer's disease (AD) and may be related to alteration of serum protein affinity for Mg relative to neurotoxic metals such as Al and enhanced entry of these proteins across the blood-brain barrier.[33]

Excitatory amino acids have been associated with the pathogenesis of ALS and AD. Experimental excess of glutamate produces neurofibrillary tangles and is toxic to neurons. A deficiency of Mg which blocks the action of glutamate by blocking NMDA receptors would exacerbate the effects of glutamate and thereby increase the potential for neurodegenerative pathology.

22.4 CONCLUSION

This review of the distribution and function of Mg and Mg-related disorders in the CNS leads to the conclusion that this mineral is essential for CNS function and has a preventive action on CNS diseases. Research in this area must continue in order to more fully understand the importance of Mg in CNS function and disease.

REFERENCES

1. Gunther T: Magnesium deficiency generally enhances cytotoxicity. *Magnesium Bull.*, 1990; 12:61–64.
2. Morris ME: Brain and CSF magnesium concentrations during magnesium deficiet in animals and humans: neurological symptoms. *Magnesium Res.*, 1992; 5:303-313.
3. Yasui M, Yano I, Yase Y, and Ota K: Distribution of magnesium in central nervous system tissue, trabecular and cortical bone in rats fed with unbalanced diets of minerals. *J. Neurol. Sci.*, 1990; 99:177–183.
4. Langley WF and Mann D: Central nervous system magnesium deficiency. *Arch. Intern. Med.*, 1991; 151:593–596.
5. Durlach J: Regulation of calcium and potassium problems caused by magnesium deficit. In Durlach J (ed): *Magnesium in Clinical Practice*, London, John Libby, 1988 pp 24–31.
6. Flatman PW: Mechanism of magnesium transport. *Annu. Rev. Physiol.*, 1991; 53:259–271.
7. Pollay M: Transport mechanism in the choroid plexus. *Fed. Proc.*, 1974; 33:2064–2069.
8. Bito LZ: Blood-brain barrier. Evidence for active cation transport between blood and the extracellular fluid of brain. *Science*, 1969; 165:81–83.
9. Hallak M, Berman RF, Irtenkauf SM, et al: Peripheral magnesium sulfate enters the brain and increases the threshold for hippocampal seizures in rats. *Am. J. Obstet. Gynecol.*, 1992; 167:1605–1610.
10. Gulya K, Kovacs GL, and Kasa P: Regulation of endogenous calcium and magnesium levels by delta opiod receptors in the rat brain. *Brain Res.*, 1991; 547:22–7.
11. DiPalma JR: Magnesium replacement therapy. *Am. Fam. Physician*, 1990; 42:173–176.
12. Meignan MI: Delirium tremens. Recent neurophysiologic concepts and therapeutic outlook. *Cah. Anesthesiol.*, 1992; 40:303–6.
13. Katz A, Kerem D, and Sherman D: Magnesium sulphate suppresses electroencephalographic manifestations of CNS oxygen. *Undersea Biomed. Res.*, 1990; 17:45–49.
14. Sommer B and Seeburg PH: Glutamate receptor channels: novel properties and new clones. *Trend Pharmacol. Sci.*, 1992; 13:291–296.
15. Janaky R, Saransaari P, and Oja SS: Release of GABA from rat hippocampal slices: involvement of quisqualate/NMDA-gated ionophores and extracellular magnesium. *Neuroscience*, 1993; 53:779–785.
16. Collingridge GL and Singer W: Excitatory amino acid receptors and synaptic plasticity. *TIPS*, 1990; 11:290–295.
17. Sarchielli P, Coata G, Firenze C, et al: Serum and salivary magnesium levels in migraine and tension-type headache. Results in a group of adult patients. *Cephalalgia*, 1992; 12:21–27.
18. Welch KM, Barkley GL, Tepley N, and Ramadan NM: Central neurogenic mechanisms of migraine. *Neurology*, 1993; 43 (Suppl. 3):S21–25.
19. Vink R: Nuclear magnetic resonance characterization of secondary mechanisms following traumatic brain injury. *Mol. Chem. Neuropathol.*, 1993; 18:279–297.
20. Vink R: Magnesium and brain trauma. *Magnesium Trace Elem.*, 1991–92; 10:1–10.
21. Vink R and McIntosh TK: Pharmacological and physiological effects of magnesium on experimental traumatic brain injury. *Magnesium Res.*, 1990; 3:163–9.
22. Lees GJ: Inhibition of sodium-potassium-ATPase: a potentially ubiquitous mechanism contributing to central nervous neuropathology. *Brain Res. Rev.*, 1991; 16:283–300.

23. McIntosh TK: Novel pharmacologic therapies in the treatment of experimental traumatic brain injury: A review. *J. Neurotrauma*, 1993; 10:215–261.

24. Altura BM, Gebrewold A, Altura BT, and Gupta RK: Role of brain [Mg^{2+}]$_i$ in alcohol-induced hemorrhagic stroke in a rat model: a ^{31}P-NMR in vivo study. *Alcohol*, 1995; 12:131–136.

25. Huang QF, Gebrewold A, Altura BT, and Altura BM: Mg^{2+} protects against PCP-induced cerebrovasospasms and vascular damage in rat brain. *Magnesium Trace Elem.*, 1990; 9:44–46.

26. Ema M, Gebrewold A, Altura BT, et al: Magnesium sulphate prevents alcohol-induced spasms of cerebral vascular effects in female rats. *Magnesium Trace Elem.*, 1991–92; 10:269–280.

27. Halt PS, Swanson RA, and Faden AI: Alcohol exacerbates behavioral and neurochemical effects of rat spinal cord trauma. *Arch. Neurol.*, 1992; 49:1178–84.

28. Yasui M, Adach K, Mukoyama M, et al: Low calcium-magnesium diet and thyrotropin releasing hormone. *Med. Sci. Res.*, 1988; 16:885–886.

29. Yasui M, Magnesium concentration in brain from multiple sclerosis patients. *Acta Neurol. Scand.*, 1990; 81:197–200.

30. Yasui M, Yase Y, Kihira T, et al: Magnesium and calcium contents in CNS tissues of amyotrophic lateral sclerosis patients from the Kii Peninsula, Japan. *Eur. Neurol.*, 1992; 32:95–98.

31. Yasui M, Kihira T, and Ota K: Calcium, magnesium and aluminum concentrations in Parkinson's disease. *Neurotoxicology*, 1992; 13:593–600.

32. Kimura K: Studies of amyotrophic lateral sclerosis in the Kozagawa district of the Kii Peninsula, Japan. (Epidemiological, genealogical and environmental studies.) *Wakayama Med. Rep.*, 1965; 9:177–192.

33. Glick JL: Dementias: the role of magnesium deficiency and an hypothesis concerning the pathogenesis of Alzheimer's disease. *Med. Hypotheses*, 1990; 31:211–225.

Section 8

Cadmium

Chapter 23

Cadmium: Acute and Chronic Neurological Disorders

Vincent A. Murphy

CONTENTS

23.1 INTRODUCTION

Neurotoxicity of cadmium (Cd) is manifested by behavioral, neurochemical, and histopathological effects. Behavioral (locomotion and conditioned responses)[1–5] and neurochemical (levels of neurotransmitter and enzyme activities)[1–3,6–8] changes have been reported after oral and parenteral exposure of Cd to animals. Histopathological effects occur after high parenteral doses and are related to cerebral and sensory ganglia hemorrhagic damage.[1,9,10] Epidemiology studies in humans have postulated a relationship between abnormal behavior or decreased intelligence in children and adults exposed to Cd.[1–3,11–13] Recent studies have examined Cd in neurodegenerative diseases.[14–16] Human studies are typically complicated by coexposure with other metals.

Over the last 20 years, there have been numerous reviews of Cd toxicity.[1–3,17,18] The major focus of most of these reviews was not neurotoxicity, because other toxic effects are more frequent in humans. Acute toxicity in humans is primarily respiratory and gastrointestinal, while chronic toxicity is primarily renal.[3,17,18] Since the focus of this chapter is the neurological effects of cadmium, the reader is referred to the above reviews for information on toxicities other than neurotoxicity. This chapter discusses factors that influence the availability of Cd to the nervous system and how Cd interacts with components of the nervous system, and then reviews the neurological effects of Cd in humans and animals with an emphasis on more recent findings.

23.2 AVAILABILITY OF CADMIUM TO THE NERVOUS SYSTEM

23.2.1 Systemic Absorption, Distribution, and Elimination

Before Cd can enter the nervous system it must enter the systemic circulation. Inhalation and oral ingestion are the major routes of entry of Cd into the body. Occupational exposures are primarily by inhalation, and nonoccupational by oral (food) intake.[3,17,18] Availability of Cd from the lungs is primarily influenced by particle size, because the major site of absorption is the alveoli.[3] Based primarily on animals studies, the oral availability of Cd is influenced by a number of factors.[18,20] Competition between Cd and calcium (Ca), copper (Cu), iron (Fe), lead (Pb), manganese (Mn), and zinc (Zn), and complexation with selenium (Se) at the site of intestinal absorption are partly responsible for alterations in the availability of Cd or the other elements after coadministration.[1,3,17,18] Cd availability after oral ingestion is increased after exposure to vitamin D, coadministration with ethanol, during dietary protein deficiency, and in younger animals.[1,3,18,19]

Once absorbed, Cd binds to plasma proteins (albumin and β_2-microglobulin).[1,3,18] Over several hours, Cd redistributes from the plasma proteins to erythrocytes and lymphocytes in the blood. Once in the blood cells, Cd does not freely exchange with the plasma. Tissue concentrations of Cd are initially highest at the site of exposure; the lung upon inhalation and the intestine after oral administration.[1,3,18] Redistribution of Cd then occurs from the site of exposure throughout the body, with the liver receiving the largest amount. In the liver, Cd becomes bound to metallothionein (see below) and preferentially distributes to the kidney. The biological half-life of Cd is about 200 days in rat and 16 to 33 years in humans; this explains why it accumulates in the body.[18] Initially, fecal excretion is the major route of elimination, primarily due to ingestion of atmospheric particles and poor intestinal absorption.[3,17,18] Some biliary excretion may occur prior to entry into the systemic circulation from the liver after oral absorption. Once in the systemic circulation, fecal and urinary elimination are roughly equal.[3,17,18]

23.2.2 Role of Metallothionein

Metallothioneins (MT) are a group of low molecular weight proteins with a high content of cysteine residues[18,21,22] which are responsible for the metal binding properties of MTs.[22] MT plays a major role in the homeostasis of Cu and Zn and modifies the toxicity and distribution of Cd.[18,22,23] Induction of MT by Cd is greatest in liver, kidney, intestine, and pancreas and lowest in brain, skeletal muscle, and testes.[18] Other metals such as Zn, bismuth (Bi), cobalt (Co), and mercury (Hg), as well as ethanol, glucocorticoids, lipopolysaccharides, and nutritional deficiencies in Ca and protein can induce MT, but to a lesser extent than Cd,[22,24] and thereby influence tissue levels of Cd. Low induction of MT in the brain is partly due to limited distribution of Cd into areas of the brain which have a blood-brain barrier.[20,23] MT protects against Cd toxicity by chelating the metal, thus preventing its interaction with key enzymes in the tissues and increasing its elimination from the circulation via the kidney.[18,22] However, Cd bound to MT is more toxic to the kidney, because the Cd-protein complex is reabsorbed and broken down by the proximal tubule cells to liberate Cd.[3,18]

Three forms of MT are present in brain: MT-I, MT-II, and MT-III.[23] MT-I and MT-II are similar to MTs found in peripheral tissues. MT-1 and MT-11 have been localized to ependymal cells, arachnoid, pia mater, endothelial cells and glia, but not neurons.[24,25] MT appears to be present in the choroid plexus of young animals, but may be absent in adults.[24] No studies examining the location of MT-III were found. MT-I and MT-II are induced in these brain areas by lipopolysaccharides, Cd, or Zn.[23,24] The Gomori-positive glia around circumventricular organs such as the area postrema, subfornical organ, and arcuate nucleus of the hypothalamus, which do not contain a blood-brain barrier contain inducible MT.[24] MT-III, which is unique to brain, differs from the other forms of MT as it contains an additional threonine residue at position 5, a six-residue insert at position 55, and eight glutamate residues.[21,23] This protein, unlike other MTs, has an acidic apoprotein, has brain cell growth inhibitory activity, and is not induced by Zn, Cd, dexamethasone, and lipopolysaccharides.[21,23]

Ebadi has proposed that MTs in the brain serve to protect the brain from elevations in free Zn and provide a means to deliver Zn to the site of synthesis of Zn-containing metalloenzymes and proteins.[23] MT in glia, endothelial, and ependymal cells may serve to bind and sequester Cd, thereby reducing the level of free Cd and preventing it from diffusing in high concentrations to other parts of the brain.[25] The induction of MT after Cd exposure in cerebral endothelial cells and the presence of MT in glia that are concentrated near the circumventricular organs support the notion that MT is involved in preventing rapid entry of toxic metals into the brain.

23.2.3 Blood-Brain Barrier and Choroid Plexus

The majority of the central nervous system (CNS) is protected from rapid entry of Cd by the blood-brain barrier (BBB).[26] The BBB consists of endothelial cells in the cerebral capillaries with tight junctions that restrict the entry of polar and protein-bound molecules. Cd in the circulation exists primarily bound to proteins and has a low permeability across the BBB.[1,20,27,28] The circumventricular organs are small areas near the ventricles of the brain containing capillaries that lack the tight junctions and are fenestrated, allowing diffusion of metal-bound proteins into the brain.[26] As mentioned above, the presence of MT in glial cells and ependymal cells near these circumventricular organs may serve to minimize Cd diffusion into the rest of the brain.

Another circumventricular organ, the choroid plexus, produces cerebrospinal fluid (CSF) and its epithelial cells are joined by tight junctions which represent the blood-cerebrospinal fluid barrier.[26,29,30] The choroid plexus epithelium, in addition to a barrier function, can accumulate heavy metals from the blood or CSF, thereby preventing them from reaching the brain parenchyma.[29,30] This accumulation does not appear to be related to high levels of thiol compounds in the choroid plexus as this tissue contains no greater concentrations of these compounds than the brain parenchyma. Cd accumulation in the choroid plexus may be energy dependent, as Na-K ATPase inhibition reduces Cd uptake into the *in vitro* choroid plexus.[29]

Newborn rats are more susceptible to acute neurotoxicity from Cd than adult rats; this difference was once thought to be due to a less well-developed blood-brain barrier in the newborn animals,[9,27,28] but it is now known that the cerebral capillaries contain tight junctions similar to the adults.[26] Increased levels of Cd in brain of young animals could result from decreased CSF bulk flow, differences in plasma protein binding, or reduced amounts of glial cells containing MT at the circumventricular organs.[24,26] The increased susceptibility of the peripheral nervous system (sensory ganglia) to Cd toxicity is in part due to the lack of a blood-tissue barrier to diffusion of protein-bound metals.[9,27]

Despite the barrier functions of the cerebral capillaries and the choroid plexus, Cd enters the CNS and produces neurotoxic effects. Because of the retention of Cd in the body, permeability barriers are only effective for acute manifestations and act primarily to prevent a rapid rise of the toxic metal into the brain. In addition, high concentrations of circulating Cd may damage the cerebral blood vessels, glia, or the choroid plexus, thus compromising their barrier or sequestering functions.[9,29,30]

23.2.4 Localization of Cadmium in the Nervous System

Radiotracer studies with Cd in rats and mice and brain regional analysis of Cd with atomic absorption have provided information on the distribution of this metal within the brain.[27,28] The circumventricular organs appear to have the most metal, these areas being the choroid plexus, hypophysis, area postrema, meninges, and pineal gland in the rat.[27] No radioactivity was found in the brain parenchyma or the spinal cord, but radioactivity was found in the sensory and autonomic ganglia, sciatic nerve, and spinal nerve roots in decreasing order. Elevated radioactivity was also found in the olfactory bulb which lies near the fenestrated capillary system of the olfactory epithelium. In the mouse, the effect of age was monitored and Cd distribution correlated with development of the vascular system.[28] Apparent movement of tracer from the blood vessels into the brain parenchyma was observed in 7- to 14-day-old mice, but not in adult mice. The highest activity was in the internal granular layer of the cerebellum, an area known to be susceptible to hemorrhage caused by Cd.[28] Regional Cd levels as measured by atomic absorption varied between 1.5- and 2-fold. Regions near circumventricular organs or fenestrated capillaries systems, such as the olfactory bulb and the hypothalamus, tended to have the higher concentrations.[20,31]

23.3 EFFECTS OF CADMIUM ON COMPONENTS OF THE NERVOUS SYSTEM

23.3.1 Transport and Enzymatic Systems Utilizing Essential Metals

Competition at the site of toxicity likely occurs between Cd and other metals. *In vitro* studies indicate that Cd competes with essential metals — primarily Zn and Ca — at metalloenzymes, proteins, and ion channels.[1,3,18] Cd can produce symptoms of Zn deficiency and administration of Zn protects against some of the acute toxicity of Cd.[1,9,18,32] Competition is less important for low-level chronic exposures than high-dose acute exposures, as most of the *in vitro* competition studies require high concentrations of Cd or low amounts of the competitive ion.

Transport proteins and enzymes important in cellular Ca homeostasis are inhibited by Cd.[1,3,9,18] Inhibition of Ca entry into neurons by Cd is due primarily to its ability to block voltage-sensitive Ca channels.[33] Cd inhibits Ca-ATPases, binds calmodulin, and inhibits mitochondrial Ca transport.[1,18,33–35] Inhibition of Ca channels involves direct competition between Ca and Cd, whereas ATPase inhibition results from interaction with thiol groups in the protein and with calmodulin.[1,34] Ca-activated channels for other ions are also influenced by Cd, primarily by displacement of Ca from its binding site.[1]

The displacement of Zn from an active or regulatory site on an enzyme, receptor, ion channel, or transcription codon by Cd can be inhibitory or stimulatory depending on the ability of Cd to behave like a Zn antagonist or agonist in a biologically active protein.[1,23,36] Both metals, in a similar fashion, modulate the responses of excitatory amino acid to their receptors.[1,23,37] Thymidine kinase, a Zn-dependent enzyme, is inhibited in the brain of young animals.[38] Carbonic anhydrase, a Zn-containing enzyme associated with glia and the choroid plexus, is inhibited by Cd.[1,36] Brain Zn-Cu superoxide dismutase is reduced in rats given Cd when the availability of the metal to the brain is increased with ethanol.[8] Lipid peroxidation in the brain is increased by either Zn deficiency or Cd toxicity.[8,19,36]

23.3.2 Neurotransmitter Levels and Reuptake

Cd exposure generally increases levels of norepinephrine and dopamine in animals.[2,7,8,18] Serotonin may be increased or decreased depending on the age, dose, duration of exposure, and brain region analyzed.[2,6–8,39] There are variable effects on acetylcholine depending on the dose and brain region examined.[2,6,7] The Zn-dependent enzyme, pyridoxal kinase, provides pyridoxal phosphate to the enzyme glutamate decarboxylase, which is essential for γ-aminobutyrate synthesis.[23] Inhibition of this enzyme could reduce γ-aminobutyrate levels, and Cd has been shown to reduce this neurotransmitter in one study.[2] The activity of acetylcholinesterase and monoamine oxidase, enzymes responsible for inactivation of neurotransmitters, is reduced after Cd exposure *in vivo*.[19,40–42] The uptake of dopamine and norepinephrine into rat synaptosomes is reduced by 25 μM of Cd, whereas the uptake of other neurotransmitters is also reduced, but at >100 μM.[1,18]

23.3.3 Nerve Conductance and Synaptic Transmission

Atchison reviewed the effects of Cd on synaptic transmission.[33] Evoked transmitter release is reduced by Cd via blockade of presynaptic Ca channels. The blockade of these channels is reversed by increasing extracellular Ca. Cd does not block the transmission of action potentials into presynaptic nerve terminals, nor does it alter spontaneous transmitter release. However, Cd can block the effect of metals which increase spontaneous transmitter release by preventing their entry into the terminal. Cd could also alter synaptic transmission by inhibition of mitochondrial metabolism, Na-K ATPases, acetylcholinesterase, and monoamine oxidase, or by increasing membrane rigidity.[1,18,34,41]

23.3.4 Second Messenger Systems, Translocating Proteins, and Receptors

The ability of brain calmodulin to stimulate purified calmodulin-dependent cyclic nucleotide phosphodiesterase was reduced after exposure to 6 mg/kg/day of Cd for 4 weeks.[35] In these same animals, the activity of calmodulin-dependent enzymes (cyclic nucleotide phosphodiesterase and synaptic Ca-ATPase) were reduced in the cerebral cortex. Cd appears to inhibit these enzymes via interaction with calmodulin by binding to Ca binding sites on the protein.

The activity of adenylate cyclase in brain synaptosomal membranes of the rat was depressed after acute injection of 4 mg/kg/day of Cd. This effect returned to normal after 18 to 24 h,[42] and adenylate cyclase activity was not found to be depressed after repeated exposure at 6 mg/kg/day for 4 weeks.[35] *In vitro*, Cd is a potent inhibitor of adenylate cyclase, K_i of 1.1 μM.[1]

Cd and Zn have similar inhibitory (*N*-methyl-D-aspartate, glutamate, and aspartate) and stimulatory actions (kainate and quisqualate) on excitatory amino acid receptors in the hippocampus of the mouse.[23,37] These metals do not compete with the excitatory amino acid or with the receptor modulator glycine.[37] Cd and Zn appear to bind near the extracellular face of the receptor. Reduced binding of ligands to muscarinic cholinergic and serotonergic receptors at micromolar concentrations and D2 dopamine receptors at higher concentrations occurs with Cd, perhaps by interacting with sulfhydryl groups on the receptor.[1,18]

23.3.5 Vascular and Supporting Elements

Acute Cd exposure results in vacuolation, endothelial gaps, and increased permeability to macromolecules in the endothelium of the vessels of the nervous system.[9,10] This type of damage occurs with endothelia in other tissues, such as the liver, uterus, ovary, placenta, and testes.[3,9] The effects of acute Cd exposure in the cerebral vessels are age related.[9] Specifically, widespread cranial hemorrhage and endothelial vacuolation occur in newborn animals, but these effects decrease with age. Unlike the cerebral vessels, endothelia in the sensory ganglia are more likely to be damaged in adults than in young animals. The susceptibility of vessels varies among strains, species, and tissues, and within tissues.[9] Within a tissue, resistant vessels proliferate and replace damaged vessels, so subsequent exposures to Cd are less damaging or produce no effect.[3,9] Distribution of Cd in nervous tissue correlates with damage in the central nervous system and sensory ganglia.[9,27,30]

The choroid plexus accumulates high levels of toxic metals relative to brain parenchyma.[27–30] Valois and Webster found dose-related degenerative changes (loss of microvilli, cytoplasmic vacuolation, rupture of apical surface, and cellular debris) in mice given 10 and 100 ppm Cd in the water (mean daily intake of 1.4 and 10.8 mg/kg) for 22 weeks, but no changes were found at 1 ppm

(0.2 mg/kg).[30] These changes were accompanied by an indication of functional damage to the epithelium (increased CSF protein). Reports of hydrocephalus in young mice and rats after acute exposure to Cd might be related to choroid plexus damage.[9,10] The choroid plexus epithelium is similar in structure to the proximal tubule cells of the kidney[29,30] and the doses that produce damage to the choroid plexus are similar to those that produce damage in the proximal tubule.[3] However, the mechanism of proximal tubule damage involves Cd-MT accumulation and degradation that has not yet been demonstrated in the choroid plexus.

Glial cells (primarily astrocytes) appear to preferentially take up metals after entering the CNS.[25] These cells contain the majority of MT in the CNS.[24,25] MT is likely involved in providing Zn to the Zn-dependent enzymes in the glial cells.[25] In culture, Cd at 1 μM decreases differentiation and function of various human and rat glial cell lines.[43] It has been postulated that damage to glia by metals results in the release of glutamate and/or enhanced lipid peroxidation that ultimately influence neuronal function.[25] Induction of MT in the Gomori-positive glia near the arcuate nucleus or the area postrema reduces damage to these circumventricular organs from gold thioglucose or cis-platin, respectively.[25] It is likely, due to the high accumulation of Cd in the circumventricular organs, that the Gomori-positive glia modify the toxic effects of Cd in these areas.[24,25]

23.3.6 Development and Growth

Effects of Cd on growth and development of the CNS are likely related to alterations in nucleic acid synthesis; Cd can inhibit Zn-dependent enzymes and compete with Zn at specific metal regulatory sites in the DNA.[3,17,23] Inhibition of microtubules and microfilaments during closure of the neural tube may also play a role.[17,32] Gupta et al. recorded decreased body weight gain, brain weight, DNA synthesis, thymidine synthetase activity, and DNA content of the young of rats exposed to Cd during gestation and lactation, but no change in total brain RNA or protein.[38] These changes were accompanied by decreases in brain Zn. In the offspring of rats exposed only during gestation, these changes were not evident except that brain weight was decreased 7 and 14 days, but not 21 days after birth.[41] It appears that exposure of Cd through both gestation and lactation alters brain growth more than exposure through gestation only.

23.4 NEUROLOGICAL AND BEHAVIORAL EFFECTS OF CADMIUM

23.4.1 Acute Animal Models

The majority of investigations examining acute neuropathological effects of Cd exposure were conducted between 1967 and 1983.[1,9,10] Injection of lethal or near-lethal doses of Cd (2 to 20 mg/kg) damages the CNS in neonatal rodents and the peripheral nervous system in adult animals.[9] Lesions in the peripheral nervous system result from damage to endothelial cells inside the spinal and trigeminal ganglia, resulting in hemorrhage and necrosis of spinal ganglion cells.[9] In the newborn CNS of rodents and rabbits, high doses of parenteral Cd produce hemorrhagic encephalopathy in the white matter and deep gray matter of the cerebrum with degenerative changes in adjacent neurons and glia, and necrosis of the cells in the granular layer of the cerebellum.[1,9,18] Hydrocephalus has also been noted.[9,10] Wong and Klaassen in 1983 examined the effect of acute Cd injection (2 to 6 mg/kg) on the brains of young (4-day-old) vs. adult rats.[10] No lesions in the brain of adult rats were evident. However, in young rats, lesions were evident 4 days after dosing in the cerebral cortex (corpus callosum and dorsal adjacent gray matter), cerebellum, and caudate putamen at 4 and 6 mg/kg. The lesions consisted of necrosis and hemorrhage with occasional inflammatory infiltrates and glial proliferation. After dosing for 19 days, the cerebral cortical lesions were replaced by a large cystic cavity with the notable absence of the corpus callosum, whereas the cerebellum and caudate putamen were found to be normal. Cd toxicity to cellular elements likely results from a combination of hemorrhagic damage and direct toxicity.[9,10,18,25]

Neurobehavioral and biochemical effects can occur after single parenteral doses of 1 to 10 mg/kg Cd and behavioral effects may occur days to weeks after the initial dose.[4,7,15,16] In neonatal animals, an increase in locomotor and social activity is generally found, but this is not always the case. Wong and Klaassen found increased exploratory and diurnal activity in 4-day-old rats 19 days after exposure to a single injection of Cd.[10] Male rats given 2 mg/kg at 5 to 6 days after birth displayed increased activity and playful interaction at 29 to 44 days of age and failed to differentiate between strange and familiar animals at 150 days of age.[5,44] Adults generally show decreases in locomotor activity and require higher doses to obtain effects. Conditioned response behaviors can be enhanced or depressed depending on the dose, the time after dosing of the measurement, and the paradigm used.[5] Inhibition of Na-K ATPase, Ca-ATPase, acetylcholinesterase, and adenylate cyclase occurs between 6 and 24 h after a single dose of 2 to 10 mg/kg intraperitoneally.[34,37]

23.4.2 Chronic and Subchronic Animal Models

A variety of neurobehavioral and biochemical effects on the nervous system are produced in rodents after repeated dosing of Cd. Levels of enzyme activity, metals, and neurotransmitters can be altered after repeated exposure to Cd. As was the case for acute exposures, neonatal animals exposed to lower doses of Cd (0.1 to 1.0 mg/kg/day intraperitoneally or orally) generally have increased locomotor and rearing activity.[1,4,7,45] Adults and adolescent animals exhibit either no change or decreases in locomotor, exploratory, and rearing activities.[1,4,7,45] Conditioned activities, such as lever pressing, have been increased after dietary exposure of Cd for 82 days.[7] Neonatal animals given repeated doses of Cd have decreased open-field inner-square exploration and improved performance in T-maze learning trials.[1,3,5] Peripheral neuropathy has been noted in rats given Cd at 4 mg/kg/day via the water for 31 months.[3]

Attempts have been made to correlate biochemical effects with behavioral effects, but no consistent conclusions have been drawn. An overview of several subchronic studies that have examined either biochemical and/or behavioral effects follows. Administration of 0.1 and 1 mg/kg/day of Cd orally to weanling rats for 30 days produced elevations of dopamine, norepinephrine, and serotonin in certain brain areas at the higher dose, but not at the lower.[1,2] However, locomotor activity was increased in both dose groups. Administration of 0.4 mg/kg/day Cd intraperitoneally for 30 days resulted in elevated levels of serotonin in the cerebellum, cerebrum, striatum, hippocampus, hypothalamus, and pons medulla in adult rats, but only in the hippocampus in weanling rats.[39] Significant decreases in ambulatory and stereotypic movement time and in the number of stereotypic or vertical movements were observed in weanling rats after this dosing regimen.[45] Adolescent rats given 0.1 mg/kg/day of Cd intraperitoneally for 51 days showed similar effects on locomotor behavior after 38, 46, and 51 days of dosing.[4] The same dose in weanling rats dosed for 33 days decreased γ-aminobutyric acid levels in the cerebellum and hypothalamus.[2] Cd at 0.5 mg/kg/day intraperitoneally increases dopamine and decreases serotonin in the adult rat brain after 30 days, decreases locomotor activity and increases failure to respond to a conditioned or unconditioned stimulus in adolescent rats after 21 days, and increases norepinephrine in the midbrain and pons medulla and serotonin in the hypothalamus in weanling rats after 33 days.[2,46] Adult rats given 100 ppm of Cd in the diet for about 82 days had decreased levels of serotonin in the cortex and brainstem along with increased levels of dopamine in the olfactory tubercule.[7] The animals were trained to press a lever to receive food over a 30-min session per day on the 60th day of exposure. The Cd-treated animals showed increased lever-pressing activity after the 5th session, which remained elevated until the 15th session, when activity returned to control levels despite being on the diet for another 6 sessions and having altered neurotransmitters at the end of the study. In another study, administration of 100 ppm of Cd in the water (~4 mg/kg/day) for 90 days in adult rats increased Zn levels, depressed activities of ATPase, monoamine oxidase, and succinic dehydrogenase, increased the level of dopamine, and decreased the level of serotonin in the whole brain.[40]

It is clearly evident that age of exposure, dose, and examination of regional or whole brain are important factors which determine what neurological effects are seen after repeated Cd exposure. However, because very few studies have examined both biochemical and behavioral changes in animals of the same age under the same study design, no consistent correlations can be derived. It is also well known that neurotransmitter levels and neurobehavioral effects can be influenced by factors outside of the CNS, such as body weight, stress, manipulation, and baseline activity. Therefore, in order to correlate neurochemical changes with neurobehavioral changes, studies which maintain consistent study design and control for outside factors are sorely needed.

23.4.3 *In Utero* Animal Models

Embryo lethality, decreased body weight, and developmental abnormalities including neural tube, limb, craniofacial, and skeletal defects, are induced by administration of Cd during gestation in experimental animals.[3,32] Effects at high parenteral doses are associated with maternal toxicity (renal, lung, and decreased body weight) or damage to the placenta.[3] Defects associated with the neural tube (abnormal cranial flexure, exencephaly, hydrocephalus and/or encephalocele) and the eye occur when animals are exposed between days 7 to 14 of gestation.[32] These defects appear to be associated with a failure of the neural tube to close. Zn given 8 to 11 h after Cd, or prior to induction of MT, can protect against these neural tube defects.[32] More subtle effects such as alterations in brain enzymes and/or neurobehavioral effects have been reported in offspring after maternal exposure to low doses of Cd during gestation.[1,3,17,38] In many of these studies, there was no evidence of maternal or overt fetal toxicity.

Details of more recent studies are described below. Pregnant rats were exposed to 50 ppm Cd (4 to 5 mg/kg/day) in the drinking water for the entire period of gestation.[41] The offspring had the same body weight as controls 7 to 21 days after birth, but brain weight was smaller after 7 and 14 days, but not 21 days. Brain protein, DNA, and RNA were the same between control and Cd-treated rats. The metal did cross the placenta, as evidenced by a twofold elevation in the tissues of Cd-treated rats. The activity of several enzymes was decreased after Cd exposure, but the decreases were seen at different times after birth: succinate dehydrogenase at 7 days, cyclic nucleotide phosphohydrolase at 14 days, and acetylcholinesterase, Na-K ATPase, and 5-nucleotidase at 21 days. A similar study to the one above was done except Cd was continued during lactation and the dose was 5 to 6 mg/kg/day during gestation and 7 to 8 mg/kg/day during lactation.[38] In this study, brain Zn and DNA content was reduced along with DNA synthesis and thymidine kinase activity, and brain weight was smaller at 21 days as well as 7 and 14 days after birth. In another study, female rats were given 0.04, 0.4, or 4 mg/kg/day for 5 weeks before mating, during mating, and during gestation.[47] Litter size, body weight gain, and viability of offspring were the same 2 months after birth, but exploratory locomotor activity and performance in the rotorod test were decreased. The offspring of rats exposed to 0.20, 0.62, or 2 mg/kg/day on days 7 to 15 of gestation showed decreased horizontal motor activity (all doses), increased immobility after amphetamine treatment during a 5-min stress swim (all doses), increased social interactions (0.62 and 2.0), retarded acquisition to a conditioned escape response, extended response latency during reacquisition (0.62 and 2.0), and reduced performance on the rotorod (2.0).[48] The birth and weaning weights of these animals were not different from control and the dams showed no signs of toxicity.

The studies above indicated that exposure to Cd *in utero* impairs performance and acquisition of learned behaviors and reduces various aspects of locomotor activity. Increases in social interactions and one-way avoidance could result from impairment of natural inhibitions or sensory activity. One needs to be cautious in attributing these changes to direct neurotoxicity of Cd, as they could result from inhibition of growth and development. It has long been recognized that behavioral changes are influenced by birth weight, which was decreased in some of the studies above. Alterations in neurotransmitters and enzymes occur after *in utero* Cd exposure. In most cases, however, these animals were not examined for behavioral changes, so how these biochemical changes relate to behavior needs further examination.

23.4.4 Observations From Occupational and Environmental Exposures

Neurological symptoms have been described in workers exposed to Cd, such as headache, vertigo, sleep disturbances, increased knee-joint reflexes, tremor, disturbances in sensory and motor function, anorexia, and anosmia.[1,3,17,18] Evaluation of 38 workers from a nickel-Cd battery factory that were exposed to Cd found 34 with headache, 16 with vertigo, 32 with general or muscle weakness, and 6 with brain atrophy by CAT scan.[49] These effects suggest some neurological component to Cd exposure, but some of these effects could have resulted from the systemic effects of toxicity on the lung and kidney rather than directly from direct neurological effects. Hart et al. examined the neuropsychologic performance among 31 workers exposed to Cd that were not experiencing renal toxicity, and those with high urinary levels of Cd performed less well than those with low urinary Cd on measures of attention, psychomotor speed, and memory.[50] Elevated levels of Cd in the hair of violent crime offenders was found in a 1981 study, but this finding was not replicated in a 1991 study.[11]

Numerous studies have examined the relationship between elevated levels of Cd and Pb in the hair of children and their neurological development.[1,3,12,13] Higher levels of Cd and Pb in the hair of children were associated with prolongation of the latency of auditory-evoked potential components and reduction in the amplitude of the potential.[13] In addition, increases in the slow wave activity and decreases in the amplitude of the EEG were related to increased hair Cd and Pb.[13] School achievement and cognitive functioning appear to be the most sensitive to the effects of these metals whereas gross and fine motor movements are the least sensitive.[13] Different aspects of cognitive functioning are altered by Pb and Cd, as Pb accounts for a greater amount of performance IQ whereas Cd accounts for a greater amount of verbal IQ.[13] Thatcher and Lester suggest these differences are related to the fact that Pb affects dopamine levels and Cd affects norepinephrine, acetylcholine and serotonin levels.[13] Cognitive deficits in children have been associated with low hair Zn and high hair Cd, but Stewart-Pinkham failed to find this relationship, instead finding decreased hair inorganic phosphorus was associated with deficits.[12] The author indicated that due to variability hair Cd was not a reliable measure of Cd pollution exposure. However, she suggested that Cd may have caused neurobehavioral changes and it decreased inorganic phosphorus, Zn, and 5-nucleotidase activity and increased hair Pb levels in these children. The correlations between metal contents in hair and cognitive functioning were continuous, making it difficult to establish a no-effect level of exposure.[13]

23.4.5 Cadmium and Neurodegenerative Disorders

Few studies have looked at Cd and Alzheimer's disease (AD) and the results of these few studies suggest Cd has a limited role in the etiology of this disease. Basun et al.[14] found elevated Cd in the plasma of AD patients, but in a smaller, better controlled study found no difference.[15] Another study found elevated Cd in the liver of AD patients, but the smoking habits of the patients was not monitored.[51] To date, no studies have found increased levels of Cd in the brain or CSF of AD patients.[14,15] Some studies have found decreased levels of Zn and Se in the brain, blood, or CSF of AD patients.[14] Decreased levels of these metals could result from Cd exposure. Although the data to date suggest a limited role for Cd in the etiology of AD, further work is needed.

Some studies examining amyotrophic lateral sclerosis (ALS) and metals have found elevated levels of Al, Ca, Hg, Mn, or Pb in spinal cord, brain, CSF, or blood.[52] Other studies have found no changes in the metals examined. Cd had not been examined in previous studies.[16,52] A recent study found elevations of MT and Cd and Zn bound to MT, but not total Cd and Zn in livers and kidneys of patients suffering from ALS.[16] The lack of reported motor neuron disease in people suffering from the effects of environmental or occupational exposure to Cd sheds doubt on Cd as a causative factor in ALS.[1–3,17,18]

23.5 SUMMARY

Histopathological changes in the nervous systems of animals generally occur with acute parenteral doses of 2 to 20 mg/kg of Cd (brain in neonates, senory ganglia in adult), but are absent after oral dosing or lower parenteral dosing. Teratogenic changes in the nervous system require parenteral doses of at least 2 mg/kg and dosing needs be done at a critical time during neural tube development (8 to 14 days gestation). The activities of enzymes and proteins important for neurotransmission, for growth and metabolism, and for protection against oxidative damage are primarily decreased in the brain of animals after doses as low as 0.4 mg/kg parenterally and 4 mg/kg orally. Depending on the age of the animal, duration of exposure, dose, and area of brain examined, neurotransmitters (acetylcholine, dopamine, norepinephrine, and serotonin) are reduced or elevated after doses as low as 0.1 mg/kg parenterally and 1 mg/kg orally. The neuropathological and neurochemical changes have occurred after acute, repeated, or *in utero* exposures, but are more prominent in animals exposed as neonates than in animals exposed as adults or *in utero*.

Behavioral alterations are the most sensitive indicators of Cd neurotoxicity in animals. Decreases in performance and activity occurred after *in utero* exposures of 0.04 mg/kg orally and increased locomotor activity occurred after neonatal oral exposures of 0.1 mg/kg. Neonatal animals exposed to Cd at low to moderate doses generally have increased locomotor activity, whereas animals exposed as adults or *in utero* tend to have decreased activity. Acquisition and performance of learned tasks are depressed in adults and animals exposed *in utero* but, as with locomotor activity, low to moderate dose in young animals can enhance these activities. As mentioned above, neurobehavioral changes need not result from direct neurotoxic effects of Cd, but could result from more generalized differences such as reduced birth weight or altered metabolism. However, the studies at low doses find little if any other effects on the animals other than the changes in behavior.

In humans, the normal exposures, even in polluted areas, are less than 10 µg/kg.[3,17] These low exposures would imply that histopathological and biochemical changes found in animals are unlikely to be detected in humans. Neurobehavioral changes, however, do occur at doses similar to human exposures. A number of neurological effects have been found in workers exposed to Cd and correlations exist between Cd exposure and intelligence and learning in children. It is interesting to speculate that changes in animal behavior are predictive of the changes seen in human behavior. However, the animal studies were not complicated by coexposure to other metals, variability in Cd exposure, nutrition, and variations in lifestyle. It is clear that exposure to heavy metals is detrimental to intelligence and behavior in children and it is possible that adult workers could also have neurological changes, but attempts to isolate Cd as a sole or primary factor remain difficult and must be done cautiously. The limited data at present suggest that Cd exposure is not involved in etiology of the neurodegenerative diseases, AD and ALS. However, additional studies are needed owing to occasional findings of altered levels of Cd, Zn, and MT in patients suffering from these diseases.

ACKNOWLEDGMENTS

The author wishes to thank Drs. Andrew S. Fix and Joseph F. Ross for providing critical review and helpful suggestions during the preparation of this chapter.

REFERENCES

1. Babitch JA: Cadmium neurotoxicity. In Bondy JC and Prasad KN (eds): *Metal Neurotoxicity.* Boca Raton, FL, CRC Press, 1988, pp 141–166.
2. Shukla GS and Singhal RL: The present status of biological effects of toxic metals in the environment: lead, cadmium, and manganese. *Can. J. Physiol. Pharmacol.*, 1984; 62: 1015–1031.
3. Taylor J and Ennever FK: Toxicological Profile for Cadmium. Atlanta, Agency for Toxic Substances and Disease Registry, United States Public Health Service, 1993.

4. Ali MM, Mathur N, and Chandra SV: Effect of chronic cadmium exposure on locomotor behaviour of rats. *Indian J. Exp. Biol.*, 1990; 28: 653–656.

5. Holloway WR and Thor DH: Social memory deficits in adult male rats exposed to cadmium in infancy. *Neurotoxicol. Teratol.*, 1988; 10: 193-197.

6. Das KP, Das PC, Dasgupta S, and Dey CD: Serotonergic-cholinergic neurotransmitters' function in brain during cadmium exposure in protein restricted rat. *Biol. Trace Elem. Res.*, 1993; 36: 119–127.

7. Nation JR, Frye GD, Von Stultz J, and Bratton GR: Effects of combined lead and cadmium exposure: changes in schedule-controlled responding and in dopamine, serotonin and their metabolites. *Behav. Neurosci.*, 1989; 5: 1108–1114.

8. Pal R, Nath R, and Gill KD: Lipid peroxidation and antioxidant defense enzymes in various regions of adult rat brain after co-exposure to cadmium and ethanol. *Pharmacol. Toxicol.*, 1993; 73: 209–214.

9. Arvidson B: Cadmium toxicity and neural cell damage. In Dreosti IE and Smith RM (eds): *Neurobiology of the Trace Elements*, Vol 2. Clifton Heights, NJ, Humana Press, 1983, pp 51-78.

10. Wong KL and Klaassen CD: Neurotoxic effects of cadmium in young rats. *Toxicol. Appl. Pharmacol.*, 1982; 63: 330–337.

11. Gottschalk LA, Rebello T, Buchsbaum MS, et al: Abnormalities in hair trace elements as indicators of aberrant behavior. *Comp. Psychiatr.*, 1991; 32: 229-237.

12. Stewart-Pinkham SM: The effect of ambient cadmium air pollution on the hair mineral content of children. *Sci. Total Environ.*, 1989; 78: 289-296.

13. Thatcher RW and Lester ML: Nutrition, environmental toxins and computerized EEG: a mini-max approach to learning disabilities. *J. Learn. Disabil.*, 1985; 18: 287–297.

14. Basun H, Forssell LG, Wetterberg L, and Winblad B: Metals and trace elements in plasma and cerebrospinal fluid in normal ageing and Alzheimer's disease. *J. Neural Transm-Park*, 1991; 4: 231–258.

15. Basun H, Lind B, Nordberg M, et al: Cadmium in blood in Alzheimer's disease and nondemented subjects: results from a population-based study. *Biometals*, 1994; 7: 130–134.

16. Sillevis Smitt PAE, van Beek H, Baars A-J, et al: Increased metallothionein in the liver and kidney of patients with amyotrophic lateral sclerosis. *Arch. Neurol.*, 1992; 49: 721–724.

17. IARC Monographs on the Evaluation of Carcinogenic Risks to Humans: Beryllium, Cadmium, Mercury and Exposures in the Glass Manufacturing Industry, Volume 58. Geneva, International Agency for Research on Cancer, World Health Organization, 1993, pp 119–237.

18. Nath R, Prasad R, Palinal VK, and Chopra RK: Molecular basis of cadmium toxicity. *Prog. Food Nutr. Sci.*, 1984; 8: 109–163.

19. Pal R, Nath R, and Gill KD: Influence of ethanol on cadmium accumulation and its impact on lipid peroxidation and membrane bound functional enzymes (Na$^+$, K$^+$-ATPase and acetylcholinesterase) in various regions of adult rat brain. *Neurochem. Int.*, 1993; 23: 451–458.

20. Murphy VA, Embrey EC, Rosenberg JM, et al: Calcium deficiency enhances cadmium accumulation in the central nervous system. *Brain Res.*, 1991; 557: 280–284.

21. Kille P, Hemmings A, and Lunney EA: Memories of metallothionein. *Biochim. Biophys. Acta*, 1994; 1205: 151–161.

22. Bremner I: Nutritional and physiologic significance of metallothionein. *Meth. Enzymol.*, 1991; 205: 25–35.

23. Ebadi M: Metallothioneins and other zinc-binding proteins in brain. *Meth. Enzymol.*, 1991; 205: 363–387.

24. Nishimura N, Nishimura H, Ghaffar A, and Tohyama C: Localization of metallothionein in the brain of rat and mouse. *J. Histochem. Cytochem.*, 1992; 40: 309-315.

25. Young JK: Glial metallothionein. *Biol. Signals*, 1994; 3: 169–175.

26. Pardridge WM: *Peptide Drug Delivery to the Brain.* New York, Raven Press, 1991, pp 52–98.

27. Arvidson, B: Autoradiographic localization of cadmium in the rat brain. *Neurotoxicology*, 1986; 7: 89–96.

28. Valois AA and Webster WS: The choroid plexus and cerebral vasculature as target sites for cadmium following acute exposure in neonatal and adult mice: an autoradiographic and gamma counting study. *Toxicology*, 1987; 46: 43–55.

29. Zheng W, Perry DF, Nelson DL, and Aposhian HV: Choroid plexus protect cerebrospinal fluid against toxic metals. *FASEB J*, 1991; 5: 2188–2193.

30. Valois AA and Webster WS: The choroid plexus as a target site for cadmium toxicity following chronic exposure in the adult mouse: an ultrastructural study. *Toxicology*, 1989; 55: 193–205.

31. Shukla GS, Srivastava RS, and Chandra SV: Glutathione status and cadmium neurotoxicity: studies in discrete brain regions of growing rats. *Fundam. Appl. Toxicol.*, 1988; 11: 229–235.

32. Pierro LJ: Cadmium and teratogenesis of the central nervous system. In Dreosti IE and Smith RM (eds): *Neurobiology of the Trace Elements*, Vol 2. Clifton Heights, NJ, Humana Press, 1983, pp 79–96.

33. Atchison WD: Effects of neurotoxicants on synaptic transmission: lessons learned from electrophysiological studies. *Neurotoxicol. Teratol.*, 1988; 10: 393–416.

34. Ahammadsahib KI, Jinna RR, and Desaiah D: Protection against cadmium toxicity and enzyme inhibition by dithiothreitol. *Cell Biochem. Func.*, 1989; 7: 185–192.

35. Vig PJS and Nath R: In vivo effects of cadmium on calmodulin and calmodulin regulated enzymes in rat brain. *Biochem. Int.*, 1991; 23: 927–934.

36. Prohaska JR: Functions of trace elements in brain metabolism. *Physiol. Rev.*, 1987; 67: 858–901.

37. Mayer ML, Vyklicky L, and Westbrook GL: Modulation of excitatory amino acid receptors by Group IIB metal cations in cultured mouse hippocampal neurones. *J. Physiol.*, 1989; 415: 329–350.

38. Gupta A, Murthy RC, and Chandra SV: Neurochemical changes in developing rat brain after pre- and postnatal cadmium exposure. *Bull. Environ. Contam. Toxicol.*, 1993; 51: 12–17.

39. Gupta A, Murthy RC, et al: Comparative neurotoxicity of cadmium in growing and adult rats after repeated administration. *Biochem. Int.*, 1990; 21: 97–105.

40. Murthy RC, Saxena B, Lal B, and Chandra SV: Chronic cadmium-ethanol administration alters metal distribution and some biochemicals in rat brain. *Biochem. Int.*, 1989; 19: 135–143.

41. Gupta A, and Chandra SV: Gestational cadmium exposure and brain development: a biochemical study. *Ind. Health*, 1991; 29: 65–71.

42. Fasitsas CD, Theocharis SE, Zoulas D, et al: Time-dependent cadmium neurotoxicity in rat brain synaptosomal plasma membranes. *Comp. Biochem. Physiol.*, 1991; 100C: 271–275.

43. Stark M, Wolff JEA, and Korbmacher A: Modulation of glial cell differentiation by exposure to lead and cadmium. *Neurotoxicol. Teratol.*, 1992; 14: 247–252.

44. Holloway WR and Thor DH: Cadmium exposure in infancy: effects on activity and social behaviors of juvenile rats. *Neurotoxicol. Teratol.*, 1988; 10: 135–142.

45. Ali MM, Shulka GS, Srivastava RS, et al: Effects of vitamin E on cadmium-induced locomotor dysfunctions in rats. *Vet. Hum. Toxicol.*, 1993; 35: 109–111.

46. Chandra SV, Murthy RC, and Ali MM: Cadmium-induced behavioral changes in growing rats. *Ind. Health*, 1985; 23: 159–162.

47. Baranski B, Stetkiewicz I, Sitarek K, and Szymczak W: Effects of oral, subchronic cadmium administration on fertility, prenatal and postnatal progeny development in rats. *Arch. Toxicol.*, 1983; 54: 297–302.

48. Lehotzky K, Ungvary G, Polinak D, and Kiss A: Behavioral deficits due to prenatal exposure to cadmium chloride in CFY rat pups. *Neurotoxicol. Teratol.*, 1990; 12: 169–172.

49. Bar-Sela S, Levy M, Westin JB, et al: Medical findings in nickel-cadmium battery workers. *Isr. J. Med. Sci.*, 1992; 28: 578-583.

50. Hart RP, Rose CS, and Hamer RM: Neuropsychological effects of occupational exposure to cadmium. *J. Clin. Exp. Neuropsychol.*, 1989; 6: 933–943.

51. Lui E, Fisman M, Wong C, and Diaz F: Metals and the liver in Alzheimer's disease: an investigation of hepatic zinc, copper, cadmium, and metallothionein. *J. Am. Geriatr. Soc.*, 1990; 38: 633–639.

52. Khare SS, Ehmann WD, Kasarskis EJ, and Markesbery WR: Trace element imbalances in amyotrophic lateral sclerosis. *Neurotoxicology*, 1990; 11: 521–532.

SECTION 9

LEAD

Chapter 24

Interactions Between Lead and Metals in Cellular Toxicity

Timothy J. B. Simons

CONTENTS

24.1 INTRODUCTION

Lead is widely distributed in the environment, largely as a result of man's activities. Ingestion of large amounts of lead has long been known to cause acute poisoning, but only in recent years has it been realized that the developing nervous system is extremely sensitive to low doses of lead. The developed nations have taken steps to limit the exposure of the population to lead, for example by introducing unleaded gasoline, and this has resulted in a reduction in blood lead levels.[1]

The toxic effects of lead obviously depend upon a large number of factors, including routes of entry into the body, absorption, transport, storage, and excretion. At the cellular and molecular level, attention has focused on the possibility that lead may produce some (but not all) of its effects by a form of "molecular mimicry", in which lead mimics the actions of calcium, although it is present at a much lower concentration. Lead-calcium interactions have been reviewed fairly recently.[2,3] This chapter will concentrate on recent developments in lead-calcium interaction, particularly in relation to the nervous system, and on emerging evidence that lead may produce some of its toxic effects by interaction with a second essential metal, zinc. Interactions between lead and magnesium are covered in Chapter 25. The author is not aware of direct interactions between lead and any other essential metal, but one cannot exclude the possibility that additional lead-metal interactions will be discovered.

24.2 GENERAL PRINCIPLES

Two main factors determine whether metals interact with each other: concentration and affinity.

Lead, calcium, and zinc exist in a variety of chemical forms in biological systems. They are all divalent cations, but they also form complexes with a number of inorganic (e.g., OH^-, $H_2PO_4^-$) and organic anions (e.g., histidine, cysteine, citrate) as well as with proteins. Protein-binding can be nonspecific (e.g., to albumin) or specific (e.g., Zn^{2+}-metallothionein). Any one of the chemical complexes of Pb^{2+} may be responsible for a particular toxic effect, so it is important to know about the speciation of lead, i.e., the concentrations of the different species present. These depend upon the concentrations of the various ligands that bind lead and competing metals, and the equilibrium constants for the metal-ligand reactions, provided those reactions are at equilibrium. Most metal-ligand reactions occur quickly in aqueous media, and can be considered to be at equilibrium. Some exceptions are known, e.g., Zn^{2+} is effectively inexchangeable when it is bound to the enzyme carbonic anhydrase, or to the plasma protein α2-macroglobulin, but Ca^{2+} and Pb^{2+} seem to be in equilibrium with their ligands, as far as is known.

The affinity is the reciprocal of the dissociation constant for a metal-ligand equilibrium, and is a measure of how strongly a metal binds to a particular site. If the metal concentration is [M] and the dissociation constant is K_d, the affinity is $1/K_d$, and the fraction of sites occupied by metal, f, will be given by $f = [M]/([M] + K_d)$. This applies to one metal on its own; a more complicated expression will apply if there is competition between two metals, or there are multisite interactions. Usually, the action of an inhibitor such as lead can not be measured unless a competing metal is also present. The actions of lead are measured by the IC_{50}, the concentration that reduces an effect by 50%. The IC_{50} can be related to the K_d for lead if measurements are made at different concentrations of the competing metal.

The toxic species of lead must be present at its site of action at a sufficient concentration and for a sufficient time to produce the toxic effect. That species may be Pb^{2+}, but even if it is not Pb^{2+}, we can use the concentration of Pb^{2+} as a surrogate for the active species, in order to test our ideas about lead toxicity. For example, if the IC_{50} of Pb^{2+} acting on an ion channel is 1×10^{-6} M, and a plasma Pb^{2+} concentration of 1×10^{-10} M is fatal, it is rather unlikely that death is caused by the 0.01% reduction in ion flow through that channel! Even if the toxic species is not Pb^{2+}, its concentration will be at a fixed ratio to $[Pb^{2+}]$ in the two cases, so this argument will still hold.

A toxic substance could exert an acute effect by inhibiting an enzyme or blocking an ion channel, or a more subtle long-term effect by modulating the synthesis of a protein, or the turnover of RNA or DNA. The principle that lead must be present at sufficient concentration and for sufficient time to produce the effect is true in either case. However, it is difficult, if not impossible, to demonstrate clear cause-effect relationships in long-term experiments because of the large number of potentially confounding factors.

24.3 SPECIATION, ION CONCENTRATION, AND DISTRIBUTION
IN BODY FLUIDS

Detailed computer simulations of the binding of lead, calcium, zinc, and other metals by the low-molecular-weight components of plasma (anions, organic acids, and amino acids) have been published.[5]

### 24.3.1	Lead

Until recently blood lead was the only readily accessible measure of lead intoxication in human subjects. Concentrations below 10 µg/dl (0.5 µM) are asymptomatic. At higher concentrations there is a steadily increasing variety of toxic effects; lead poisoning, with encephalopathy, occurs at around 80 to 100 µg/dl (4 to 5 µM) in children. Lead readily enters cells, and 98 to 99% of the lead in blood is bound to erythrocytes.[6] Of the 1% or so that is in plasma, about 40% is bound to protein (mainly albumin) and about 60% to cysteine.[5,7] Free Pb^{2+} has been estimated to be 0.02% of the total plasma lead, so this would be about 1×10^{-12} M at 10 µg/dl blood lead, and 1×10^{-11} M at 100 µg/dl blood lead.[7] Lead is much more strongly bound to cytoplasmic constituents than in extracellular fluid: 99% of blood lead is in the erythrocytes, and the cytoplasmic concentration of Pb^{2+} has been estimated to be 6×10^{-12} M at 10 µg/dl blood lead (note that all these estimates involve some uncertainty due to extrapolation).[8] Cytoplasmic Pb^{2+} concentrations have also been measured in lead-exposed tissues *in vitro*, using fura-2 fluorescence[9] or ^{19}F NMR with 5F-BAPTA.[10,11] Bovine adrenal chromaffin cells showed elevated norepinephrine release when Pb^{2+} was increased to 1.4×10^{-11} M,[9] and values of about 3×10^{-11} M have been measured in lead-intoxicated osteoblastic[11] and neuroblastoma × glioma cells.[10] Taken together, these observations suggest that such intracellular free Pb^{2+} concentrations can be regarded as toxic.

Lead readily enters other body tissues, especially bone, brain, and kidney. The blood-brain barrier is ineffective at keeping lead out of brain, so any principles of lead toxicity in other tissues will be expected to apply also to brain. At steady-state the total lead concentration in brain is one to three times that in blood.[12] There is no information on the binding of lead in brain, but there are no other obvious mechanisms whereby Pb^{2+} concentrations could be sustained in brain that are significantly higher than those present in the blood.

### 24.3.2	Calcium

There is about 1 mM free Ca^{2+} in plasma, which is about half of the total. The remainder is reversibly bound to protein, mainly albumin, and anions such as hydrogen phosphate and citrate.[5] The cytoplasmic Ca^{2+} concentration is generally around 1×10^{-7} M. It can increase to around 1×10^{-6} M in many cells where Ca^{2+} is used as a "second messenger". The increase in Ca^{2+} leads to the activation of a Ca^{2+}-dependent protein such as troponin C (in muscle), calmodulin, protein kinase C, or to the activation of neurosecretion. At 1×10^{-11} M Pb^{2+} (putatively toxic) the ratio of concentrations $[Pb^{2+}]:[Ca^{2+}]$ will be around $1:10^8$ in extracellular space and $1:10^4$ inside cells.

### 24.3.3	Zinc

The total zinc concentration in plasma is about 15 µM, of which 3 to 5 µM is tightly bound to α2-macroglobulin. The remainder is exchangeable: about 9 to 11 µM bound to albumin and 1 µM to transferrin. Only about 0.15 µM is bound to low molecular weight components, mainly to cysteine and histidine.[13] It is extremely difficult to measure the free Zn^{2+} concentration; a value of 2×10^{-10} M has been estimated for horse plasma.[14]

Cellular total zinc concentrations are typically in the 100 to 500 µM range, higher than total zinc in extracellular fluids. Zinc is a component of many enzymes, and in many cells is bound to a storage protein, metallothionein. Little is known of the intracellular speciation of zinc; in human erythrocytes the intracellular free Zn^{2+} concentration has been estimated to be 2×10^{-11} M.[15] If one can generalize from these very limited observations, at a "toxic" level of Pb^{2+} of 1×10^{-11} M, the ratio of $[Pb^{2+}]:[Zn^{2+}]$ will be around 1:20 in extracellular space and 1:2 inside cells.

24.4 LEAD-METAL INTERACTIONS

Lead could produce an effect in two different ways, direct and indirect. In the direct mechanism, lead binds at a site normally occupied by another metal, and produces either a positive (stimulatory) or negative (inhibitory) effect. Lead could also displace a metal (e.g., calcium) from a binding site, increasing the free Ca^{2+} concentration, which would then produce an effect at a second site.

24.4.1 Lead-Calcium Interactions

The topics discussed here are not intended to be an exhaustive list of lead-calcium interactions, but areas of current research interest. For a more complete list the reader is referred to References 2 and 3.

24.4.1.1 Ion Channels

Lead inhibits current flow through a variety of ion channels, including L- and T-type Ca channels, Na channels, and nicotinic acetylcholine-gated channels. In most cases the IC_{50} is around 1×10^{-6} M Pb^{2+}, although some channels are more sensitive. The most sensitive channels are L-type Ca channels in rat E18 hippocampal neurones[16] and nicotinic acetylcholine-gated channels in mouse neuroblastoma cells,[17] with IC_{50}s of 2 to 3×10^{-8} M. Even these channels are too insensitive to lead to be affected significantly by the toxic (extracellular) concentration of 1×10^{-11} M Pb^{2+}.

More information about the actions of lead on ion channels can be found in a recent review.[18]

24.4.1.2 Calmodulin

Pb^{2+} can replace Ca^{2+} as an activator of calmodulin, but the affinity for Pb^{2+} is about the same as Ca^{2+}.[3] 1×10^{-11} M Pb^{2+} would therefore be expected to have little effect on calmodulin-dependent processes when the Ca^{2+} concentration is 10^{-7} to 10^{-6} M. However, intracellular free $[Ca^{2+}]$ increases when cells are exposed to lead,[10,11] and it has recently been shown that exposure of rats to lead *in vivo* (50 mg/kg intragastrically for 8 weeks) causes a 50% increase in the calmodulin content of the cerebral cortex as well as increasing $[Ca^{2+}]$ in synaptosomes.[19] Both effects would be expected to stimulate calmodulin-dependent processes in an additive manner. This reopens the possibility that relatively low doses of lead may be able to interfere with calmodulin-dependent processes *in vivo*. However, it should be pointed out that the lead content of the brain tissue (2 µg/g, about 1×10^{-5} M) was considerably higher than in other recent studies of the developmental effects of lead on the brain,[20,21] which may have affected the results.

24.4.1.3 Protein Kinase C

Protein kinase C, like calmodulin, is a component of Ca^{2+} signaling pathways in cells. Considerable interest was aroused by the report that 1×10^{-12} M Pb^{2+} is as effective as 1×10^{-6} M Ca^{2+} in activating the enzyme from rat brain.[22] This was confirmed in a much more careful study, in which the enzyme was shown to be 50% activated by 5.5×10^{-11} M Pb^{2+} or 2.6×10^{-7} M Ca^{2+}.[23] Toxic doses of lead *in vivo* would therefore be expected to mimic Ca^{2+} in activating protein kinase C, but all the evidence available relates to experiments carried out *in vitro*.

Lead inhibits endothelial cell growth in coculture with astroglia.[24] This is associated with protein kinase C activation in the astroglial cells, but the possibility remains that lead is acting in some other way on the endothelial cells, not via protein kinase C. (The inhibition of endothelial cell growth may be related to the encephalopathy of lead poisoning, in which blood-brain barrier permeability increases.)

Extracellular lead stimulates the contraction of vascular smooth muscle *in vitro*.[25] The effect is enhanced by phorbol ester and mezerein, indicating an involvement of protein kinase C, and inhibited by blocking Ca entry through L-type channels with verapamil.[25] In these experiments there was no direct control of intracellular free Pb^{2+} or Ca^{2+}. After entering the cells, lead might be stimulating protein kinase C directly, or indirectly by raising intracellular Ca^{2+}, by displacement from binding sites or from mitochondria.

Very recently, direct measurements of protein kinase C activity in permeabilized bovine adrenal medullary cells have shown that Pb^{2+} stimulates the enzyme in the concentration range from 10^{-12} to 10^{-9} M, but inhibits it between 10^{-9} and 10^{-6} M Pb^{2+}.[26] When protein kinase C is stimulated with Ca^{2+}, Pb^{2+} only has an inhibitory effect. This helps to explain an earlier report that lead failed to stimulate protein kinase C at all.[27]

24.4.1.4   Neurosecretion

In the past, lead has been shown either to increase or decrease transmitter release from brain synaptosomes and the neuromuscular junction, depending upon the circumstances. Neurosecretion depends upon calcium entry through calcium channels at the nerve terminal, which is blocked by lead, although with relatively low affinity (see above). Attention has now focused on the intracellular effects of Pb^{2+}, using bovine adrenal chromaffin cells as a model system. Lead readily enters these cells, partly at least through calcium channels, and stimulates norepinephrine release when the cytoplasmic $[Pb^{2+}]$ is above 1×10^{-11} M.[9] In permeabilized cells Pb^{2+} can replace Ca^{2+} and stimulate secretion. The norepinephrine release process is half-maximally activated at 4.6×10^{-9} M Pb^{2+}, compared with 2.4×10^{-6} M Ca^{2+}.[28] Activation of protein kinase C (by phorbol esters) enhances catecholamine release in response to Ca^{2+} or Pb^{2+}. Staurosporine, an inhibitor of protein kinase C, blocks the enhancement of norepinephrine release caused by phorbol esters, but has no effect on basal release caused by Pb^{2+}.[28] This indicates that the stimulation of secretion by Pb^{2+} is not dependent on protein kinase C, although Pb^{2+} can modify secretion by stimulating protein kinase C.

Intracellular Pb^{2+} also activates outward K^+ currents in bovine adrenal chromaffin cells, with half-maximal effect at 5×10^{-10} M Pb^{2+}.[29] This is another Ca^{2+}-like action of Pb^{2+}. In the intact cell, activation of K^+ currents would be expected to hyperpolarize the cell, decrease Ca^{2+} entry, and decrease neurosecretion, although it is clear from studies with intact cells that low concentrations of lead do stimulate norepinephrine secretion.[9]

24.4.1.5   Developmental Aspects of Lead Neurotoxicity

The developing nervous system is much more sensitive to lead than the adult. Cognitive deficits occur in children exposed to lead in the first 2 years of life, even at blood lead levels as low as 15 to 20 μg/dl (about 1×10^{-6} M), which are otherwise asymptomatic.[30] Recently, animal models have begun to be developed to explore cellular changes resulting from perinatal exposure to low doses of lead. Using permeabilized neurones from rat cerebral cortex, the increase in intracellular Ca^{2+} induced by inositol 1,4,5-trisphosphate (IP_3) was reduced when the rats had been exposed to lead perinatally, but not when exposed as adults, even though the lead content of the brain was similar (0.1 to 0.15 μg/g) in both cases.[20] The effect was associated with a reduction in the number of IP_3 receptors in the cells. This is an exciting result, because it is opposite from the usual acute effect of lead on calcium metabolism, which is to increase Ca^{2+} signaling, and also because the effect is specific to the developing brain. However, the overall effect on cellular function is unclear, because there is also evidence that lead increases cellular IP_3 transients in astroglia,[31] which would have an opposite effect.

In another study, a reduction in the number of muscarinic receptors was seen in the septum, but not in the hippocampus, in perinatally lead-exposed rats.[21] This is significant because such changes have been associated with cognitive deficits, and emphasizes that neurotoxic effects of lead are not necessarily associated with lead-metal interactions.

24.4.1.6 Summary and Further Work

A number of calcium-dependent processes are sensitive to lead. Protein kinase C is the most sensitive one currently known, but no cellular effect of lead has been clearly shown to be dependent upon protein kinase C. Lead has many effects on calcium homeostasis and calcium signaling pathways, but isolated observations are not necessarily a good guide to cellular function. There is a need for integration, to show how lead affects the cell as a whole, and also to demonstrate that relevant effects occur at lead concentrations that are toxic to the whole organism. These are often much lower than the concentrations applied to isolated cells. However, very low concentrations of lead, acting over a long time, may produce cumulative effects that are similar to much larger concentrations acting for a short time. This possibility has not yet been tested.

24.4.2 Lead-Zinc Interactions

This is a relatively new area and there is little published work.

24.4.2.1 δ-Aminolevulinate Dehydratase

This enzyme condenses two molecules of δ-aminolevulinate to make porphobilinogen. It is on the pathway of heme synthesis, and is present in most cells, including neurones. It is probably the enzyme most sensitive to inhibition by lead. The activity of the enzyme in erythrocytes can be used as an index of past lead exposure in humans.[32] It contains a zinc ion at the catalytic center, coordinated by two cysteine and two histidine residues.[33] Recent work in the author's laboratory has shown that the Zn^{2+} ion is labile at 37°C, and can readily be removed or replaced by Pb^{2+}. Neither Ca^{2+} nor Mg^{2+} are involved at all in the enzyme's activity.

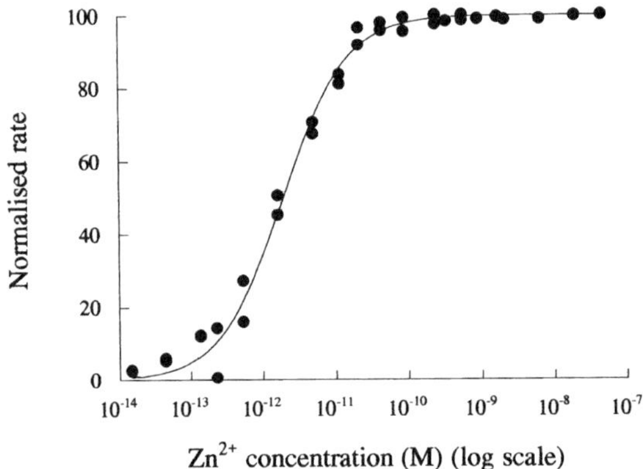

FIGURE 24.1 Variation of human erythrocyte δ-aminolevulinate dehydratase rate with Zn^{2+} concentration. The enzyme rate was measured using erythrocyte lysate at pH 7.2 and 37°C. The free $[Zn^{2+}]$ concentration was controlled by buffering Zn^{2+} with EDTA, HEDTA, EGTA, or NTA, and the enzyme was preincubated for 40 min at each $[Zn^{2+}]$, before addition of 4 mM δ-aminolevulinate and measuring product formation in 20 min. The results are fitted by the Michaelis-Menten equation (note logarithmic scale).

Figure 24.1 shows the enzyme rate as a function of Zn^{2+} concentration, in the absence of Pb^{2+}. The rate is half-maximal at 1.6×10^{-12} M Zn^{2+}. This compares with an estimate of $<10^{-7}$ M for the affinity for Zn^{2+} by equilibrium dialysis.[34] Pb^{2+} is a competitive inhibitor of Zn^{2+} ions (Figure 24.2). The IC_{50} for Pb^{2+} varies linearly with the Zn^{2+} concentration (Figure 24.2B). Extrapolation to 0-Zn gives a value of 6.5×10^{-14} M for the K_i for Pb^{2+}. This does not mean that concentrations of Pb^{2+} as low as 10^{-13} M would be inhibitory, because Pb^{2+} is a competitive inhibitor of Zn^{2+} and Zn^{2+} is always present inside cells. In human erythrocytes the Zn^{2+} concentration is about 2.4×10^{-11} M,[15] so the enzyme would be 95% activated in the complete absence of Pb^{2+}. The effective IC_{50} for Pb^{2+} would be expected to be 1×10^{-12} M at this Zn^{2+} concentration. This is compatible with previous observations *in vivo* and *in vitro*.[32,35] δ-Aminolevulinate is neurotoxic, and its plasma concentration increases during lead poisoning, so it might play a role in the neurotoxicity of lead.[35]

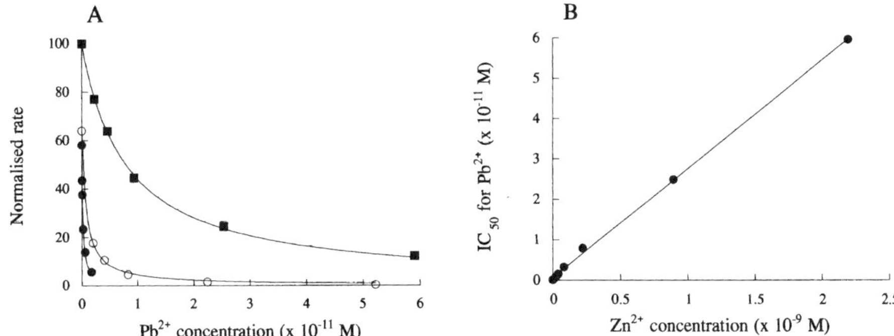

FIGURE 24.2 Inhibition of human erythrocyte δ-aminolevulinate dehydratase synthesis by Pb^{2+}. A: Enzyme rate is shown as a function of $[Pb^{2+}]$ while $[Zn^{2+}]$ was kept fixed at 2.3×10^{-10} M (■), 2.0×10^{-11} M (○), or 1.6×10^{-12} M (●). The results are fitted by the Michaelis-Menten equation with competitive inhibition: rate $= V_{max}[S]/(K_m\{1 + ([I]/K_i\} + [S])$, which reduces to the equation rate $= V'/(IC_{50} + [Pb^{2+}])$, when rate is plotted against inhibitor concentration ($I = Pb^{2+}$) at constant substrate concentration ($S = Zn^{2+}$). In turn, $IC_{50} = K_i(1 + [S]/K_m)$, and Figure 24.2B shows the variation of IC_{50} with Zn^{2+} concentration from the data in Figure 24.2A, and at 5 other Zn^{2+} concentrations.

24.4.2.2   Cyclic GMP-Phosphodiesterase

Lead inhibits cGMP-specific phosphodiesterase in bovine retinal rods.[36] The analogous enzyme in bovine lung has been shown to bind zinc and contain amino acid sequences characteristic of zinc hydrolases.[37] Further work will be needed to show whether this is another example of lead-zinc competition.

24.4.2.3   Metallothionein

Metallothionein acts as a cytoplasmic store for zinc. It binds seven zinc or cadmium ions per molecule, and is synthesized in response to an increase in cellular zinc or cadmium content. The apparent dissociation constants at pH 7 are about 5×10^{-13} M for Zn^{2+} and 5×10^{-17} M for Cd^{2+}.[38] The dissociation constant for Pb^{2+} is intermediate between Zn^{2+} and Cd^{2+}.[39] Corresponding to this observation, the Zn^{2+} form of metallothionein is able to protect rat liver δ-aminolevulinate dehydratase from inhibition by Pb^{2+}, while the Cd^{2+} form is not.[40] Thus, even if cellular effects of Pb^{2+} are not mediated directly by competition with Zn^{2+}, there could be an indirect effect. Zn^{2+}-metallothionein inside cells should act as a buffer to absorb Pb^{2+} and protect against its toxic effect.

24.4.2.4   Summary and Further Work

Studies of lead-zinc interactions are in their infancy; there is a need for much more work. Zinc is present in large quantities in some parts of the brain, is released during nervous activity, and can modulate amino acid receptors, especially *N*-methyl-D-aspartate (NMDA) receptors.[41] The potential actions of lead deserve investigation.

24.5 SUMMARY AND CONCLUSIONS

The goal of studies of lead toxicology must be not just to understand the acute effects of lead (lead poisoning) but also the longer-term chronic effects, particularly those which occur at low levels of lead exposure. At present some progress has been made towards the first goal, but little towards the second.

The developing nervous system is much more sensitive to lead than the adult system. Cognitive deficits occur in children exposed to lead in the first 2 years of life, even at blood lead levels as low as 15 to 20 µg/dl (about 1×10^{-6} M), which are otherwise asymptomatic.[30] It is now possible to set up animal models and demonstrate long-term effects of perinatal exposure to lead on the nervous system. However, it is not possible to provide an explanation for these effects. They might be due to lead-metal interactions, for example long-term disruption of calcium homeostasis, or to an effect on protein synthesis. Lead is known to act as a specific catalyst in the hydrolysis of RNA.[42,43] This effect does not occur with other metals, and has only been demonstrated *in vitro* at lead concentrations around 1×10^{-3} M. Much more work needs to be done to establish whether significant RNA breakdown occurs over a biological time-scale at doses of lead known to be toxic *in vivo*, and also whether such a generalized mechanism is capable of producing the specific effects of lead on the developing nervous system.

REFERENCES

1. Brody DJ, Pirkle JL, Kramer RA, et al: Blood lead levels in the U.S. population. Phase 1 of the third national health and nutrition examination survey (NHANES III, 1988 to 1991). *J. Am. Med. Assoc.*, 1994; 272:277–283.
2. Clarkson TW: Molecular and ionic mimicry of toxic metals. *Annu. Rev. Pharmacol. Toxicol.*, 1993; 32: 545–571.
3. Simons TJB: Lead-calcium interactions in cellular lead toxicity. *NeuroToxicology*, 1993; 14:77–86.
4. Srivastava D and Fox DA: Molecular interactions between lead and magnesium ions. This volume, Chap. 25.
5. May PM, Linder PW, and Williams DR: Computer simulation of metal-ion equilibria in biofluids: Models for the low-molecular weight complex distribution of calcium (II), magnesium (II), manganese (II), iron (III), copper (II), zinc (II) and lead (II) ions in human blood plasma. *J. Chem. Soc. Dalton Trans.*, 1977; 588–596.
6. Manton WI and Cook JD: High accuracy (stable isotope dilution) measurements of lead in serum and cerebrospinal fluid. *Br. J. Ind. Med.*, 1984; 41:313–319.
7. Al-Modhefer AJA, Bradbury MWB, and Simons TJB: Observations on the chemical nature of lead in human blood serum. *Clin. Sci.*, 1991; 81:823–829.
8. Simons TJB: Lead transport and binding by human erythrocytes in vitro. *Pflüegers Arch.*, 1993; 423:307–313.
9. Tomsig JL and Suszkiw JB: Pb^{2+}-induced secretion from bovine chromaffin cells: fura-2 as a probe for Pb^{2+}. *Am. J. Physiol.*, 1990; 259:C762–C768.
10. Schanne FAX, Moskal JR, and Gupta RK: Effect of lead on intracellular free calcium ion concentration in a presynaptic neuronal model: ^{19}F-NMR study of NG108-15 cells. *Brain Res.*, 1989; 503: 308-311.
11. Schanne FAX, Dowd TL, Gupta RK, et al: Lead increases free Ca^{2+} concentration in cultured osteoblastic bone cells. Simultaneous detection of intracellular free Pb^{2+} by ^{19}F NMR. *Proc. Natl. Acad. Sci. U.S.A.*, 1989; 86:5133–5135.
12. Bradbury MWB and Deane R: Permeability of the blood-brain barrier to lead. *NeuroToxicology*, 1993; 14:131–136.
13. Harris WR and Keen C: Calculations of the distribution of zinc in a computer model of human serum. *J. Nutr.*, 1989; 119:1677–1682.
14. Magneson GR, Puvathingal JM, and Ray WJ: The concentrations of free Mg^{2+} and free Zn^{2+} in equine blood plasma. *J. Biol. Chem.*, 1987; 262:11140–11148.
15. Simons TJB: Intracellular free zinc and zinc buffering in human red blood cells. *J. Membrane Biol.*, 1991; 123: 63–71.
16. Audesirk G and Audesirk T: The effects of inorganic lead on voltage-sensitive calcium channels differ among cell types and among channel subtypes. *NeuroToxicology*, 1993; 14:259–266.
17. Oortgiesen M, Van Kleef RGDM, Bajnath RB, et al: Nanomolar concentrations of lead selectively block neuronal nicotinic acetylcholine responses in mouse neuroblastoma cells. *Toxicol. Appl. Pharmacol.*, 1990; 103:165–174.
18. Kiss T and Osipenko ON: Toxic effects of heavy metals on ionic channels. *Pharmacol. Rev.*, 1994; 46:245–267.
19. Sandhir R and Gill KD: Alterations in calcium homeostasis on lead exposure in rat synaptosomes. *Mol. Cell. Biochem.*, 1994; 131:25–33.
20. Singh AK: Age-dependent neurotoxicity in rats chronically exposed to low levels of lead: calcium homeostasis in central neurons. *NeuroToxicology*, 1993; 14:417–428.
21. Bielarczyk H, Tomsig JL, and Suszkiw JB: Perinatal low-level lead exposure and the septo-hippocampal cholinergic system: selective reduction of muscarinic receptors and choline acetyltransferase in the rat septum. *Brain Res.*, 1994; 643:211–217.

22. Markovac J and Goldstein GW: Picomolar concentrations of lead stimulate brain protein kinase C. *Nature*, 1988; 334:71–73.

23. Long GJ, Rosen JF, and Schanne FAX: Lead activation of protein kinase C from rat brain. Determination of free calcium, lead and zinc by ^{19}F NMR. *J. Biol. Chem.*, 1994; 269:834–837.

24. Laterra J, Bressler JP, Indurti RR, et al: Inhibition of astroglia-induced endothelial differentiation by inorganic lead: a role for protein kinase C. *Proc. Natl. Acad. Sci. U.S.A.*, 1992; 89:10748–10752.

25. Watts SW, Chai S, and Webb RC: Lead acetate-induced contraction in rabbit mesenteric artery: interaction with calcium and protein kinase C. *Toxicology*, 1995; 99:55–65.

26. Tomsig JL and Suszkiw JB: Multisite interactions between Pb^{2+} and protein kinase C and its role in norepinephrine release from bovine adrenal cells. *J. Neurochem.*, 1995; 64:2667–2673.

27. Murakami K, Feng G, and Chen SG: Inhibition of brain protein kinase C subtypes by lead. *J. Pharm. Exp. Ther.*, 1993; 264:757–761.

28. Tomsig JL and Suszkiw JB: Intracellular mechanism of Pb^{2+}-induced norepinephrine release from bovine chromaffin cells. *Am. J. Physiol.*, 1993; 265:C1630–C1636.

29. Sun LR and Suszkiw JB: Pb^{2+} activates potassium currents in bovine adrenal chromaffin cells. *Neurosci. Lett.*, 1994; 182:41–43.

30. Needleman HL: The current status of childhood low-level lead toxicity. *NeuroToxicology*, 14: 1993; 161–166.

31. Dave V, Vitarella D, Aschner JL, et al: Lead increases inositol 1,4,5-trisphosphate levels but does not interfere with calcium transients in primary rat astrocytes. *Brain Res.*, 1993; 618:9–18.

32. Astrin KH, Bishop DF, Wetmure JG, et al: δ-Aminolevulinic acid dehydratase isozymes and lead toxicity. *Ann. N.Y. Acad. Sci.*, 1987; 514:23–29.

33. Tsukamoto I, Yoshinaga T, and Sano S: The role of zinc with special reference to the essential thiol groups in δ-aminolevulinic acid dehydratase of bovine liver. *Biochem. Biophys. Acta*, 1979; 570:167–178.

34. Jaffe EK, Abrams WR, Kaempfen HX, and Harris KA: 5-Chlorolevulinate modification of porphobilinogen synthase identifies a potential role for the catalytic zinc. *Biochemistry*, 1992; 31:2113–2123.

35. Moore MR, Meredith PA, and Goldberg A: Lead and heme biosynthesis. In Singhal RL and Thomas JA (eds): *Lead Toxicity*. Baltimore, Urban & Schwarzenberg, 1980 pp 79–118.

36. Fox DA, Srivastava D, and Hurwitz RL: Lead-induced alterations in rod-mediated visual functions and cGMP metabolism: new insights. *NeuroToxicology*, 1994; 15:503–512.

37. Francis SH, Colbran JL, McAllister-Lucas LM, et al: Zinc interactions and conserved motifs of the cGMP-specific phosphodiesterase suggest that it is a zinc hydrolase. *J. Biol. Chem.*, 1994; 269:22477–22480.

38. Vasak M and Kagi JHR: Spectroscopic properties of metallothionein. In Sigel, H (ed): Zinc and its role in biology and nutrition, *Metal Ions in Biological Systems*, Vol. 15. New York, Marcel Dekker, 1983, pp 213–273.

39. Nielson KB, Atkin CL, and Winge DR: Distinct metal-binding configurations in metallothionein. *J. Biol. Chem.*, 1985; 260:5342–5350.

40. Goering PL and Fowler BA: Metal constitution of metallothionein influences inhibition of δ-aminolaevulinic acid dehydratase (porphobilinogen synthase) by lead. *Biochem. J.*, 1987; 245:339–345.

41. Frederickson CJ and Moncrieff DW: Zinc-containing neurons. *Biol. Signals*, 1994; 3:127–139.

42. Pan T, Dichtl B, and Uhlenbeck OC: Properties of an *in vitro* selected Pb^{2+} cleavage motif. *Biochemistry*, 1994; 33:9561-9565.

43. Otzen DE, Barciszewski J, and Clark BFC: Dual hydrolytic role for Pb(II) ions. *Biochimie*, 1994; 76:15-21.

Chapter 25

Molecular Interactions Between Lead and Magnesium Ions

Devesh Srivastava and Donald A. Fox

CONTENTS

25.1 INTRODUCTION

Lead poisoning is the most common environmentally related disease of young children. Three to four million children in the U.S. are exposed to environmental sources of lead that place them at risk of adverse health effects. The tissue most sensitive to the adverse effects of lead is the central nervous system.[1] To determine and elucidate the cellular and molecular mechanisms of lead neurotoxicity, we have used the neural retina as a model system for the central nervous system.

The physiology and biochemistry of the retinal rod photoreceptor is relatively well understood. Similar to most signal transduction systems, the light signal is converted to a cellular signal via a receptor → G-protein → target enzyme → second messenger pathway. In the rod photoreceptor, this phototransduction cascade is composed of rhodopsin → transducin → cGMP phosphodiesterase (PDE) → cGMP.[2]

In vitro and *in vivo* electrophysiological investigations have demonstrated that lead alters the ability of rod photoreceptors to respond to light.[3–6] The abnormal electrophysiological responses were consistent with an inhibition of the activation process in phototransduction. A lead-induced elevation in retinal cGMP levels due to decreased retinal cGMP hydrolysis supported this hypothesis.[5,6] In this chapter we identify a site of action and describe a novel biochemical mechanism of lead neurotoxicity that is entirely consistent with the electrophysiological findings. That is, nanomolar concentrations of free Pb^{2+} inhibit the rod photoreceptor cGMP PDE by directly competing with millimolar concentrations of free Mg^{2+} within the catalytic site.[7,8]

25.2 CHEMICAL PROPERTIES OF LEAD AND MAGNESIUM

Metal ion substitution provides an important basis for metal ion toxicity. Size similarity is generally more important than charge similarity in permitting metal ion substitutions.[9] Except for charge, however, Pb^{2+} and Mg^{2+} appear to be two dissimilar divalent cations. Pb^{2+} is a Group 14 metal whereas Mg^{2+} is a Group 2 metal (new IUPAC notation system). The effective ionic radii for the two metals, which should be defined in reference to a specific coordination number, are different. For a coordination number of 6 (preferred coordination number for Mg^{2+}), the effective ionic radius for Pb^{2+} is 1.19 Å and for Mg^{2+} is 0.72 Å.[9] This is in contrast to the relatively similar effective ionic radii for Pb^{2+} and Ca^{2+} at a coordination number of 8 (preferred by Ca^{2+}): 1.29 Å and 1.12 Å, respectively.[9] Moreover, the metal ion stabilities for two standard bidentate ligands, glycine and 1,2-diaminoethane, are several log units greater for Pb^{2+} than Mg^{2+} or Ca^{2+}.[10] Furthermore, and possibly most relevant for the Pb^{2+}-Mg^{2+} interaction, the preferred ligands for Pb^{2+} and Mg^{2+} are also significantly different. Pb^{2+} shows a strong tendency to form bonds with amines and sulfhydryls on small molecules and proteins whereas Mg^{2+} forms only weak complexes with these ligands.[9–11] Interestingly, the association of Mg^{2+} with amines and sulfhydryls is greater than that for Ca^{2+} and these ligands.[9,10,12] Although, at first glance, there does not appear to be a strong chemical basis for Pb^{2+} acting at a Mg^{2+} site on a protein, Pb^{2+} and Ca^{2+} appear to interact at common biochemical sites without sharing many common chemical properties.[11,13]

25.3 CELLULAR LEVELS OF MAGNESIUM

To assess the physiological function of Mg^{2+} and to determine the pathophysiological consequences when its concentration is altered, it is essential to know the intracellular free Mg^{2+} concentration ($[Mg^{2+}]_i$). A wide variety of different methodologies, species, and tissue types have been utilized to determine the $[Mg^{2+}]_i$. They all indicate that the $[Mg^{2+}]_i$ is between 0.4 and 0.6 mM.[14] Calculations reveal that this represents about 5% of the total magnesium within a cell.[14]

The rod photoreceptor $[Mg^{2+}]_i$ has not yet been measured although it can be estimated from published works. A value of 2.9 fmol of total magnesium has been reported for a single mouse photoreceptor,[15] which has an average volume of 234 fl.[16] A simple calculation reveals a total magnesium concentration of 12.4 mM for a mouse photoreceptor: 5% of which represents 0.6 mM $[Mg^{2+}]_i$. Moreover, the total concentration of magnesium in a frog rod outer segment from dark-adapted and light-adapted retinas is 11 mM and 9 mM, respectively;[17] 5% of these values would be 0.55 and 0.45 mM $[Mg^{2+}]_i$, respectively. Thus, the rod photoreceptor $[Mg^{2+}]_i$ falls within range of other tissues.

25.3.1 Effects of Lead on Cellular Levels of Magnesium

In vivo and *in vitro* lead exposure both decreased the cellular Mg^{2+} concentration. Moderate lead exposure to adult male rats, resulting in blood lead values of 45 μg/dl, slightly decreased the tissue concentrations of magnesium.[18] Exposure to micromolar concentrations of Pb^{2+} significantly decreased (–21%) the $[Mg^{2+}]_i$ of cultured osteoclastic bone cells.[19] These studies suggest that lead exposure may indirectly affect cell function by reducing the $[Mg^{2+}]_i$, an essential metal in energy metabolism as well as other fundamental biochemical processes.[20]

25.4 DIRECT INHIBITION OF THE ROD PHOTORECEPTOR cGMP PHOSPHODIESTERASE BY LEAD

Following developmental lead exposure, biochemical studies revealed a dose-dependent increase in retinal cGMP content that was due to an inhibition of retinal cGMP hydrolysis and not due to an increase in cGMP synthesis.[5,6,8] A similar concentration-dependent inhibition of cGMP hydrolysis was observed with direct Pb^{2+} exposure to adult control retinal homogenates, although the site and mechanism of action were not identified.[5,6,8]

To test the hypothesis that Pb^{2+} directly inhibits the rod cGMP PDE, experiments were conducted using purified, trypsin-activated rod photoreceptor cGMP PDE isolated from frozen, dark-adapted bovine retinas. The detailed methodology has been described previously.[7,8] All reported concentrations of Pb^{2+} and Mg^{2+} represent free concentrations of cations that were calculated using a computer program (Calcon, Version 9.4: graciously provided by Joseph Tash) that is based on the critical stability constants of Martell and Smith and is corrected for temperature, pH, and ionic strength.[10] The concentrations of Pb^{2+} were verified, within range of detection, using a Pb^{2+}-selective electrode (Model 948200, Orion Research, Inc., Boston, MA) and an Orion 701A digital ion analyzer.

Picomolar to nanomolar concentrations of Pb^{2+} directly inhibited the bovine rod cGMP PDE. The half-maximal inhibitory concentration (IC_{50}) of Pb^{2+} varied as a function of the concentration of the substrate, cGMP, and the cofactor, Mg^{2+} (Table 25.1). For a given concentration of cGMP, the IC_{50} of Pb^{2+} *increased* as the concentration of Mg^{2+} increased from 10 μM to 10 mM. For a given concentration of Mg^{2+}, the IC_{50} of Pb^{2+} *decreased* as the concentration of cGMP increased from 1 μM to 1 mM. Previous results suggested that both cGMP and Mg^{2+} act within the catalytic pocket of the rod cGMP PDE.[7] Since the IC_{50} of Pb^{2+} varied with the concentration of both cGMP and Mg^{2+}, it appears that Pb^{2+} acts within the same catalytic pocket.

Table 25.1 Half-Maximal Inhibitory Concentrations (IC_{50}) of Pb^{2+} (nM) for the Rod cGMP Phosphodiesterase[a]: Effects of cGMP[b] and Mg^{2+c}

	10 μM Mg^{2+}	500 μM Mg^{2+}	10 mM Mg^{2+}
1 μM cGMP	2.13 ± 0.17	12.23 ± 2.55	67.0 ± 14.0
5 μM cGMP	0.80 ± 0.07	7.86 ± 1.09	44.0 ± 4.5
1 mM cGMP	0.45 ± 0.07	0.80 ± 0.05	3.86 ± 0.44

[a] The IC_{50} values represent the mean ± SEM of triplicate samples from 3 to 5 separate experiments. The IC_{50} values were derived using nonlinear least-squares analysis.
[b] At each concentration of cGMP, the mean IC_{50} values at each concentration of Mg^{2+} were significantly different from each other at $p < .05$.
[c] At each concentration of Mg^{2+}, the mean IC_{50} values at each concentration of cGMP were significantly different from each other at $p < .05$.

Reprinted from Srivastava, D., Hurwitz, R.L., and Fox, D.A., *Toxicol. Appl. Pharmacol.*, 134: 43–52, 1995. With permission.

As previously noted, Pb^{2+} was a potent inhibitor of cGMP PDE. Under physiologically relevant conditions (500 μM Mg^{2+} and 1 to 5 μM cGMP), 1 nM Pb^{2+} significantly inhibited the rod PDE and the IC_{50} values of Pb^{2+} ranged from 7 to 12 nM. Under pathophysiological conditions, such as those encountered in lead-exposed rats with elevated levels of retinal cGMP[5,6] and possible lead-induced decrease in $[Mg^{2+}]_i$, as observed in other tissues,[18,19] the IC_{50} of Pb^{2+} may be as low as 800 pM Pb^{2+}.[8] These concentrations of free Pb^{2+} are consistent with our animal models of lead exposure (mean peak whole blood lead values of 19 and 59 μg/dl[5,6]), since ≤5 nM free Pb^{2+} would be found at a serum lead concentration of 10 μg/dl.[21] In addition, exposure of developing rats to lead results in a 25 to 35% decrease in the amount of cGMP PDE mRNA and protein found in the rod photoreceptors (Srivastava, Hurwitz, and Fox: unpublished data). Moreover, lead, like calcium,[22,23] appears to be concentrated in photoreceptors of lead-exposed rats relative to the remainder of the retina (Fox: unpublished data). These results suggests that an IC_{50} value of 7 to 12 nM Pb^{2+} is consistent with the *in vivo* data.

25.5 MECHANISM OF THE DIRECT INHIBITION OF THE ROD PHOTORECEPTOR cGMP PHOSPHODIESTERASE BY LEAD

Magnesium is an essential cofactor for the rod cGMP PDE. The concentration of Mg^{2+} that results in half-maximal stimulation ($K_{0.5}$) of cGMP PDE activity is dependent on the concentration of the substrate cGMP.[7] For example, using 5 μM cGMP the apparent $K_{0.5}$ of Mg^{2+} was 24 μM (Figure 25.1A). In the presence of 750 pM Pb^{2+} and 5 nM Pb^{2+}, the apparent $K_{0.5}$ of Mg^{2+} increased 2- to 29-fold to 52 μM and 687 μM, respectively. Despite the presence of Pb^{2+}, the apparent maximal velocity (V_{max}) was attained in the presence of 10 mM Mg^{2+}. This result suggested that there was competition between Pb^{2+} and Mg^{2+}.

The next set of experiments defined the nature of the Pb^{2+}-induced inhibition of cGMP PDE. The rod cGMP PDE was assayed using 3.5 μM cGMP and the indicated concentrations of Mg^{2+} and Pb^{2+} (Figure 25.1B). A double reciprocal plot of the data revealed that the apparent $K_{0.5}$ of Mg^{2+} increased over twofold in the presence of increasing concentrations of Pb^{2+} (from 29 to 67 μM Mg^{2+}) while the apparent V_{max} was unchanged. This result demonstrated that Pb^{2+} is a competitive inhibitor of the rod cGMP PDE with respect to Mg^{2+}.

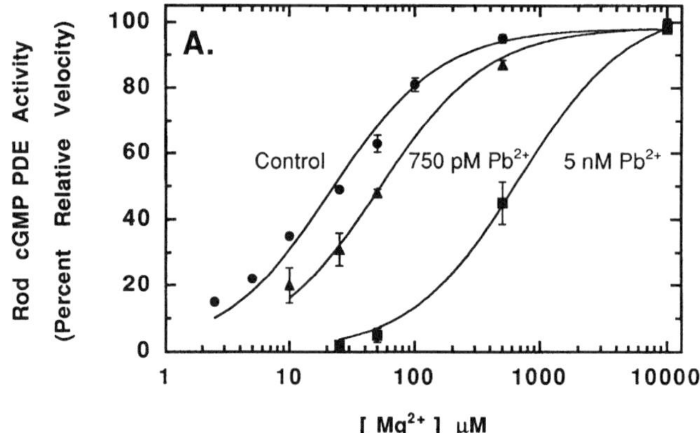

FIGURE 25.1   The isolated, trypsin-activated rod photoreceptor cGMP PDE was assayed in the presence of the indicated concentrations of cGMP, Mg^{2+}, and Pb^{2+}. **(A) Lead increases the apparent $K_{0.5}$ of magnesium for the rod cGMP PDE:** 50 pM of rod photoreceptor cGMP PDE was assayed using 5 μM cGMP at the indicated concentrations of Mg^{2+} and Pb^{2+}. PDE activity was normalized to 100% (19.62 μmol cGMP hydrolyzed min^{-1} mg $protein^{-1}$). Each data point represents the mean ± SEM of 3 to 6 separate experiments. Curves were fit and apparent $K_{0.5}$ values were determined using nonlinear least-squares analysis. There was no PDE activity observed in the absence of Mg^{2+} or in the absence of Mg^{2+} and the presence of Pb^{2+} (data not shown). **(B) Lead is a competitive inhibitor of the rod cGMP PDE with respect to magnesium:** each data point represents the mean of 3 to 5 separate experiments. Curves were fit using linear regression analysis. The point of convergence on the y-axis is consistent with a competition between Pb^{2+} and Mg^{2+}. **(C) Magnesium reverses the lead-induced inhibition of the rod cGMP PDE:** 430 pM rod photoreceptor cGMP PDE was assayed using 500 μM cGMP and 500 μM Mg^{2+} in the absence or presence of 5 nM Pb^{2+}. The hydrolysis of cGMP was linear with respect to time in both conditions. In the presence of Pb^{2+}, however, the rate of hydrolysis was inhibited 78% (closed circles). Each data point represents the mean of 3 to 5 separate experiments. At the 5-min time point, an additional 10 mM Mg^{2+} was added to a set of tubes containing 500 μM cGMP, 500 μM Mg^{2+}, and 5 nM Pb^{2+} (arrow). From this time point onwards, the rate of cGMP hydrolysis was not affected significantly by Pb^{2+} (open circles). (Figures 25.1B and C are reprinted from Srivastava, D., Hurwitz, R.L., and Fox, D.A., *Toxicol. Appl. Pharmacol.*, 134: 43–52, 1995. With permission.)

By definition, the competitive inhibition by Pb^{2+} should be reversible by the addition of excess Mg^{2+}. To examine this experimentally, the hydrolysis of cGMP by the rod cGMP PDE was measured as a function of time using 500 μM cGMP and 500 μM Mg^{2+} in the absence (control) or presence of 5 nM Pb^{2+} (Figure 25.1C). cGMP hydrolysis was inhibited 78% by 5 nM Pb^{2+}. This inhibition was reversed at the 5-min time point following the addition of 10 mM Mg^{2+} (see arrow). That is, the rate of hydrolysis in the presence of 5 nM Pb^{2+} plus 10 mM Mg^{2+} was not significantly different from that of the control. This result clearly suggests that Mg^{2+} may be useful in reversing the inhibition of the rod cGMP PDE by Pb^{2+}.

A detailed kinetic analysis using various concentrations of cGMP, Mg^{2+}, Pb^{2+}, and IBMX (a competitive inhibitor of PDE with respect to cGMP) was performed.[7,8] The experiments allowed the generation of secondary and tertiary double reciprocal plots from which binding constants for each step of the reaction were derived (Figure 25.2).[7,8] These plots also resulted in the derivation of binding constants for the steps involved with the Pb^{2+}-induced inhibition of cGMP PDE.[8] The constant α represents the factor by which the prior binding of one coreactant, cGMP or Mg^{2+}, to the PDE affects the binding of the second co-reactant, Mg^{2+} or cGMP, respectively, to the PDE. The constant β represents the factor by which the prior binding of cGMP or Pb^{2+} to the PDE affects

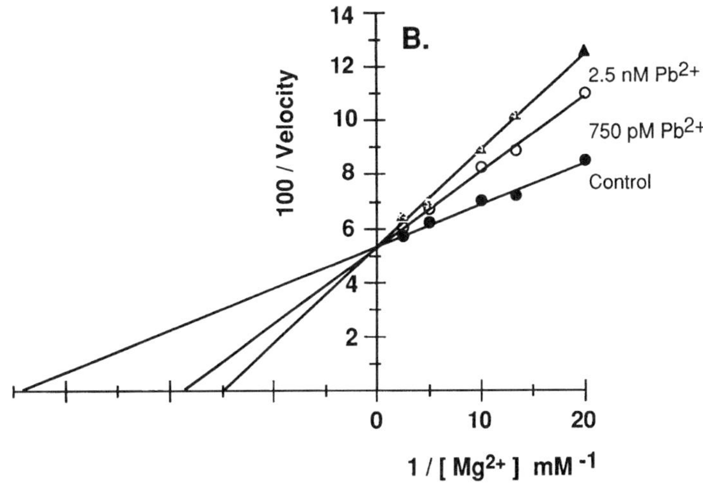

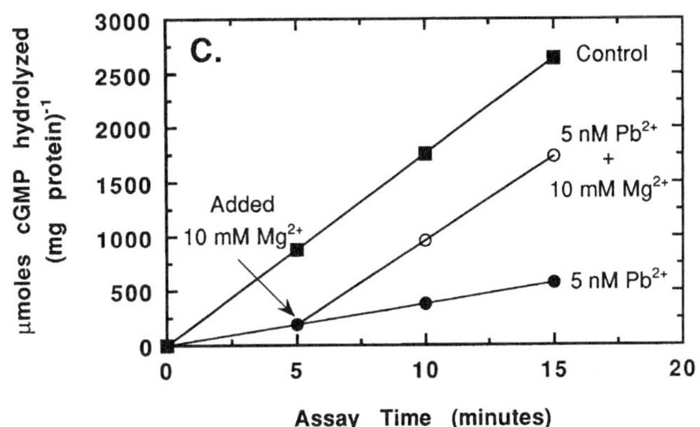

FIGURE 25.1 B and C

the binding of Pb^{2+} or cGMP, respectively, to the PDE. A value less than 1 implies that the binding of the first reactant increased the affinity for the second reactant while a value greater than 1 implies that the affinity for the second reactant was reduced by the binding of the first reactant. The factors, α and β, were different for Mg^{2+} and Pb^{2+}, respectively. The prior binding of cGMP to the PDE decreased the affinity of the PDE for Mg^{2+} ($\alpha > 1$) but increased the affinity of the PDE for Pb^{2+} ($\beta < 1$). Similarly, the prior binding of Mg^{2+} to the PDE decreased the affinity of the PDE for cGMP ($\alpha > 1$) but increased the affinity of the PDE for IBMX ($\beta < 1$). However, the prior binding of Pb^{2+} to the PDE increased the affinity of the PDE for cGMP ($\beta < 1$). These differences in α and β, coupled with the competition between Pb^{2+} and Mg^{2+}, suggest that Pb^{2+} and Mg^{2+} may be binding at proximal, yet distinct sites. Pb^{2+}, whose effective ionic radius is almost twice the size of Mg^{2+},[9] may bind to the PDE and obscure the Mg^{2+} binding site, thus giving the appearance of competition.

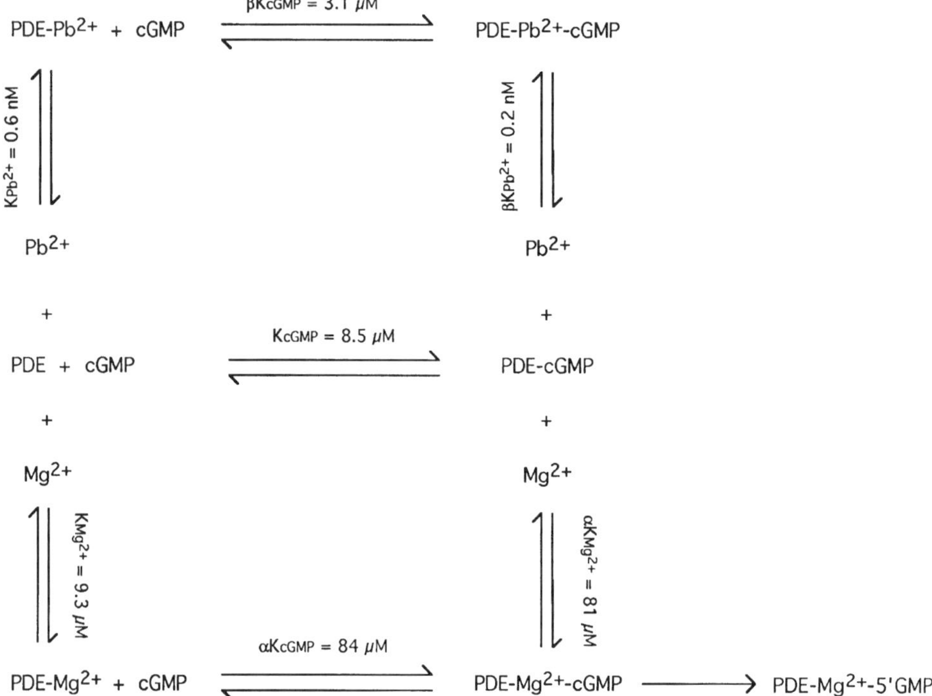

FIGURE 25.2   Proposed mechanism of the lead-induced inhibition of the rod cGMP PDE. The hydrolysis of cGMP by the rod PDE may be modeled with a random binding order of Mg^{2+} and cGMP to the PDE.[7] Based on the results illustrated in Figure 25.1B,[8] Pb^{2+} may substitute for Mg^{2+} and ultimately form a ternary complex with cGMP and the rod PDE. This ternary complex will not produce the product 5′GMP. Secondary and tertiary double reciprocal plots were used to derive the binding constants for each step of the reaction.[7,8] The results of this kinetic analysis are summarized here. The derived binding constants K_{cGMP}, $K_{Mg^{2+}}$, $K_{Pb^{2+}}$, represent the dissociation constants for cGMP, Mg^{2+}, and Pb^{2+}, respectively, and the PDE. The constants α ($\alpha = 8.7–9.9$) and β ($\beta = 0.3$) represent the factor by which the prior binding of one coreactant affects the binding of the second coreactant. (Reprinted from Srivastava, D., Hurwitz, R.L., and Fox, D.A., *Toxicol. Appl. Pharmacol.*, 134: 43–52, 1995. With permission.)

25.6   LEAD AND CALCIUM INTERACTIONS

Classical investigations of the mechanisms of lead toxicity have focused on Pb^{2+} and Ca^{2+} interactions.[11,13,24–26] For example, Pb^{2+} substitutes for Ca^{2+} in calmodulin-dependent processes and troponin C-dependent processes. Pb^{2+} inhibited $^{45}Ca^{2+}$ binding to calmodulin and, based on tyrosine fluorescence, bound to calmodulin.[24] Pb^{2+} displaced $^{45}Ca^{2+}$ from calmodulin, the vitamin D-dependent calcium binding proteins CaBP type I and II, and, to a lesser extent, troponin C and parvalbumin in a dose-dependent manner.[25] Pb^{2+} and Ca^{2+} were equally efficacious in stimulating troponin C to activate myofibrillar ATPase although the binding of Pb^{2+} to troponin C was undetectable when using tyrosine fluorescence as an indicator.[26] The results presented in this chapter strongly suggest that Pb^{2+} may also produce its toxic effects at the site of another essential divalent cation: Mg^{2+}.

25.7 MAGNESIUM AS AN ESSENTIAL ION FOR ENZYMES

Mg^{2+} serves as an essential cofactor in a wide range of enzymatic reactions.[20,27] These reactions may be divided into two general categories. First, there are the reactions that involve ATP, where the true substrate is an MgATP complex. Second, Mg^{2+} may be required by enzymes either as a structural component or as an allosteric activator. Both situations are discussed below.

25.7.1 Lead and Na+,K+-ATPase

The Na+,K+-ATPase is an excellent example of an enzyme that utilizes MgATP as a substrate. Pb^{2+} is a competitive inhibitor of the retinal Na+,K+-ATPase with respect to its substrate MgATP.[28] Under conditions where ATP was saturated with Mg^{2+}, the addition of more Mg^{2+} increased the inhibition of retinal Na+,K+-ATPase activity by Pb^{2+}. Previous studies demonstrated that the phosphorylation of Na+,K+-ATPase by ATP occurs at high-affinity binding sites. When Mg^{2+} binds to capacity-regulating sites on the Na+,K+-ATPase, it induces a conformational change which results in additional low-affinity nucleotide binding sites.[14] With respect to the potentiating effect of Mg^{2+} on the Pb^{2+}-induced inhibition of Na+,K+-ATPase, perhaps Mg^{2+} caused conformational changes in the Na+,K+-ATPase such that Pb^{2+} was able to occupy the additional low-affinity nucleotide binding sites and thus increase its inhibition. Although Pb^{2+} is a competitive inhibitor of the retinal Na+,K+-ATPase with respect to MgATP, it is unclear if Pb^{2+} acts at a specific Mg^{2+} site. Further investigation in this area should yield greater understanding of the basic mechanism of the Na+,K+-ATPase as well as its Pb^{2+}-induced inhibition.

25.7.2 Lead and Other Phosphodiesterases

There are many examples of enzymes that require Mg^{2+} as a cofactor. Among this large group are enolase, pyrophosphatase, glutamine synthetase, alkaline phosphatase, and phosphodiesterase.[27] The mechanism by which Pb^{2+} inhibits the rod photoreceptor cGMP PDE has been elucidated[7,8] and reviewed in this chapter. This interaction appears to occur within the catalytic domain of the PDE, whose amino acid sequence is highly conserved among PDEs.[29] Therefore, it is tempting to speculate that Pb^{2+} may disrupt cyclic nucleotide hydrolysis by other PDEs via a similar mechanism. One PDE that is of particular interest is the brain type IV cAMP-specific PDE.[29] A mutation in the gene for this enzyme results in defective memory and learning in *Drosophila*.[30] Theoretically, a direct inhibition of the mammalian homologue of this cAMP PDE by Pb^{2+} at the Mg^{2+} site may contribute to the long-term learning deficits produced by lead exposure during development.[1]

25.8 CONCLUDING REMARKS

To date, most of the studies exploring the ionic basis of lead's effects have focused on lead-induced alterations in calcium homeostasis, calcium-mediated functions, or direct Pb^{2+}-Ca^{2+} interactions.[6,11,13,22,23] The results presented here and by Simons in Chapter 24 reveal that Pb^{2+} also interacts with Mg^{2+} and Zn^{2+} sites, respectively. Based on the molecular mechanism of the Pb^{2+}-Mg^{2+} interaction (i.e., competitive inhibition), it may be prudent to conduct studies using magnesium (and maybe zinc) supplementation as possible therapeutic or prophylactic treatments in lead-exposed animals and children.

ACKNOWLEDGMENT

This work was supported partially by National Institute of Environmental Health Science Grant RO1-ES03183 (DAF).

REFERENCES

1. Needleman HL and Bellinger D: The health effects of low level exposure to lead. *Annu. Rev. Public Health*, 1991; 12:111–140.
2. Pugh EN, Jr and Lamb TD: Amplification and kinetics of the activation steps in phototransduction. *Biochim. Biophys. Acta*, 1993; 1141:111–149.
3. Fox DA and Sillman AJ: Heavy metals affect rod, but not cone, photoreceptors. *Science*, 1979; 206:78–80.
4. Tessier-Lavigne M, Mobbs P, and Attwell D: Lead and mercury toxicity and the rod light response. *Invest. Ophthalmol. Vis. Sci.*, 1985; 26:1117-1123.
5. Fox DA: Visual and auditory system alterations following developmental or adult lead exposure. A critical review. In Needleman HL (ed): *Human Lead Exposure*. Boca Raton, FL, CRC Press, 1992; pp 105–123.
6. Otto DA and Fox DA: Auditory and visual dysfunction following lead exposure. *NeuroToxicology*, 1993; 14:191–208.
7. Srivastava D, Fox DA, and Hurwitz RL: Effects of magnesium on cyclic GMP hydrolysis by the bovine retinal rod cyclic GMP phosphodiesterase. *Biochem. J.*, 1995; 308:653–658.
8. Srivastava D, Hurwitz RL, and Fox DA: Lead- and calcium-mediated inhibition of bovine rod cGMP phosphodiesterase: interactions with magnesium. *Toxicol. Appl. Pharmacol.*, 1995; 134: 43–52.
9. Martin RB: Bioinorganic chemistry of metal ion toxicity. In Sigel H (ed): *Metal Ions in Biological Systems: Concepts on Metal Ion Toxicity*, Volume 20. New York; Marcel Dekker, 1986; pp 21–65.
10. Martell AE and Smith RM: *Critical Stability Constants*. New York; Plenum Press; 1974; pp 27, 109, 182, 269, and 496.
11. Simons TJB: Cellular interactions between lead and calcium. *Br. Med. Bull.*, 1986; 42:431–434.
12. Martin RB: Bioinorganic chemistry of magnesium. In Sigel H and Sigel A (eds): *Metal Ions in Biological Systems: Compendium on Magnesium and its Role in Biology, Nutrition, and Physiology*, Volume 26. New York; Marcel Dekker, 1990; pp 1–13.
13. Pounds JG: Effect of lead intoxication on calcium homeostasis and calcium-mediated function. A review. *Neuro-Toxicology*, 1984; 5:295–332.
14. Heaton FW: Distribution and function of magnesium within the cell. In Birch NJ (ed): *Magnesium and the Cell*. London; Academic Press, 1993; pp 121–136.
15. Farber DB and Lolley RN: Calcium and magnesium content of rodent photoreceptor cells as inferred from studies of retinal degeneration. *Vision Res.*, 1976; 22:219–228.
16. Lolley RN, Lee RH, Chase DG, and Racz E: Rod photoreceptor cells dissociated from mature mice retinas. *Invest. Ophthalmol. Vis. Sci.*, 1986; 27:285–295.
17. Somlyo AP and Walz B: Elemental distribution in *Rana pipiens* retinal rods. Quantitative electron probe analysis. *J. Physiol.*, 1985; 358:183–195.
18. Flora SJS, Kumar D, Sachan SRS, and Gupta SD: Combined exposure to lead and ethanol on tissue concentration of essential metals and some biochemical indices in rat. *Biol. Trace Elem. Res.*, 1991; 28:157–164.
19. Dowd TL, Rosen JF, and Gupta RK: ^{31}P NMR and saturation transfer studies of the effect of Pb^{2+} on cultured osteoblastic bone cells. *J. Biol. Chem.*, 1990; 265:20833–20838.
20. Heaton FW: Role of magnesium in enzyme systems. In Sigel H and Sigel A (eds): *Metal Ions in Biological Systems: Compendium on Magnesium and its Role in Biology, Nutrition, and Physiology*, Volume 26. New York; Marcel Dekker, 1990; pp 119–133.
21. Al-Modhefer AJA, Bradbury MWB, and Simons TJB: Observations on the chemical nature of lead in human blood serum. *Clin. Sci.*, 1991; 81:823–829.
22. Katz LM and Fox DA: Developmental lead exposure selectively alters the scotopic ERG component of dark and light adaptation and increases rod calcium content. *Vision Res.*, 1992; 32:249–255.
23. Medrano CJ and Fox DA: Substrate-dependent effects of calcium on rat retinal mitochondrial respiration. Physio-logical and toxicological studies. *Toxicol. Appl. Pharmacol.*, 1994; 125:309–321.
24. Chao S-H, Suzuki Y, Zysk JR, and Cheung WY: Activation of calmodulin by various metal cations as a function of ionic radius. *Mol. Pharmacol.*, 1984; 26:75–82.
25. Richardt G, Federolf G, and Habermann E: Affinity of heavy metal ions to intracellular Ca^{2+}-binding proteins. *Biochem. Pharmacol.*, 1986; 35:1331–1335.
26. Chao S-H, Bu C-H, and Cheung WY: Activation of troponin C by Cd^{2+} and Pb^{2+}. *Arch. Toxicol.*, 1990; 64:490–496.
27. Williams RJP: Magnesium: An introduction to its biochemistry. In Birch NJ (ed): *Magnesium and the Cell*. London; Academic Press, 1993; pp 15–30.
28. Fox DA, Rubinstein SD, and Hsu P: Developmental lead exposure inhibits adult rat retinal, but not kidney, Na^+,K^+-ATPase. *Toxicol. Appl. Pharmacol.*, 1991; 109:482–493.

29. Beavo JA, Conti M, and Heaslip RJ: Multiple cyclic nucleotide phosphodiesterases. *Mol. Pharmacol.*, 1994; 46:399–405.
30. Davis RL and Dauwalder B: The Drosophila dunce locus. Learning and memory genes in the fly. *Trends Genet.*, 1991; 7:224–229.

Chapter 26

Inorganic Lead, Neurotransmitters, and Neuropeptides

Nicole Pages and Roger Deloncle

CONTENTS

26.1 INTRODUCTION

Lead is ubiquitous in the human environment as a result of industrialization and, in recent years, traffic. It has no known physiological values. The blood lead level considered to indicate lead poisoning has fallen steadily since the 1970s. As yet, no threshold has been identified for the harmful effects of lead, suggesting that blood levels of lead currently prevalent in industrialized societies (4 to 30 µg/100 ml) and previously thought to be safe may signify some degree of risk for brain injury.[1–4]

Clinically, lead neurotoxicity is insidious, progressing from subtle psychological effects to death, relative to the importance of exposure and rate. High-dose acute exposure (blood lead levels >80 µg/100 ml) is associated with signs of encephalopathy mainly in childhood (drowsiness, ataxia, paralysis, convulsions, coma) and leads either to death or to major irreversible neurotoxic sequelae such as neurological damage and mental retardation. This degree of lead poisoning has become uncommon in recent years, probably as a result of screening and treatment of children and workers at risk. At lower levels, in children, toxicity may be insidious, does not appear readily reversible, and affects both neurological and sensory processes. Current screening programs have revealed that most poisoned children have no distinctive symptoms. However, blood lead levels as low as 10 µg/100 ml of whole blood are associated with decreased intelligence and impaired neurobehavioral development (as manifested by hyperactivity), decreased stature or growth, decreased hearing acuity, and decreased ability to maintain a steady posture.[1] In adults, lead induces subtle central alterations as manifested by neuropsychiatric changes.[2] Lead also exerts a neurotoxic effect on the peripheral system function both in children and adults by reducing motor nerve conduction velocities and impairing fine motor coordination. This decreased nervous conduction velocity was suggested experimentally to be due to a segmental demyelination and axonal degeneration of peripheral motor nerves and to the interaction of lead with the kinetics of acetylcholine release at the myoneural junction.[5]

During the past several years, there has been a renewed interest in the basic mechanisms underlying the neurotoxic effects caused by acute or chronic exposure to lead. Experimental studies have corroborated performance changes in response to low-level lead exposure and showed increased spontaneous motor activity in mice, rats, and monkeys, thus allowing, together with recent clinical data, the development of various hypotheses on the mechanisms of inorganic lead neurotoxicity.[5] Many of the biological alterations produced by lead appear related to the ability of this heavy metal to either inhibit or mimic the action of calcium.[7,8] Other mechanisms were also implicated, involving mainly inhibition by lead of either cell energetics or heme synthesis at the level of δ-aminolevulinate dehydratase, resulting in depletion of heme-dependent proteins (including cytochromes) and in accumulation of heme precursors such as δ-aminolevulinic acid that may be neuroactive.[2,9] More recently, the inhibitory effect of lead on brain physiological development was also reported.[10,11] In the present review, we will focus our attention on the neurochemical changes induced by lead on neurotransmitter processes.

26.2 LEAD AND NEUROTRANSMISSION

Lead is not uniformly distributed within the nervous tissue.[4] The cerebellum is particularly susceptible to large doses of lead. At lower dose, another important target of lead is the hippocampus, which plays a major role in the consolidation of memory.[11] During the increased locomotor activity period in rats, (8 weeks of age) the highest lead contents were observed in the hypothalamus and striatum, regions of the brain believed to play a fundamental role in the elaboration of behavior and in the regulation of motor activity, respectively.[6] However, there seems to be little correlation between the amount of lead present in various brain regions and the degree of neurochemical changes observed.[6] Lead acts as a neurotoxin, at very low doses, and the mechanisms involved are probably neurochemical, altering biochemical and electrophysical properties of nerve cells as well as the behavior of animals in the absence of morphological damage which appears at higher dose, thus reflecting the great sensitivity of the CNS to cell injury. It probably acts nonspecifically, in the submolecular range, at any site to which it gains access by mechanisms of uptake and distribution which are not completely understood but involving, at least in children, a disruption of blood-brain barrier function.[2,10,11]

26.3 BASIS OF LEAD-INDUCED BRAIN DYSFUNCTION: *EX VIVO* AND *IN VITRO* LEAD EFFECTS

The mechanism by which lead interferes with the normal process of transmitter release has been particularly studied. A number of neurotransmitter systems have been implicated in lead toxicity; however, no single system has emerged as a selective or primary site of insult. Anyway, *ex vivo* and *in vitro* results were consistent in spite of the variability in the experimental systems, varying from neuromuscular junction and cervical ganglia preparations (PNS) to slices and synaptosomes, i.e., suspensions of isolated nerve endings isolated from brain (CNS). In some experiments, the animals were exposed to lead prior to tissue preparation (*ex vivo*) while in others, the tissue under study was suspended in or superfused with a buffer (*in vitro*) containing lead.[7]

26.3.1 Cholinergic Processes[3,5,13–16]

26.3.1.1 Acetylcholine (ACh) Release

In the peripheral nervous system, lead has been shown to impair cholinergic function in several different experimental models. **Spontaneous ACh release** was enhanced in both the PNS and CNS. Electrophysiologic measurements on peripheral synapses showed that lead enhanced the frequency of spontaneous miniature end-plate potentials.[7] Similarly, the spontaneous release of ACh, but not of choline (Ch), was increased in minces from lead-treated mice, while *in vitro* removal of calcium did not change spontaneous release of ACh.[5] In contrast, electrophysiological studies of peripheral synapses on frog neuromuscular or mouse phrenic nerve-diaphragm junctions have shown that relatively low concentrations of Pb attenuate the nerve-evoked end-plate potential, i.e., the **depolarization-evoked release.**[12] Lead had an inhibitory effect on the release of ACh from the superfused superior ganglia of cats during stimulation of the preganglionic fiber; the addition of extracellular calcium has been shown to reverse, or by pretreatment, to protect against this inhibition.[12] Concurrently, the presence of lead ions did not change the contractile mechanism of postganglionic nerve stimulation and the sensitivity of ganglionic cells to injected ACh.[3] This presynaptic blocking action of lead on evoked ACh release, which is calcium dependent,[13] has been confirmed in brain minces[5] where Ch-evoked release was also decreased and in synaptosomes prepared from mice brain,[13] or rat hippocampus and cortex.[14] Reduced calcium concentration potentiated this effect.[3]

26.3.1.2 Choline (Ch) Uptake

The high-affinity Ch uptake is proposed to be a regulatory part of cholinergic neurotransmission, as a means of partially supplying precursor for the synthesis of readily releasable acetylcholine.[13] Increased Pb levels inhibited **high-affinity Ch uptake** by synaptosomes prepared from caudate nucleus or forebrain, but not from hippocampus of lead-treated mice. This lead-induced inhibition was mimicked by reduced *in vitro* calcium concentrations, thus suggesting again a competitive interaction between lead and calcium. In contrast, the low-affinity Ch uptake was not changed.[15] However, conflicting results were obtained in cortical minces of chronic lead-treated mice where both high and low activity processes of Ch transport were not changed.[5]

26.3.1.3 Conclusion

Lead, *in vitro* and *ex vivo*, increases ACh spontaneous release. In contrast, it inhibits evoked ACh release and Ch uptake through a competitive replacement of calcium at presynaptic sites, thus resulting in an inhibition of cholinergic function.[5,10,13] It is possible that the inhibitory effects of lead for choline uptake may indirectly result from Pb-induced inhibition of ACh-evoked release, since high-affinity uptake of Ch is affected by changes in ACh release induced by drugs or potassium stimulation.[10] Finally, the inhibition of energy production appears to limit availability of acetyl-coenzyme A, which is essential for ACh production.[16]

26.3.2 Catecholaminergic Processes[3,13,16,20]

26.3.2.1 Catecholamine Synthesis and Uptake

In mouse forebrain synaptosomes, the **high-affinity uptake** for the catecholamine precursor amino acid, tyrosine, was found to be increased by lead while no change in high-affinity uptake of noradrenaline (NA) was found.[15] In contrast, the high-affinity uptake of dopamine (DA) was reduced by both *in vitro* and *ex vivo* lead exposure.[13,16] The effects of lead on dopaminergic function are not directly calcium dependent: Silbergeld has demonstrated that in synaptosome preparations, *in vitro*-decreased calcium concentration does not affect DA uptake;[13] moreover, addition of calcium does not reverse lead-induced inhibition of dopamine uptake. However, **DA synthesis** and tyrosine hydroxylase, the initial and rate-limiting enzyme in the catecholaminergic pathway remained unchanged in rat forebrain synaptosomes after exposure to lead.[17]

26.3.2.2 Catecholamine Release

Lead (25 µM) was found to decrease **spontaneous release** or to induce no change (100 µM) of DA release.[18,19] In fact, lead alone (1 to 100 µM) appeared to have no effect on release of DA, but it nevertheless potentiated calcium-induced release in a dose-dependent manner (1 to 30 µM range).[13,20] This effect may have resulted from lead's ability to increase uptake of calcium into caudate dopaminergic nerve terminals. In this regard, it is noteworthy that lead in concentration between 1 to 100 µM increases calcium uptake in a dose-dependent manner.[13] However, **potassium-evoked DA release** from forebrain slices of lead-treated animals was decreased.[17] Similarly, striatal synaptosomes of lead-treated rats had a diminished DA-evoked release.[20]

26.3.2.3 Conclusion

Lead, *in vitro* and *ex vivo*, has no direct effect on dopamine spontaneous release. In contrast, it decreases evoked dopamine release and inhibits dopamine uptake. The uptake inhibition thus results in an increased dopaminergic function.

26.3.3 GABAergic Processes[3,9,16,21]

In the first studies, a dual effect of lead was reported. *Ex vivo*, chronic high lead exposure inhibits both uptake and release of γ-aminobutyric acid (GABA) from nerve terminals of all areas but the cerebellum. However, when added *in vitro*, lead has no effect on both the incorporation and liberation of GABA in forebrain synaptosomes in lead-treated mice. This lack of effect suggested to Silbergeld and Lamon that lead was not directly responsible for the neurotoxic effect, but indirectly through the accumulation of ALA.[9] ALA and GABA have a very close structure and ALA could act as a GABA partial agonist. Indeed, *in vitro*, ALA (100 µM) weakly displaces GABA from its receptors. It was also reported to directly inhibit GABA uptake and release from synaptosomes even though Minnema and Michaelson failed to observe such effect.[21] However, more recent studies, using "purified" cortical synaptosomes to eliminate potential confounding influences by extraneuronal tissues and/or organelles reported that in fact, lead (1 to 30 µM) increased spontaneous release. In contrast, it attenuated depolarization-evoked GABA release, which is dependent on extraneuronal calcium.[21] So, lead acts on GABA in a similar manner as on the other neurotransmitters.

26.3.4 Miscellaneous

High-affinity uptake for serotonin, another important putative neurotransmitter, and for various aminoacids, phenylalanine, glycine, and leucine from forebrain synaptosomes was not altered in lead-treated mice.[15]

26.3.5 Conclusion

Most studies reported for ACh, DA, and GABA suggest that lead *in vitro* affects the release of all neurotransmitter systems in a similar manner, both quantitatively and qualitatively, indicating that lead acts at sites basic to the transmitter release process.[21] First, lead appears to competitively block the opening of voltage-sensitive calcium channels. Consequently calcium cannot enter the cell resulting in inhibition of evoked neurotransmitter release from presynaptic nerve endings. In contrast, the delayed onset (15 to 30 s) of lead-induced spontaneous neurotransmitter release and the apparent competition between lead and calcium suggest that lead may enter the nerve ending through calcium channels to exert its releasing action intraneurally. However, lead may also enter through other routes than calcium, sodium, and/or potassium channels since their blockade by inhibitors is ineffective in blocking lead-induced release.[21] Second, inside the nerve terminal lead may act either by increasing intraneuronal ionized calcium or by stimulating calcium-activated molecules mediating transmitter release.[3,8,21] A support for the later hypothesis was first provided by the observation that Pb can elicit acetylcholine release from digitonin-permeabilized synaptosomes in the absence of Ca^{2+}. This last effect could be due to activation of protein kinase II in the nerve endings. This enzyme is widespread in the neural tissue and is proposed to have a role in the release of neurotransmitters. It is activated by a calcium-calmodulin complex. Many of the agents which regulate neurotransmitter release also activate both calmodulin protein kinase II and increase the state of phosphorylation of a synaptic vesicle protein, synapsin I. Consequently, it has been hypothesized that activated protein kinase II would phosphorylate the synapsin I, thus allowing it to separate from the synaptic vesicle. The vesicle is then more likely to fuse with the synaptic membrane and to release neurotransmitters. This sequence may explain how lead, by mimicking calcium as an intracellular second messenger inside the nerve terminal, increases the basal rate of neurotransmitter release.[7,8] The other hypothesis that lead may increase spontaneous neurotransmission by inducing intrasynaptosomal free calcium accumulation as a result of alterations of intracellular calcium homeostasis is controversial. Indeed, various mechanisms which were proposed to explain that accumulation (inhibition of Ca-Na ATPase-mediated exchange, inhibition of the calcium flux into mitochondria or endoplasmic reticulum, or calcium displacement from these nerve terminal organelles) were refuted successively.[21]

Finally, the biphasic response of neurotransmitter release to lead, observed *in vitro* with stimulation of the basal rate and inhibition of the depolarization-induced fraction, may have special relevance to the immature nervous system in the absence of overt pathological damage by interfering with the physiological pruning of excess synapses.[7,10,11]

26.4 *IN VIVO* LEAD EFFECTS

The biological analysis of nervous tissue from lead-exposed animals might be expected to provide insights on lead neurotoxicity. Indeed, lead-induced brain dysfunction involves an overall derangement of neural processing concerning the main pathways of neurotransmission: ACh, DA, NA, and GABA. However, neurochemical studies of lead-exposed animals, while numerous, are too variable in treatment and other parameters to provide a comprehensive or systemic analysis of brain chemistry or a systematic hypothesis for transmitter-specific substrate for the effects of lead.[16]

An experimental model with morphological alterations closely resembling those occurring in humans with lead toxicity has been described first in rats. Lead was provided in drinking water or food to nursing mothers or directly to pups right after birth and before weaning. In the initial model (acute toxicity), rats developed paraplegia, extensive cerebellar vascular damage, and reduced growth rate. It has been suggested that most observed effects could be attributable to undernutrition occurring at blood lead levels in neonates of about 100 µg/100 ml.[16] Consequently, from the 1980s, in further models (chronic toxicity) in either mice, rats, or monkeys, the doses were decreased so as to reflect low-level lead exposure and to avoid differences in growing rate. Some of the more predominant behavioral symptoms displayed by developing animals were hyperactivity, increased aggressiveness, and stereotyped behavior as manifested by excessive self grooming at 6 and 8 weeks of age. Since lead is present in the lipid membranes, it was reported that lead could act either presynaptically by interfering with neurotransmitter processes, as reported *in vitro*, or may postsynaptically induce receptor alterations either by altering their characteristics or as a consequence of lead-induced changes in either presynaptic neurons or other factors governing receptor efficiency.[23]

26.4.1 Cholinergic Processes[3,16]

It has been suggested that lead-induced hyperactivity in mice is associated, at least in part, with a decrement in central cholinergic function. Indeed, administration of several drugs to the lead-induced hyperactive mice influences both the hyperactive state and central ACh metabolism. For example, the muscarinic blocking agents, atropine and benztropine, exacerbate the hyperactive state and block the effect of ACh at postsynaptic receptor sites. Conversely, the anticholinesterase agent, physostigmine, suppresses the hyperactive state probably by causing ACh accumulation at postsynaptic receptor sites.[2,5]

The steady state levels of the neurotransmitter ACh and its prescursor, Ch, were reported to be unchanged in mouse forebrain and in the following rat brain regions: cerebellum, hippocampus, midbrain, medulla pons, and striatum.[5,15,16] A significant elevation of 20% was found in the diencephalon of lead-treated animals.[16] By observing suitable precautions against post-mortem enzymatic degradation, a regional diminution of mouse brain ACh was reported.[22] Conflicting results were reported in rat cortex with either a 32 to 48% increase or no change at all.[3,16] The steady state level of Ch was significantly reduced in rat midbrain, but no changes were found in cortex, hippocampus, and striatum of lead-treated rats.[3,16] Finally, a consistent inhibitory effect of ACh turnover was observed in all investigated areas.[3]

In mouse forebrain, acetylcholinesterase (AChE) and choline acetyltransferase (ChAT) activities were not altered by lead treatment.[3] However, in more discrete brain regions, lead changed the degradative and synthetic enzymatic activities in opposite directions. In lead-treated rats, AChE had significantly lower activity in the diencephalon (where ACh levels increased), medulla-pons, and midbrain, while ChAT activity increased in the cerebral cortex, hippocampus, and medulla-pons regions.[3]

Affinity and density of muscarinic receptors were evaluated by radioligand binding studies on rats treated by lead at different periods: during pregnancy, lactation, and after weaning. The rats treated during pregnancy showed decreases both in affinity and density of striatal muscarinic receptors. The density was reduced only in rats treated during pregnancy while no change was observed in rats treated after weaning. The apparent reduction in the first two groups reveals that antenatal and neonatal exposure to lead interferes with the physiological peripostnatal increase in muscarinic receptor density.[27] This may account for the deficit in central cholinergic function consecutive to lead exposure.[23]

While effects on cholinergic metabolism have been reported in lead-treated rodents, the doses of lead used were high, resulting in undernutrition that may be implicated in the observed cholinergic alterations.[16] However, Carroll et al. pointed out that the release in lead-exposed animals was consistent with the *in vitro/ex vivo* effects of lead on PNS, so that it is unlikely that the undernutrition is the cause of the altered neurochemistry observed *in vivo*.[5]

26.4.2 Catecholaminergic Processes[3,6,16,21,24]

Since the catecholamines are thought to regulate motor activity in rodents, the levels of the amines were among the first neurochemical parameters to be investigated in experimental lead poisoning.[2,6,23] Moreover, the frequent finding that increased motor activity was suppressed by the administration of amphetamines, methylphenidate, and aminergic antagonists, and was exacerbated by aminergic agonists, and that lead induced increased serum prolactin levels, which are mainly under hypothalamic dopaminergic inhibitory control, has remained a continuing stimulus of research in lead.[2,3,15,16,23] So, alterations in functioning of catecholamines have been postulated to play a role in toxic effects of lead on the central nervous system.[15,25]

Lead-induced brain dysfunction has been proposed to involve overactivation of biogenic amine systems (mainly at high lead levels), imbalance among such systems, or inhibition of dopamine function.[25] Most of the first studies, using rather high lead doses, were conducted on whole brains or forebrains and measured the endogenous levels of catecholamines, NA, or DA, and their amino acid precursor, tyrosine. Mouse whole-brain levels of tyrosine were not changed.[3] It was also reported first that rat whole-brain NA endogenous level was unchanged. Subsequently they were found to be increased either in the rat whole brain, midbrain and brainstem, or mouse forebrain.[6,15,26,28–30] Levels of DA were not found by several authors to be changed in mouse whole brain or forebrain.[15,26,29,30] Others have found a decrease in DA levels in whole brain.[25,28] However, when measuring neurotransmitters in the whole brain, regional changes in the concentrations of neurotransmitters may have easily been masked by measurements in larger portions of the CNS.[24] The involvement of brain biogenic amines was then examined in discrete brain regions of rats that had been chronically exposed to low levels of lead from birth by measuring either steady state or turnover via catecholamine metabolites. Results revealed that each region has its own characteristic pattern of responses to lead toxicity on the two neurotransmitters.

For example, in the **hypothalamus**, the first studies reported no change of NA;[6,28–31] no change of DA;[23,30] or a reduction of DA;[6] although in another paper Dubas et al. reported an increase in DA level.[31] In a further study, NA and DA levels were reported to be significantly increased, indicating that catecholamine projections were susceptible to lead.[32] After discontinuation of the consumption of lead, the abnormality of DA but not of NA was relieved, indicating that the NA system seems to be very susceptible to lead in this area. In the **striatum**, Grant et al. observed no changes of NA while Dubas et al. reported a reduction of NA and an increase in DA.[30,31] In contrast to Dubas et al., Jason and Kellog observed a reduction of DA levels.[28] In the **striatum-accumbens**, only NA was increased and did not return to normal after rehabilitation while no changes occurred in the levels of DA. Thus there is some basis to assume a correlation between the hyperactive behavior and the increase in NA in the striatum-accumbens since it is known that the balance in the aminergic transmitter actions determines the levels of locomotor activity.[32] In the **hippocampus**, NA was significantly elevated while DA showed no changes, which is a minor point since DA innervation is very poor in this area. The NA abnormality recovered after rehabilitation.[32] The **motor cortex**, which receives NA projections but not DA, did not show changes in NA level.[28–30,32,33] However, Dubas and Hrdina observed a reduction in NA and DA in the cerebral cortex.[6] In the **brainstem**, conflicting results were reported: NA level was either increased in various studies,[28,32] or unchanged.[29,30,33] DA levels were decreased,[6,32] or increased.[31] The DA decrease did not recover after rehabilitation whereas increased NA returned to normal.[32] In the **cerebellum**, NA level was found either unchanged[29,33] or increased, returning to normal after rehabilitation.[32] In the **midbrain**, NA and DA levels were reported to be first increased,[31] whereas subsequently DA levels were found decreased.[6]

Besides the high sensitivity of the noradrenergic system, it is noteworthy that the lesion is not permanent in the majority of the brain regions, even with a very high dose of lead consumption, when abnormal levels of lead are relieved. One possible explanation for the elevated levels of NA seen during lead treatment is that lead might alter the activities of the enzymes involved in noradrenaline metabolism.[32] The effect of lead on the DA system is highly region-specific, suggesting the possibility that different mechanisms may be operant in the dopaminergic systems.

NA turnover in rat whole brain was increased whereas the DA turnover was either increased or unchanged in rat whole brain.[16,24,32] Govoni et al. observed decreased dihydroxyphenylacetic acid (DOPAC) levels in the striatum and the hypothalamus that seemed to be associated with a reduced DA turnover. DOPAC is a product of MAO action on DA, either synthesized or taken up primarily inside the axon terminal. Since chronic lead treatment might lead to inhibition of both type A and B MAO, DOPAC levels could be expected to decrease, except for any special regional differences as reported in nucleus accumbens and frontal cortex. The decrease in DA turnover was not accompanied by a compensatory supersensitive response of the DA receptors.[24] In the nucleus accumbens dopamine levels were unchanged, but increased DOPAC concentrations were found probably reflecting an enhancement of DA turnover. This may be correlated with the hypermotility that follows lead intoxication. In fact, a bilateral administration of DA into the nucleus accumbens enhances locomotor activity. An increase in DOPAC levels in the frontal cortex was also reported, suggesting an increase in dopamine turnover in this area.[24]

Changes in metabolites indicate changes in the synthesis and/or release of DA, which in turn determine dopamine receptor numbers (sulpiride binding) which increased in striatum and decreased in nucleus accumbens, in acute lead exposure. However, Winder and Lazareno, in a less severe intoxication schedule (30 or 1000 ppm Pb for 3 weeks), reported an absence of any dopaminergic D2 receptor change in striatum and limbic forebrain (including the nucleus accumbens) thus suggesting that the neurochemical effects of lead in the dopaminergic system would appear to be minimal.[35] It was also reported that semichronic but not acute lead exposure affected responsiveness of noradrenergic receptors.[2] Alternatively, lead postsynaptic effects can be estimated by measuring the activation of adenylate cyclase by NA or DA resulting in the intracellular formation of the second messenger, cAMP. Lead *in vivo* and *in vitro*, in conditions of high exposure, directly inhibited transmitter-sensitive (dopamine-stimulated) adenyl cyclase in striatal tissue or did not alter it.[16,17,24]

The discrepancy in the results might be due to differences in doses of lead, duration of exposure, when and how lead was administered to the suckling animals, the general conditions of the animals, and the age of animals when the studies were performed.[3] However, the above comparative assessment brings out an important suggestion that the effect of lead is more related to some local or regional factors rather than depending on a given neuron-type projection. Among these factors, regional differences in the metabolic patterns, multiple transmitter composition of the NA axon terminals, and the interstitial environment were evoked.[32]

26.4.3 GABAergic Processes[3,23,32]

The importance of GABA as a regulator of overall brain activity and mood is well known, particularly in light of its major role in regulating cortical electrical activity and its intricate involvement with benzodiazepine-mediated pathways. GABAergic neurotransmission is very sensitive to lead exposure, under both acute and chronic conditions of exposure. This hypothesis that lead is a GABAergic neurotoxin provides an interesting perspective on its behavioral effects which have been described experimentally and clinically. Indeed, a GABA involvement in lead intoxication is consistent with the clinical and experimental signs of excitability, hyperreactivity, hyperactivity, and at high levels of exposure, convulsions. In addition, lead-exposed rats showed a lower seizure threshold for convulsant agents and auditory evoked seizures.[2,4] Finally, many of the behavioral effects of lead can be characterized as failures of appropriate inhibition of response. This may range from the general level of motor activity, an increase which is frequently (although not always) observed in animals, to the constellation of off-task and inattentive behaviors described in classroom observation of lead-exposed children.[2] Under conditions of either high- or low-level lead exposure, a reduction of the GABA level was found in rat cerebellum, but not in rat whole brain, cerebral cortex, or brainstem.[2,32] Lead acts also postsynaptically by increasing the cerebellar GABA receptor density.[23] In addition, the specific (^{3}H)-GABA binding was reduced in striatum, increased in cerebellum, while it was not modified in substantia nigra, nucleus accumbens, hypothalamus, and cerebral cortex in conditions of low-level lead exposure.[23] Finally, cGMP levels, a second messenger which may be related with the function of GABAergic neurons, decreased in striatum and increased in cerebellum but not in the other areas.[23] These results raise the question as to why only the cerebellar and striatal GABAergic transmissions were affected and not other regions also known to contain GABAergic neurons and terminals.[33] Finally, Silbergeld et al. reported in high-dose exposed rats a decrease in cortical GABA transaminase and a decrease in glutamic acid decarboxylase in the substantia nigra while Govoni et al. using lower lead doses found no differences in the activity of these metabolic enzymes.[23]

It is noteworthy that the effect of lead on GABA could be an indirect one since lead, by inhibiting ALA-dehydratase, stimulates the production of ALA, which may be neuroactive by competing with GABA at the synaptic level.[9] However, the concentration of ALA necessary to mimic lead-induced GABA release from synaptosomes is seen only at concentrations of ALA far exceeding those that might occur in the range of likely human exposure to lead.[21]

26.4.4 Serotoninergic Processes[6,34]

The serotoninergic pathway was less studied. The steady state levels of serotonin (5HT) were not altered by lead treatment in mouse,[3] or rat whole brain, or brain regions (cortex, brainstem, cerebellum).[29,33] Neither was the level of its metabolite 5-hydroxyindolacetic acid, 5HIAA, found to be altered in rat brain regions (cortex, brainstem, cerebellum).[29,33] However, rats administered a high dose of lead acetate, responsible for malnutrition, showed increases in 5HT levels in the motor cortex, the hippocampus, and after a 40-day delay, the brainstem.[32] In contrast, low-level lead exposure was responsible for an increase in 5HT level in the hippocampus,[32] and for a decrease in 5HT and 5HIAA in the hypothalamus, the cortex,[6] and the brainstem, which contains both serotoninergic cell body neurons and terminals.[34] In the 5HT terminal area of the striatum, 5HT and 5HIAA were found to be either decreased or unchanged.[6,34] Finally, Dubas et al. reported an increase in 5HT levels in the midbrain and a corresponding decline in the concentration of 5HIAA, suggesting a decrease in the neurotransmitter utilization in this area.[31] The serotoninergic changes persisted both in rats withdrawn from lead,[31,34] and in rats whose motor activity returned to normal by 12 weeks of age.[6] So, it appears that changes in 5HT metabolism may result from a nonspecific toxic effect of lead. However, the hippocampal increase in 5HT as well as in NA is noteworthy, suggesting that the major amine pathways in the hippocampus are significantly affected by lead.

26.4.5 Glutaminergic Processes

Lead blocks, in the hippocampus, N-methyl-D-aspartate (NMDA)-sensitive glutaminergic synapses. In training experiments, Pb-exposed rats learned the stimulus properties of NMDA faster than controls. These results raise the possibility of a lead-induced supersensitivity of glutaminergic systems which justifies further evaluation in light of the role of this system in learning and memory functions.[2,16]

26.4.6 Neuropeptides

Attempts have been made to integrate some of the findings in lead toxicity involving imbalance and feedback regulations probably mediated by other interneurons and neurotransmitters. However, little information is readily available.

26.4.6.1 Neuropeptide Y

NPY is reported to be a noradrenergic neuromodulator or a neurotransmitter. In conditions of high-level exposure, we observed during the hyperactive period a decrease in hypothalamic and cortical

NPY levels. This may account for the reduced food intake and the hyperactivity, since NPY was shown to enhance food intake and to exert behavioral sedation. Interestingly, NPY level returned to normal in the hypothalamus but not in the cortex at 12 weeks of age, a period where rats were more quiet and returned to a higher diet consumption (unpublished data).

26.4.6.2   Enkephalins

An increase in striatal enkephalins has been observed following a 3-month period of lead exposure. Studies using lower and shorter lead exposures have shown a depression of enkephalin levels, in parallel with dopaminergic turnover in the striatum (which are both implicated in reward process) and a delay in the ontogeny of the neurotransmitter.[16,23]

26.5 CONCLUSION

Recognition of the sensitivity of children to lead toxicity and concern about environmental lead contamination have prompted generation of various experimental systems and of experimental animal models of lead exposure during development. The wide variability in experimental conditions makes it difficult to determine the mechanism by which lead induces nervous system impairment, because of the lack of sensitive neurochemical markers of lead neurotoxicity. Lead modifies different neurotransmitter systems and various functional steps of neurotransmission (synthesis, transport, release, metabolism, receptor function, and second messenger formation).[23] In addition, results obtained at low doses did not always appear at a higher dose.[2]

Instead of their variability, recent *in vitro* and *ex vivo* studies resulted in converging evidence for the cellular mechanism underlying lead toxicity on neurotransmitters. Lead was shown to increase spontaneous release of either cholinergic, catecholaminergic, and GABAergic systems, probably by stimulating calcium-activated molecules mediating transmitter release. Conversely, lead decreased the potassium-evoked release of the previous transmitters, probably by competitively interacting with calcium at sites involved in the release of neurotransmitters, resulting in inhibition of Ca-dependent stimulus-coupled release. Alterations of both spontaneous and evoked release in opposite directions could result, if the same occurs *in vivo,* to alterations of the physiological brain maturation in children.[7] In contrast, conventional animal toxicology models have provided limited information about lead effects on neural processes. Marked effects on behavior, mainly hyperactivity, increased aggressiveness, impaired motor coordination, and reduced learning ability have been reported in rodents and monkeys following exposure to inorganic lead during the prenatal or developing neonatal periods at lead doses lower than those necessary to cause encephalopathy. These models generally use low or medium lead exposure to avoid the nonspecific effect of undernutrition generated by high-dose lead exposure. Unfortunately, many studies done in the 1970s used high doses, thus complicating the interpretation of the neurotransmitter alteration pattern induced by lead. In most of these studies, the exposure schedule was reported but the resulting blood lead levels were not measured, thereby preventing assessment of the dose-effect relationship. In addition to the dose heterogeneity, other methodological problems contributed to the conflicting results previously reported. The exposure periods were variable concerning either gestation or the lactating period or both or the postweaning period. Other authors stopped the lead exposure at the end of the hyperactive period.[16] In addition, either the whole brain was studied, thereby missing small regional studies, or more discrete areas were examined, but in that case, with few exceptions, all the brain areas were not studied at the same time so that simultaneous changes of neurotransmitter levels were not assessed. However, lead exposure in rodents has been associated with selective actions of lead on neurotransmitters in discrete brain areas.[3,23] The vulnerability of axon terminals of any given type to lead may be dependent on some regional factors, although the projections of the different regions originate from an apparently similar category of neurons in the brainstem.[32] These effects have been hypothesized to result from competitive interactions between lead and calcium at sites involved in uptake and release of neurotransmitters and their precursors.[8] Some attempts have been made to correlate behavioral findings with those reported at the neurochemical level.[3] For instance, the existence of an interrelated hypocholinergic-hyperaminergic dysfunction in lead-induced hyperactive mice was hypothesized.[3] Similarly, it is currently agreed that the GABAergic system is the most sensitive to lead neurotoxicity, and is the site involved in the earliest expression of lead neurotoxicity. Such a hypothesis is consistent with the other neurochemical effects of lead exposure since reduced GABAergic function can itself be associated with disinhibition of adrenergic and dopaminergic function.[2] Finally, it may be assumed that, like enkephalins and NPY, some peptides may also vary in some areas, thus increasing the complexity of the mechanisms involved in lead-induced altered neurotransmission.

REFERENCES

1. CDC: Preventing lead poisoning in young children, October 1991. Centers for Disease Control, Atlanta, GA.
2. Silbergeld EK: *Neurotoxicology of Lead*. Marcel Dekker, New York, K Blum and L Manzo, Eds., 1985, pp 299–322.
3. Shih TM and Hanin I: Chronic lead exposure in immature animals. Neurochemical correlates. *Life Sci.*, 1978; 23:877–888.
4. Silbergeld EK: Neurological perspective on lead toxicity in *Human Lead Exposure*. CRC Press, Boca Raton, FL, Needleman HL. Ed., 1992, pp 90–103.
5. Carroll PT, Silbergeld EK, and Goldberg AM: Alteration of central cholinergic function by chronic lead acetate exposure. *Biochem. Pharmacol.*, 1977; 26:397–402.
6. Dubas TC and Hrdina PD; Behavioural and neurochemical consequences of neonatal exposure to lead in rats. *J. Environ. Pathol. Toxicol.*, 1978; 2:473–484.
7. Bressler JP and Goldstein GW: Mechanisms of lead neurotoxicity. *Biochem. Pharmacol.*, 1991; 41:479–484.
8. Pounds J: Effects of lead on calcium homeostasis and calcium-mediated cell function: a review. *Neurotoxicology*, 1984; 5:295–332.
9. Silbergeld EK and Lamon JM: The role of altered haem synthesis in the neurotoxicity of lead. *J. Occup. Med.*, 1980; 25:680–684.
10. Goldberg GW: Developmental neurobiology of lead toxicity in *Human Lead Exposure*. CRC Press, Boca Raton, FL, Needleman HL. Ed., 1992, pp 125–135.
11. Verity MA: Comparative observations on inorganic and organic lead neurotoxicity. *Environ. Health Perspect.*, 1990; 89:43–48.
12. Silbergeld EK, Fales JT, and Goldberg AM: Evidence for a junctional effect of lead on neuromuscular junction. *Nature*, 1974; 247:49–50.
13. Silbergeld HD: Interaction of lead and calcium on the synaptosomal uptake of dopamine and choline. *Life Sci.*, 1977; 23:309–318.
14. Minnema DJ, Michaelson IA, and Cooper GP: Calcium efflux and neurotransmitter release from hippocampal synaptosome exposed to lead. *Toxicol. Appl. Pharmacol.*, 1988; 92:351–357.
15. Silbergeld EK and Goldberg AM: Pharmacological and neurochemical investigations of lead-induced hyperactivity. *Neuropharmacology*, 1975; 14:431–444.
16. Winder C and Kitchen I: Lead neurotoxicity: a review of the biochemical, neurochemical and drug induced behavioral evidence. *Prog. Neurobiol.*, 1984; 22:59–87.
17. Wince LC, Donavan CA, and Azzaro AJ: Alterations in the biochemical properties of central dopamine synapses following chronic postnatal PbCO$_3$ exposure. *J. Pharmacol. Exp. Ther.*, 1980; 214:642–650.
18. Ramsay P, Krigman M, and Morell P: Developmental studies on the uptake of choline, GABA and dopamine by crude synaptosomal preparations after *in vivo* and *in vitro* lead treatment. *Brain Res.*, 1980; 187:383–402.
19. Komulainen H and Tuomisto J: Effect of heavy metals on dopamine, noradrenaline and serotonin uptake and release in rat brain synaptosomes. *Acta Pharmacol. Toxicol.*, 1981; 48:199–20.
20. Minnema DJ, Greenland RD, and Michaelson IA: Effect of *in vitro* inorganic lead on dopamine release from superfused rat striatal synaptosomes. *Toxicol. Appl. Pharmacol.*, 1986; 84:400–411.
21. Minnema DJ and Michaelson IA: Differential effects of inorganic lead and δ–aminolevulinic acid *in vitro* on synaptosomal γ-aminobutyric acid release. *Toxicol. Appl. Pharmacol.*, 1986; 86:437–447.
22. Modak AT, Purdy RH, and Stavinhoa WB: Changes in acetylcholine concentration in mouse brain following ingestion of lead acetate in drinking water. *Drug Chem. Toxic.*, 1978; 1:373–389.
23. Govoni S, Memo M, Lucchi L, et al: Brain neurotransmitters and chronic lead intoxication. *Pharmacol. Res. Commun.*, 1980; 12:447-460.
24. Govoni S, Memo M, Spano PF et al: Chronic lead treatment differentially affects dopamine synthesis in various brain areas. *Toxicology*, 1979, 12, 343–349.

25. Sauerhoff MW and Michaelson IA: Hyperactivity and brain catecholamines in lead exposed developing rats. *Science*, 1973; 182:1022–1024.

26. Golter M and Michaelson IA: Growth, behavior, and brain catecholamines in lead exposed neonatal rats. A reappraisal. *Science*, 1975; 187:359–361.

27. Rossouw J, Offermeier J, and Van Rooyen JM: Apparent central neurotransmitter receptor changes induced by low-level lead exposure during different developmental phases in the rat. *Toxicol. Appl. Pharmacol.*, 1987; 91: 132–139.

28. Jason KM and Kellog CK: Neonatal lead exposure. Effects on development of behavior and striatal dopamine neurones. *Pharmacol. Biochem. Behav.*, 1981; 15: 641–649.

29. Sobotka TJ, Brodie RE, and Cook MP: Psychophysiologic effects of early lead exposure. *Toxicology*, 1975; 5:175–191.

30. Grant LD, Kimmel CA, Martinez-Vargas CM, et al: Assessment of developmental toxicity associated with chronic lead exposure. *Environ. Health Perspect.*, 1976; 17:290–302.

31. Dubas TC, Stevenson A, Singhal RL, et al.: Regional alterations of brain biogenic amines in young rats following chronic lead exposure. *Toxicology*, 1978; 9:185–190.

32. Shailesh Kumar MV and Desiraju T: Regional alterations of brain biogenic amines and GABA/glutamate levels in rats following chronic lead exposure during neonatal development. *Arch. Toxicol.*, 1990; 64:305–314.

33. Sobotka TJ and Cook MP: Postnatal lead acetate exposure in rats; possible relationship to minimal brain dysfunction. *Am. J. Ment. Defic.*, 1974; 79:5–9.

34. Widmer HR, Bûtikofer EE, Schlumpf M, et al: Pre- and postnatal lead exposure affects the serotoninergic system in the immature rat brain. *Experientia*, 1991, 47:463–466.

35. Winder C and Lazareno S: Effect of lead exposure on dopaminergic D2 receptor binding in the 21-day-old rat. *Toxicol. Lett.*, 1985; 24:209–214.

Chapter 27

Epidemiologic Approaches to Characterizing the Developmental Neurotoxicity of Lead

David C. Bellinger

CONTENTS

27.1 GENERAL ISSUES IN THE EPIDEMIOLOGIC STUDY OF LEAD NEUROTOXICITY

Lead is unlike most other metals in that knowledge about human health effects is not based primarily on extrapolations across species or from isolated high-dose exposures incurred under unusual circumstances. Indeed, for no other metal are nationally representative estimates of population levels available.[1] Because the range of population exposures includes levels known to be neurotoxic, empirical data support the risk assessment analyses that form the foundation of regulatory policy.

Children are one of the population segments considered to be most sensitive to lead toxicity. Their behaviors make some lead sources and pathways more accessible to them than to adults (e.g., paint, dust, soil), while their physiology makes them more vulnerable than adults to lead's toxicity, particularly its central nervous system effects. Much of the concern over lead toxicity has focused on the effects of lead on children's growth and development, especially their cognitive or intellectual skills.

In the following section, key elements of an epidemiologic study design are discussed from the perspective of a study of lead neurotoxicity.

27.1.1 Sampling Frame and Strategy

In early studies of lead neurotoxicity, inferences about exposure-outcome associations were often based on samples of convenience consisting of children with clinical lead poisoning or clinically defined syndromes of cognitive or behavioral dysfunction. Generalizations about disease etiology and natural history that are based on highly selected samples generated by poorly characterized referral mechanisms may be biased since the patients are not likely to be representative of all individuals with a particular disease.[2] In the 1970s, efforts were made to assemble population-based samples that permit inferences with external validity. Several sampling strategies have been employed. In some studies the aim was to assemble a study cohort that represents a random sampling of some population (e.g., registrants at prenatal care clinics serving a particular catchment area, children attending certain grades within a school system). The advantage of such an approach, if properly implemented, is that the distribution of lead exposures in the cohort reflects the distribution of exposures in the underlying population, permitting inferences about population-attributable risk. A drawback of this approach is that the number of individuals in the cohort with very high or very low lead levels (with respect to the underlying population) is relatively small, with the levels of most clustering around the mean. Some investigators oversampled individuals in certain portions of the exposure distribution (e.g., very high, very low), allowing more precise estimation of the form of the dose-effect relationship at the extremes of the distribution.[3]

27.1.2 Measurement of Exposure

The concentration of lead at the critical target organ (e.g., brain) cannot be measured for most adverse health effects, requiring reliance on surrogate measures of target organ dose. Because of the complex toxicokinetics of lead, inferring brain concentration from a surrogate biomarker entails a risk of dose misclassification and thus the possibility of, at best, lack of precision or, at worst, systematic bias in estimation of dose-response relationships.

Although blood lead concentration is the exposure index most commonly used, only a small percentage of total body burden resides in blood. The relatively brief biological residence time of lead in blood is the basis for the oft-cited view that blood lead measures only relatively recent exposure. Moreover, although nearly all of lead in whole blood is in red cells, the small fraction in plasma is more bioavailable. The large portion of total body burden in bone (~95% in adults) was regarded as metabolically inert, but is now recognized as potentially toxicologically active, especially the lead in trabecular bone. Efflux of bone lead under physiologic conditions is now thought to contribute substantially to blood lead. Reliable and sensitive methods for measuring bone lead concentration have recently become available for use in field epidemiological studies[4] and provide new opportunities for exploring exposure-outcome relationships.[5]

In many studies, children's lead dose was estimated from the concentration in shed deciduous teeth. The usefulness of this tissue for epidemiologic studies is limited by the relatively brief interval in which naturally shed teeth can be collected and by limitations in the information about dose history conveyed by a tooth lead level. Important advances are being made in this regard, however.[6]

To study the impact of prenatal exposure, investigators have relied on the concentration of lead in a variety of tissues as indirect indices of dose delivered to the fetus (e.g., maternal blood level during pregnancy or at delivery, umbilical cord blood, neonatal blood, placental or cord tissue, and hair). These indices are not always highly correlated, nor do they always bear the same relationship to indices of neonatal status.[7] Development of a valid model of lead biokinetics in the maternal-fetal unit will significantly increase our ability to study the impact of prenatal lead exposure.

27.1.3 Measurement of Outcome

The endpoints used to assess lead's developmental toxicity have evolved considerably. Many early studies of "subclinical" lead exposures examined whether higher exposures were associated with increased risk of clinical diagnoses such as mental retardation. In the early 1970s, studies began to employ more sensitive continuously distributed indices of outcome, such as scores on group intelligence tests. In the middle 1970s, investigators began to use individually administered tests of intelligence, neuropsychologic function, and academic achievement. Relatively little has since changed in assessment strategy. Because much of the impetus for this research has been to inform public health policy, the major endpoint used has remained the familiar index IQ. As a result, however, available data provide relatively little insight into the neuropsychological mechanism(s) of the IQ effects observed. Adapting for use with children process-oriented experimental tasks that have identified lead-related deficits in animal learning and performance[8] might facilitate the extrapolation of conclusions across species and perhaps clarify the neuropsychological bases of the IQ deficits. A recent demonstration of differences in the performance of twins discordant for lead poisoning on such a task (discrimination reversal) illustrates this approach.[9]

27.1.4 Statistical Analysis

Because most questions about the cognitive and behavioral sequelae of lead toxicity in children must be addressed using an observational design, much of the controversy in this field concerns the validity of the statistical methods used in a study to disentangle the impact of lead from the impacts of other, often colinear developmental risks. In early studies, hypothesis-testing often consisted of unadjusted two-group comparisons (i.e., "exposed" vs. "unexposed"). Acknowledging the importance of statistical adjustment for confounding bias, later studies employed analysis of covariance or multiple regression. Path analysis or latent structure models have recently been used to describe the direct and indirect relationships among exposure, outcome, intermediate, and confounding or effect-modifying variables.

Specifying a valid model for an outcome with complex determinants, such as intellectual functioning, remains a hurdle not yet fully overcome. Increasing efforts have been made to assess the impact of measurement errors on the precision and accuracy of parameter estimation. Interest has also increased in methods that assess the impact of missing data on estimation by modeling the "nonresponse mechanism". This is especially important for prospective studies, which are vulnerable to attrition that may be differential (i.e., nonrandom).

27.2 APPLICATION OF EPIDEMIOLOGIC STRATEGIES

The evolution in the designs used to study lead neurotoxicity is characteristic of the sequence used to study other exposure-disease associations: clinical case series, case-control, cross-sectional or retrospective cohort, prospective cohort, experimental or quasi-experimental studies. The following section provides generic descriptions of the key elements of each design prototype as applied to the study of lead neurotoxicity, as well as a brief description of an example of each design.

27.2.1 Clinical Case Series

Descriptions of the clinical characteristics and outcomes of convenience samples of lead-poisoned children began to appear early in this century. From an epidemiologic standpoint, these reports left much to be desired. Unknown selection factors generated the available patients from an unspecified population base. The contributions of these case series were nevertheless important. They provided information about aspects of the range of the late outcomes of lead-poisoned children, although not necessarily valid or generalizable estimates of the average outcome. They also generated hypotheses for investigation in studies employing rigorous population-based sampling mechanisms.

27.2.1.1   Example

The case series of Byers and Lord[10] remains a classic in the lead poisoning literature. They reported on a group of 20 children, none of whom had a severe acute lead encephalopathy, who were discharged from hospital as "cured". These 20 were selected from 128 children treated over a 10-year period at the Children's Hospital (Boston). Only one, or perhaps two, children were judged to have succeeded in school, although most had an IQ score in the average range. Behavioral difficulties were also prevalent, notably impulsivity, irritability, and distractibility. The major contribution of this report was the suggestion that "silent" lead poisoning may cause learning failure in children. Byers and Lord concluded that, "It seems probable that lead poisoning of the sort here discussed can at present be recognized in only a small percentage of cases."[10]

27.2.2 Case-Control Studies

To evaluate hypotheses suggested by case reports, studies were undertaken in which the lead exposure histories of children with diagnosed intellectual or behavioral handicaps ("cases") were compared to the histories of children free of handicap ("controls"). The handicaps studied included mental retardation, hyperactivity, learning disability or dyslexia, autism, and other neuropsychiatric disorders.

27.2.2.1   Example

Gittelman and Eskenazi [11] compared the lead burdens of 3 groups of 6- to 12-year-old children: (1) children with cross-situational hyperactivity (N = 103), (2) nonhyperactive learning-disabled children (N = 31), and (3) normal siblings of children in the hyperactive group (N = 33). A child's lead burden was inferred from urinary lead concentration in response to a single 250 mg challenge dose of penicillamine, an oral chelating agent. Group comparisons were adjusted for socioeconomic status, race, age, and sex. Matched sibling analyses were also conducted. The mean urinary lead level did not differ among the three groups, nor did the percentages of children with levels greater than 0.08 mg/l, the concentration presumed to reflect a "toxic" level. Hyperactive children had significantly higher levels than their paired siblings, although no relationship was found between the urinary lead level and severity of behavioral disturbance in the hyperactive children. The authors concluded that their findings, "...suggest that lead is not salient, but not negligible, as a contributor to hyperactivity and possibly to learning disability, in nondisadvantaged children."[11]

The case-control method has serious limitations as a strategy for studying lead neurotoxicity. First, the diagnosis of a neuropsychiatric disorder is usually made in middle childhood, long after children typically experience peak lead exposures. Biological markers of exposure readily measurable at this time (e.g., blood lead) may not provide an accurate index of exposure of either the cases or controls during this critical early period, increasing the risk of exposure misclassification in both groups. Second, because of the retrospective nature of a case-control study, the assumption that lead exposure preceded and thus caused the adverse outcome, rather than being a result of that outcome, can be difficult to substantiate. Third, the only outcome that can be evaluated in relation to lead exposure is the one that is the basis of case status classification. Fourth, many aspects of the case-control design are difficult to implement, including the identification of appropriate controls.[12]

27.2.3 Cross-Sectional or Retrospective Cohort Studies

In studies using this design, classification is based on exposure rather than outcome (cf. case-control studies). In some applications, information on children's lead exposure and their outcome status is collected at the same time (i.e., cross-sectional design). In others, exposure classification is based on a biomarker, such as tooth lead level, that reflects past exposure (retrospective).

27.2.3.1 Example

Fulton and colleagues[13] selected as their sampling frame all children in two classes (6- to 9-year-olds) in 18 primary schools within an area of Edinburgh. To increase statistical efficiency, they oversampled children with venous blood lead levels in the top quartile within each school, and randomly selected approximately one-third of the remaining children. The 501 children enrolled were administered a battery of cognitive and educational achievement tests and a test of reaction time. Parents and teachers rated children's behavior. Adjusting for potential confounders such as child health status, sociodemographic factors, and parental IQ, blood lead levels were inversely related to scores on tests of ability (e.g., BAS subtests of Matrices, Similarities, and Digit Span) and educational achievement (Numbers, Reading). Although only a small percentage of outcome variance was associated with blood lead level, the relationship seemed to be linear over the blood lead range of 3.3 to 34.0 µg/dl, without evidence of a threshold. Blood lead level was also associated with children's performance on the reaction time task[14] and both parents' and teachers' ratings of behavior, especially aggressive/anti-social and hyperactive behavior.[15]

A cohort study design has several advantages over a case-series or case-control design. The first concerns the enhanced external validity of study inferences. Cross-sectional and cohort studies can be population based, and depending on the specific sampling strategy used, permit the derivation of effect measures with important public health implications, such as population-attributable risk. Second, because children are selected for participation on the basis of exposure rather than outcome status, the association between lead and many outcomes can be evaluated in a cohort study.

The cross-sectional/retrospective cohort design has limitations, however. As with case-control studies, assigning exposure status on the basis of a blood lead level measured at the time of outcome assessment (e.g., school age) risks misclassification of a child's exposure at an earlier perhaps toxicologically critical period of development. The bias introduced is most likely differential, increasing the risk of a Type II error (i.e., accepting the null hypothesis of "no association"). The risk of error in estimating quantitative aspects of dose-effect/response relationships is also increased. Finally, the temporal characteristics of exposure history are poorly characterized, preventing the exploration of the relationship between the age at exposure and the severity or nature of neurobehavioral toxicity.

27.2.4 Prospective Cohort Studies

Incorporating longitudinal measurements of children's lead burden and development, prospective cohort studies can address several questions that case series and cross-sectional studies cannot, including the threshold(s) associated with neurobehavioral toxicity, age-dependent variation in vulnerability to lead, the time course of associations between exposure and specific expressions of outcome, and correlates of persistence and reversibility.

27.2.4.1 Example

In a study conducted in Boston,[16] 3 groups of children were enrolled: (1) those with umbilical cord blood lead levels below the 10th percentile for the delivery population sampled (<3 µg/dl); (2) those with levels around the mean (6 to 7 µg/dl); and (3) those with levels greater than the 90th percentile ($\geq$10 µg/dl but <25 µg/dl). Children's development and postnatal blood lead levels were evaluated at 6, 12, 18, 24, and 57 months, and at 10 years of age. The confounding impact of risks for poor development that often co-occur with elevated lead exposure were reduced by selecting as the base population a group of mostly middle- and upper-middle-class women. In infancy, the cord blood lead level was inversely associated with adjusted scores on the Mental Development Index of the Bayley Scales of Infant Development. By age 57 months, this association had attenuated, although the extent of attenuation varied with features of a child's rearing environment. At 57 months and 10 years of age, blood lead levels measured postnatally, especially at around age 2 years, were inversely associated with covariate-adjusted intellectual performance.

Although exposure biomarkers are measured repeatedly, the intervals between measures are typically long enough that the exposure history compiled may not reflect features of neurobehavioral significance. Moreover, exposure classification in a prospective cohort study is still based on surrogate measures of the concentration of lead in the brain, the critical target organ for neurotoxicity. This requires the strong assumption, for which little evidence currently exists, that blood lead level is linearly, or at least monotonically related to brain lead under typical exposure scenarios. Finally, cohort attrition over the course of a longitudinal study may introduce bias if loss-to-followup or data "missingness" is not random with respect to both exposure history and neurobehavioral status.

27.2.5 Experimental Studies/Clinical Trials

Ethical considerations constrain the extent to which experimental studies involving random assignment to exposure groups can be conducted in humans. As a result, the methods used to study the impact of lead on children's cognitive function are usually observational and subject to the inferential limitations that apply when the possibility of residual confounding cannot be eliminated. Experimental studies of lead-exposed rodents[17] and primates[8] thus are a vital piece of the context within which epidemiological studies of lead neurotoxicity should be interpreted.

Experimental studies of alternative medical treatments or source abatement strategies may be feasible under certain circumstances. The impact of various drug therapies on children's cognitive functioning has been evaluated in several quasi-experimental studies, involving nonrandom assignment to treatment.[18] The U.S. National Institute of Environmental Health Sciences is currently sponsoring a multicenter randomized placebo-controlled clinical trial assessing the efficacy of the oral chelator 2,3-dimercaptosuccinic acid in limiting the adverse neurodevelopmental impact of blood lead levels of between 20 and 40 µg/dl. Weitzman et al.[19] reported the results of a randomized community-based trial of leaded soil abatement as a public health intervention for high-risk children. Removal of lead-contaminated soil reduced the blood lead levels of preschool-aged children over the following year, although the decline over and above that observed among children in the two control groups was only about 1 µg/dl.

27.3 ASPECTS OF THE DOSE-EFFECT RELATIONSHIP DESCRIBING LEAD NEUROTOXICITY IN CHILDREN

The final section summarizes the contributions of epidemiologic data to our understanding of four critical aspects of lead neurotoxicity in children.

27.3.1 Magnitude of Effect

Recent quantitative syntheses of epidemiologic data have concluded that a modest inverse association holds between lead exposure biomarkers (e.g., blood or tooth lead level) and indices of global intellectual function at school age (e.g., IQ). One meta-analysis conducted under the auspices of the World Health Organization[20] concluded that, "...the size of the apparent IQ effect (at age 4 and above) is a deficit between 0 and 5 points (on a scale with a standard deviation of 15) for each

10 µg/dl increase in blood lead level, with a likely apparent effect size of between 1 and 3 points." This conclusion is phrased in terms of a lead-associated shift in central tendency (i.e., mean IQ). For a normally distributed variable, a small shift in the mean may be accompanied by more dramatic changes in the percentages of individuals in the group with either very low or very high scores.[21]

Despite the statistical significance of this aggregate estimate of the decline in children's intellectual function with increasing lead burden, the magnitude of the estimated lead effect varies across studies. Identifying the exposure, host, or contextual factors that alter this effect, either enhancing or diminishing its magnitude, remains an important research need.[22]

Current data are inconclusive regarding the shape of the dose-effect relationship at blood lead levels below 10 to 15 µg/dl. The WHO report attributes this to uncertainties in analytical and psychometric measurement. However, a significant covariate-adjusted inverse linear association between blood lead level at 2 years of age and IQ at 5 and 10 years of age was observed in a cohort in which nearly all children had blood lead levels below 15 µg/dl.[23]

27.3.2 Critical Periods of Vulnerability

Drawing strong inferences about the age(s) at which exposure to lead is especially harmful is difficult unless exposures at different ages are uncorrelated. In many cohorts, especially those living in high-risk environments, blood lead levels tend to "track", demonstrating intraindividual stability. This permits conclusions about the impact of cumulative exposure but not of age-specific exposures. Moreover, the blood lead level does not remain stable but tends to peak at around 2 years of age, perhaps due to developmental changes in patterns of mobility and hand-to-mouth activity. This confounding of age and maximum blood lead level further constrains inferences about their relative neurotoxicologic significance. Nevertheless, blood lead levels measured in the 1- to 3-year age range seem to be the best predictors of later intellectual function (i.e., IQ). Primate studies suggest, however, that different behavioral endpoints are sensitive to exposures at different ages, or to exposure biomarkers that average exposure over different durations (i.e., acute vs. chronic exposure). Because blood lead level measured at a particular age integrates exogenous (external) and endogenous (internal) inputs, study findings regarding the age(s) at which the blood lead level has greatest prognostic significance should not be strictly interpreted as the age(s) on which exposure prevention policies should focus.

27.3.3 Nature and Neuropsychological Mechanism

Primate models of lead toxicity implicate "perseveration/distractibility/inability to inhibit inappropriate responding" as a possible behavioral mechanism of lead neurotoxicity.[8] In contrast, human epidemiological studies do not provide a clear picture of what it is that underlies the lower IQ scores of more highly exposed children, or indeed, on which types of tasks more highly exposed children tend to do poorly. One approach to data synthesis that has not been extensively exploited with regard to human epidemiological studies is consideration of the influence that a study's "system" may have on the nature and magnitude of the effects observed. Among the factors that contribute to the "system" in a lead study are cohort characteristics (e.g., rearing history, other toxicant exposures, family characteristics, distribution of genetic determinants of variation in susceptibility), exposure parameters (e.g., dose, timing, chronicity), and testing context. Slight differences in these parameters can alter the behavioral presentation of lead-exposed animals.[22]

Even in cohorts manifesting a dose-related decline in global functioning (IQ), attempts to use standard neuropsychological tests of specific domains to identify its neuropsychological underpinnings have been unsuccessful.[24] No lead-specific neuropsychological "signature" has yet been identified. It is not clear whether this reflects insensitivity of the measures of neuropsychological function, interstudy differences in exposure parameters and environmental factors, or both.

27.3.4 Reversibility of Cognitive Effects

Determining whether persisting lead-associated deficits reflect true permanence or are the impact of ongoing exposure requires a precision in the characterization of delivered dose that can rarely be achieved in an observational study. Again it is necessary to consider animal models, in which exposure is experimentally manipulated and the role of potentially confounding factors eliminated. The set of long-term studies by Rice provide compelling evidence that behavioral deficits are still evident at age 10 years in animals whose blood lead level peaked at 25 µg/dl at 300 days of age and whose steady-state levels remained below 15 µg/dl thereafter.[8]

The prospective studies have fulfilled at least some of their promise by providing relatively clear evidence about the time course of the association between low-level prenatal exposure and early developmental deficit. This association appears to be transient, attenuating during the preschool period to an extent that varies with level of postnatal exposure, sex, and socioeconomic status.[25] A similar "catch-up" phenomenon has been noted for lead's apparent suppression of early postnatal growth.[26] The absence of detectable effects does not mean, however, that subtle residual deficits remain which may be expressed under certain conditions.[27]

The behavioral correlates of early postnatal exposures appear to be more persistent.[23,28,29] In one study, growth curve analyses were used to model the change over time (from 8 to 12 years of age) in the association between children's tooth lead levels and word reading skills.[30] A model postulating a simple constant lead-associated decrement fit the data as well as or better than models postulating either "catch up" or deterioration of performance over time.

Whether medical treatments or environmental interventions that reduce children's blood lead levels reverse or limit exposure-related decrements in cognitive performance remains uncertain.

27.4 CONCLUSION

The amount of epidemiologic data on lead far exceeds that of any other heavy metal. The fact that many questions about the neurobiologic and neuropsychologic mechanisms of lead's neurotoxicity remain unanswered bespeaks the difficulty of obtaining clear answers to complex questions on the basis of observational studies. The proper interpretation of epidemiologic data requires close consideration of research on experimental animal models and mechanistic studies of cellular and subcellular toxicity.

REFERENCES

1. Brody D, Pirkle J, Kramer R, et al: Blood lead levels in the US population. Phase 1 of the Third National Health and Nutrition Examination Survey (NHANES III, 1988 to 1991). *J. Am. Med. Assoc.*, 1994; 272, 277–283.
2. Ellenberg J and Nelson K: Sample selection and the natural history of disease. *J. Am. Med. Assoc.*, 1980; 243:1337–1340.
3. Bellinger D, Needleman H, Leviton A, et al: Early sensory-motor development and prenatal exposure to lead. *Neurobehav. Toxicol. Teratol.*, 1984; 6:387–402.
4. Todd A and Chettle D: *In vivo* X-ray fluorescence of lead in bone: Review and current issues. *Environ. Health Perspec.*, 1994; 102:172–177.
5. Hu H, Watanabe H, Payton M, et al: The relationship between bone lead and hemoglobin. *J. Am. Med. Assoc.*, 1994; 272:1512–1517.
6. Gulson B and Wilson D: History of lead exposure in children revealed from isotopic analyses of teeth. *Arch. Environ. Health,* 1994; 49:279–283.
7. Bellinger D: Teratogen update: Lead. *Teratology*, 1994; 50:367–373.
8. Rice D: Lead-induced changes in learning. Evidence for behavioral mechanisms from experimental animal studies. *Neurotoxicology*, 1993; 14:167–178.
9. Evans H, Daniel S, and Marmor M: Reversal learning tasks may provide rapid determination of cognitive deficits in lead-exposed children. *Neurotoxicol. Teratol.*, 1994; 16:471–477.
10. Byers R and Lord E: Late effects of lead poisoning on mental development. *Am. J. Dis. Child.*, 1943; 66:471–494.
11. Gittelman R and Eskenazi B: Lead and hyperactivity revisited. An investigation of nondisadvantaged children. *Arch. Gen. Psychiatr.*, 1983; 40:827–833.

12. Kelsey J, Thompson W, and Evans A: *Methods in Observational Epidemiology.* New York: Oxford University Press, 1986.
13. Fulton M, Thomson G, Hunter R, et al: Influence of blood lead on the ability and attainment of children in Edinburgh. *Lancet*, 1987; 1: 1221–1225.
14. Raab G, Thomson G, Boyd L, et al: Blood lead levels, reaction time, inspection time and ability in Edinburgh children. *Br. J. Dev. Psychol.*, 1990; 8:101–118.
15. Thomson G, Raab G, Hepburn W, et al: Blood-lead levels and children's behaviour. Results from the Edinburgh Lead Study. *J. Child Psychol. Psychiatr.*, 1989; 30:515–528.
16. Bellinger D: Lead and children's neuropsychologic development. Problems and progress in establishing brain-behavior relationships. In Tramontana M and Hooper S (eds): *Advances in Child Neuropsychology*, Vol. 3 New York: Springer-Verlag, 1995, pp 12–47.
17. Cory-Slechta D: The lessons of lead for behavioral toxicology. In: Smith M, Grant L and Sors A (eds): *Lead Exposure and Child Development: An International Assessment.* Boston: Kluwer Academic, 1989, pp 399–413.
18. Ruff H, Bijur P, Markowitz M, et al: Declining blood lead levels and cognitive changes in moderately lead-poisoned children. *J. Am. Med. Assoc.*, 1993; 269:1641–1646.
19. Weitzman M, Aschengrau A, Bellinger D, et al: Lead-contaminated soil abatement and urban children's blood lead levels. *J. Am. Med. Assoc.*, 1993; 269:1647–1654.
20. International Programme on Chemical Safety. *Environmental Health Criteria 165.* Inorganic Lead. Geneva: WHO, 1995.
21. Weiss B: Risk assessment. The insidious nature of neurotoxicity and the aging brain. *Neurotoxicology*, 1990; 11:305–314.
22. Bellinger D: Interpreting the literature on lead and child development. The neglected role of the "experimental system". *Neurotoxicol. Teratol.*, 1995; 17: 201–212.
23. Bellinger D, Stiles K, and Needleman H: Low-level lead exposure, intelligence, and academic achievement. A long-term follow-up study. *Pediatrics*, 1992; 90:855–861.
24. Stiles K and Bellinger D: Neuropsychological correlates of low-lead exposure in children. A prospective study. *Neurotoxicol. Teratol.*, 1993; 15:27–35.
25. Bellinger D, Leviton A, and Sloman J: Antecedents and correlates of improved cognitive performance in children exposed to low levels of lead. *Environ. Health Perspect.*, 1990; 89:5–11.
26. Shukla R, Dietrich K, Bornschein R, et al: Lead exposure and growth in the early preschool child. A follow-up report from the Cincinnati Lead Study. *Pediatrics*, 1991; 88:886–892.
27. Cory-Slechta D: Exposure duration modifies the effects of low level lead on fixed-interval performance. *Neurotoxicology*, 1990; 11:427–442.
28. Baghurst P, McMichael A, Wigg N, et al: Environmental exposure to lead and children's intelligence at the age of 7 years. The Port Pirie Cohort Study. *N. Engl. J. Med.*, 1992; 327:1279–1284.
29. Dietrich K, Berger O, Succop P, et al: The developmental consequences of low to moderate prenatal and postnatal lead exposure. Intellectual attainment in the Cincinnati Lead Study cohort following school entry. *Neurotoxicol. Teratol.*, 1993; 15:37–44.
30. Fergusson D and Horwood J: The effects of lead on the growth of word recognition in middle childhood. *Int. J. Epidemiol.*, 1993; 22:891–897.

Chapter 28

Efforts to Reduce Lead Exposure in the United States

J. Michael Davis, Robert W. Elias, and Lester D. Grant

CONTENTS

28.1 INTRODUCTION

Scientific understanding of the health effects of lead has flourished over the past two decades. Advances in this area of research have spawned a series of efforts by governmental agencies to enhance the protection of public health from the adverse effects of lead. Underlying this dynamic history is the fact that lead has been repeatedly found to induce adverse health effects at successively lower and lower levels of exposure. In addition, as regulatory efforts have succeeded in reducing lead exposure involving one source or pathway, the importance of other sources or pathways has changed.

This chapter updates earlier work by the authors[1] and provides a recent historical perspective on the contributions of air, water, food, paint, soil, and dust to lead exposure in the U.S. and on some of the regulatory and policy actions directed at reducing these and other sources of exposure. Although the frame of reference is the U.S., the discussion is likely to have broad relevance to other countries, for within the U.S. the relative importance of different sources of lead exposure may vary from one locale or region to another, even as it may vary from one nation to another.

The extensive database on lead health effects is too massive to summarize adequately within the limited scope of this chapter (see reviews by U.S. Environmental Protection Agency[2,3]). Even the more specialized area of lead neurotoxicity, which has been a primary public health concern and basis for regulatory efforts to reduce lead exposure, especially in the developing fetus and child, is best summarized by referring the reader to other reviews[1,4,5] (also see Chapter 27). Suffice it to say that the overall pattern of evidence from several prospective studies of neurobehavioral development in children has provided a major part of the rationale for the U.S. Environmental Protection Agency's conclusion that a blood lead concentration of 10 to 15 µg/dl, and possibly lower, constitutes a level of concern for adverse health effects of lead.[2,3] However, it is also important to realize that epidemiological findings do not constitute the only basis for concluding that low-level lead exposure poses the risk of developmental neurotoxicity. An even larger body of evidence of lead neurotoxicity comes from experimental studies of laboratory animals. These studies complement the epidemiological evidence by controlling or eliminating many variables that tend to confound epidemiological studies. Taken together, the literature points to notable parallels in the nature of the neurotoxic effects of lead in children, primates, and rodents.[6]

Given the breadth, depth, and coherence of the available evidence, it is difficult to escape the conclusion that low-level lead exposure (as indicated by blood lead levels of 10 to 15 µg/dl, and possibly lower) induces neurobehavioral effects in young children. This evidence has been extensively reviewed by independent scientific panels and essentially the same conclusion has been affirmed by other U.S. federal agencies, including the Agency for Toxic Substances and Disease Registry,[7] the U.S. Department of Health and Human Services,[8] and the U.S. Centers for Disease Control and Prevention.[9]

Efforts on the part of EPA and other agencies charged with protecting the public health have in fact succeeded in reducing lead exposure in the U.S. Recent data from the National Health and Nutrition Survey (NHANES) III indicate that U.S. nationwide mean blood lead levels of persons 1 to 74 years of age have decreased from 12.8 to 2.8 µg/dl from 1976 to 1991.[10] For children under 6 years of age, the nationwide mean blood lead levels have declined from 13.7 to 3.2 µg/dl in the case of non-Hispanic white children and from 20.2 to 5.6 µg/dl in the case of non-Hispanic black children. The prevalence of blood lead levels of 10 µg/dl or greater decreased from 85 to 5.5% in non-Hispanic white children and from 97.7 to 20.6% in non-Hispanic black children.

Despite these notable percentage reductions, in 1988-1991 approximately 1.7 million U.S. children under 6 years of age still had blood lead levels of 10 µg/dl or higher.[11] Thus, the above figures tell a story of success mixed with remaining challenges, for although average U.S. blood lead levels have declined dramatically, higher blood lead levels are still associated with certain sociodemographic factors such as low income level, black race/ethnicity, and male gender. These successes and challenges are discussed in terms of sources and pathways of lead exposure in the following sections.

28.2 AIR

Although breathing the ambient air was once a dominant pathway of lead exposure in the U.S., reductions in air lead concentrations over the past two decades have greatly reduced the contribution of polluted air to total lead exposure. Much of this reduction has been related to the decrease in the amount of lead in gasoline, as suggested by Figure 28.1. This trend began around 1973 with the introduction of catalytic converters on U.S. automobiles. The reason for introducing catalytic converters was not to alleviate lead pollution but to reduce other pollutants from automobile engine exhaust emissions; unleaded gasoline was needed simply to avoid fouling the catalysts. Nevertheless, the benefits of this action in reducing the blood lead levels of the U.S. population were soon evident (Figure 28.2). The recognized health benefits of reduced lead exposure led to a phase-down of the allowable concentration of lead in gasoline to 0.1 g/gal by EPA in the late 1980s. Similar patterns have been observed in other countries as well.[13] The 1990 amendments to the *Clean Air Act* called for EPA to ban the manufacture, sale, or introduction after 1992 of any engine that requires leaded gasoline and the prohibition after 1995 of all leaded gasoline for highway use.[14]

In addition to these reductions in mobile sources of lead, decreases in lead emissions from stationary sources have also contributed to an improvement in general population exposures to lead. Since 1978, the national ambient air quality standard (NAAQS) for lead has been 1.5 µg/m³ on a quarterly average basis, which each state determines how to achieve through its own respective state implementation plan (SIP). Substantial reductions in industrial sources of lead emissions to

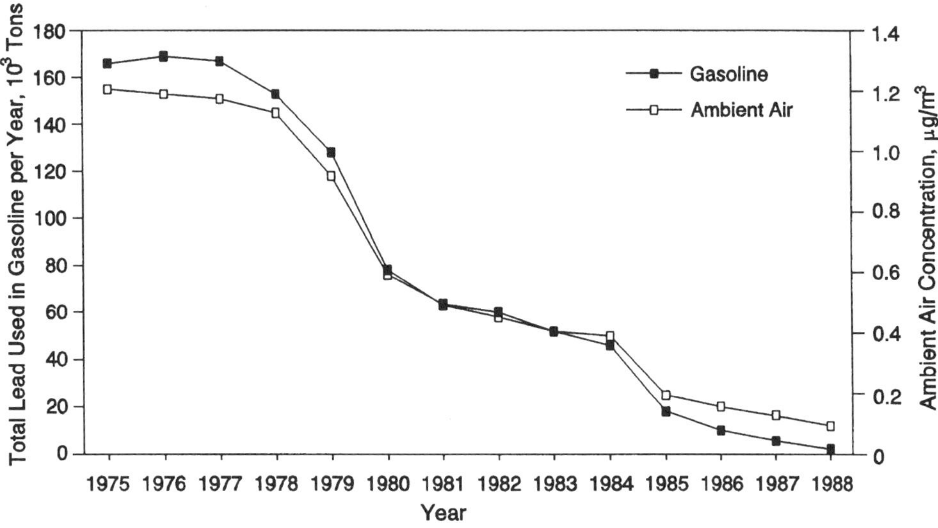

FIGURE 28.1   Gasoline lead vs. air lead levels in the U.S. (From U.S. Environmental Protection Agency, 1986, with updating.)

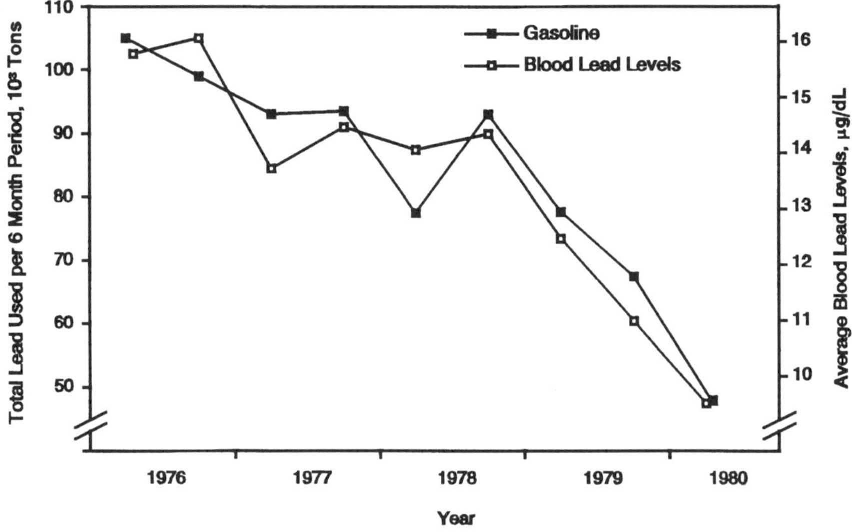

FIGURE 28.2   Gasoline lead vs. blood lead levels in the U.S. (From Annest, J.L., *Lead vs. Health: Sources and Effects of Low Level Lead Exposure*, Rutter, M. and Russell Jones, R., Eds., John Wiley & Sons, New York, 1983.)

the atmosphere have resulted from SIP and other types of regulatory controls for lead and for particulate matter in general.

Consideration has been given to revising the existing NAAQS for lead to a concentration lower than 1.5 $\mu g/m^3$, with attention centering on 0.75 $\mu g/m^3$ on a monthly (vis-a-vis quarterly) average basis. However, the number of major point sources likely to be affected by a revised lead NAAQS is relatively small, i.e., around 30 primary and secondary lead smelters nationwide. The limited number and circumscribed geographical areas of these sources mean that the estimated number of children with blood lead levels above 10 $\mu g/dl$ expected to be affected by a revised NAAQS is small in comparison to past figures or in comparison to exposures to some other current sources (such as lead-based paint).

Because ambient air lead problems tend to occur around a few point sources, in 1990 EPA began to concentrate on major point sources of lead through aggressive implementation of the existing NAAQS with expanded monitoring, enforcement, and regulatory actions. Approximately one-third of these sources have never recorded a violation of the lead NAAQS, and of the remaining sources with violations, 15 have shown improvements in air lead levels.[15]

As part of a regulatory strategy to reduce lead exposure, EPA has also recently turned to provisions for maximum achievable control technology (MACT) under the 1990 Amendments to the *Clean Air Act*.[14] As the term suggests, MACT standards are intended to provide "the maximum degree of reduction in emissions of hazardous air pollutants (including a prohibition on such emissions, where achievable) that the [EPA] Administrator, taking into consideration the cost of achieving such emission reductions, and any non-air quality health and environmental impacts and energy requirements, determines is achievable."[16] As recently promulgated by EPA, MACT standards for secondary lead smelters require such sources to implement various procedures and facility modifications, including construction of specified enclosures, controlled ventilation equipment, dust-control measures such as paving of roadways, and regular cleaning of pavement and vehicles.[16]

28.3 FOOD

Although lead in food has been an important pathway and source of exposure, it has typically originated from two other major sources: the deposition of lead from the atmosphere and the lead solder used to seal metal containers of food. Lead deposited from the atmosphere onto vegetables and fruits may adhere through harvesting, processing, and distribution and ultimately end up being ingested. With the reduction of air lead levels, concentrations of lead in food have also declined. Perhaps even more importantly, the use of lead solder to seal cans has also resulted in varying degrees of contamination of the foodstuffs packaged in cans and has thereby contributed significantly to the intake of lead by consumers. Since the early 1980s, the Food and Drug Administration (FDA) and food processors in the U.S. have cooperated in a voluntary program to eliminate the use of lead-soldered cans. In 1979, approximately 90% of domestically produced cans used lead solder; currently, no domestic cans are made with lead solder.[17] However, the importation of lead-soldered cans from foreign suppliers remains problematic. To help address this potential health hazard, the FDA has issued emergency "action levels" (lead concentrations in various types of canned foods) as well as a total ban on the use of lead-soldered cans.

Other actions, such as the use of lead in pesticides for crops, may have also contributed to the contamination of food in the past, but lead has been largely eliminated from such products in recent years.[18] The FDA has concluded that the largest single source of lead in the U.S. diet is currently lead-glazed ceramic ware such as plates, coffee mugs, bowls, and similar frequently used articles. The FDA has had guidelines since 1971 pertaining to lead glazing on ceramic ware used with food. Recently, the guidelines on levels of lead leaching from ceramic ware of various types were lowered to less than one-half to one-tenth of the levels that were last revised in 1979. The FDA is also directing efforts toward eliminating the use of lead foil on wine bottles and reducing other potential sources of lead exposure such as bottled water, food additives, and leaded crystal ware.[17]

28.4 DRINKING WATER

It has been estimated that, on average, approximately 20% of a child's lead exposure in the U.S. comes from drinking water.[18] The problem of lead in drinking water arises primarily because of the contamination of water by lead-containing components of the plumbing system. Source water itself seldom contains excessive levels of lead; rather, lead is typically introduced within the distribution system and individual households by virtue of the chemical reaction of corrosive water with lead connectors, lead service lines, and other lead-bearing plumbing materials such as solder, brass, and possibly some plastics. The concentration of lead in drinking water varies greatly from one system to the next and even from one residence to the next, depending on variables such as the pH, alkalinity, and temperature of the water, as well as the configuration of the plumbing system and the amount of time water stands in the system.

In 1975 EPA set a national primary drinking water regulation (NPDWR) for lead at 50 µg/l. To limit further use of lead in drinking water systems, a ban was issued in 1986 on the use of lead pipes and lead solder that come in contact with potable water,[19] and a recall and ban on water coolers containing lead components was issued in 1988.[20] In 1991, EPA established a revised NPDWR that required, for the first time, monitoring for lead at household taps in homes identified as high risk for lead contamination.[21] Because lead levels in tap water vary significantly depending on standing time and volume of the sample, as well as plumbing materials in the home, the rule requires collection of first-draw samples from high-risk homes (e.g., those having relatively new lead solder or lead service connections) so as to capture worst-case conditions. First-flush water refers to that which is obtained from a faucet after at least 6 h of nonuse and is collected without flushing the tap. Public water systems that have lead levels above 15 µg/l in more than 10% of samples (defined as the action level) are required to install and maintain optimal corrosion control treatment.

The NPDWR for lead also requires water systems that exceed the action level to regularly deliver public education materials to their consumers, as well as replace lead service connections if corrosion control treatment does not reduce lead levels below the action level. As part of the rule, EPA also established a maximum contaminant level goal (MCLG) of zero as a nonenforceable target.

It is difficult to characterize the extent of lead exposure from drinking water on a national basis. Based on the most recent compliance data, between 15 and 20% of public water systems exceed the action level, which means that at least 10% of the high-risk homes in each system have greater than 15 µg/l in first-draw water. The actual percentage of the population exposed to elevated lead levels in drinking water is very likely much lower because (1) not all the high-risk homes in those communities exceed the action level, (2) most homes are not high-risk candidates, and (3) most water consumed is not first-draw water.

EPA estimated that approximately 2% of the children in the U.S. not living in deteriorated lead-painted housing and not exposed to highly lead-contaminated soil have blood lead levels above 10 µg/dl.[17] It was also estimated that this percentage could be reduced to approximately 1.4% with effective implementation of the NPDWR for lead.

28.5 PAINT

As lead exposures through air, food, and drinking water have declined in recent years in the U.S. and are expected to decline further, the relative importance of exposure to lead in paint has increased. Lead-based paint is currently viewed as the predominant contributor to elevated blood lead levels (>10 µg/dl) in U.S. children.[18] Based on a survey conducted by the Department of Housing and Urban Development (HUD), it has been estimated that approximately 64 million (± 7 million, 95% confidence interval) or 83% (± 9%) of the privately owned housing units built before 1980 in the U.S. contain lead-based paint, with about 12 million (±1 million) of these homes occupied by families with children under the age of 7 years.[22] In addition, approximately 782,000 or 86% of the public housing units contain lead-based paint.

Efforts to deal with paint lead began in the 1950s with voluntary efforts by U.S. paint manufacturers to reduce interior paint lead content to 1%. A limit of 0.5% was made mandatory for residential paint lead content by the *Lead-Based Paint Poisoning Prevention Act* in 1971. The current maximum of 0.06% was issued by the Consumer Product Safety Commission (CPSC) in 1977 and became effective in 1978.

More recent federal actions related to lead-based paint include the "Guidelines for the Evaluation and Control of Lead-Based Paint Hazards in Housing" issued in 1995 by the Department of

Housing and Urban Development (HUD).[23] These guidelines specify that abatement efforts are to be triggered by a lead concentration equal to or greater than 1.0 mg/cm^2 (as measured by X-ray fluorescence) or 0.5% by weight (5000 µg/g as measured by atomic absorption spectrometry), unless state or local regulations are more stringent. Various dust lead levels (by wipe sampling) have been specified as federal lead standards for risk assessment and clearance purposes: 100 µg/ft^2 for carpeted and uncarpeted floors, 500 µg/ft^2 for interior window sills, and 800 µg/ft^2 for window troughs and for exterior concrete surfaces. (The guidelines also specify lead levels for bare residential soil, which is addressed below.) These guidance levels are subject to periodic updating and revision.

The main thrust of current efforts to deal with the paint lead problem are directed at abatement of *in situ* lead paint and dust.[24] Blood lead levels have been shown to decline after abatement procedures, at least for children with blood lead levels above 20 µg/dl, but the success of such interventions depends very much on the methods employed and their proper implementation. At this time, inadequate information is available to identify a particular abatement strategy as markedly more effective than others.[24] Because of the importance of the proper implementation of clean-up methods, EPA has been developing programs for training accreditation and contractor certification for abatement of lead hazards.

28.6 DUST AND SOIL

As in the case of paint lead, the issue of lead in household dust and urban soil has grown in relative importance in recent years as other sources of childhood lead exposure have declined. Indeed, the problem of lead in dust and soil originates to a considerable extent with leaded paint because of the deterioration that painted surfaces (both interior and exterior) commonly undergo with aging, thereby producing lead dust inside and contaminating soil adjacent to exterior painted walls. Because lead-contaminated soil tracked into the home can result in increased dust lead levels, the distinction between soil and dust as exposure sources is not clear-cut. In addition to paint, industrial and mobile sources contribute to soil lead levels in some areas.

Exposure to lead in dust and soil is primarily of concern from the standpoint of infants and young children, who frequently come into contact with soil and floors and who typically engage in hand-to-mouth activities that allow ingestion of lead from soil- or dust-contaminated objects. Based on the HUD-sponsored survey of U.S. housing, 17% of pre-1980 housing units have excessive levels of dust lead and 21% of such homes have excessive soil lead levels.[22] Where the painted surfaces are not intact, nearly half of the units have excessive dust and soil levels.

EPA is now completing an Urban Soil Lead Abatement Demonstration Project[25] to investigate whether reductions in soil lead concentrations result in reduced blood lead levels in inner-city children. Soils with elevated lead levels from study areas in three cities (Boston, Baltimore, and Cincinnati) were replaced with soils of approximately 100 ppm lead. Changes in lead concentrations of street dust, house dust, and children's hand dust were monitored, along with children's blood lead levels, to assess the effectiveness of soil abatement. In addition, these areas were compared to areas where treatments (e.g., containment of exterior lead-painted surfaces) other than soil removal have been applied. This project is expected to provide information that will help guide future, larger-scale abatement efforts.

Apart from residential sources of dust and soil lead, contaminated industrial sites are another potential source of lead exposure, particularly when such sites have been abandoned or are no longer effectively restricted from the general population. The EPA Office of Solid Waste and Emergency Response (OSWER) has issued interim guidance on soil lead cleanup levels at Superfund sites, specifying 400 ppm as a screening level for consideration of cleanup efforts at residential sites.[26] These levels constitute general guidance or recommendations, but they are not intended to be binding or enforceable standards. EPA expects to issue further guidance for assessing dangerous levels of lead in soil in late 1995.

One of the difficulties in setting an absolute standard for lead in soil is illustrated by the lack of a Reference Dose (RfD) for lead. Because of the practical inability to identify a threshold for lead health effects, given its pervasive presence and the lack of a true "control" or "reference" group in epidemiological studies of sensitive populations such as young children, no RfD can be derived for lead at present. This does not necessarily mean that no threshold exists for lead toxicity — only that if it does exist, it is presumably at or below the levels of exposure already commonly encountered.[27]

As an alternative to an RfD for lead, the U.S. Environmental Protection Agency has developed an integrated exposure uptake biokinetic (IEUBK) model to use in assessing the exposure risks of lead in various media.[28] As the name implies, the IEUBK model considers the concentrations of lead in air, water, food, and soil and dust, and incorporates intake, absorption, elimination, and other parameters to estimate average blood lead levels in children up to age 7 years. Alternatively, one may specify a desired blood lead value and back-calculate requisite lead concentrations in one or more environmental media to achieve the target value. This model provides a much more refined approach to assessing lead-contaminated sites than is possible using a general soil lead standard or a universal RfD for lead.

28.7 EPA LEAD STRATEGY

In 1991, EPA formally issued its "Strategy for Reducing Lead Exposures"[18] in an attempt to address the problem of reducing lead exposure, particularly in children, in a fashion that recognizes the multiple sources and pathways of lead exposure. The specific objectives of this strategy were to: (1) significantly reduce the incidence of blood lead levels above 10 µg/dl in children, and (2) significantly reduce the amount of lead introduced into the environment.

The EPA Lead Strategy contains several elements. Part of the strategy involves the use of regulatory approaches under different legal authorities such as the *Clean Air Act*,[14] the *Safe Drinking Water Act*,[19] and the *Toxic Substances Control Act*.[29] The vigorous enforcement of current standards and the development of additional regulations, as appropriate, underlie the regulatory facet of the EPA Lead Strategy. In addition, the Strategy includes the development of methods to deal with existing lead contamination. Accurate and efficient technical means to identify and compare "hot spots" of lead are essential to the success of abatement efforts. Another key focus of the EPA Lead Strategy is the prevention of lead pollution. Through a combination of market-based incentives, regulatory mechanisms, and technology development, efforts are being made to decrease future exposure to lead by reducing lead production and consumption. Recycling of lead in an environmentally sound manner is part of this aspect of the Strategy, not only to prevent such objects as car batteries from being improperly discarded but to reduce the need for mining and smelting new lead.

The value of public education is also recognized in the EPA Lead Strategy. Several outreach efforts, including telephone "hot lines," informational pamphlets, and other public education activities, are supported by EPA so that individuals can be better informed about what they can do to reduce lead exposure risks for themselves, their children, or others for whom they may be responsible.

The term "strategy" implies a coordinated effort, and indeed EPA recognizes the need to coordinate its actions both internally and with other agencies in a manner that contributes to the goal of reducing lead exposure overall rather than competing or interfering with other efforts. With the numerous and sundry legislative mandates pertaining to lead, some actions could easily result in simply shifting the problem from one part of the environment to another. For example, if secondary smelters were eliminated directly or indirectly through extremely stringent air standards, the reduced capacity to recycle lead-acid batteries might well result in more batteries being disposed of improperly and contaminating the environment. Similarly, efforts to remove lead from residential structures because of concerns about the health effects of lead in children should not be conducted in a manner that results in adult workers being excessively exposed to lead. A broad perspective on the lead problem is necessary if undesirable trade-offs between different segments of the environment or the population are to be avoided.

28.8 FUTURE DIRECTIONS

Given the environmental persistence of lead and the history of our evolving understanding of its public health significance, continuing efforts to investigate various facets of the lead problem are needed. Apart from remaining questions about the basic science of lead toxicity and exposure, solutions are needed for the difficult technical problems involved in measuring and abating lead in the environment. Because of the pervasiveness of lead, the costs associated with remediation could be enormous. Therefore, part of the challenge of dealing with lead is to find cost-effective methods for abating lead exposure through further research and development.

ACKNOWLEDGMENTS

We thank Doug Fennell and Marianne Barrier for assistance in producing this paper and Michael Bolger, Jeff Cohen, John Haines, Denise Kearns, Brian Lee, Laura McKelvey, Lori Saltzman, Mark Schmidt, John Schwemberger, and Chon Shoaf for comments or information used in preparing the manuscript. The views expressed in this paper are solely those of the authors and do not necessarily reflect the views or policies of the U.S. Environmental Protection Agency. The U.S. Government has the right to retain a nonexclusive royalty-free license in and to any copyright covering this article.

REFERENCES

1. Davis JM, Elias RW, and Grant LD: Current issues in human lead exposure and regulation of lead. *NeuroToxicology*, 1993; 14:15–28.
2. U.S. Environmental Protection Agency: Air Quality Criteria for Lead. Research Triangle Park, NC, Office of Health and Environmental Assessment, Environmental Criteria and Assessment Office, EPA Report no. EPA-600/8-83-028aF-dF, 1986.
3. U.S. Environmental Protection Agency: Air Quality Criteria for Lead: Supplement to the 1986 Addendum. Research Triangle Park, NC, Office of Health and Environmental Assessment, Environmental Criteria and Assessment Office, EPA Report no. EPA/600-8-89-049F, 1990.
4. Davis JM and Svendsgaard DJ: Lead and child development. *Nature*, 1987; 329:297–300.
5. Grant LD and Davis JM: Effect of low-level lead exposure on paediatric neurobehavioural development: current findings and future directions. In Smith M, Grant LD, and Sors A (eds): *Lead Exposure and Child Development: An International Assessment*. London, Kluwer Academic, 1989, pp 49–115.
6. Davis JM, Otto DA, Weil DE, and Grant LD: The comparative developmental neurotoxicity of lead in humans and animals. *Neurotoxicol. Teratol.*, 1990; 12:215–229.
7. Agency for Toxic Substance and Disease Registry: The Nature and Extent of Lead Poisoning in Children in the United States: A Report to Congress, Atlanta, GA, U.S. Department of Health and Human Services, 1988.
8. U.S. Department of Health and Human Services: Strategic Plan for the Elimination of Childhood Lead Poisoning. Developed for the Risk Management Subcommittee, Committee to Coordinate Environmental Health and Related Programs, U.S. Department of Health and Human Services. Atlanta, GA, Public Health Service, Centers for Disease Control, 1991.
9. U.S. Centers for Disease Control: Preventing Lead Poisoning in Young Children: a Statement by the Centers for Disease Control— October, 1991. Atlanta, GA, U. S. Department of Health and Human Services, Public Health Service, 1991.
10. Pirkle JL, Brody DJ, Gunter EW, et al: The decline in blood lead levels in the United States: the National Health and Nutrition Examination Surveys (NHANES). *J. Am. Med. Assoc.*, 1994; 272: 284–291.
11. Brody DJ, Pirkle JL, Kramer RA, et al: Blood lead levels in the US population: phase 1 of the third National Health and Nutrition Examination Survey (NHANES III, 1988 to 1991). *J. Am. Med. Assoc.*, 1994; 272:277–283.
12. Annest JL: Trends in the blood lead levels of the U.S. population: the second National Health and Nutrition Examination Survey (NHANES II) 1976-1980. In: *Lead vs. Health: Sources and Effects of Low Level Lead Exposure*, Rutter M and Russell Jones R, eds. New York, John Wiley & Sons, 1983, pp 33–58.
13. Wietlisbach V, Rickenbach M, Berode M, and Guillemin M: Time trend and determinants of blood lead levels in a Swiss population over a transition period (1984-1993) from leaded to unleaded gasoline use. *Environ. Res.*, 1995; 68:82–90.

14. Clean Air Act, as amended by PL 101-549, November 15, 1990. U.S. Code 42:7401–7626.

15. U.S. Environmental Protection Agency: Lead NAAQS attainment strategy management report. Research Triangle Park, NC, Office of Air Quality Planning and Standards, June 30, 1995.

16. U.S. Environmental Protection Agency: National emission standards for hazardous air pollutants from secondary lead smelting (40 CFR 63.545). Federal Register (June 23, 1995) 60:32587–32601. U.S. Government Printing Office, Washington, D.C.

17. Bolger PM, Yess NJ, Gunderson EL, et al: Identification and reduction of sources of dietary lead in the United States. *Food Additives and Contaminants* 1996; 13: 53–60.

18. U.S. Environmental Protection Agency: Strategy for Reducing Lead Exposures. Washington, DC, U.S. Environmental Protection Agency, February 21, 1991.

19. Safe Drinking Water Act Amendments: PL 99-339, June 19, 1986, U.S. Code 42:300f et seq. U.S. Government Printing Office, Washington, D.C.

20. Lead Contamination Control Act of 1988: U.S. Code 42:300j-21 - 300j-26. U.S. Government Printing Office, Washington, D.C.

21. U.S. Environmental Protection Agency: Maximum contaminant level goals and national primary drinking water regulations for lead and copper. Federal Register (June 7, 1991) 56:26460–26564. U.S. Government Printing Office, Washington, D.C.

22. U.S. Environmental Protection Agency: Report on the national survey of lead-based paint in housing, base report. Washington, DC, Office of Pollution Prevention and Toxics, EPA report no. EPA 747-R95-003, 1995.

23. U.S. Department of Housing and Urban Development: Guidelines for the evaluation and control of lead-based paint hazards in housing. Washington, DC, Office of Lead-Based Paint Abatement and Poisoning Prevention, document no. HUD-1539-LBP, 1995.

24. U.S. Environmental Protection Agency: Review of studies addressing lead abatement effectiveness. Washington, DC, Office of Pollution Prevention and Toxics, report no. EPA 747-R-95-006, 1995.

25. U.S. Environmental Protection Agency: Urban soil lead abatement demonstration project, vols. I-IV [draft]. Washington, DC, Office of Research and Development, report no. EPA/600/AP093/001a-d, July 1993.

26. Laws EP: Revised interim soil lead guidance for CERCLA sites and RCRA corrective action facilities [memorandum from EPA's Assistant Administrator for Solid Waste and Emergency Response to regional administrators I-X]. Washington, DC, U. S. Environmental Protection Agency, Office of Solid Waste and Emergency Response, OSWER directive no. 9355.4-12; report no. EPA/540/F-94/043; July 14, 1994. (Available from NTIS, Springfield, VA; PB94–963282).

27. Davis JM: Risk assessment of the developmental neurotoxicity of lead. *NeuroToxicology*, 1990; 11:285–292.

28. U.S. Environmental Protection Agency: Guidance manual for the integrated exposure uptake biokinetic model for lead in children [report] and the integrated exposure uptake biokinetic model for lead in children (IEUBK) version 0.99d [computer diskette]. Washington, DC: Office of Solid Waste and Emergency Response, report no. EPA/540-R-93-081, publication no. 9285.7-15-1, 1994 (Available from: NTIS, Springfield, VA: PB93-963510 [report] and PB94-501517 [diskette]).

29. Toxic Substances Control Act, as amended. 1990, U.S. Code 15:2601 et seq. U.S. Government Printing Office, Washington, D.C.

SECTION 10

MANGANESE

Chapter 29

Relationships Between Manganese and Other Minerals

James C.K. Lai, Alex W.K. Chan, Margaret J. Minski, and Louis Lim

CONTENTS

29.1 INTRODUCTION

Although metals play many roles in brain function and behavior, the underlying biochemical and molecular mechanisms are poorly understood.[1–3] This minireview introduces several key and testable hypotheses that have been proposed to account for certain roles of metals in brain function and development and discusses the advances in methodology and approaches that have prompted a new generation of multidisciplinary studies to investigate such hypotheses.[2,4,5] The relationships between brain manganese and other metals are discussed in light of such critical conceptual and methodological advances.[1-6] (For other related topics not covered in this minireview, see reviews by Lai et al.,[1-4] Lai,[5] Wedler,[6] and references cited therein; also see other chapters in this volume.) Some directions and prospects for future studies are briefly entertained.

29.2 ROLES OF MANGANESE AND OTHER ESSENTIAL MINERALS IN BRAIN FUNCTION

29.2.1 Roles of Metals in Neurotransmitter Metabolic Cycle

The concept of a neurotransmitter metabolic cycle has been synthesized to account for the normal roles of metals in neurotransmitter function and metabolism.[3] Nine such roles have been identified — roles of metals in (1) precursor and neurotransmitter uptake, (2) neurotransmitter synthesis, (3) neurotransmitter storage and compartmentation, (4) neurotransmitter release, (5) neurotransmitter binding, (6) neurotransmitter-induced or mediated changes in ion fluxes, (7) other neurotransmitter-linked events (e.g., activation of cyclases), (8) neurotransmitter catabolism, and (9) transport of neurotransmitter metabolites. (see Lai et al.[3] for a full discussion and additional references.) The neurotransmitter metabolic cycle provides the most appropriate theoretical framework upon which one can systematically formulate testable hypotheses concerning the mechanisms of actions of essential and neurotoxic metals on neurotransmitter function and metabolism.[3]

29.2.2 Roles of Metals in Brain Development

Because many different metals are closely associated with nucleic acids in their intracellular microenvironments and many biological functions of nucleic acids are critically dependent on metals, metals have been postulated to play some important roles in regulating neuronal and glial development and differentiation.[2,5] However, these roles are poorly understood.

Some metals may control blood-brain barrier maturation. During development, the age-related decreases in the brain regional contents of electrolytes correlate well with the age-dependent lowering of brain water[8] and are inversely related to the age-dependent rise in activities of brain regional Na-K- and Mg-ATPases,[7] reflecting the rates of maturation of the blood-brain barrier in the different brain regions.[7,8] These observations are consistent with the proposal that metals regulate the maturation and function of the blood-brain barrier.[2,5,8]

Many enzymes related to energy metabolism (e.g., cytochrome oxidase and superoxide dismutase being cuproenzymes; glutamate dehydrogenase being a Zn metalloenzyme) are metalloproteins, and in metal deficiency states activities of these enzymes are decreased.[2,5] Thus, the metals involved may play structure-and-function roles in the development of these metalloenzymes, and hence the maturation of brain energy metabolism.[2,5]

29.2.3 Roles of Metals in Brain Aging

The blood-brain barrier may be compromised in aging: this may be one consequence of changes in metal metabolism in aging. For example, since increased vascular permeability is often associated with cerebrovascular diseases, the increased incidence of cerebrovascular diseases found in aging implies that the blood-brain barrier is functionally disturbed in aging.[2,5] Brain calcification is one form of cerebral atrophy associated with aging.[2,5] There is some speculation that the increases in brain levels of other metals such as Cu and Fe in aging may be responsible for the metal-mediated generation of free radicals which induce neuronal cell death.[9]

The changes in endocrine functions in aging are associated with alterations in trace metal metabolism. However, whether the increases in brain Ca, Fe, and Cu, and decreases in brain Mn and Se can be directly attributed to the hormonal changes in aging remains to be elucidated.[2,5]

29.3 ARE THERE RELATIONSHIPS BETWEEN MANGANESE AND OTHER METALS AND ELECTROLYTES IN BRAIN?

29.3.1 Suggestive Evidence From Peripheral Tissue Studies

Early studies indicate that, in animals fed high dietary manganese, decreased levels of Fe were found in liver, kidney, and spleen.[10] Thus, manganese may interact with other metals and electrolytes in absorption and organ distribution.[10] However, the early studies did not address the targeting of manganese to the central nervous system. Furthermore, the absence of suitable multidisciplinary approaches greatly hampered the progress in this field.

29.3.2 Advances in Methodologies for Addressing Key Issues

Several major advances in methodologies and approaches in the last decade have allowed the development of versatile multidisciplinary approaches that greatly facilitate the elucidation of the relationships between manganese and other metals and electrolytes in the brains of normal and manganese-treated animals.[3,4] On the one hand, the development of instrumental neutron activation analysis (INAA) technique played a critical role in shaping the studies of the interactions between multiple metals in small biological samples such as discrete regions of mammalian brain.[4] On the other hand, parallel advances in the development of subcellular fractionation and other biochemical and cell biological techniques have stimulated the applications of these techniques to investigate biochemical and physiological functions in nervous tissues.[11] As a result of these critical developments, a new era of studying the functions of trace metals and their interactions in neural tissues was initiated (see below).

29.4 METHODOLOGICAL CONSIDERATIONS IN ELUCIDATING RELATIONSHIPS BETWEEN MANGANESE AND OTHER MINERALS

29.4.1 Multidisciplinary Approach

By suitable combinations of anatomical, biochemical, toxicokinetic, physiological, and analytical techniques, different variations of a multidisciplinary approach can be designed to address specific issues pertaining to the functions of trace metals and their interactions in neural tissues. For example, by combining the versatile analytical technique of INAA with the anatomical method of brain regional dissection, new and insightful information has been obtained concerning the distribution of some 30 elements in 7 rat brain regions.[12] The results of such studies have led to a broad classification of elements into three distinct categories: (1) elements that show great regional variability (e.g., Ca, Se, Cu), (2) elements that show moderate regional variability (e.g., V, Mo, Zn), and (3) elements that exhibit little or no regional variability (e.g., Cl, Na, K). Other variations of such a multidisciplinary approach have also proven to be extremely instrumental in the design of seminal studies that elucidate important aspects of trace metal functions and interactions in brain (see below).

29.4.2 Design and Use of Animal Models

Good animal models simulate and resemble the human condition. Animal models suitable for studying the relationships between Mn and other minerals have to satisfy two essential criteria. First, they allow the entry of peripherally administered metal ions (such as manganese) into brain across the blood-brain barrier. Second, the models provide adequate brain tissue for regional and subcellular studies.[3,4]

Our model involves continuously exposing rats to $MnCl_2$ (added in the drinking water) throughout pre- and postnatal development till the Mn-treated animals are used for experiments.[4,7,8] In this model, Mn crosses the blood-brain barrier and Mn accumulation in different brain regions shows dose-dependent increases.[4,7,8] Furthermore, the model does provide adequate brain tissue samples for regional and subcellular studies.[4,7,8]

29.5 ELUCIDATION OF INTERACTIONS BETWEEN MANGANESE AND OTHER MINERALS VIA DEVELOPMENTAL AND TOXICOLOGICAL STUDIES

29.5.1 Stages of Brain Development and Essential Minerals Interactions

These correlations and interactions are best illustrated by examining manganese and a chemically closely related essential metal, namely, magnesium. Relevant to this consideration are two hypotheses: (1) chemically similar metal ions show developmental and regional correlations in their brain levels; (2) the correlations with stages of brain development may suggest that they — the chemically similar metals — share common or very similar metabolic mechanisms during brain development. These hypotheses have been investigated in a series of studies by Lai and colleagues;[5,7,8,13] the results of some of the studies are summarized below.

In general, the age-related decreases in brain regional Mn and Mg levels parallel the decreases in brain water and increases in tissue weight, suggesting that the maturation of the blood-brain barrier in the various brain regions examined may play important role(s) in brain Mn and Mg homeostasis.[8,13] For example, the age-dependent Mn accumulation shows regional variations: at day 5 postnatal this accumulation is most pronounced in striatum (12.05 μ/g wet weight) but least marked in cerebral cortex (0.85 μ/g wet weight). By day 10 postnatal, pons and medulla and hypothalamus are regions with, respectively, the highest (4.73 μ/g wet weight) and the lowest (0.52 μ/g wet weight) Mn levels. By contrast, brain regional Mn variations are less pronounced in weanling and adult rats. The age-dependent Mg accumulation shows regional variation at postnatal day 5, being most marked in pons and medulla (720 μ/g wet weight) and least marked in cerebral cortex (295 μ/g wet weight). Mg levels in all regions decrease after day 5; by day 120, only Mg level in cerebral cortex is lower than levels in other regions (the latter being very similar).[8] Thus, these observations are consistent with the two hypotheses, i.e., (1) and (2) mentioned above.

Chronic Mn treatment (10 mg $MnCl_2 \cdot 4H_2O$ per ml of drinking water) gives rise to significant increases in Mn accumulation in all regions during development, the only exception being the 5-day-old cerebellum.[8] This treatment also alters the regional Mn distribution in development as well as the developmental pattern of Mn levels within a particular region.[8] For example, in the treated animals, Mn levels in all regions demonstrate that Mn accumulation in brain is higher postweaning than preweaning.[8] The results of these studies[8] strongly suggest that Mn intake modulates its accumulation and metabolism in brain during development.

Compared to the corresponding effects on brain Mn, the effects of chronic Mn treatment on brain Mg are much less extensive. One consequence of the treatment is a general decrease in Mg levels in all regions at day 5 postnatal such that the marked regional variation in Mg levels observed in control animals is abolished in Mn-treated rats.[8] Significant Mn-induced decreases in Mg levels are also noted in most regions at day 22, in pons and medulla at day 10, and in cerebral cortex at day 120.[8] Moreover, Mn treatment also modifies the developmental pattern of Mg levels within a particular region.[8] Consistent with the two hypotheses, i.e., (1) and (2) discussed above, the results of these studies[8] indicate that Mn interacts with Mg metabolism in a region-specific manner during brain development.

Other ongoing studies demonstrate that Mn also interacts with the metabolism of other essential metals (e.g., Se, Co, Fe) in a region-specific manner during brain development (Lai, J.C.K., Chan,

A.W.K., and Minski, M.J., unpublished observations). These and other studies will further elucidate how Mn may interact with essential metals during brain development and shed some light on the underlying physiological mechanism(s).

29.5.2 Stages of Brain Development: Manganese and Nonessential and/or Toxic Metals

There is a paucity in the literature concerning these correlations. Our ongoing studies indicate that Mn may interact with nonessential metals. For example, chronic manganese treatment leads to significant decreases in Al accumulation in all regions of the 5-day-old rat brain, implying that Mn may interact with Al (Chan, A.W.K., Lai, J.C.K., Minski, M.J., and Lim, L., unpublished data) during brain development. Clearly, future studies will elucidate if this type of interaction can be generalized to other nonessential metals.

29.5.3 Brain Manganese and Minerals in Aging

These aspects have not been investigated extensively. We have noted that among the brain electrolytes examined, only Ca levels are increased (+85%) in aging.[2] Of the trace metals investigated, brain levels of Fe (+58%) and Cu (+100%) are increased but those of Mn (−37%) and Se (−48%) are decreased in aging. Brain levels of Al, Zn, Co, and Rb are unaffected.[2] Interestingly, in rats that had been chronically treated with Mn (1 mg $MnCl_2 \cdot 4H_2O$ in the drinking water) throughout their life span, the aging-induced changes in brain Ca, Fe, and Mn levels are abolished (Lai, J.C.K., Chan, A.W.K., Minski, M.J., and Lim, L., unpublished data). Thus, these results are consistent with the hypothesis that aging induces changes in trace metal metabolism. Furthermore, these observations suggest that chronic and life span Mn treatment exerts some modulatory effects on the interactions between brain minerals in aging. This important topic certainly merits further investigation.

29.6 ELUCIDATION OF INTERACTIONS BETWEEN MANGANESE AND OTHER MINERALS VIA CELLULAR AND SUBCELLULAR APPROACHES

29.6.1 Using Cellular Approaches

Though primary cultures of neurons and glial cells have been exploited to investigate the effects of toxic metals on neural cell types,[1,3] interactions between Mn and other metals have only been investigated to a very limited extent in these cells. For example, Zn inhibits but Cu enhances the uptake of Mn into glial cells.[6] Since neurons are known to contain higher Ca, Cu, and Mn than glial cells,[14] one might speculate that neurons may behave differently compared to glial cells in the interactions between Mn and other metals. Only future studies, however, will clarify these mechanisms.

29.6.2 Employing Subcellular Approaches

Lai et al.[4] were the first to employ the sensitive technique of instrumental neutron activation analysis for the determinations of over 15 elements in subcellular fractions that had been biochemically characterized.[11] However, a full discussion of the results of these studies is beyond the scope of this minireview. A brief discussion (see below) of some of the key findings will serve to illustrate the importance of such multidisciplinary studies. The major conclusion resides in the fact that each metal has its own subcellular distribution pattern, which is modulated by chronic Mn treatment *in vivo*.[4]

In rat brain, the subcellular distributions of Mn, Cu, and Al are not uniform.[4] For example, Mn levels are highest in mitochondria, capillaries, and cytosol, intermediate in microsomes and synaptosomes, and lowest in myelin and nuclei.[4] Cu levels are highest in cytosol and capillaries, intermediate in mitochondria, nuclei, and microsomes, but lowest in synaptosomes and myelin.[4] For Al, a nonessential trace metal, the rank order of distribution differs yet again, being highest in synaptosomes, nuclei, and mitochondria, intermediate in myelin and capillaries, but lowest in cytosol and microsomes.[4]

Previous studies by Rajan et al.[15] who investigated five metals (Cu, Fe, Zn, Mg, Ca) in only three subcellular fractions (myelin, synaptosomes, mitochondria) derived from whole brain and selected brain regions from normal rats yielded results that were comparable with those of Lai et al.[4]

Chronic Mn treatment *in vivo* gives rise to a 93% increase in Mn in brain homogenate but does not markedly change the relative distribution of Mn in various fractions.[4] This treatment, however, alters the subcellular distributions of Cu and Al.[4] Cu in brain homogenate is increased by 36%; nuclear and synaptosomal Cu levels are increased by 102 and 70%, respectively, whereas capillary Cu level is decreased by 52%, Cu levels in other fractions being unaltered.[4] The decrease in Al level in homogenate (–37%) is matched by a decrease in Al level in synaptosomes (–64%) but an increase in Al (+191%) in microsomes.[4]

Chronic Mn treatment *in vivo* also influences the subcellular distributions of other metals in brain. This treatment gives rise to increases in the levels of cobalt in mitochondria (+230%), synaptosomes (+75%), and myelin (+110%) but decreases in cobalt in cytosol (–48%). On the other hand, this treatment results in increases in the levels of chromium in mitochondria (+160%), synaptosomes (+110%), and a decrease in the level of chromium in cytosol (–48%) (Chan, A.W.K., Minski, M.J., and Lai, J.C.K., unpublished data). Additional studies are in progress to determine if chronic Mn treatment also induces changes in the levels of other metals (e.g., Rb, Se, V, etc.) in subcellular fractions derived from brain.

29.7 INTERACTIONS BETWEEN MANGANESE AND OTHER MINERALS IN BRAIN: IMPLICATIONS IN NEUROLOGICAL AND PSYCHIATRIC DISEASES

Evidence has been accumulating that metal toxicity may play pathogenetic and/or pathophysiological roles in a variety of neurological and psychiatric diseases.[2,3] For example, Al, Fe, Mn, and Zn have been implicated in Alzheimer's disease.[2,3,16,17] Chronic manganese toxicity in humans shows an acute phase of psychosis (known as locura manganica), followed by a chronic phase whose signs and symptoms are characteristic of those of Parkinson's disease and dystonia.[7,8,13] Furthermore, metal ions-mediated oxidative damage that leads to neuronal cell death has been postulated to be one of the key mechanisms underlying many neurodegenerative diseases including Parkinsonism and Alzheimer's disease.[9] (For full discussions and references, see the reviews by Lai et al.,[1–4] Vatassery,[9] Wedler,[6] Connor,[16,17] and Lai.[5])

Since more than one metal may be implicated in the pathogenetic and/or pathophysiological mechanisms underlying the above-mentioned neurological and psychiatric diseases, it is reasonable to speculate that interactions between trace metals may be at least one of the complicating factors involved in such mechanisms. Thus, a better understanding of the mechanisms of interactions between metals — such as those between brain Mn and other minerals — will elucidate the roles of metals in the pathogenesis and/or pathophysiology of many neurological and psychiatric diseases.

29.8 CONCLUSIONS AND PROSPECTS FOR FUTURE STUDIES

This minireview has highlighted the methodological advances that were — and continue to be — critical in opening new areas for investigating the roles of metals in brain function. These advances in methodology and approaches allow one to address hypotheses that focus on mechanisms of interactions between metals in brain function and dysfunction. In particular, this new generation of studies has yielded important and novel hypotheses concerning the roles of Mn in brain development and aging and the interactions between Mn and other minerals underlying such roles.

As emphasized above, the prior successes of multidisciplinary approaches point to the utility and importance of new combinations in the design of such studies. Thus, in future studies, such approaches can be directed to address mechanistic issues implicating the interactions between Mn and other minerals in discrete cellular and/or subcellular compartments. Similarly, clinical studies (employing human brain) are needed to directly address pathogenetic and pathophysiological mechanisms underlying neurological and psychiatric diseases in which metal toxicity is implicated.

REFERENCES

1. Lai JCK, Wong PCL, and Lim L: Structure and function of synaptosomes and mitochondrial membranes: elucidation using neurotoxic metals and neuromodulatory agents. In Sun GY, Bazan N, Wu J-Y, Porcellati G, and Sun AY (eds): *Neural Membranes*. Clifton Heights, NJ, Humana Press, 1983, pp 355–374.
2. Lai JCK, Chan AWK, Minski MJ, and Lim L: Roles of metal ions in brain development and aging. In Gabay S, Harris J, and Ho BT (eds): *Metal Ions in Neurology and Psychiatry*. New York, Alan R. Liss, 1985, pp 49–67.
3. Lai JCK, Leung TKC, and Lim L: Effects of metal ions on neurotransmitter function and metabolism. In Gabay S, Harris J, and Ho BT (eds): *Metal Ions in Neurology and Psychiatry*. New York, Alan R. Liss, 1985, pp 177–197.
4. Lai JCK, Chan AWK, Minski MJ, Leung TKC, Lim L, and Davison AN: Application of instrumental neutron activation analysis to the study of trace metals in brain and metal toxicity. In Gabay S, Harris J, and Ho BT (eds): *Metal Ions in Neurology and Psychiatry*. New York, Alan R. Liss, 1985, pp 323–343.
5. Lai JCK: Metal ions and brain. In Berthon G (ed): *Handbook on Metal-Ligand Interactions in Biological Fluids*, Vol. 2. New York, Marcel Dekker, 1995, in press.
6. Wedler FC: Biological significance of manganese in mammalian systems. In Ellis GP and Luscombe DK (eds): *Progress in Medicinal Chemistry*, Vol. 30. Amsterdam, Elsevier Science, 1993, pp 89–133.
7. Lai JCK, Leung TKC, Lim L, et al.: Effects of chronic manganese treatment on rat brain regional sodium-potassium-activated and magnesium-activated adenosine triphosphatase activities during development. *Metab. Brain Dis.*, 1991; 6:165–174.
8. Chan AWK, Minski MJ, Lim L, and Lai JCK: Changes in brain regional manganese and magnesium levels during postnatal development: modulations by chronic manganese administration. *Metab. Brain Dis.*, 1992; 7:21–33.
9. Vatassery GT: Vitamin E: neurochemistry and implications for neurodegeneration in Parkinson's disease. *Ann. N.Y. Acad. Sci.*, 1992; 669:97–110.
10. Underwood EJ: Manganese. In *Trace Elements in Human and Animal Nutrition*, 4th ed. New York, Academic Press, 1977, pp 170–195.
11. Lai JCK and Clark JB: Isolation and characterization of synaptic and non-synaptic mitochondria from mammalian brain. In Boulton AA, Baker GB, and Butterworth (eds): *NeuroMethods*, Vol. 11. Clifton Heights, NJ, *Humana Press*, 1989, pp 43–98.
12. Chan AWK, Minski MJ, and Lai JCK: An application of neutron activation analysis to small biological samples: simultaneous determination of thirty elements in rat brain regions. *J. Neurosci. Meth.*, 1983; 7:317–328.
13. Lai JCK, Chan AWK, Leung TKC, et al.: Neurochemical changes in rats chronically treated with a high concentration of manganese chloride. *Neurochem. Res.*, 1992; 17:841–847.
14. Tholey G, Ledig M, Kopp P, et al.: Levels and sub-cellular distribution of physiologically important metal ions in neuronal cells cultured from chick embryo cerebral cortex. *Neurochem. Res.*, 1988; 13:1163–1167.
15. Rajan KS, Colburn RW, and Davis JM: Distribution of metal ions in the subcellular fractions of several rat brain areas. *Life Sci.*, 1976; 18:423–432.
16. Connor JR: Iron acquisition and expression of iron regulatory proteins in the developing brain: manipulation by ethanol exposure, iron deprivation and cellular dysfunction. *Dev. Neurosci.*, 1994; 16:233–247.
17. Connor JR: Iron regulation in the brain at the cell and molecular level. In Hershko et al. (eds): *Progress in Iron Research*. New York, Plenum Press, 1994, pp 229–238.

Chapter 30

Manganese Dynamics in Brain Mitochondria

Claire E. Gavin and Thomas E. Gunter

CONTENTS

30.1 INTRODUCTION

Manganese, like iron, is a transition metal whose redox properties are both valuable and dangerous to biological systems. Unlike iron, intracellular levels of manganese are primarily controlled by tight homeostatic regulation of absorption and excretion rather than by sequestration within storage molecules. Mn^{2+} levels in brain are further controlled by the blood-brain barrier.[1] Because Mn^{2+} shares with Mg^{2+} and Ca^{2+} an affinity for phosphate, carboxyl, and nitrogen groups, it binds strongly to membrane phospholipids and glycoproteins, to some carboxylic acids, and to adenosine triphosphate (ATP). Clearance of Mn^{2+} from brain is extremely slow.[2]

Manganese poisoning continues to occur in industry and agriculture.[3,4,5] The resulting neurological disorder is characterized by rigidity, which is related to depletion of striatal dopamine,[6] and by dystonia, which is related to destruction of the globus pallidus.[7] The focal degeneration found in striatum of manganese-intoxicated brains may be explained by the synergistic cytotoxicity of Mn^{2+} and L-dopa or dopamine (see Reference 8 and citations therein). However, the outstanding histopathological finding in severe cases of manganism is the virtual obliteration of the globus pallidus (see Reference 9 and citations therein).

The literature in the late 1980s suggested to us that some toxic effects of manganese may involve interference with mitochondrial function or Ca^{2+} homeostasis:

1. After *in vivo* administration, manganese appeared in the mitochondrial fraction of liver[10] and of striatum.[11] Mn^{2+} is sequestered *in vitro* by isolated liver mitochondria via the Ca^{2+} uptake mechanism, the uniporter.[12]

2. At physiological levels of Mg^{2+}, Mn^{2+} enhances mitochondrial Ca^{2+} uptake,[13] raising the possibility of interference with other Ca^{2+} fluxes and ultimately with intracellular Ca^{2+} homeostasis.

3. Manganese binds tightly to the inner mitochondrial membrane,[14] the location of the electron transport chain and oxidative phosphorylation, raising the possibility of interference with mitochondrial energy production.

30.2 METHODS

Rat brain mitochondria were isolated as described in Reference 15, and suspended in either mannitol/sucrose or KCl medium. Endogenous Ca^{2+} concentration ($[Ca^{2+}]$) was measured spectrophotometrically with the dye Arsenazo III and typically equalled 5 to 17 nmol/mg protein. Ion fluxes were studied using $^{54}Mn^{2+}$ and $^{45}Ca^{2+}$ as described in Reference 15 and Figure 30.1. Oxygen consumption was measured in isolated liver mitochondria incubated in mannitol/sucrose medium with 10 mM KCl, as described in Reference 9 and in Figure 30.2.

30.3 KINETICS OF Mn^{2+} TRANSPORT IN BRAIN MITOCHONDRIA

The *in vivo* evidence for rapid mitochondrial accumulation of Mn^{2+}[10] appears to conflict with Mn^{2+} uptake kinetics and binding characteristics *in vitro*. First, sequestration of Mn^{2+} by isolated liver mitochondria is much slower than that of Ca^{2+}. Second, not only does Mn^{2+} bind strongly to cellular membranes, but it also binds to ATP with an affinity higher than that of Mg^{2+}. We hypothesized that mitochondrial accumulation might nevertheless occur if Mn^{2+} efflux is even slower than its influx and if Mn^{2+} uptake is facilitated *in vivo* by one or more soluble brain constituents normally omitted from *in vitro* preparations. A likely candidate for Mn^{2+} uptake enhancement was the polyamine spermine, present in millimolar amounts and reported to activate mitochondrial sequestration of Ca^{2+}.[16]

Results showed that, as predicted, ATP strongly inhibits Mn^{2+} sequestration by brain and liver mitochondria, even in the presence of physiological $[Mg^{2+}]$s.[9] However, Mn^{2+} uptake in the presence of ATP was significantly enhanced both by physiological levels of spermine (0.5 to 1.0 mM) and by $[Ca^{2+}]$s within the range of cytosolic $[Ca^{2+}]$ "spikes" following depolarization (i.e., 4 to 5 μM). The latter observation suggested that the mitochondria of tonically active neurons might sequester significantly more Mn^{2+} than resting neurons.

Mn^{2+} efflux from brain mitochondria proved to be extremely slow. We expected that it would be transported over the two Ca^{2+} efflux mechanisms, one of which involves exchange with Na^+ (for reviews see References 12 and 17). However, although Na^+-dependent efflux is the primary efflux mechanism for Ca^{2+} in mitochondria of brain and other excitable tissues, and is a full order of magnitude faster than Na^+-independent efflux, Mn^{2+} is not transported via this mechanism.[15] Moreover, at physiological $[Mg^{2+}]$ and in the presence of ATP, the maximum velocity of Mn^{2+} efflux is ~1.0 nmol/mg·min vs. 1.5 to 3.0 nmol/mg·min for Ca^{2+}. More importantly, the concentration of Mn^{2+} at half-maximal transport is nearly twice that of Ca^{2+}.

Mn^{2+} is thus capable of accumulating within brain mitochondria, a phenomenon that may explain its slow clearance from brain.[15]

30.4 EFFECT OF Mn^{2+} ON MITOCHONDRIAL Ca^{2+} FLUXES

At physiological $[Mg^{2+}]$, Mn^{2+} enhanced Ca^{2+} uptake into brain mitochondria by 6–15%, with an EC_{50} of 11.0 μM Mn^{2+}. In contrast to this modest facilitation of Ca^{2+} influx, Na^+-independent Ca^{2+} efflux was reduced to ~60% of control values by intramitochondrial (but not extramitochondrial) Mn^{2+}. The inhibition was competitive, with an apparent K_i of 8.0 nmol Mn^{2+}/mg protein in brain and about 34 nmol Mn^{2+}/mg protein in liver. Intramitochondrial Mn^{2+} inhibited Na^+-dependent Ca^{2+}

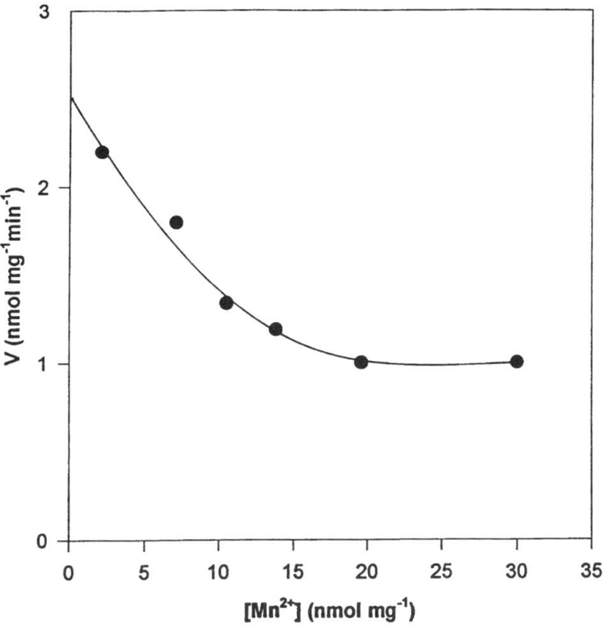

FIGURE 30.1 Mn^{2+} inhibition of Na^+-dependent Ca^{2+} efflux. Brain mitochondria were incubated at 2 to 4 mg/ml in 125 mM KCl, 10 mM K-Hepes, 0.2 mM ATP, 0.05 mM K-phosphate, 5.0 mM K-glutamate, 5.0 mM K-malate, pH 7.2 at 20°C. After 4 min, $^{45}Ca^{2+}$ (2 µCi/ml) was added, followed 1 min later by $^{54}Mn^{2+}$ (1.6 µCi/ml). After 10 min, equal portions of uptake samples were added to medium containing either EGTA/ruthenium red (RR) or EGTA/RR plus NaCl (final [Na$^+$] = 4.0 mM). Aliquots were removed at 0.75 to 1 min intervals and centrifuged for 30 s, after which the supernatants were poured off and counted for radioactivity as described.[15] Na^+-dependent efflux rates were obtained from the difference between the rates with and without Na^+. Intramitochondrial [Mn^{2+}] (nmol/mg mitochondrial protein) was determined by subtracting the amount of Mn^{2+} present in the supernatant from the total amount added. The [Ca^{2+}] in the experiment shown here was 27.3 nmol/mg protein; at 13.8 nmol Ca^{2+}/mg protein, maximum inhibition was 78%. Equivalent results were obtained in medium containing 2 mM MgCl$_2$.

FIGURE 30.2 Mn^{2+} inhibition of ADP-stimulated respiration in liver mitochondria with succinate or glutamate/malate as substrates. Mn^{2+} was added to isolated rat liver mitochondria suspended in a lucite chamber at 1 mg/ml in 195 mM mannitol, 65 mM sucrose, 10 mM KCl, 10 mM K-Hepes, 2.0 mM K-phosphate, 1.0 mM MgCl$_2$, 3.0 mM succinate (Δ) or 5.0 mM each glutamate and malate (●), pH 7.2 at 25°C. In some preparations Mn^{2+} uptake was stimulated by spermine, in others by 5.0 µM Ca^{2+}. ADP was added after 3 min to initiate State 3 respiration, after which oxygen consumption was recorded with an oxygen electrode, as described in Reference 9. Each data point represents the mean and standard error of 2 to 5 samples. (From Gavin, C.E., Gunter, K.K., and Gunter, T.E., *Toxicol. Appl. Pharmacol.*, 115, 1, 1992. With permission.)

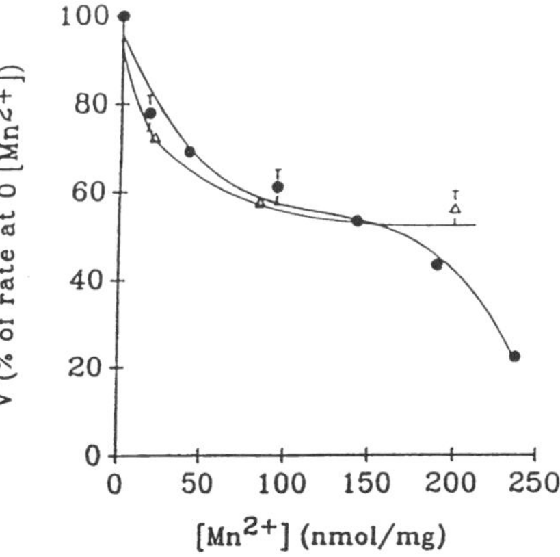

transport as well, with an IC$_{50}$ of 5.0 nmol Mn^{2+}/mg protein and a maximum inhibition of 60 to 80% (Figure 30.1). This combination of enhanced Ca^{2+} influx and diminished Ca^{2+} efflux suggests that Mn^{2+} is capable of altering Ca^{2+} fluxes enough to increase the level of intramitochondrial Ca^{2+}, particularly when cellular Ca^{2+} influx is unusually high, as in neurons that are undergoing prolonged stimulation of synaptic activity.

Elevated intramitochondrial [Ca^{2+}] has been associated with depletion of glutathione,[18] inhibition of oxidative phosphorylation,[19] and deterioration of mitochondrial function.[20] Perhaps most importantly, elevated intramitochondrial $[Ca^{2+}]$ increases the probability of the irreversible opening of a proteinaceous pore in the inner mitochondrial membrane (for a review see Reference 17). This permeability transition collapses the mitochondrial membrane potential, with concomitant loss of ATP production, followed by colloid-osmotic swelling of the mitochondrial matrix. Grossly swollen mitochondria are characteristic of a variety of pathological states, including ischemia or hypoxia and subsequent reperfusion.[21,22] Moreover, severe depletion of ATP has been shown to trigger cell death.[23,24]

Results of these studies suggest that the more Mn^{2+} accumulates in brain mitochondria, the more Ca^{2+} is likely to accumulate there as well, with a resulting risk of Ca^{2+} overload.

30.5 EFFECT OF Mn²⁺ ON MITOCHONDRIAL RESPIRATION

Once we had established sound kinetic reasons why elevated brain $[Mn^{2+}]$ might accumulate within brain mitochondria and might injure cells indirectly by facilitating the elevation of intramitochondrial $[Ca^{2+}]$, we asked whether elevated mitochondrial manganese might be able to interfere with oxidative phosphorylation. Mn^{2+} could affect mitochondrial function in at least two ways: first, Mn^{2+} could simply substitute for Mg^{2+} or Ca^{2+} at binding sites on enzymes of the electron transport system or on the $F_1ATPase$, inhibiting enzyme activity. Alternatively, Mn^{2+} may become oxidized by superoxide to the powerful oxidant Mn^{3+},[25] with resulting membrane damage and secondary loss of ATP production. The latter event would be expected to trigger either increased oxygen consumption in the absence of ADP (due to proton leaks through a damaged membrane) or decreased oxygen consumption in uncoupled mitochondria (due to damaged enzymes).

Because the yield of brain mitochondria is low, oxygen consumption experiments were carried out in liver mitochondria. Figure 30.2 shows that intramitochondrial Mn^{2+} decreased the rate of ADP-stimulated respiration with an IC_{50} of about 15 nmol/mg.[9] The inhibition ranged in various preparations from 30–50% with either succinate or the NADH-linked substrates glutamate and malate. With glutamate and malate as substrates, extremely high $[Mn^{2+}]$s further decreased respiration (Figure 30.2); this additional decrease was due to inhibition of uncoupled respiration (data not shown). The absence of an effect on uncoupled respiration at $[Mn^{2+}]$s that inhibit ADP-stimulated respiration suggests that Mn^{2+} may act at the level of ADP phosphorylation: the relatively low IC_{50} is consistent with binding to a high-affinity, saturable site on either ATP, the ADP/ATP translocase, or the $F_1ATPase$ itself. A site on the $F_1ATPase$ that binds Mn^{2+} with high affinity has already been identified.[9]

The relatively mild decrease observed in ATP production would likely be significant only under prolonged or sudden energy demand. However, according to Bradford,[26] the energy demand of synaptically active brain tissue is extremely high even under normal conditions, because of the need to maintain ion gradients. This enormous energy demand is an important aspect of the model proposed in the following section, which assumes that the effect of Mn^{2+} on respiration of brain mitochondria is similar to our observations in liver mitochondria.

30.6 PROPOSED MODEL FOR THE MECHANISM OF SEVERE MANGANESE TOXICITY

The clinical signs of manganese intoxication all indicate dysfunction of the basal ganglia, a group of brain nuclei which together function to suppress involuntary movements. These nuclei include the striatum and the globus pallidus (GP) as well as two associated nuclei, the substantia nigra (SN) and the subthalamic nucleus (STN). The basal ganglia receive excitatory input from cortical neurons synapsing in the striatum, while a major portion of their output is from the internal segment of the GP (GP_i) projecting to thalamic motor nuclei. Like Parkinsonism, manganism depletes the striatum of dopamine, producing rigidity and bradykinesia. Severe manganism is also characterized by dystonia and its histopathological correlates, severe injury to the GP and often to the STN.[7]

The following model links mitochondrial Mn^{2+} accumulation to empirical observations of GP_i destruction (Figure 30.3). Depletion of striatal dopamine is reported to cause a pathologic tonic increase in the synaptic activity of GP_i neurons.[27,28] This increase is due not only to disinhibition from lack of dopamine, but also to an increase in excitatory input from the STN, which is itself disinhibited by low dopamine.[29] The increased synaptic activity of pallidal neurons necessarily increases both their Ca^{2+} influx and their energy demand, just as manganese levels are increasing throughout the basal ganglia and most notably within the GP.[30] As we have seen, manganese is likely to accumulate within the mitochondria of these hyperactive neurons, where it may inhibit their ATP production directly and cause Ca^{2+} accumulation by inhibiting Ca^{2+} efflux while enhancing Ca^{2+} uptake. Elevated intramitochondrial $[Ca^{2+}]$ carries the risk of inducing the permeability transition and thereby eliminating oxidative phosphorylation. A decrease in Mn^{2+}-binding extramitochondrial ATP would accelerate both Mn^{2+} and Ca^{2+} accumulation within the remaining intact mitochondria. When a critical number of mitochondria become irretrievably permeabilized, ATP levels fall to a level incompatible with cell survival. The death of a critical number of GP_i and STN neurons results in dystonia.

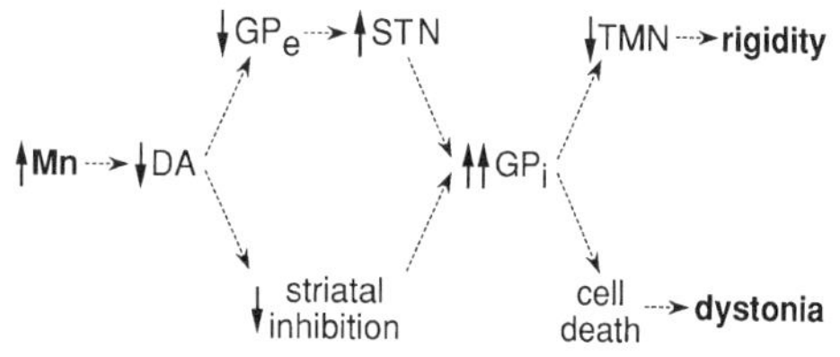

FIGURE 30.3 A proposed model for the mechanism of manganese toxicity. Manganese depletes striatal dopamine (DA), thereby decreasing striatal inhibition of the internal segment of the globus pallidus (GP_i). Loss of DA also decreases the activity of the external pallidum (GP_e), disinhibiting the subthalamic nucleus (STN) and increasing its excitatory input to GP_i. The resulting pathological increase in GP_i activity enhances inhibition of thalamic motor nuclei (TMN), causing rigidity. Meanwhile, the presence of manganese in GP_i mitochondria eventually results in the death of GP_i neurons, producing dystonia.

The model outlined in Figure 30.3 is meant to stimulate efforts to elucidate the mechanism(s) underlying the distinguishing neurological sign and singular pathology of severe manganese toxicity, both of which appear to be consistent with toxicity to mitochondria of the globus pallidus.

ACKNOWLEDGMENTS

This work was supported in part by the National Institute of General Medical Sciences, Grant GM-35550. We thank Dr. Michael Zuscik for redrawing Figure 30.3.

REFERENCES

1. Aschner M and Aschner JL: Manganese neurotoxicity: cellular effects and blood-brain barrier transport. *Neurosci. Biobehav. Rev.*, 1991; 15:333–340.
2. Dastur DK, Manghani DK, and Rachavendran KV: Distribution and fate of ^{54}Mn in the monkey. Studies of different parts of the central nervous system and other organs. *J. Clin. Invest.*, 1971; 50:9–20.
3. Lu CS, Huang CC, Chu NS, and Calne DB: Levodopa failure in chronic manganism. *Neurology*, 1994; 44:1600–1602.
4. Calne DB, Chur NS, Huang CC, Lu CS, and Olanow W: Manganism and idiopathic Parkinsonism: similarities and differences. *Neurology*, 1994; 44:1583–1586.
5. Megler D, Huel G, Bowler R, et al: Nervous system dysfunction among workers with long-term exposure to manganese. *Environ. Res.*, 1994; 64:151–180.
6. Bernheimer H, Birkmayer W, Hornykiewicz O, et al: Brain dopamine and the syndromes of Parkinson and Huntington. Clinical, morphological and neurochemical correlations. *J. Neurol. Sci.*, 1973; 20:415–425.
7. Barbeau A: Mn and extrapyramidal disorders. *Neurotoxicology*, 1984; 5:13–36.
8. Montine TJ, Underhill TM, Linney E, and Graham DG: Fibroblasts that express aromatic amino acid decarboxylase have increased sensitivity to the synergistic cytotoxicity of L-Dopa and manganese. *Toxicol. Appl. Pharmacol.*, 1994; 128:116–122.

9. Gavin CE, Gunter KK, and Gunter TE: Mn²⁺ sequestration by mitochondria and inhibition of oxidative phosphorylation. *Toxicol. Appl. Pharmacol.*, 1992; 115:1–5.

10. Maynard LS and Cotzias GC: Partition of Mn among organs and organelles of the rat. *J. Biol. Chem.*, 1955; 214:489–495.

11. Liccione J and Maines M: Selective vulnerability of glutathione metabolism and cellular defense mechanisms in rat striatum to manganese. *J. Pharmacol. Exp. Ther.*, 1988; 247:156–161.

12. Gunter TE and Pfeiffer D: Mechanisms by which mitochondria transport calcium. *Am. J. Physiol.*, 1990; 258:C755–C786.

13. Konji V, Montag A, Sandri G, et al: Transport of Ca²⁺ and Mn²⁺ by mitochondria from rat liver, heart and brain. *Biochimie*, 1985; 67:1241–1250.

14. Gunter TE, Puskin JS, and Russell PR: Quantitative magnetic resonance studies of manganese uptake by mitochondria. *Biophys. J.*, 1975; 15:319–333.

15. Gavin CE, Gunter, KK, and Gunter TE: Manganese and calcium efflux kinetics in brain mitochondria. *Biochem. J.*, 1990; 266:329–334.

16. Nicchitta CV and Williamson JR: Spermine. *J. Biol. Chem.*, 1984; 259:12,978–12,983.

17. Gunter TE, Gunter KK, Sheu S-S, and Gavin CE: Mitochondrial calcium transport: physiological and pathological relevance. *Am. J. Physiol.*, 1994; 267:C313–C339.

18. Olafsdottir K, Pascoe GA, and Reed DJ: Mitochondrial GSH status during Ca-ionophore-induced injury to isolated hepatocytes. *Arch. Biochem. Biophys.*, 1988; 263:226–235.

19. Fagian MM, Da Silva LP, and Vercesi AE: Inhibition of oxidative phosphorylation by Ca²⁺ or Sr²⁺: a competition with Mg²⁺ for the formation of adenine nucleotide complexes. *Biochim. Biophys. Acta*, 1986; 852:262–268.

20. McCormack JG and Denton RM: Influence of calcium ions on mammalian intramitochondrial dehydrogenases. In Fleischer S and Fleischer B (eds): *Methods in Enzymology*, Vol 174; Biomembranes part U: Cellular and subcellular transport: eukaryotic (nonepithelial) cells. San Diego, Academic Press, 1989, pp 95-118.

21. Meldrum B: Cell damage in epilepsy and the role of calcium in cytotoxicity. *Adv. Neurol.*, 1986; 44:849–855.

22. Crompton M and Andreeva L: On the involvement of a mitochondrial pore in reperfusion injury. *Basic Res. Cardiol.*, 1993; 88:513–523.

23. Lemasters JJ, Gores GJ, Nieminen A-L, et al: Multiparameter digitized video microscopy of toxic and hypoxic injury in single cells. *Environ. Health Perspect.*, 1990; 84:83–94.

24. Jurkowitz-Alexander MS, Altschuld RA, Hohl CM, et al: Cell swelling, blebbing, and death are dependent on ATP depletion and independent of calcium during chemical hypoxia in a glial cell line (ROC-1). *J. Neurochem.*, 1992; 59:344–352.

25. Archibald FS and Tyree C: Manganese poisoning and the attack of Mn³⁺ upon catecholamines. *Arch. Biochem. Biophys.*, 1987; 256:638–650.

26. Bradford HF: *Chemical Neurobiology*. W.H. Freeman, New York, 1986.

27. Miller WC and DeLong M: Altered tonic activity of neurons in the globus pallidus and subthalamic nucleus in the primate MPTP model of parkinsonism. In Carpenter MB and Jayarman A (eds): *The Basal Ganglia*, Vol 2. New York, Plenum Press, 1986, pp 415–427.

28. Mitchell IF, Cross AJ, Sambrook MA and Crossman AR: Neural mechanisms mediating 1-methyl-4-phenyl-1,2,3,6-tetrahydropyridine-induced Parkinsonism in the monkey. Relative contributions of the striatopallidal and striatonigral pathways as suggested by 2-deoxyglucose uptake. *Neurosci Lett.*, 1986; 63:61–66.

29. Klockgether T and Turski L: Excitatory amino acids and the basal ganglia: implications for the therapy of Parkinson's disease. *Trends Neurosci.*, 1989; 12:285–86.

30. Newland MC, Ceckler RL, Kordower JH, and Weiss B: Visualizing manganese in the primate basal ganglia with magnetic resonance imaging. *Exp. Neurol.*, 1989; 106:251–258.

Chapter 31

Implications of Manganese in Disease, Especially Central Nervous System Disorders

Edith Heilbronn and Håkan Eriksson

CONTENTS

31.1 MANGANESE IN BIOLOGICAL SYSTEMS

Manganese is a normal constituent of many biological systems[1] and an essential trace element. It is well recognized that manganese deficiency can result in a wide variety of structural and physiological defects.[2] It is necessary for the formation of connective tissue and bone, and for growth, carbohydrate and lipid metabolism, the embryonic development of, e.g., the inner ear and reproductive organs.[1] Manganese is a necessary constituent of several enzymes, including glutamine synthetase, calmodulin-dependent phosphatases, mitochondrial superoxide dismutase, phosphoenolpyruvate carboxykinase, pyruvate carboxylase, arginase, and galactosyl transferase.

Under physiological conditions, glutamine synthetase accounts for around 80% of all available manganese in the brain and is fairly uniformly distributed in the brain,[3] but after toxic overexposure manganese accumulates in the basal ganglia,[4] structures that have a high non-heme iron content.[5]

31.2 MANGANESE AND EPILEPSY

Hurley and co-workers[6] first suggested a link between epilepsy and manganese (Mn^{2+}) when they observed that Mn^{2+}-deficient rats were more susceptible to seizures than after Mn^{2+}-repletion. They also showed that the Mn^{2+}-deficient animals exhibited an epileptiform EEG. In addition, blood Mn^{2+} levels in epileptic patients were found to be lower than normal.[7] The lowest levels were seen in patients with the most frequent seizures.

Thus, while a low Mn^{2+} concentration is attributed by some to the seizure activity associated with epilepsy, others propose that the low Mn^{2+} concentrations may be secondary to genetic mechanisms underlying the epilepsy. Studies concerning Mn^{2+}-binding enzymes of brain and liver, i.e., glutamine synthetase and arginase, indicate that seizures affect liver arginase activity through changes in liver Mn^{2+} concentration, whereas genetically epilepsy-prone rats show abnormalities, apparently dependent on the seizure activity.[8]

The glutamine synthetase activity is lower in brains from epileptic animals than in those from nonepileptic controls, paralleling lower Mn^{2+} levels in brains from epilepsy-prone animals.[8] As lowering of the glutamine synthetase activity by inhibitors causes seizures in animals, one might speculate that there is a link between genetic forms of epilepsy and a diminished capacity of the brain to metabolize both the excitatory transmitter glutamate and ammonia. The accumulation of these compounds should increase the spread of excitation. Whether there is a general deficit of astrocyte functions in detoxifying glutamate, i.e., uptake and metabolism, or a restricted capacity to regulate enzymatic functions that underlies certain epileptic conditions has yet to be demonstrated.

31.3 OBSERVATIONS FROM HUMAN CASES OF MANGANESE INTOXICATION AND FROM EXPERIMENTALLY INDUCED MANGANESE INTOXICATION

The possible role of environmental neurotoxic compounds in the etiology of neurodegenerative disorders is an important research issue. Manganese is a neurotoxic agent in humans, known to induce psychiatric and extrapyramidal symptoms after chronic exposure in mining or industry.[9,10] In addition to adverse effects of inorganic manganese dust or vapor among steel-manufacturing workers or welders,[11,12] health risks of exposure to organic manganese compounds, including the widely used pesticide manganese ethylenebis (dithiocarbamate)[13] and the antiknock agent methyl-cyclopentadienyl manganese tricarbonyl[14] in unleaded gasoline, have created concern. It is also notable that new detergents with bleaching agents containing organic manganese complexes have been introduced for public use.[15]

Acute manganese intoxication includes irritability, speech disturbance, compulsive actions, and hallucinations. In the chronic stage of manganism neurological signs resemble a Parkinson's syndrome and dystonia. Neuropathological changes associated with manganese intoxication are mainly localized to the basal ganglia. Neuronal degeneration is found in the putamen and globus pallidus.[16] Decreased dopamine concentrations in the striatum are found in intoxicated patients; and also in experimentally intoxicated monkeys[17] and rodents. Post-mortem studies in humans[16] and chronic studies in nonhuman primates[4,17–19] and rodents[14,20] showed that manganese intoxication produces neuropathological changes in basal ganglia, especially the globus pallidus, caudate, and putamen, with lesions localized both pre- and postsynaptically to the dopaminergic nigrostriatal pathway.

Chronic occupational exposure to manganese-containing dust has been demonstrated to produce a slowly progressive, largely irreversible extrapyramidal movement disorder that may include psychiatric dysfunction early in its course.[21] Only a few autopsy studies of patients who suffered from manganism have been reported (reviewed in References 10, 16, and 22). The typical finding of basal ganglia atrophy is characterized by neuronal loss, gliosis, and sometimes pigment deposition. Lesions tend to focus in the pallidoluysian system. However, the striatum seems frequently to be involved. Occasionally, an involvement of the substantia nigra has also been observed.

These clinical and pathological changes have recently been found to be duplicated in monkeys after chronic exposure to $Mn(IV)O_2$.[4,18,19] Neurochemical analysis of the brain from these monkeys showed that:

1. Mn levels were elevated most in the pallidum and putamen;
2. Dopamine and DOPAC but not homovanillic acid levels were significantly reduced in the lentiform nuclei;
3. Mn had the greatest toxic effect on dopaminergic neurons in these brain regions.

There have been numerous reports of manganese-mediated dopaminergic neurotoxicity in other animal models (reviewed in References 4, 9, and 10). Although specific results have varied because of experimental design, the conclusions have been broadly consonantal.

31.3.1   Findings in Mn-Exposed Humans

A recent examination, using magnetic resonance imaging and positron emission tomography,[23] of four patients with manganism from a Taiwanese ferromanganese plant suggested that mild neurological features of manganism might result from functional disturbance of postsynaptic striatal or pallidal neurons, rather than from a dopamine depletion that has been reported subsequent to postmortem examination. There was a widespread decline in cortical and subcortical glucose consumption, but no change in 6-fluorodopa utilization was observed. The symptoms of intoxication among the workers became progressively worse, and were only occasionally and transiently improved by L-dopa treatment.[21] This is partly in contradiction to what was originally described by Cotzias and co-workers[24] who found that L-dopa yielded even better benefits to patients (miners) with chronic manganese poisoning than to patients with Parkinson's disease. There are, however, many differences between the two studies, such as type, level, and duration of exposure, as well as treatment protocols.

31.3.2   Findings in Low-Level Mn-Exposed Humans

A study by Mergler and collaborators[25] on workers in a manganese alloy production plant with low levels of airborne manganese dust compared to the plant in the previously mentioned study, presented data on neurofunctional tests. Significantly higher blood Mn but unaffected urinary Mn levels were found in exposed workers than in matched controls. Exposed persons also differed from controls in motor functions (particularly those which require alternating and rapid movements), emotional state, cognitive flexibility, and olfactory perception threshold. On the whole, intellectual functions were similar in the two groups. Other studies have indicated an increased irritability, trembling of fingers, and fatigue among exposed workers, compared to matched controls.[11] In general, when effects of occupational exposure are studied, results may be ambiguous due to lack of information concerning exposure by other neurotoxic agents as well as by lack of knowledge about exact degrees of exposure.

31.4   TRANSPORT AND DISTRIBUTION OF MANGANESE IN THE CNS

Within the brain manganese is distributed heterogeneously, with highest concentrations in the pallidum and putamen and the lowest concentration in the frontal cortex.[26]

The level of manganese in brain and brain capillaries exceeds that in plasma by more than a hundredfold,[2] which implies some form of selective binding or active transport. Various routes have been suggested through which manganese may reach the brain. Manganese in serum is thought to circulate as Mn^{3+} bound tightly to transferrin ($K_a \sim 10^{10}$ M^{-1}).[27] It has been suggested that Mn^{3+} is taken up into the brain via a transferrin-dependent mechanism.[28] Brain capillaries are also known to contain high levels of transferrin receptors[29] and Mn^{3+} has been shown to be taken up and internalized in cells by a transferrin-dependent mechanism. However, manganese is also reported to exist in plasma as Mn^{2+} circulating as the free ion or bound to plasma proteins such as albumin and α_2–macroglobulin or to low-weight solutes. These other species may contribute to brain uptake as well. Murphy and co-workers[30] reported that Mn^{2+} may be transported into brain by a saturable high-affinity mechanism. Several transport systems are reported to accept Mn^{2+} as a ligand, including Ca^{2+} channels, the Na^+/Ca^{2+} exchanger, the active Ca^{2+} uniporter in mitochondria, and the Na^+/Mg^{2+} antiporter. In addition, specific Mn^{2+} transport systems have been identified in a number of cells (see Reference 31). Others[31] suggest that Mn^{2+} rapidly crosses the blood-brain barrier by a facilitated mechanism and that uptake is critically influenced by plasma protein binding. The relative contribution of the various pathways probably differs depending on the chemical form, amounts, and administration or exposure route of manganese.

The transport of manganese after intracerebral Mn^{2+} injections in rat was recently analyzed.[32] Dopaminergic neurons were found to take up and transport the manganese ions in an anterograde direction, but not in a retrograde direction. GABAergic neurons also took up and transported manganese in the anterograde direction. Manganese was found to be localized in both glial cells and neurons. Manganese binding seemed nonsaturable, suggesting multiple binding sites, binding in a ferritin-like core or intracellular precipitation. Appearance of manganese in brain areas remote from the injection site was predominantly due to axonal transport, but only a relatively small proportion of ^{54}Mn was transported intraneuronally, possibly as a transferrin complex. Larger quantities are probably bound by glial cells, i.e., astrocytes, oligodendroglia, and activated microglial cells, and stored as a ferritin complex. The rationale for this was based on several studies during recent years.[28,33] (For a review of iron homeostasis see Reference 34).

It has been established that the distribution of manganese after overexposure is predominantly confined to brain areas containing high levels of non-heme iron,[5] including the caudate-putamen, globus pallidus, substantia nigra, and subthalamic nucleus. Further, transferrin receptors are found predominantly in the projection areas of the non-heme iron-rich areas. The Fe-transferrin and Mn-transferrin complexes have both been shown to be internalized by the transferrin receptor, and subsequently some of the metal ions were found as ferritin complexes. Other important sites of manganese accumulation are mitochondria, as is discussed in Chapter 30 of this volume. The cell types responsible for manganese accumulation in the brain have not yet been defined, although ferritin-containing cells like microglia and the oligodendrocytes might accumulate manganese as well as iron.

31.5 POSSIBLE MECHANISMS OF MANGANESE NEUROTOXICITY

Manganese neurotoxicity is possibly caused by the affinity of the manganese ions to areas with high levels of neuromelanin, e.g., the nigrostriatal tract, as well as by the ability to assume multioxidative states in areas which offer a suitable biochemical milieu for the transformation of Mn^{2+} to Mn^{3+}, thus favoring (auto)oxidation of dopamine to semiquinone and the production of toxic free radicals.[35–37]

It has been hypothesized[35,36] that the distribution of lesions in the CNS following chronic exposure to manganese could be explained by manganese interaction with dopamine. Catechols present in dopaminergic neurons may undergo a series of oxidations and nucleophilic additions that culminate in neuromelanin formation. Following the observation that trivalent manganese rapidly and efficiently oxidizes catechols *in vitro* it has been hypothesized that Mn^{3+}-induced oxidation of dopamine may explain the topographic distribution and neurochemical changes observed in chronic manganese intoxication. In support of this proposal, one group of investigators has observed a correlation between cytotoxic effects in PC12 cells exposed to $MnCl_2$ and increased malondialdehyde formation.[38]

In recent experiments, the hypothesis was tested that L-dopa, dopamine, and DOPAC serve as endogenous "protoxins" for manganese ions, with dopamine being the most potent.[39] Aromatic amino acid decarboxylase was transfected into a nonneuronal cell line (Chinese hamster ovary cells; CHO). With this procedure it was possible to have a cell-based system able to generate dopamine, but free from the other metabolic pathways for catechols present in neuronal systems. The experiments allowed the conclusion that manganese and L-dopa are potent synergistic cytotoxicants in CHO (wild-type) cultures and that the increased sensitivity of CHO/AADC (aromatic amino acid decarboxylase) cells to manganese and L-dopa was accounted for by greater total intracellular concentrations of catechols and not by differences between the cytotoxicity of L-dopa and dopamine. The results indicate that in the presence of manganese, it was the total amount of catechols within the cell that determined the degree of cytotoxicity, with no significant difference between L-dopa and dopamine. Side-chain decarboxylation of L-dopa to dopamine was biologically significant because it altered the cellular handling of the catechol moiety, resulting in significantly increased intracellular concentrations and therefore a greater range of manganese-induced cytotoxicity. Extracellular addition of dopamine to the medium resulted in low intracellular levels in CHO cells and was toxic only at high concentrations of manganese. Assays of cellular and medium fractions indicated that addition of Mn induced oxidation of intracellular and extracellular catechols and that all L-dopa and dopamine was consumed within hours. These results are supported by another study[40] showing that intracellular nonvesicular dopamine or L-dopa are more potent cytotoxins than extracellular catechols in the presence of manganese.

Recently, the role of endogenous antioxidant systems in the striatum and striatal synaptosomes (as models of neuronal terminals) of the rat was assessed during manganese-induced oxidative stress.[41,42] Dopamine metabolism may be associated with an increased production of H_2O_2 and increased oxidative damage of the dopaminergic system. It was found that manganese treatment increased both the dopamine turnover and the uric acid level in the tissue. The uric acid might originate from xanthine oxidase acting on xanthine or hypoxanthine, a pathway that also generates superoxide radical anion. Xanthine/hypoxanthine are preferentially produced during oxidative stress, suggesting a mitochondrial interference (see below) or increased energy demands during high manganese load.

The hypothesis that manganese neurotoxicity may result from an impairment of cellular antioxidant systems has also been advanced[9,36,37,43] on the basis of decreased GSH levels and GSH peroxidase activity. An important *in vivo* function of GSH is to maintain appropriate tissue ascorbic acid concentrations, which may be needed for reducing functions not efficiently performed by GSH. In Parkinson's disease a regional depletion of GSH was found to correlate well with the disease severity.[44] In addition, a consistent decrease in regional ascorbic acid levels was found (–30% in putamen, –33% in globus pallidus, –75% in amygdala).

Oxidative DNA damage causes mutations and cell death. L-dopa, dopamine, and 3-*O*-methyl-L-dopa cause extensive oxidative DNA-damage in the presence of H_2O_2 and traces of copper ions (Cu^{2+}), as recently shown.[45] 8-Hydroxyguanine is the major product. Ferric ions (Fe^{3+}) were much less effective and manganous ions (Mn^{2+}) did not catalyze DNA damage. The authors propose that copper ion release, in the presence of L-dopa and its metabolites, might be an important mechanism of neurotoxicity, e.g., in Parkinson's disease and amyotrophic lateral sclerosis. The question whether manganese, as manganic ions (Mn^{3+}) or in combination with other metal ions or during stressful conditions, is also able to cause oxidative DNA damage, however, needs to be evaluated before similar mechanisms of manganese action can be ruled out.

Mitochondria are natural parts of manganese toxicity discussions (see Chapter 30) as manganese accumulates and otherwise interferes with mitochondrial functions. As discussed previously, several mitochondrial enzymes use manganese as essential cofactors. Recently it was suggested that impaired oxidative energy metabolism by manganese causes additional excitotoxicity.[46] The manganese toxicity after intrastriatal injections of $MnCl_2$ was shown[46] to be manifested in lesions suggesting an excitotoxic damage. These lesions were blocked by MK-801, a noncompetitive NMDA-receptor antagonist. Although this rapidly leads to neurotoxic lesions after intracerebral manganese injections, it is questionable whether the toxic manifestations are comparable to those attainable after chronic exposure of manganese compounds (usually airborne oxides). Nevertheless, the hypothesis of excitotoxic lesions is very interesting and deserves more attention and continued research efforts.

31.6 CONCLUSIONS

Several pathogenic mechanisms for manganese-induced neurodegeneration of the dopamine and GABAergic pathways in the basal ganglia have been proposed. The most widely cited is (auto)oxidation of brain catechols by (Mn^{2+})/Mn^{3+} to produce cytotoxic quinone species and partially reduced oxygen species.[35–37,47] This cytotoxic mechanism may be augmented by compensatory increases in dopamine turnover[10] or by impairment of cellular protective mechanisms by manganese.[43,48,49] These results certainly point to an involvement of the cellular antioxidant system in the control of neuronal damage caused by manganese-induced oxidative stress.

Other pathogenic mechanisms that have been proposed include increased activity of mitochondrial cytochrome P450[50] and disruption of mitochondrial calcium regulation with inhibition of oxidative phosphorylation,[51] which might lead to excessive amounts of extracellular glutamate, leading to increased NMDA receptor stimulation and finally to excitotoxic lesions. The relative importance and interrelationship of these potential targets are unknown.

The interrelationship between intracellular catechol oxidation and other pathogenic processes, like mitochondrial dysfunction, excitotoxic mechanism, and limitations of free radical scavenging should therefore be a focus of future experiments. The techniques provided by modern molecular biology may be of great help as they make it possible to design cellular systems containing or lacking specific enzymes, receptors, ion channels, or other relevant proteins. Such techniques may be helpful for elucidating the mechanisms and the cellular specificity of manganese toxicity and how to counteract or prevent this toxicity.

REFERENCES

1. Schramm VL and Wedler FC (eds): *Manganese in Metabolism and Enzyme Function*. New York; Academic Press; 1986.
2. Keen CL: Metabolism and biochemistry of manganese: In Frieden E (ed) *Biochemistry of the Essential Ultra-Trace Elements*. New York, Plenum Press, 1984.
3. Wedler FC and Denman RB: Glutamine synthetase: the major Mn(II) enzyme in mammalian brain. *Curr. Top. Cell Regul.*, 1984; 24:153–169.
4. Eriksson H, Mägiste K, Plantin LO, et al: Effects of manganese oxide on monkeys as revealed by a combined neurochemical, histological and neurophysiological evaluation. *Arch. Toxicol.*, 1987; 61:46–52.
5. Hill JM and Switzer RC: The regional distribution and cellular localization of iron in the rat brain. *Neuroscience*, 1984; 11:595–603.
6. Hurley LS, Woolley DE, Rosenthal F, et al: Influence of manganese on susceptibility of rats to convulsions. *Am. J. Physiol.*, 1963; 204:493–496.
7. Carl GF, Keen CL, Gallagher BB, et al: Association of low blood manganese concentrations with epilepsy. *Neurology*, 1986; 36:1584–1587.
8. Carl GF, Blackwell LK, Barnett FC, et al: Manganese and epilepsy: brain glutamine synthetase and liver arginase activities in genetically epilepsy prone and chronically seizured rats. *Epilepsia*, 1993; 34:441–446.
9. Donaldson J: The physiopathologic significance of manganese in brain: its relation to schizophrenia and neurodegenerative disorders. *Neurotoxicology*, 1987; 8:451–462.
10. Barbeau A: Manganese and extrapyramidal disorders (a critical review and tribute to Dr. George C. Cotzias). *Neurotoxicology*, 1984; 5:13–35.
11. Roels H, Lauwerys R, Buchet JP, et al: Epidemiological survey among workers exposed to manganese: effects on lung, central nervous system, and some biological indices. *Am. J. Ind. Med.*, 1987; 11:307–327.
12. Wang JD, Huang CC, Hwang YH, et al: Manganese-induced parkinsonism: an outbreak due to an unrepaired ventilation control system in a ferromanganese smelter. *Br. J. Ind. Med.*, 1989; 46:856–859.
13. Ferraz HB, Bertolucci PHF, Pereira JS, et al: Chronic exposure to the fungicide maneb may produce symptoms and signs of CNS manganese intoxication. *Neurology*, 1988; 38:550–553.
14. Gianutsos G and Murray MT: Alterations in brain dopamine and GABA following inorganic or organic manganese administration. *Neurotoxicology*, 1982; 3:75–81.
15. Hage R, Iburg JE, Kerschner J, et al: Efficient manganese catalysts for low-temperature bleaching. *Nature*, 1994; 369:637–639.
16. Yamada M, Ohno S, Okayasu I, et al: Chronic manganese poisoning: a neuropathological study with determination of manganese distribution in the brain. *Acta Neuropathol. (Berl.)*, 1986; 70:273–278.
17. Bird ED, Anton AH, and Bullock B: The effect of manganese inhalation on basal ganglia dopamine concentrations in rhesus monkey. *Neurotoxicology*, 1984; 5:59–65.
18. Eriksson H, Gillberg PG, Aquilonius SM, et al: Receptor alterations in manganese intoxicated monkeys. *Arch. Toxicol.*, 1992; 66:359–364.
19. Eriksson H, Tedroff J, Thuomas KA, et al: Manganese-induced brain lesions in *Macaca fascicularis* as revealed by positron emission tomography and magnetic resonance imaging. *Arch. Toxicol.*, 1992; 66:403–407.
20. Autissier N, Rochette L, Dumas P, et al: Dopamine and norepinephrine turnover in various regions of the rat brain after chronic manganese chloride administration. *Toxicology*, 1982; 24:175–182.

21. Huang CC, Lu CS, Chu NS, et al: Progression after chronic manganese exposure. *Neurology*, 1993; 43:1479–1483.
22. Barbeau A, Inoue N, and Cloutier T: Role of manganese in dystonia. *Adv. Neurol.*, 1976; 14:339–352.
23. Wolters EC, Huang CC, Clark C, et al: Positron emission tomography in manganese intoxication. *Ann. Neurol.*, 1989; 26:647–651.
24. Cotzias GC, Papavasiliou PS, Ginos J, et al: Metabolic modification of Parkinson's disease and of chronic manganese poisoning. *Annu. Rev. Med.*, 1971; 22:305–326.
25. Mergler D, Huel G, Bowler R, et al: Nervous system dysfunction among workers with long-term exposure to manganese. *Environ. Res.*, 1994; 64:151–180.
26. Larsen NA, Pakkenberg H, Damsgaard E, et al: Topographical distribution of arsenic, manganese, and selenium in the normal human brain. *J. Neurol. Sci.*, 1979; 42:407–416.
27. Scheuhammer AM and Cherian MG: Binding of manganese in human and rat plasma. *Biochim. Biophys. Acta*, 1985; 840:163–169.
28. Aschner M and Aschner JL: Manganese neurotoxicity: cellular effects and blood-brain barrier transport. *Neurosci. Biobehav. Rev.*, 1991; 15:333–340.
29. Pardridge WM, Eisenberg J, and Yang J: Human blood-brain barrier transferrin receptor. *Metabolism*, 1987; 36:892–895.
30. Murphy VA, Wadhwani KC, Smith QR, et al: Saturable transport of manganese(II) across the rat blood-brain barrier. *J. Neurochem.*, 1991; 57:948–954.
31. Rabin O, Hegedus L, Bourre JM, et al: Rapid brain uptake of manganese(II) across the blood-brain barrier. *J. Neurochem.*, 1993; 61:509–517.
32. Sloot WN and Gramsbergen JBP: Axonal transport of manganese and its relevance to selective neurotoxicity in the rat basal ganglia. *Brain Res.*, 1994; 657:124–132.
33. Suárez N and Eriksson H: Receptor-mediated endocytosis of a manganese complex of transferrin into neuroblastoma (SHSY5Y) cells in culture. *J. Neurochem.*, 1993; 61:127–131.
34. Connor JR: Cellular and regional maintenance of iron homeostasis in the brain — normal and diseased states. In *Iron in Central Nervous System Disorders*. Springer-Verlag, New York, 1993, pp 1–18.
35. Donaldson J, McGregor D, and LaBella F: Manganese neurotoxicity: a model for free radical mediated neurodegeneration? *Can. J. Physiol. Pharmacol.*, 1982; 60:1398–1405.
36. Graham DG: Catecholamine toxicity: a proposal for the molecular pathogenesis of manganese neurotoxicity and Parkinson's disease. *Neurotoxicology*, 1984; 5:83–95.
37. Archibald FS and Tyree C: Manganese poisoning and the attack of trivalent manganese upon catecholamines. *Arch. Biochem. Biophys.*, 1987; 256:638–650.
38. Sun AY, Yang WL, and Kim HD: Free radical and lipid peroxidation in manganese-induced neuronal cell injury. In *Markers of Neuronal Injury and Degeneration*. New York Academy of Science, New York, 1993, pp 358–363.
39. Montine TJ, Underhill TM, Linney E, et al: Fibroblasts that express aromatic amino acid decarboxylase have increased sensitivity to the synergistic cytotoxicity of L-Dopa and manganese. *Toxicol. Appl. Pharmacol.*, 1994; 128:116–122.
40. Vescovi A, Facheris L, Zaffaroni A, et al: Dopamine metabolism alterations in a manganese-treated pheochromocytoma cell line (PC12). *Toxicology*, 1991; 67:129–142.
41. Desole MS, Miele M, Esposito G, et al: Monoaminergic systems activity and cellular defense mechanisms in the brainstem of young and aged rats subchronically exposed to manganese. *Neurosci. Lett.*, 1994; 177:71–74.
42. Desole MS, Miele M, Esposito G, et al: Dopaminergic system activity and cellular defense mechanisms in the striatum and striatal synaptosomes of the rat subchronically exposed to manganese. *Arch. Toxicol.*, 1994; 68:566–570.
43. Liccione JJ and Maines MD: Selective vulnerability of glutathione metabolism and cellular defense mechanisms in rat striatum to manganese. *J. Pharmacol. Exp. Ther.*, 1988; 247:156–161.
44. Riederer P, Sofic E, Rausch WD, et al: Transition metals, ferritin, glutathione, and ascorbic acid in parkinsonian brains. *J. Neurochem.*, 1989; 52:515–520.
45. Spencer JPE, Jenner A, Aruoma OI, et al: Intense oxidative DNA damage promoted by L-DOPA and its metabolites: implications for neurodegenerative disease. *FEBS Lett.*, 1994; 353:246–250.
46. Brouillet EP, Shinobu L, McGarvey U, et al: Manganese injection into the rat striatum produces excitotoxic lesions by impairing energy metabolism. *Exp. Neurol.*, 1993; 120:89–94.
47. Halliwell B: Manganese ions, oxidation reactions and superoxide radical. *Neurotoxicology*, 1984; 5:113–118.
48. Vescovi A, Gebbia M, Cappelletti G, et al: Interactions of manganese with human brain glutathione-S-transferase. *Toxicology*, 1989; 57:183–191.
49. Shi XL and Dalal NS: The glutathionyl radical formation in the reaction between manganese and glutathione and its neurotoxic implications. *Med. Hypotheses*, 1990; 33:83–87.
50. Liccione JJ and Maines MD: Manganese-mediated increase in the rat brain mitochondrial cytochrome P-450 and drug metabolism activity: susceptibility of the striatum. *J. Pharmacol. Exp. Ther.*, 1989; 248:222–228.
51. Gavin CE, Gunter KK, and Gunter TE: Mn^{2+} sequestration by mitochondria and inhibition of oxidative phosphorylation. *Toxicol. Appl. Pharmacol.*, 1992; 115:1–5.

Chapter 32

Manganese-Induced Parkinsonism

Erik Ch. Wolters

CONTENTS

32.1 INTRODUCTION

Manganese is an essential trace element in the human body, being a constituent of connective tissue as well as of many enzymes, in particular glutamine synthetase. After iron, aluminum, and copper, it is the fourth most widely used metal in the world. Manganese oxide (ore) is converted by electrolytic high-temperature reduction to ferromanganese or silicomanganese alloys. Yearly, about 8 tons of manganese metal are extracted and mainly employed for the manufacture of steel, while a minor portion is alloyed with aluminum and copper. Manganese dioxide is used in the manufacture of dry batteries, chlorine gas, paints, varnish, and enamel. Other manganese-containing compounds are potassium permanganate, with bactericidal and fungicidal properties, the pesticide manganese ethylenebis, and methylcyclopentadienyl manganese tricarbonyl (MMT), applied as an antiknock agent in lead-free gasoline — 90% of the world's manganese is supplied by Australia, Brazil, Gabon, and the Republic of South Africa.

Heavy metals are often involved in the development of neurological disorders. It is assumed that zinc plays a role in the etiology of epilepsy,[1] aluminum in Alzheimer's disease,[2] mercury in Minimata's disease,[3] and copper in the hepatolenticular degeneration of Wilson disease.[4] Manganese is also known to be involved in various cerebral disease processes, such as manganese deficiency leading to epilepsy. It was only 20 years after the publication of the classic paper of James Parkinson on Parkinson's disease (PD) in 1817 that the mask-like features of workers employed in the grinding of manganese ores (especially braunite — Mn_2O_3 and $MnSiO_3$) drew Couper's attention to the possibility that manganese could be a causative agent in the development of a bizarre disorder with neurological as well as behavioral components.[5] He described Parkinson-like features in five patients, working in a manganese ore-crushing plant. Von Oettingen,[6] and especially Cotzias et al.,[7,8] expanded on the toxic symptomatology. It was found that neurological symptoms of intoxication were exclusively related to the inhalation of massive amounts of "fresh" manganese dust or fumes when manganese was present in particle sizes of less than 5 μm. Patients, mainly miners and smelter workers as well as manufacturers of dry batteries, were described in Chile, Morocco, Cuba, India, Japan, Australia, the former U.S.S.R., and the U.S. It was established that an acute intoxication could lead to behavioral disturbances with involuntary laughing and auditory hallucinations (the so-called manganese madness or locura manganica), and a chronic intoxication to extrapyramidal symptoms resembling those seen in Parkinson's and Wilson diseases.

32.2 ACUTE MANGANESE INTOXICATION: LOCURA MANGANICA

Manganese madness is prevalent in manganese miners in Chile, but not in manufacturers. The main characteristics of this mild and mostly transient syndrome are disorientation, memory impairment, anxiety, compulsive behavior, emotional instability, and hallucinations. The hallmarks are reminiscent of schizophrenia as well as amphetamine-induced psychosis. Dopamine receptors are involved in both conditions. It was suggested therefore that manganese may possess a specific affinity for dopaminergic receptors. Manganese is known to oxidize dopamine, resulting in quinones and free ·OH radicals.

32.3 CHRONIC MANGANESE INTOXICATION: MANGANESE PARKINSONISM

In chronic manganese intoxication for a period of 6 months or longer, marked asthenia and extrapyramidal features such as hypokinesia, rigidity, and tremor, and occasionally dystonia, siallorrhea, impotence, and insomnia may occur. As long as exposure continues, the symptoms will progress. Recovery is rare,[9] and symptoms may continue to progress even after cessation of manganese exposure.[10] Because the clinical hallmarks of this disease closely resembled the manifestations of idiopathic PD, and as these symptoms were reported to respond favorably to levodopa therapy, it was thought that the dopaminergic system was involved and that the vulnerability of this system to manganese underlies the clinical syndrome. Hence the syndrome was denominated manganese parkinsonism.

The main neuropathological feature in PD is an idiopathic degeneration of dopaminergic cells in the ventral mesencephalon in combination with the occurrence of so-called Lewy bodies (predominantly in the brain stem). Lewy bodies are eosinophilic intraneuronal inclusions containing filamentous material. The etiology of the degenerative process in idiopathic PD is still an enigma. Although it is generally accepted that PD results primarily from a loss of dopaminergic neurons in the mesencephalon, the resulting alterations in the activity of the basal ganglia are still rather poorly characterized.

32.4 ROLE OF BASAL GANGLIA IN THE REGULATION OF MOTOR BEHAVIOR

Dopaminergic neurons in the ventral mesencephalon are distributed over different cell groups, including the substantia nigra pars compacta (A9 cell group), the ventral tegmental area (A10 cell group), and the retrorubral area (A8 cell group). Recent neuroanatomical and functional studies have revealed that the dopaminergic projection can influence extensive basal ganglia-thalamocortical circuits which are intimately involved in the regulation of motor and complex behavioral activity.[11,12] The following brief description of the functional neuroanatomy of these basal ganglia-thalamocortical circuits is useful in understanding the pathophysiology of Parkinson-like symptoms, the result of presynaptic dopaminergic deficiency as well as postsynaptic events in the so-called motor loops.

The basal ganglia include the striatum (caudate nucleus, putamen, and nucleus accumbens), the pallidum (external and internal segments of the globus pallidus and ventral pallidum), the subthalamic nucleus and the substantia nigra (pars compacta and pars reticulata). The striatum is the input structure of the basal ganglia, receiving afferents from the entire cerebral cortex, the midline and intralaminar thalamic nuclei, and midbrain serotonergic and dopaminergic cell groups. The output structure of the basal ganglia consists of the internal segment of the globus pallidus (GP$_i$) and the pars reticulata of the substantia nigra (SNR) which project, also in a topographical manner, to different medial and ventral thalamic nuclei, the deep layers of the superior colliculus, and the reticular formation of the mesencephalon. The various thalamic nuclei that are innervated by these output structures of the basal ganglia project to different cortical areas of the frontal lobe, including motor, premotor, and prefrontal cortical areas. The topography in the projections from different frontal cortical areas through the basal ganglia and the thalamus by subsequent corticostriatal, striatopallidal (or striatonigral), pallidothalamic (or nigrothalamic), and thalamocortical projections is such that a number of parallel, functionally segregated basal ganglia-thalamocortical circuits can be conceived. Whereas motor functions are dealt with in the circuit that originates in the (pre)motor cortex (and that, at the striatal level, also receives inputs from the somatosensory cortex), other circuits are involved in complex motor/behavioral, cognitive, and affective processes.[11]

The input (striatum) and output structures (GP$_i$ and SNR) of the basal ganglia (see Figure 32.1A) are connected with each other by means of two pathways, i.e., a "direct" and an "indirect" pathway. The direct pathway consists of the GABA/substance P/dynorphin-containing striatopallidal (GP$_i$) and striatonigral (SNR) projections. The indirect pathway comprises the GABA/enkephalin-containing striatopallidal (external segment = GP$_e$), the GABAergic pallido-subthalamic, and the glutamatergic subthalamo-pallidal (GP$_i$), and subthalamo-nigral (SNR) projections.

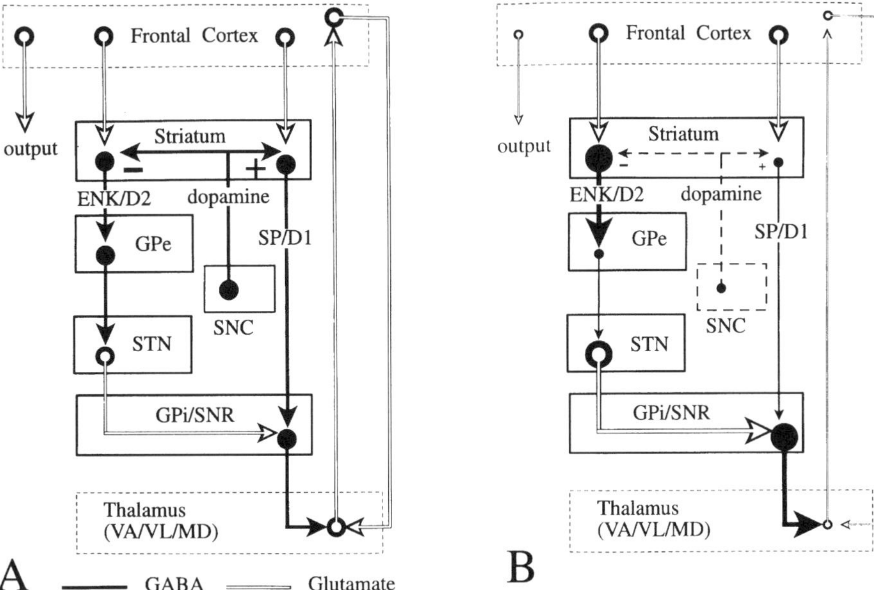

FIGURE 32.1   Schematic representation of the main connections in a basal ganglia-thalamocortical circuit as a "model" for multiple, parallel organized, functionally segregated loops with the same design. A: Normal status: the two output routes ("indirect" and "direct") are in "balance" at the level of the output structures. B: Presumed disturbance in Parkinson's disease: depletion of dopamine in the striatum leads to dysbalance in two output routes, and a suppression of thalamocortical activity. Abbreviations: GP, globus pallidus; GP$_e$, external segment of GP; GP$_i$, internal segment of GP; MD, mediodorsal thalamic nucleus; SNR, reticular part of the substantia nigra; SNC, pars compacta of the substantia nigra (A9); SP, substance P; STN, subthalamic nucleus; VA/VL, ventral anterior/ventral lateral thalamic nuclei.

At the level of the output structures of the basal ganglia these direct and indirect pathways have opposite effects on the GABAergic output neurons that project to the thalamic nuclei, the superior colliculus, and the reticular formation.

A "balance" between the two striatal output pathways seems to be essential for the normal regulation of movement. Although recent data show a more complex picture of the distribution of dopamine receptors, it has previously been suggested that dopamine, through different dopaminergic receptors, has opposing effects on the direct and indirect pathways:[11] dopamine acting via D1 receptors would have a stimulating effect on the direct pathway, whereas it would have an inhibitory effect on the indirect pathway via D2 receptors (see also Figure 32.1A). The consequence of the loss of dopamine is thought to be an increase in the output from the GP_i and SNR to the thalamus[13] (see also Figure 32.1B), ultimately resulting (supposedly at a loss of about 75% of striatal DA) in a reduction of cortical activation, which accounts for (most of) the parkinsonian signs, including bradykinesia, rigidity, tremor, and postural imbalance.

32.5 PATHOLOGY AND PATHOPHYSIOLOGY IN MANGANESE PARKINSONISM

In humans, neuropathologic changes in manganese intoxication are characterized by degeneration especially of the medial segment of the pallidum, to a lesser extent of the caudate nucleus and putamen, and only rarely of the substantia nigra.[14] Severe lesions of subthalamic and pallidal nuclei were also reported in monkeys following manganese dust aerosol inhalation.[15] Increasing the levels of manganese,[16,17] however, induced neuronal death with dopamine and neuromelanin loss in the nigral substance. Probably dose dependently, both pre- and postsynaptic lesions to the dopaminergic nigrostriatal projection system may be induced.

Functional imaging offers the possibility to estimate the functional integrity of the central dopaminergic system *in vivo*. In mild manganese-induced parkinsonian patients, fluorodopa PET studies showed, in contrast to PD patients, a fairly normal striatal uptake[18] in the caudate-putamen complex. Raclopride, a ligand for dopamine receptors, however, evidenced a reduced binding.[19] These results suggest a down regulation of dopamine receptors with an intact presynaptic dopaminergic system, in concordance with the pathological findings in manganese-intoxicated patients.

In idiopathic PD, however, a significant reduction of striatal fluorodopa uptake is seen, whereas raclopride shows a normal to slightly increased binding, which could be inferred to be the result of natural compensation processes in (dopaminergic) denervation. Based on the pathology and functional imaging, it may be conceived that pathophysiology in mild manganese intoxication, however, is based on postsynaptical, probably GABAergic,[20] dysfunction within the basal ganglia circuitry rather than presynaptical dopaminergic deficiency, although leading to the same increased thalamic inhibition and decreased thalamo-cortical output (see Figure 32.1B). Indeed, accordingly, in a double-blind controlled study in these mildly affected patients, levodopa was found to lack clinical benefit.[21] Earlier reports, however, mentioned some clinical efficacy of this drug in manganese-induced parkinsonian patients. These observations may be explained by the earlier-suggested dose-dependent toxic effect of manganese both postsynaptically and presynaptically to the nigrostriatal dopaminergic projection system.

32.6 ETIOLOGY OF MANGANESE PARKINSONISM

The underlying mechanisms whereby the inhalation of manganese dust results in cerebral disease remain a mystery. Hypotheses include an increased oxidation of dopamine with the production of quinones and ·OH free radicals.

Manganese circulates in the blood and enters the central nervous system as Mn^{++} as well as Mn^{+++}. In the central nervous system, manganese is found in glial cells as well as predominantly dopaminergic and GABAergic neurons.[20] The neurotoxic effect of manganese is hypothesized to be related to the several valency forms of manganese. During manganese intoxication, the rate of oxidation of Mn^{++} may exceed the rate of reduction of higher Mn valencies. Cotzias et. al. found a correlation between tissue manganese levels and the presence of the pigmented polymer melanin.[22] Both in manganese-induced Parkinsonism and PD, neuromelanin, formed by the nonenzymatic oxidation of dopamine, is found to be decreased.[23] It was concluded therefore that accumulation of manganese in the melanin-containing tissues was the result of the unique predisposition of manganese for binding sites on this polymer. Moreover, manganese, in its trivalent form, is a powerful oxidant of dopamine, and it has been suggested that this dopamine oxidation-enhancing property of manganese may explain its ability to induce a toxic lesion in strategic regions of the brain, as dopamine oxidation byproducts, such as free radicals, are extremely cytotoxic.[24]

More recently, Brouillet et al. noted that the damage associated with manganese intoxication is similar in distribution to that seen with carbon monoxide, cyanide, and other mitochondrial toxins. They suggested that manganese might also cause damage by impairing mitochondrial activity.[25] This could result in a loss of the voltage-dependent magnesium blockade in NMDA receptors, permitting an excitatory amino acid-induced rise in cytosolic free Ca^{++} and a cascade of events leading to a disrupted oxidative metabolism and cell death. As manganese toxicity can be prevented either by NMDA receptor antagonists or by removal of excitatory input to the basal ganglia by prior decortication, it may be inferred that oxidant stress may contribute to manganese toxicity.

32.7 THERAPEUTIC CONSIDERATIONS IN MANGANESE PARKINSONISM

Levodopa as well as disodium calcium edetate are said to bring some relief in manganese Parkinsonism. However, the one and only double-blind levodopa study[20] failed to show any significant therapeutic effect in (mild) manganese parkinsonism. Based on the hypothesized etiology of manganese-induced parkinsonism, other therapeutic strategies may bring clinical relief, such as glutamate receptor antagonists and free-radical scavenging agents. Clinical applications, however, have not been reported yet.

32.7.1 Glutamate Receptor Antagonists

The striatum is extensively innervated by glutamatergic projections from the cortex, the thalamus, and the amygdala. The medium-sized spiny neurons (most of which contain GABA as a neurotransmitter) are considered to be the main receiving neuronal elements in the striatum. Consequently, it can be inferred that glutamate receptors are involved in the regulation of the activity of these GABAergic cells. As can be seen from Figure 32.1, glutamatergic projections occur elsewhere in the basal ganglia circuitry. We would like to focus especially on the subthalamic nucleus (STN) which seems to play an essential role in controlling the output from the striatum and the GP_e. Glutamatergic cells project from the STN to the GP_i and the SNR. The experiments performed by Bergman et al.[26] are indicative of the essential role of the STN. Bergman discovered that the parkinsonian symptoms in MPTP-lesioned rhesus monkeys disappeared upon lesioning of the STN. Instead of lesioning the STN, one would expect that a pharmacological blockade of the glutamate receptors in the GP_i/SNR region would provide similar results.

The occurrence of both AMPA and NMDA receptors (glutamate receptor subtypes) has been demonstrated in these regions and both receptor types are involved in synaptic transmission processes in this brain region. Indeed, as has recently been demonstrated in several animal models for PD (reserpinized mouse, 6-OHDA rat, MPTP-lesioned monkey), NMDA and AMPA receptor antagonists are able to potentiate the behavioral effects of L-DOPA and to induce clear stimulating effects on motor behavior in the absence of L-DOPA. For instance, both the competitive (3-[2-carboxypiperazin-4-yl]-propyl-1-phosphonic acid or CPP) and the noncompetitive (MK-801) NMDA receptor antagonist were able to restore motor activity in these animal models.[27] Therefore, blockade of glutamate receptors provides a novel therapeutic strategy for the treatment of manganese-induced parkinsonism as well as idiopathic PD. However, the almost ubiquitous presence of the glutamatergic system in the brain poses the risk that the side effects of the glutamate receptor blockers may be even more serious than the symptoms of those diseases. On the other hand, the ongoing "proliferation" of glutamate receptor subtypes, which can be expected in the forthcoming years, might allow us to develop drugs which can be used as "smart weapons", seeking their own discrete target in the brain.

32.7.2 Free-Radical Scavengers

Free radicals are highly reactive entities (atoms or molecules) that can damage a variety of critical biological molecules. Enzymatic breakdown of dopamine as well as autooxidation of dopamine generates hydrogen peroxide, which under normal circumstances is detoxified by glutathione and catalyzed by glutathione peroxidase. In patients with manganese-induced parkinsonism and/or idiopathic PD, however, a consistently decreased level of striatal glutathione peroxidase and an increased level of free iron have been demonstrated, which promote the formation of hydroxyl radicals by the so-called Fenton reaction.[25,28] Therefore, it has been hypothesized that free radicals might be implicated in the pathogenesis of both conditions. If this were the case one might expect a therapeutic effect of free-radical scavengers such as tocopherol. Whereas the aforementioned pharmacotherapeutic interventions are all symptomatic treatments, treatment with free-radical trapping agents might slow the progression of the disease. In a recent multicenter trial[29] with 800 patients it turned out, however, that there was no beneficial effect of tocopherol in patients with early idiopathic PD. In the same study a beneficial effect was reported from deprenyl. Since deprenyl also inhibits the enzymatic breakdown of dopamine, it is still unclear whether this compound acts by interfering with oxidative stress mechanisms or solely by increasing the dopamine content.

REFERENCES

1. Barbeau A and Donaldson J: Zinc, taurine and epilepsy. *Arch. Neurol.*, 1974; 30:52-58.
2. Crapper DR, Krishnan SS, and Dalton AJ: Brain aluminum in Alzheimer's disease and experimental neurofibrillary degeneration. *Science*, 1973; 180:511–513.
3. McAlpine D and Araki S: Minimata disease: an unusual neurological disorder caused by contaminated fish. *Lancet*, 1958; 2:629–631.
4. Cummings JN: *Heavy Metals and the Brain*. Blackwell Scientific, Oxford, 1959.
5. Couper J: On the effects of black oxide of manganese when inhaled into the lungs. *Br. Ann. Med. Pharmacol.*, 1837; 1:41–42.
6. Von Oetingen WF: Manganese: its distribution, pharmacology and health hazards. *Physiol. Rev.*, 1935; 2:292–295.
7. Cotzias G: Manganese in health and disease. *Physiol. Rev.*, 1958; 38:503–532.
8. Cotzias CG, Hoiriuchi K, Fuenzalida S, Mena I: Chronic manganese poisoning — clearance of tissue manganese concentration with persistence of the neurological picture. *Neurology*, 1968; 18:376–382.
9. Huang CC, Chu NS, Lu CS, et al: Chronic manganese intoxication. *Arch. Neurol.*, 1989; 46:1104–1106.
10. Huang CC, Lu CS, Chu NS, et al: Progression after chronic manganese exposure. *Neurology*, 1993; 43:1479–1483.
11. Alexander GE and Crutcher MD: Functional architecture of basal ganglia circuits: neural substrates of parallel processing. *TINS*, 1990; 13:266–271.
12. Groenewegen HJ, Roeling TAP, Voorn P, and Berendse HW: The parallel arrangement of basal ganglia-thalamocortical circuits: a neuronal substrate for the role of dopamine in motor and cognitive functions? In: Wolters ECh, Scheltens Ph, eds. *Mental Dysfunction in Parkinson's Disease*. ICG Printing, Dordrecht, 1993: pp. 3–18.
13. DeLong MR: Primate models of movement disorders of basal ganglia origin. *Trends Neurosci.*, 1990; 13:281–285.
14. Yamada M, Ohno S, Okayasy I, et al: Chronic manganese poisoning: a neuropathological study with determination of manganese distribution in the brain, *Acta Neuropathol. (Berl.)*, 1986; 70:723–728.
15. Pentschew A, Ebner FF, and Kovatch RM: Experimental manganese encephalopathy in monkeys. A preliminary report. *J. Neuropathol. Exp. Neurol.*, 1963; 22:488–499.
16. Neff NH, Barrett RE, and Costa E: Selective depletion of caudate nucleus dopamine and serotonin during chronic manganese dioxide administration to squirrel monkeys. *Experientia*, 1969; 25:1140–1141.
17. Gupta SK, Murthy RC, and Chandra SV: Neuromelanin in manganese-exposed primates. *Toxicol. Lett.*, 1980; 6:17–20.

18. Wolters EC, Huang CC, Clark C, et al: Positron emission tomography in manganese intoxication. *Ann. Neurol.*, 1989; 26:647–651.

19. Shinotoh H, Snow BJ, Huang CC, et al: Presynaptic and postsynaptic striatal dopaminergic function in manganese intoxication studied by positron emission tomography. *Can. J. Neurol. Sci.*, 1993; 20:S236.

20. Sloot WN and Gramsbergen JBP: Axonal transport of manganese and its relevance to selective neurotoxicity in the rat basal ganglia. *Brain Res.*, 1994; 657:127–131.

21. Lu CS, Huang CC, and Calne DB: Levodopa failure in chronic manganism. *Neurology*, 1994; 44:1600–1602.

22. Cotzias CG, Papavasiliou S, and Miller ST: Manganese in melanin. *Nature*, 1964; 201:1228–1229.

23. Das KC, Abramson MB, and Katzman R: Neuronal pigments: spectroscopic characteristics of human brain melanin. *J. Neurochem.*, 1978; 30:601–605.

24. Donaldson J and Barbeau A: Metal ions in neurology and psychiatry. Manganese neurotoxicity: Possible clues to the etiology of human brain disorders. *Neurol. Neurobiol.*, 1985; 15:259–285.

25. Brouillet BP, Shinobu L, McGarvey U, Hochberg F, and Beal MF: Manganese injection into the striatum produces excitotoxic lesions by impairing energy metabolism. *Exp. Neurol.*, 1993; 120:89–94.

26. Bergman H, Wichmann T, and DeLong MR: Reversal of experimental parkinsonism by lesions of the subthalamic nucleus. *Science*, 1990; 249:1436–1439.

27. Klockgether T and Turski L: Toward an understanding of the role of glutamate in experimental Parkinsonism. Agonist-sensitive sites in the basal ganglia. *Ann. Neurol.*, 1993; 34:585–593.

28. Jenner P, Dexter DT, Sian J, Schapira AHV, and Marsden CD: Oxidative stress as a cause of nigral cell death in Parkinson's disease and incidental Lewy Body Disease. *Ann. Neurol.*, 1992; 32:S82–S87.

29. The Parkinson Study Group. Effects of tocopherol and deprenyl on the progression of disability in early Parkinson's disease. *N. Engl. J. Med.*, 1993; 328:176–183.

SECTION 11

CALCIUM

Chapter 33

The Involvement of Intracellular Calcium in Amyotrophic Lateral Sclerosis

Mohammed I. Hasham and Charles Krieger

CONTENTS

33.1 INTRODUCTION

Amyotrophic lateral sclerosis (ALS) is a progressive neurodegenerative disorder of unknown cause primarily affecting anterior horn cells and descending motor pathways. Three forms of ALS have been defined: a familial form with autosomal dominant inheritance (familial ALS, FALS); a more frequent, nonhereditary "sporadic" form; and a form seen in the Mariana Islands in the Western Pacific which is often associated with Parkinson's disease and dementia. Although mutations in the Cu^{2+}/Zn^{2+} superoxide dismutase (SOD) gene have been detected in some pedigrees with FALS, FALS patients constitute less than 10% of cases seen worldwide. Furthermore, the number of FALS pedigrees having the SOD mutation is likely around 15 to 20%.[1] Patients with the far more common sporadic (i.e., nonfamilial) form do not demonstrate abnormalities in the SOD gene. The pathogenic mechanisms underlying the sporadic form of ALS are unknown; however, some evidence supports a role for changes in concentrations of intracellular Ca^{2+} ($[Ca^{2+}]_i$) in ALS.[2]

33.2 EPIDEMIOLOGICAL EVIDENCE FOR A ROLE FOR Ca^{2+} IN ALS

Circumstantial evidence linking alterations in Ca^{2+} and ALS have been provided by environmental surveys of possible triggering factors for the high incidence of ALS in the Western Pacific.[3] Detailed analysis of soil and water samples from Guam and the Kii Peninsula of Japan have shown a high content of Mn^{2+} and Al^{3+} with low levels of Ca^{2+} and Mg^{2+}.[3] It has been postulated that patients in regions with a high incidence of ALS have a chronic nutritional deficiency of Ca^{2+} which leads to secondary hyperparathyroidism with increased absorption of Al^{3+} and the intraneuronal deposition of Ca^{2+} and Al^{3+}. This hypothesis has been supported by nutritional studies of rodents and primates fed Ca^{2+}-deficient diets.[4,5] These studies have revealed altered bone mineralization and intracellular deposition of Ca^{2+} in neurons. In addition, several case reports indicate that occasionally patients with hyperparathyroidism clinically resemble those with ALS and will respond favorably with treatment of the hyperparathyroidism.[6]

33.3 PUTATIVE ROLE OF $[Ca^{2+}]_i$ IN CELL DEATH

On the basis of numerous observations using models of ischemia and excitatory amino acid toxicity, it has been suggested that elevated $[Ca^{2+}]_i$ may play a significant role in mediating neuron death. Evidence supporting the role of altered $[Ca^{2+}]_i$ homeostasis as a mediator of cell death during ischemia in experimental models includes observations of Ca^{2+} accumulation in tissues exposed to ischemic conditions and elevated $[Ca^{2+}]_i$ in cultured cells following exposure to metabolic inhibitors. Some forms of excitatory amino acid (EAA) toxicity in cultured neurons are clearly associated with elevated $[Ca^{2+}]_i$ and alterations in the activities of several Ca^{2+}-dependent enzymes.[7]

Neurons have several Ca^{2+} influx pathways and buffering mechanisms which are capable of regulating $[Ca^{2+}]_i$. The principal routes of Ca^{2+} entry include voltage-dependent Ca^{2+} channels (VDCC) and receptor-operated channels such as the N-methyl-D-aspartate (NMDA) subclass of EAA receptor-gated ion channels (Figure 33.1). In addition, $[Ca^{2+}]_i$ can increase by Ca^{2+} influx through the Na^+/Ca^{2+} exchanger, by Ca^{2+} release from internal stores, or as a result of decreased efflux through the Ca^{2+}-ATPase (Figure 33.1).[8] Pharmacological treatments which block Ca^{2+} influx or intracellular Ca^{2+} release, or which enhance Ca^{2+} efflux, have been shown to be protective against cell death in studies using cultured cells. However, not all reports have confirmed an association between increased $[Ca^{2+}]_i$ and cell death under ischemic or other cytotoxic conditions.[9] It is likely that differing observations on changes in $[Ca^{2+}]_i$ under different circumstances depend upon the cell type, the period of cellular development, [10] and/or the source and nature of the Ca^{2+} influx.[11] Additionally, decreased cellular energy stores during periods of ischemia likely compromise the cell's ability to compensate for a Ca^{2+} influx.[12]

33.3.1 Ca^{2+} in Excitatory Amino Acid-Mediated Cell Death

The EAAs, glutamate and aspartate, act as neurotransmitters at the majority of excitatory synapses in the central nervous system. In some experimental models of ischemia and hypoxia, excessive glutamate release has been demonstrated which may be causally linked to neuronal death. It is well established that NMDA receptor activation by glutamate will produce Ca^{2+} influx.[13] This Ca^{2+} influx is thought to be responsible for mediating cell death, as removal of Ca^{2+} from the extracellular medium prior to NMDA application decreases the extent of cell death in some culture models.[14] Blockade of non-NMDA subtypes of glutamate receptors has also been reported by some investigators to be protective against EAA-mediated cell death in culture, an effect which may result from attenuated Ca^{2+} influx.[15] In experiments where $[Ca^{2+}]_i$ was made to rise transiently and to approximately equal values by activation of either VDCCs or NMDA-gated ion channels, the death of cultured spinal neurons was greater following Ca^{2+} influx through NMDA-gated channels than with influx through VDCCs.[11] Thus the source, rather than the magnitude of the Ca^{2+} influx, may be an important factor in determining cytotoxicity. If Ca^{2+} influx through a specific pathway is a critical determinant of cell fate, it is possible that Ca^{2+} interacts with distinct signaling pathways such as protein kinases.

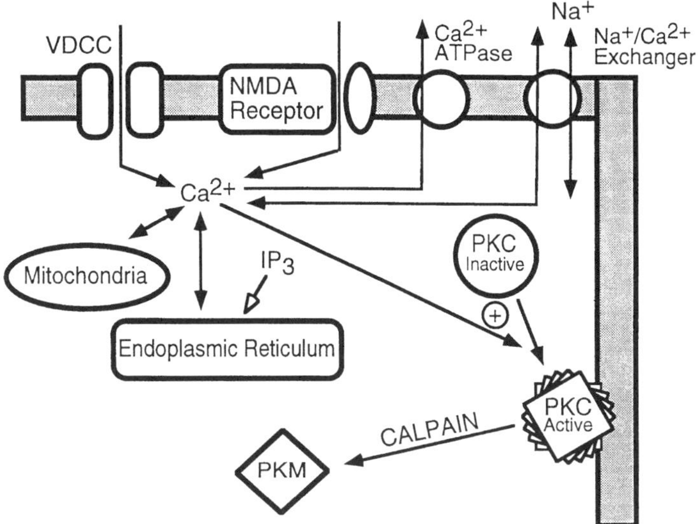

FIGURE 33.1 Role of Ca^{2+} in cell death. Ca^{2+} influx pathways include the NMDA receptor-gated channel, voltage-dependent calcium channels (VDCCs), and reverse operation of the Na^+/Ca^{2+} exchanger. Release of calcium from internal stores can be produced by inositol trisphosphate (IP_3; open arrow) or other stimuli. Elevated $[Ca^{2+}]_i$ can be achieved through reductions in Ca^{2+} efflux through the ATP-dependent Ca^{2+}-pump. Raised $[Ca^{2+}]_i$ can lead to the activation of protein kinases, such as protein kinase C (PKC; plus sign). The actions of Ca^{2+}-stimulated proteases (calpains) can convert PKC to the Ca^{2+}-independent form, PKM, which can phosphorylate a large variety of intracellular components, possibly leading to cell death. Calpains can also catalyze the breakdown of cytoskeletal elements, compromising membrane integrity.

33.3.2 Ca^{2+} and Downstream Signaling Pathways

Ca^{2+} can stimulate a variety of intracellular signaling enzymes including protein kinases, protein phosphatases, and proteases.[16,17] Perhaps the best evidence that Ca^{2+}-dependent processes mediate cell death comes from studies which show protein kinase C (PKC) activity is increased following NMDA-stimulated Ca^{2+} influx. PKC activation is dependent upon the presence of Ca^{2+} as well as membrane phospholipids and diacylglycerol.[18] Studies in tissue culture models have shown that NMDA-associated neuronal death can be attenuated by the addition of inhibitors of PKC.[7] Alternatively, cell survival increases if PKC is downregulated prior to NMDA application to cultured cells.[19] The activation of PKC leads to its translocation to the membrane and a conversion into the Ca^{2+}-independent cleavage product, PKM, by the action of the Ca^{2+}-dependent proteases (calpains, Figure 33.1). Thus, potentially, $[Ca^{2+}]_i$ can result in augmented phosphorylating activity of PKC and PKM which may persist even if $[Ca^{2+}]_i$ returns to normal. In addition, IP_3-mediated Ca^{2+} release from internal stores can also increase PKC activity in the absence of Ca^{2+} influx from the outside (Figure 33.1).

33.4 CELLULAR MECHANISMS UNDERLYING ALTERED $[Ca^{2+}]_i$ IN ALS

As indicated above, raised $[Ca^{2+}]_i$ can be produced by a number of mechanisms. Ca^{2+} influx can increase, buffering of $[Ca^{2+}]_i$ can be impaired, or Ca^{2+} extrusion decreased. Alternatively, raised $[Ca^{2+}]_i$ can be associated with Ca^{2+} release from intracellular pools or by reverse operation of a Ca^{2+} extrusion mechanism. Two well-studied mechanisms leading to increased Ca^{2+} influx include: (1) Ca^{2+} entry through VDCCs, and (2) activation of EAA receptor-gated channels.

33.4.1 Immune-Mediated Changes in Voltage-Dependent Calcium Channels

Evidence in favor of increased Ca^{2+} influx through VDCC in ALS has been reported by Appel and colleagues.[20–23] Initial studies by this group demonstrated that immunoglobulins from the sera of ALS patients stimulated an increase in muscle miniature endplate potential frequency in an *in vitro* mouse muscle preparation, presumably by increasing Ca^{2+} influx into the presynaptic terminals of motoneurons leading to increased quantal release of acetylcholine.[20] Later studies reported that IgG fractions from ALS patients reduced the amplitudes of the dihydropyridine-sensitive (L-type) currents of skeletal muscle.[21,22] In more recent work, however, reports from the same group indicate that ALS IgG will increase Ca^{2+} influx through P-type VDCCs.[23]

An independent group has reported that ALS sera reduces the amplitude of currents due to VDCCs in rat cerebellar granule cells.[24] Serum fractions containing either IgG, IgM, or other serum proteins from three of four ALS patients produced a 30 to 100% reduction in currents due to VDCCs, compared to fractions from two control sera. Reductions of Ca^{2+} current were seen in both the low- and high-threshold currents, although these currents could not be distinguished pharmacologically. This study by Zhainazarov and colleagues is hampered by the small number of serum samples studied. Furthermore, the results are confusing as Ca^{2+} currents were modified not only by ALS IgG, but also by serum proteins other than IgG.[24] Although work on the relation between ALS immunoglobulins and Ca^{2+} influx is of interest, there has been almost no confirmation of the effects of ALS IgG on VDCCs by other investigators. Furthermore, it is puzzling that ALS IgG is capable of so many distinct actions on VDCCs.

A study of $[Ca^{2+}]_i$ in cultured fibroblasts obtained from muscle biopsies of ALS patients and controls was similar with regard to the basal $[Ca^{2+}]_i$ levels between groups.[25] Lower values of $[Ca^{2+}]_i$ were achieved in fibroblasts from ALS patients following stimulation with neurotensin and brady-kinin, compared to controls. These data were interpreted by the authors to suggest an abnormality in G protein-related signal transduction in ALS. However, evidence for this possibility is scant.[25]

Cell lines having some properties similar to those of motoneurons have been used to elucidate the pathogenesis of ALS.[26] A recent report by Smith et al. [26] demonstrates that one type of motoneuron clone, VSC 4.1, has reduced viability when incubated with ALS IgG. The cytotoxicity associated with ALS IgG was dependent on the extracellular $[Ca^{2+}]$ and could be blocked by preincubation of immunoglobulins with L-type VDCCs or the purified α subunit of the L-type VDCC.[26] Selective blockers of N-type VDCCs (ω-conotoxin) and P-type VDCCs (Aga-IVa) atten-uated cell death in this clonal line. Viability was not affected by incubation with an antagonist of L-type VDCCs (nifedipine) or with Bay K8644 (an L-type VDCC activator). These data on VSC 4.1 further substantiate the role of ALS IgG as enhancing various VDCC currents in neurons. A recent report has claimed that the Ca^{2+}-induced cell death in this motoneuron cell line is due to apoptotic mechanisms.[27]

A confounding aspect to the general hypothesis that ALS IgG acts on VDCCs is that since many neuronal types have P- and N-type VDCCs, why are motoneurons affected in a relatively selective manner in ALS? One possible explanation for the selectivity of the Ca^{2+}-induced cyto-toxicity in ALS could be the absence of Ca^{2+}-binding proteins in motoneurons and other susceptible cells. One could hypothesize that although ALS may be characterized by increased Ca^{2+} influx through N- and P-type VDCCs in many neurons in the neuraxis, many of these neurons possess sufficiently high levels of calcium-binding proteins to effectively buffer the increased Ca^{2+} influx and protect against Ca^{2+}-induced cytotoxicity. It might be possible that motoneurons do not possess appreciable amounts of calcium buffering proteins.[28,29] Appel and colleagues have recently provided support for the view that reduced calcium buffering could be present in ALS with evidence that motoneuron populations which were lost early in ALS had lower levels of calbindin D-28K (CaBP) and parvalbumin (PV) immunoreactivity than motoneurons lost later in the disease.[28] Motoneuron-hybrid cell lines which have lower levels of CaBP and PV have reduced viability following exposure to ALS IgG than other cell lines with higher amounts of CaBP and PV immunoreactivity.[28] One limitation of this study is that it is difficult to compare cell lines with regard to cell survival and conclude that the difference in viability is related to a single factor, such as the presence of CaBP or PV, as cell lines may differ in many respects from one another.

In spite of the attractiveness of Appel's model of IgG-induced changes in Ca^{2+} influx producing motoneuron death, several problems remain. First, in the event that ALS IgG is responsible for increasing Ca^{2+} influx, one might expect that plasmaphoresis or immunosuppression would alter the course of ALS. There is very good evidence, however, that total body irradiation or other immunosuppressant techniques do not alter the clinical deterioration in ALS. It might be argued that IgG-induced cell damage might be present early in the course of ALS and by the time symptoms developed immunosuppression would be too late to change the clinical course. This would imply that calcium cytotoxicity could occur many months, or possibly years, prior to the progressive motoneuron death. In spite of this tantalizing possibility, we have no real evidence in favor of a "delayed Ca^{2+}-induced cytotoxicity" occurring over such long time periods. Secondly, the assumption that motor neurons and other cells might be more susceptible to calcium-induced cytotoxicity because of reduced calcium binding proteins or other buffering proteins is compelling, but need not be correct. Independent groups have recently demonstrated that the presence of CaBP or other intracellular Ca^{2+} buffers does not necessarily predict which cells will survive a Ca^{2+}-induced cytotoxic insult (Baimbridge, K.G., personal communication, 1995). Thirdly, the fundamental assumption that increased Ca^{2+} influx results in higher levels of $[Ca^{2+}]_i$ and neuron death in ALS is unproven. Potentially, an animal model of ALS created using antibodies to VDCCs might clarify if this is the case.

33.4.2 Excitatory Amino Acid-Mediated Elevation of $[Ca^{2+}]_i$

At least six independent lines of evidence have been presented to suggest that "glutamatergic dysfunction", or EAA receptor-mediated toxicity with presumptively augmented Ca^{2+} influx, may be responsible for the development of ALS. These lines of evidence include:

1. The association between exposure to known excitotoxins (e.g., domoic acid) and the development of motoneuronopathies in humans and primates.
2. Observations that elevated concentrations of glutamate and possibly other EAAs are present in the plasma and/or cerebrospinal fluid of patients who have died with ALS.
3. Observations of reduced tissue contents of glutamate and aspartate in the spinal cords and brains of patients who have died with ALS.
4. Evidence of decreased glutamate uptake by synaptosomes obtained from spinal cords and brains of ALS patients.
5. Autoradiographic evidence for altered numbers of EAA receptors in ALS patients.
6. Evidence of clinical improvement with therapies aimed at altering glutamate levels (e.g., Riluzole).

Some of the data supporting these claims have been reviewed recently (see Reference 30).

33.4.3 Role of Ca^{2+} Buffering, Intracellular Ca^{2+} Release or Altered Efflux

There have been no reports of altered release of Ca^{2+} from intracellular stores or impaired Ca^{2+} efflux in ALS as potential mechanisms for elevation of $[Ca^{2+}]_i$. As noted above, several studies have reported that comparatively low levels of CaBP and PV immunoreactivity are present in motoneurons of ALS and control subjects.[28,29] Low levels of Ca^{2+} binding proteins may contribute to the susceptibility of motoneurons to develop Ca^{2+}-induced toxicity. However, levels of these Ca^{2+} binding proteins do not appear to differ between ALS patients and controls.[29]

33.5 ACTIVATION OF PROTEIN KINASE C

A consequence of raising $[Ca^{2+}]_i$ is the activation of Ca^{2+}-dependent protein kinases such as the families of Ca^{2+}- and lipid-dependent protein kinases (PKC) and Ca^{2+}- and calmodulin-dependent protein kinases.

Evidence for altered activity of PKC in ALS was provided by a study evaluating the regulation of NMDA receptor binding in ALS and control spinal cord tissue.[31] In baseline studies it was found that [³H]MK-801 binding was reduced in spinal cords from ALS patients compared to controls.[31] To evaluate NMDA receptor regulation, spinal cord sections from ALS patients and controls were exposed to phorbol ester (an activator of PKC) before incubation with [³H]MK-801 to determine levels of NMDA receptor channel binding. Following incubation in phorbol ester, [³H]MK-801 binding was increased in both ALS and control spinal cord tissue to almost identical levels of specific binding for both groups.[31] The increase in [³H]MK-801 binding could be completely blocked by concurrent exposure of spinal cord sections to H-7, a general protein kinase inhibitor.

To further evaluate the involvement of PKC in ALS, PKC activity (histone H1 phosphotransferase activity) was determined in cytosolic and membrane-derived fractions from ALS and control spinal cord tissue following liquid chromatography. Phosphotransferase activity in one peak (Peak I) of mono Q fractions from both cytosolic and membrane-associated extracts from ALS and control spinal cord tissue was due to PKC. Significantly higher PKC activity was observed in spinal cords from ALS patients compared to controls; cytosolic and membrane-derived extracts from ALS patients showed 330% and 100% increases in PKC compared to control, respectively. A second fractionation peak (Peak II) had phosphotransferase activity which was largely independent of Ca^{2+} and lipid and resembled the properties of PKM, a 50-kDa protein generated by the limited proteolysis of PKC by calpain. PKM activity was also increased in cytosolic and membrane-derived extracts from ALS spinal cords compared to control, but did not achieve statistical significance.[32] Part of the elevated PKC activity could be accounted for by an increase in the amount of PKC protein.[32]

These data suggest that PKC is hyperactive in ALS. The differences in PKC activity between ALS and control tissue do not result from variability in the post-mortem delay or handling of the tissue, as these parameters were similar for ALS and control patients. Based on these observations, we presume that PKC activity is increased in ALS, likely as a result of raised $[Ca^{2+}]_i$. Elevated $[Ca^{2+}]_i$ could be generated by some of the mechanisms outlined above. Once persistently elevated, PKC may be capable of phosphorylating a number of cell surface receptors such as NMDA,[31] glutathione,[33] or other neurotransmitter receptors. Possibly, growth factor action could be affected by PKC, either at the level of the receptor or at subsequent steps in the signal transduction cascade. Other secondary effects of PKC would likely include activation of "downstream kinases" such as the mitogen-activated protein kinases (MAP kinases).[34] Also of possible importance in this scheme is the Ca^{2+}-dependent enzyme, calpain, which is responsible for the limited proteolysis of PKC to PKM. Our current studies are aimed at evaluating the extent of changes in PKC and PKM activities at different sites in the nervous system of ALS patients and controls.

33.6 CONCLUSION

Indirect evidence suggests that $[Ca^{2+}]_i$ may be altered in ALS. This evidence is broadly derived from data demonstrating:

1. That immunoglobulins from ALS patients can modify the amplitude of Ca^{2+} currents through VDCCs,
2. That impaired regulation of EAAs occurs in ALS which can potentially augment $[Ca^{2+}]_i$,
3. That motoneurons possess low levels of the calcium-binding proteins CaBP and PV, and
4. That increased activity of PKC is seen in ALS tissues.

This work must be supplemented by further studies aimed at evaluating the amplitude and time course of the change in $[Ca^{2+}]_i$ necessary to produce motoneuron dysfunction and cell death.

ACKNOWLEDGMENTS

Work described in this chapter is supported by the Amyotrophic Lateral Sclerosis Society of Canada and the Heart and Stroke Foundation of British Columbia. CK holds a scholarship from the Medical Research Council of Canada.

REFERENCES

1. Orrell RW and deBelleroche JS: Superoxide dismutase and ALS. *Lancet*, 1994; 344:1651–1652.
2. Krieger C, Jones K, Kim SU, and Eisen AA: The role of intracellular free calcium in motor neuron disease. *J. Neurol. Sci.* (Suppl.), 1994; 124:27–32.
3. Yase Y: Environmental contribution to the amyotrophic lateral sclerosis process. In Serratrice G, Cros D, Desnuelle C, et al. (eds): *Neuromuscular Diseases*. New York, Raven Press, 1984, pp 335–339.
4. Garruto RM, Shankar SK, Yanagihara R, et al: Low-calcium, high-aluminum diet-induced motor neuron pathology in cynomolgus monkeys. *Acta Neuropathol. (Berl.)*, 1989; 78:210–219.
5. Yasui M, Yase Y, and Ota K: Distribution of calcium in central nervous system tissues and bones of rats maintained on calcium-deficient diets. *J. Neurol. Sci.*, 1991; 105:206–210.
6. Melmed S and Braunstein GD: Endocrine function in amyotrophic lateral sclerosis. A review. *Neurol. Clin.*, 1987; 5:33–42.
7. Maiese K, Boniece IR, Skurat K, and Wagner, JA: Protein kinases modulate the sensitivity of hippocampal neurons to nitric oxide toxicity and anoxia. *J. Neurosci. Res.*, 1993; 36:77–87.
8. Exton JH: Mechanisms of action of calcium-mobilizing agonists: some variations on a young theme. *FASEB J.*, 1988; 2:2670–2676.
9. Dubinsky JM: Examination of the role of calcium in neuronal death. *N.Y. Acad. Sci.*, 1993; 679:34–42.
10. Mattson MP, Rychlik B, You JS, and Sisken JE: Sensitivity of cultured human embryonic cerebral cortical neurons to excitatory amino acid-induced calcium influx and neurotoxicity. *Brain Res.*, 1991; 542:97–106.
11. Tymianski M, Charlton MP, Carlen PL, and Tator CH: Source specificity of early calcium neurotoxicity in cultured embryonic spinal neurons. *J. Neurosci.*, 1993; 13:2085–2104.
12. Beal MF: Does impairment of energy metabolism result in excitotoxic neuronal death in neurodegenerative diseases? *Ann. Neurol.*, 1992; 312:119–130.
13. Millani D, Guidolin D, Facci L, et al: Excitatory amino acid-induced alterations of cytoplasmic free Ca^{2+} in individual cerebellar granule neurons: role in neurotoxicity. *J. Neurosci. Res.*, 1991; 28:434–441.
14. Dubinsky JM and Rothman SM: Extracellular calcium concentrations during "chemical hypoxia" and excitotoxic neuronal injury. *J. Neurosci.*, 1991; 11:2545–2551.
15. Michaels L and Rothman SM: Glutamate neurotoxicity *in vitro*: antagonist pharmacology and extracellular calcium concentrations. *J. Neurosci.*, 1991; 10:283–292.
16. Mattson MP: Evidence for the involvement of protein kinase C in neurodegenerative changes in cultured human cortical neurons. *Exp. Neurol.*, 1991; 112:95–103.
17. Fukunaga K, Soderling TR, and Miyamoto E: Activation of Ca^{2+}/calmodulin-dependent protein kinase II and protein kinase C by glutamate in cultured rat hippocampal neurons. *J. Biol. Chem.*, 1992; 267:22522–22527.
18. Kikkawa U and Nishizuka Y: The role of protein kinase C in transmembrane signalling. *Ann. Rev. Cell. Biol.*, 1986; 2:149–178.
19. Favaron M, Manev H, Siman R, et al: Downregulation of protein kinase C protects cerebellar granule neurons in primary culture from glutamate-induced neuronal death. *Proc. Natl. Acad. Sci. U.S.A.*, 1990; 87:1983–1987.
20. Uchitel OD, Appel SH, Crawford F, and Sczcupak L: Immunoglobulins from amyotrophic lateral sclerosis patients enhance spontaneous transmitter release from motor-nerve terminals. *Proc. Natl. Acad. Sci. U.S.A.*, 1988; 85:7371–7374.
21. Delbono O, Garcia J, Appel SH, and Stefani E: Calcium current and charge movement of mammalian muscle. Action of amyotrophic lateral sclerosis immunoglobulins. *J. Physiol.*, 1991; 444:723–742.
22. Smith RG, Hamilton S, Hofmann F, et al: Serum antibodies to L-type calcium channels in patients with amyotrophic lateral sclerosis. *N. Engl. J. Med.*, 1992; 327:1721–1728.
23. Llinas R, Sugimori M, Cherksey BD, et al: IgG from amyotrophic lateral sclerosis patients increases current through P-type calcium channels in mammalian cerebellar Purkinje cells and in isolated channel protein in lipid bilayer. *Proc. Natl. Acad. Sci. U.S.A.*, 1993; 90:11743–11747.
24. Zhainazarov AB, Annunziata P, Toneatto S, et al: Serum fractions from amyotrophic lateral sclerosis patients depress voltage-activated Ca^{2+} currents of rat cerebellar granule cells in culture. *Neurosci. Lett.*, 1994; 172:111–114.

25. Witt MR, Gredal O, Dekermendjian K, et al: Calcium homeostasis in fibroblasts from patients with amyotrophic lateral sclerosis. *J. Neurol. Sci.*, 1994; 126:206–212.

26. Smith RG, Alexianu ME, Crawford G, et al: Cytotoxicity of immunoglobulins from amyotrophic lateral sclerosis patients on a hybrid motoneuron cell line. *Proc. Natl. Acad. Sci. U.S.A.*, 1994; 91:3393–3397.

27. Alexianu ME, Mohamed AH, Smith RG, et al: Apoptotic cell death of a hybrid motoneuron cell line induced by immunoglobulins from patients with amyotrophic lateral sclerosis. *J. Neurochem.*, 1995; 63:2365–2368.

28. Alexianu ME, Ho BK, Mohamed AH, et al: The role of calcium-binding proteins in selective motoneuron vulnerability in amyotrophic lateral sclerosis. *Ann. Neurol.*, 1994; 36:846–858.

29. Ince P, Stout N, Shaw P, et al: Albumin and calbindin D-28k in the human motor system and in motor neuron disease. *Neuropathol. Appl. Neurobiol.*, 1993; 19:291–299.

30. Rowland LP: Riluzole for the treatment of amyotrophic lateral sclerosis — too soon to tell? *N. Engl. J. Med.*, 1994; 330:592–596.

31. Krieger C, Wagey R, Lanius RA, and Shaw CA: Activation of PKC reverses apparent NMDA receptor reduction in ALS. *Neuroreport*, 1993; 4:931–934.

32. Lanius RA, Paddon HB, Mezei M, et al: A role for amplified PKC activity in the pathogenesis of amyotrophic lateral sclerosis. *J. Neurochem.*, 1995; 65:927–930.

33. Lanius RA, Shaw CA, Wagey R, and Krieger C: Characterization, distribution, and protein kinase C-mediated regulation of [^{35}S] glutathione binding sites in mouse and human spinal cord. *J. Neurochem.*, 1994; 63:155–160.

34. Saitoh T, Masliah E, Jin LW, et al: Protein kinases and phosphorylation in neurological disorders and cell death, *Lab. Invest.*, 1991; 64:596–616.

Chapter 34

Spinal Canal Stenosis Secondary to Calcium Crystal Deposition Disease: Relationship Between Neurologic Symptoms and Locations of Crystal Deposition

Munehito Yoshida, Tetsuya Tamaki, and Yasunori Taniguchi

34.1 INTRODUCTION

Neuropathy caused by calcification of the spinal canal was first reported in 1976 by Nanko et al., who reported on spinal root disease caused by calcification of the cervical ligamentum flavum (CLF).[1] Reports on this type of neuropathy have been increasing in recent years, but few have examined neuropathy caused by calcification of structures other than the ligamentum flavum.[2–5] Calcification of extremity joints was previously recognized as chondrocalcinosis or pseudogout and was collectively referred to by Ryan and McCarty as calcium pyrophosphate dihydrate crystal deposition disease (CPPD-cdd).[6] As will be described below, cases of CLF complicated by calcification of articular cartilage are frequently encountered. It is therefore necessary to analyze calcification of the spinal soft tissue from the viewpoint of systemic calcium crystal deposition disease.[6] In the present study, we examined spinal canal stenosis caused by calcium crystal deposition and the resultant neurologic symptoms and pain in 25 cases of calcium crystal deposition disease of the spinal ligaments.

34.1.1 Subjects

All of the 25 patients studied here were identified for fulfilling the diagnosis of calcification of the spinal ligaments; 13 patients had calcification of the cervical spine and 12 had calcification of the lumbar spine. The cervical spine group was composed of 2 males and 11 females, with a mean age of 74.7 years (67 to 79 years). The lumbar spine group was composed of 2 males and 10 females with a mean age of 72.4 years (63 to 81 years). Four of the 25 patients had disease of both the cervical and lumbar spine. The locations and features of calcification were first examined using CT scans. In operated cases, the calcium crystal-deposited areas of the spine-supportive tissue were observed macroscopically and by microscopy of removed tissue samples. Surgically removed tissues were histologically examined, using HE and Kossa staining. Transmission electron microscopy was carried out on some samples. Crystal components were identified using X-ray diffraction analysis or corrective polarizing microscopy.

34.1.2 Results

Neurologic symptoms were observed in 24 cases, excluding 1 patient whose chief complaint was nuchal pain. The neurologic symptoms observed were cervical myelopathy (8 cases), cervical radiculopathy (4 cases), cauda equina syndrome (5 cases), and sciatica (7 cases); of these cases, 22 were treated surgically. The surgical techniques used for the cervical spine group were laminectomy (1 case) and expansive laminoplasty(11 cases). In the lumbar spine group, laminectomy (3 cases) and fenestration (7 cases) were carried out. When neurologic conditions were evaluated using the JOA scoring system,[7] before and after surgery, the average score for the cervical spine group was 6.4 before surgery and 10.8 after surgery, with a postoperative improvement of 41.5%. In the lumbar spine group in which objective symptoms were evaluated (full score = 15) by using the JOA scoring system of the lumbar spine,[8] the average score was 5.4 before surgery and 11.2 after surgery (an 80% improvement after surgery). Thus, postoperative improvement was less marked in the cervical spine group, probably because this group included more elderly patients and more severe cases, and because patients in this group were often complicated by calcification-induced arthropathy of the extremities.

34.2 FEATURES OF X-RAY IMAGES

Disc narrowing, vacuum phenomenon, sclerosis of the subchondral bone of the vertebral body, and changes similar to spinal bone proliferation were often revealed by X-rays. Spondylolisthesis and degenerative scoliosis were not uncommon. One patient showed destructive spondylopathy, accompanied by marked degenerative spondylolisthesis. Calcification of the disc was also visible, although it was not found in any patient with severe disc degeneration.

34.3 CT SCANS OF SPINAL CANAL STENOSIS

On CT images, calcium crystal deposition was clearly visible and had various forms (nodular, granular, linear, etc.). Calcification of the ligamentum flavum appeared to be responsible for spinal canal stenosis. However, when observed in more detail, it was rare that spinal canal stenosis at the cervical segments was caused only by calcification of the ligamentum flavum. In many cases, marked spinal canal stenosis of the cervical segments appeared to have been caused by the following factors: (1) posterior osteophytes due to spondylopathic changes, (2) granular calcification and thickening of the posterior longitudinal ligament, (3) bone proliferation accompanied by calcification of the intervertebral joint, and (4) others (Figure 34.1). Nodular calcification of the ligamentum flavum, which was often seen in the cervical spine group, was rare in the lumbar spine group. In the lumbar spine group, hypertrophy of the ligament, accompanied by granular calcification of various sizes, was often observed (Figure 34.2A). In this group, dural canal compression from the anterior direction was seen, accompanied by calcification of the disc and the posterior longitudinal ligament (Figure 34.2B). Degenerative spondylolisthesis was revealed in three cases, accompanied by calcification and proliferative changes of the intervertebral articular capsule. Disc calcification was noted in two cases from the cervical spine group and six from the lumbar spine group.

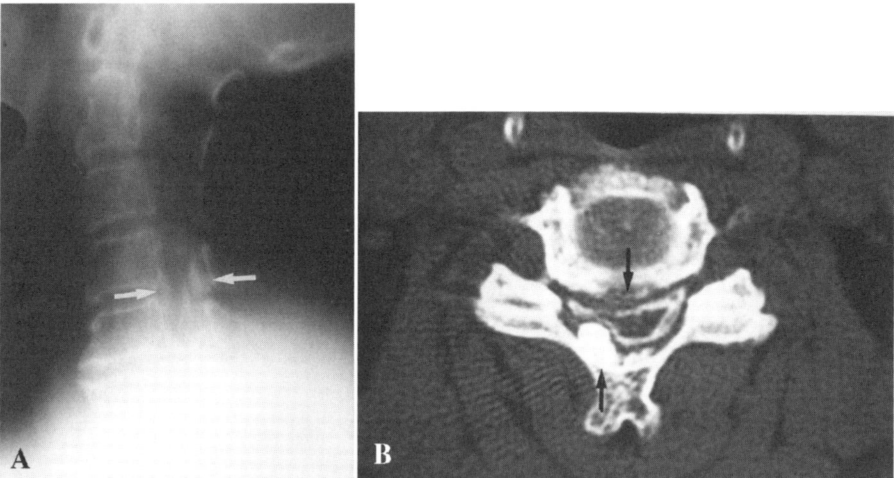

FIGURE 34.1   (A): 71-year-old female; tomograph of the cervical spine showing calcification of both the posterior longitudinal ligament and ligamentum flavum. (B): CT scan demonstrating the nodular calcification of the ligamentum flavum and calcification of the posterior longitudinal ligament.

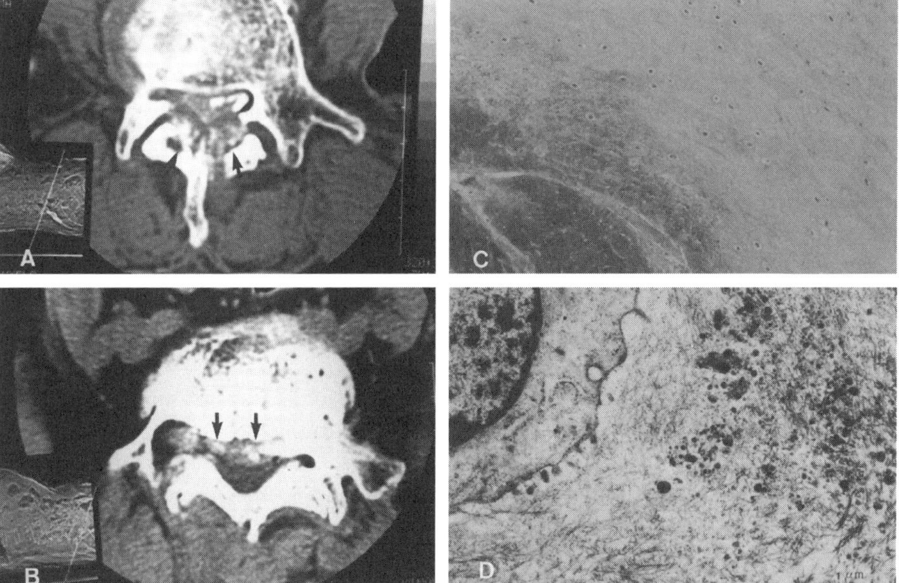

FIGURE 34.2   (A): 76-year-old female; CT scan showing the massive granular calcium deposition in the ligamentum flavum. (B): CT scan showing spinal canal stenosis secondary to calcification of the posterior longitudinal ligament. (C): Microphotograph demonstrating a number of chondrocytes around the calcium crystal depositions. (Hematoxylin-eosin staining, × 50). (D): Electromicroscopic photograph showing numerous matrix vesicles containing deposits with a high electron density (× 5000).

Calcification of the posterior longitudinal ligament was detected in six cases from the cervical group and six from the lumbar group. Calcification of the ligamentum flavum was seen in 11 cases from the cervical group and 10 from the lumbar group. Calcification of the intervertebral articular capsule was present in one case from the cervical group and four cases from the lumbar group. Calcification of the transverse ligament of the atlas was observed in one case (Table 34.1).

Table 34.1 Localization of Calcium Crystal Deposition in Histopathologic Findings and CT Scan

	Cervical (n = 13)		Lumbar (n = 12)	
	Histologic (n = 12)	CT (n = 13)	Histologic (n = 10)	CT (n = 12)
Intervertebral disc	—	2	6	6
Posterior longitudinal lig.	/	6	5	6
Lig. flavum	12	11	10	10
Capsule of facet joint	3	1	7	4
Interpinosus lig.	/	0	1	1
Transvers atlas lig.	/	1	/	/
Fibrous tissue	3	/	4	/

34.4 HISTOPATHOLOGY AND CALCIUM-DEPOSITED SITES

Surgical specimens were obtained in 12 cases from the cervical spine group and 10 cases from the lumbar spine group. When examined histopathologically, the calcium crystal-deposited sites were stained black with the Kossa stain. Crystals were clearly visible under a polarizing microscope. In the cervical spine group, only the posterior component of the cervical spine was collected in all cases. Calcium crystal deposition were seen in the ligaments flavum of 12 cases and the interspinous ligaments of 3 cases. In the lumbar spine group, calcium crystal deposition in the disc or the posterior longitudinal ligament was seen in five cases. Calcium crystals were present in the ligamentum flavum of all cases. Deposition in the capsule of the facet joint was noted in three cases from the cervical spine group and seven cases from the lumbar spine group. Another characteristic finding was fibrous bands, which were detected on the dura mater or the nerve root in three cases from the cervical spine group and four cases from the lumbar spine group. The area showing fibrous bands had signs of chronic inflammation (including macrophages) and calcium crystal deposits (Table 34.1). When each ligament was examined histopathologically, the ligamentum flavum was found to contain many calcium crystal deposits and to show intense degenerative hypertrophy at a low magnification. At a higher magnification, this ligament was observed to contain characteristic granulation tissue including foreign-body giant cells around the calcified site and showed signs of chronic inflammation, including lymphocytes. The posterior longitudinal ligament also showed a number of calcium crystal deposits even at a low magnification. The disc and the capsule of the facet joint were also found to have calcium crystal deposits and granulation tissue around the deposits. A number of chondrocytes were visible around the calcium crystals of each tissue (Figure 34.2C) When the areas showing chondrocytes were observed under an electron microscope, numerous matrix vesicles were seen around the chondrocytes, and the vesicles contained deposits with a high electron density (Figure 34.2D).

34.5 JOINTS SHOWING CHONDROCALCINOSIS

In 24 cases, the joints of the extremities were examined for chondrocalcinosis, using radiography. Chondrocalcinosis was most frequently seen in the knee meniscus (18 cases, 75%), followed by the wrist joint (6 cases, 37.5%), and the pubic symphysis (6 cases, 37.5%). Other areas which frequently showed chondrocalcinosis include the hip joint, the shoulder, and the elbow joint. In total, the incidence of chondrocalcinosis was as high as 83.3% (20/24).

34.6 COMPOSITION OF CALCIUM CRYSTALS

The composition of calcium crystals was identified in 16 cases, using X-ray diffraction analysis and corrective polarizing microscopy: the crystals were composed of calcium pyrophosphate dihydrate (CPPD) in 10 cases, hydroxyapatite (HAP) in 3 cases, and both in 3 cases.

34.7 DISCUSSION

It is believed that the first report on calcium crystal deposition in spine-supportive tissue was published in 1858 by von Luschka.[9] Later, Taylor and Little[5] demonstrated HAP crystal deposition in the disc. Bywaters et al.[10] found CPPD crystal deposits in the disc for the first time. Richards and Hamilton[11] examined in detail changes in the spinal vertebrae associated with chondrocalcinosis. Regarding cases complicated by neurologic symptoms, Nanko et al.[1] reported cases of spinal root disease associated with calcification of the cervical yellow ligament for the first time. Dehais et al.[12] reported cases of lumbar spinal canal stenosis caused by CPPD-cdd of the lumbar yellow ligament. In the same year, Taillard and Lagier[13] reported a case of degenerative spondylolisthesis of the L4 caused by chondrocalcinosis. Thereafter, more than 100 cases of yellow ligament calcification have been reported.[14] In addition, cases of CPPD-cdd of the syndesmo-odontoid joint of the craniocervical junction have also been reported, although this condition is rare.[3] A general evaluation of our CT findings, macroscopic findings during surgery, and histopathological findings indicates that calcium crystal deposition occurs not only in the disc and the yellow ligament but also in the posterior longitudinal ligament, the capsule of the facet joint, the interspinous ligament, and the epidural fibrous tissue, leading to the induction of spinal canal stenosis.[15] Resnick and Pined[16] also reported a detailed study of 1000 autopsied cases, and found CPPD crystal deposition in each supportive tissue of the spinal column. Their study, however, only suggested that neuropathy can be induced by CPPD crystal deposition. It did not present such a case.

In the present study, evident neurologic symptoms were seen in 24 cases. The conditions of these cases suggest that hypertrophy of the calcium crystal-deposited ligament is responsible for neurologic symptoms. It has been reported that reactive cells (related to chronic inflammation) and foreign-body type giant cells infiltrated the calcium crystal-deposited ligaments, resulting in hypertrophy of these ligaments, accompanied by granulation.[17] It is therefore likely that this sequence of changes induces spinal canal stenosis and the appearance of symptoms of nerve compression, nerve stimulation, or denervation.

Two phenomena can develop following calcium deposition of the disc: (1) tissue destruction, which can progress into disc degeneration;[11] and (2) pain, resembling the pain-associated subacute disc inflammation as reported by Cottin et al.[18] We cannot rule out that the first of these two phenomena can lead to the appearance of characteristic vacuum phenomenon, sclerosis of the subchondral bone, degenerative scoliosis, ankylosing spondylopathy, etc.[19–21] The second phenomenon seems to serve as an origin of the onset of nuchal and back pain as seen in cases of acute calcification of the disc. The transfer of calcium crystals to the epidural cavity seems to be another mechanism for the induction of pain, i.e., pain caused by calcium crystal-induced inflammation. In seven cases, we demonstrated the presence of calcium crystal deposition and signs of chronic inflammation in the epidural fibrous tissue. In view of one of our cases, with a chief complaint of nuchal pain, and the cases reported by Yasukawa et al.,[22] this finding suggests that the calcified lesion can spontaneously reduce in size. We speculate that the calcium crystals which flow into the epidural cavity can originate from the disc, the posterior longitudinal ligament, the yellow ligament, and even the intervertebral joint. Calcium crystal deposition of the intervertebral joint is thought to be responsible not only for facet pain but also for the onset of spondylolisthesis secondary to crystal-caused articular destruction.[13] Le Goff et al.[23] reported a case showing meningitis-like symptoms due to CPPD-cdd of the atlantoaxial articulation. It needs to be borne in mind that destructive spondylopathy can develop, although rarely, in cases of diffuse deposition.[24]

A significant question is whether or not we can regard calcium crystal deposition of the spinal vertebrae as being equal to category CPPD-cdd used by Ryan and McCarty.[6] To solve this question, we reviewed 100 reported cases of calcification of the yellow ligament (the most frequent type of calcification).[14] Crystals had been identified in 48 of these 100 cases, including 23 cases of CPPD, 14 cases of HAP, 8 cases of CPPD + HAP, and 3 other cases. This result was similar to that of the analysis of our cases. The incidence of chondrocalcinosis of the extremity joints was higher in the present study (83%) than that reported in the literature (32%). This is probably because we additionally carried out radiography of the extremity joints on 24 cases to determine the incidence of chondrocalcinosis more accurately. Our figure suggests that the incidence of complication by chondrocalcinosis is higher than has been reported previously.

These results allow us to conclude that calcium crystal deposition should be regarded as a systemic disease, and that spinal lesions are a part of such a systemic disease. Because calcium crystals were found to be composed either of CPPD alone, HAP alone, or CPPD + HAP, this disease can be divided into the following three types if the terms used in the literature are employed: (1) CPPD-cdd,[6] (2) basic calcium phosphate-cdd,[25] and (3) mixed crystal deposition disease.[26] Ryan and McCarty[6] divided CPPD-cdd into four subtypes: familial, idiopathic, abnormal metabolism, and posttraumatic/postoperative types. All of our cases seem to correspond to the idiopathic type, because none had any abnormalities in their family history, biochemical parameters, metabolism, or hormones.

It was recently reported that CPPD crystals in the ligaments can change into the more stable hydroxyapatite.[27,28] This report allows us to explain why some cases have crystals composed of both CPPD and HAP. However, since there are still a number of open questions, we cannot at present use the term CPPD-cdd to collectively indicate the condition of our cases, and hence use the term calcium crystal deposition disease, tentatively.

Our cases had numerous chondrocytes in the vicinity of calcified areas and a number of matrix vesicles around these chondrocytes. These findings allow us to propose a mechanism of crystal formation. That is, we assume that calcification begins with these matrix vesicles and then spreads to degenerative collagen fibers.

Following the recent increase in the percentage of elderly people, this disease has been increasing in significance because it is often seen in elderly women. Its pathogenesis, however, remains unknown. Systemic studies of ligament calcification and other phenomena seem to be indispensable to clarify its etiology.

34.8 CONCLUSION

1. Clinical symptoms and pathophysiological features of 25 cases of calcium crystal deposition disease (of the cervical vertebrae in 13 cases and the lumbar vertebrae in 12 cases) were reported.
2. CT scans and histological examination revealed calcification not only in the yellow ligament but also in the disc, posterior longitudinal ligament, the intervertebral articular capsule, the interspinous ligament, the atlantic transverse ligament, and the epidural fibrous band.
3. By these findings, it is concluded that the calcium crystal deposition in a variety of vertebral structures is responsible for the pathogenesis of spinal canal stenosis, which results in spinal symptoms and signs.

REFERENCES

1. Nanko S, Takagi A, Mannen T, et al: Myelo-radiculopathy secondary to calcification of the ligamentum flavum in the cervical spine. *Shinkeinaika* (in Japanese) 1976; 4:205-210.
2. Berghausen E, Balogh K, Landis WJ, et al: Cervical myelopathy attributable to pseudogout. *Clin. Orthop.*, 1987; 214:217–221.

3. Dirheimer Y, Bensimon C, Christmann D, and Wackenheim C: Syndesmo-odontoid joint and calcium pyrophosphte dihydrate deposition disease(CPPD). *Neuroradiology*, 1983; 5:319–321.

4. Ellman MH, Vazquez T, Ferguson L, and Mandel N: Calcium pyrophosphate deposition in ligamentum flavum. *Arthritis Rheum.*, 1978; 21:611–613.

5. Taylor TKF and Little K: Calcification in the intervertebral disc. *Nature*, 1963; 199:612.

6. Ryan LM and McCarty DJ: Calcium pyrophosphate crystal deposition disease; pseudogout; articular chondrocalcinosis. *Arthritis and Allied Condition*. 11th ed. Lea & Febiger, Philadelphia, 1989, pp 1711–1736.

7. Hirabayashi K: Scoring system for cervical myelopathy (Japanese Orthopaedic Association). *J. Jpn. Orthop. Assoc.*, 1994; 68:134–147.

8. Inoue S: Assessment of treatment for low back pain (Japanese Orthopaedic Association). *J. Jpn. Orthop. Assoc.*, 1986; 60:391–394.

9. Weinberger A and Myers A: Intervertebral disc calcification in adults: a review. *Semin. Arthritis Rheum.*, 1978; 8:69–75.

10. Bywaters EGL, Hamilton EBD, Williams R, et al: The spine in idiopathic haemochromatosis. *Ann. Rheum. Dis.*, 1971; 30:453–465.

11. Richards AJ and Hamilton BD: Spinal changes in idiopathic chondrocalcinosis artiocularis. *Rheumatol. Rehabil.*, 1976; 15:138–142.

12. Dehais J, Senegas J, Bauduceau B, et al: Stenose du canal Rachidien lumbaire au cours d'une chondrocalcinose articulaire diffuse. *Rev. Rhum. Mal. Osteoartyic.*, 1977; 44:585–588.

13. Taillard W and Lagier R: Pseudo-spondylolisthesis et chondrocalcinose. *Rev. Chir. Orthoped.*, 63:149–156, 1977.

14. Yoshida M, Funaoka N, Taniguchi Y, et al: Calcification of the ligamentum flavum of the cervical spine. *Central Jpn. J. Orthopaed. Surg. Traumatol.*, (in Japanese) 1991; 34:1333–1334.

15. Arnordi CC: Lumbar spinal stenosis and nerve root entrapment syndromes; definition and classification. *Clin. Orthop.*, 1976; 115:4–5.

16. Resnick D and Pined C: Vertebral involvement in calcium. pyrophosphate dihydrate crystal deposition disease. *Radiology*, 1984; 153:55–60.

17. Yoshida M, Shima K, Taniguchi Y, et al: Hypertrophied ligamentum flavum in lumbar canal stenosis — pathogenesis and morphologic and immunohistochemical observation. *Spine*, 1992; 17:1353–1360.

18. Cottin S, Le Gall G, Lanoiselee JM, and Rault D: Les pseudospondlodiscites de la chondrocalcinose articulaire diffuse. *Nouv. Presse Med.*, 1980; 9:1827–1830.

19. Layfer LF, Katz R, and Golden H: Chondrocalcinosis simulating ankylosing spondylitis. *J. Am. Med. Assoc.*, 1978; 240:55–56.

20. Okazaki T, Saito T, Mitomo T, and Siota Y: Pseudogout. Clinical observation and chemical analyses of deposits. *Arthritis Rheum.*, 1976; 9:293–305.

21. Zitnan D and Sitaj S: Chondrocalcinosis articularis. Section 1, clinical and radiological study. *Ann. Rheum. Dis.*, 1963; 22:142–169.

22. Yasukawa Y, Akizuki S, Wada T, and Takizawa T: Calcification of ligamentum flavum of cervical spine with unusual clinical symptoms and course. *Seikeigeka*, (in Japanese) 1990; 41:1968–1969.

23. LeGoff P, Pennec Y, and Youinou P: Signes cervicaux aigus, pseudo-meninges, revelateurs de la chondrocalcinose articulaire. *Semin. Hop. Paris*, 1980; 56:1515–1518.

24. Doherty M and Dieppe P: Crystal deposition disease in the elderly. *Clin. Rheum. Dis.*, 1986; 12:97–116.

25. Halverson PB and McCarty DJ: Basic calcium phosphate (apatite, octacalcium, tricalcium phosphate) crystal deposition diseases. *Arthritis and Allied Conditions*. 11th ed., Lea & Febiger, Philadelphia, 1989, pp 1737–1755.

26. Dieppe PA: Mixed crystal deposition disease and osteoarthritis. *Br. Med. J.*, 1978; 1:150.

27. Kawano N, Matsuno T, Miyazawa S, et al: *Nousinkeigeka* (Japanese) 1987; 15:181–190.

28. Kubokura T, Sanno N, Yagishita S, and Ito Y: Ultrastructural study of calcification of the ligamentum flavum of the cervical spine. Report 1. *J. Clin. Electron Microsc.*, 1988; 21:267–275.

Chapter 35

Epilepsy, Calcium, and Minerals

Elizabeth J. Waterhouse and Robert J. DeLorenzo

CONTENTS

35.1 INTRODUCTION

A number of minerals have been implicated in the pathophysiology of seizures and the development of epilepsy. Due to its multiple roles in neurotransmission and intracellular signaling mechanisms, calcium is the most important among them.[1] Thus, this chapter will review neuronal mechanisms of calcium regulation and homeostasis, intracellular calcium-dependent second messenger systems, and calcium systems as they relate to epilepsy and antiepileptic drug therapy, and evaluate the role of other minerals in the pathophysiology of epilepsy.

35.2 ROLE OF CALCIUM IN NEURONAL FUNCTIONING

35.2.1 Calcium and Excitatory Amino Acids

Over the past 10 years, numerous advances have been made in understanding the role of excitatory amino acids (EAA) in neurotransmission. EAA systems play a major role in the pathophysiology of excitotoxic injuries, and offer promise for pharmacologic intervention. The EAAs glutamate and aspartate are the major excitatory neurotransmitters in the central nervous system. They bind to several distinct receptors, classified according to the agonists that selectively activate them. The major EAA receptor subtypes include postsynaptic AMPA, kainate, and N-methyl-D-aspartate (NMDA) and metabotropic receptors,[4–6] as well as kainate and LAP-4 receptors which may be located presynaptically. Multiple isoforms of the AMPA and kainate receptor subunits have been cloned, and a differential expression of subunits between anatomic regions has been demonstrated.[7,8] Interestingly, NMDA and kainate receptors are highly enriched in the hippocampus,[6] an area of the brain that has been implicated in the generation of some types of seizures.

Each of these EAA receptors is associated with a cation channel that opens in response to agonist binding, allowing the cell to depolarize. The AMPA receptor is coupled to a monovalent cation channel that conducts sodium inward and potassium outward. The NMDA receptor acts as an ionophore, and is an important mechanism of entry of both extracellular calcium and sodium. It is activated by binding of an agonist ligand, and the resulting partial membrane depolarization removes tonic blockade of the cation channel by magnesium. The channel is then highly permeable to calcium ions as well as monovalent cations.

The metabotropic glutamate receptor generates calcium flux by liberating calcium from intracellular stores such as the endoplasmic reticulum. Activation of this receptor initiates a metabolic cascade in which phosphoinositol bis-4,5-phosphate is cleaved to diacylglycerol (DAG) and inositol polyphosphates such as inositol 1,4,5-triphosphate (IP_3).

The kainate and LAP-4 receptors may be situated presynaptically. The kainate receptor controls a monovalent cation channel, but less is known about LAP-4 receptors, which may act as autoreceptors that inhibit glutamate release when activated.

35.2.2 Calcium Channels

In recent years, calcium channels have generated considerable interest because of the important second messenger properties of calcium. There are two major categories of calcium channels: voltage regulated and receptor regulated. Three major types of voltage-regulated calcium channels have been well characterized: the T, L, and N channels. These channels differ in their location, pharmacological antagonists, and channel properties.

L-type calcium channels are sensitive to dihydropyridine calcium channel blockers, such as nimodipine and nitrendepine. They are also strongly blocked by inorganic divalent cations such as cadmium and cobalt. They are located postsynaptically, predominantly on the soma. L-type calcium channels are opened by strong depolarizations and inactivate slowly. They have relatively large conductances. L-type calcium channel activities are thought to be modulated by internal accumulation of calcium ions as well as by cyclic AMP-dependent phosphorylation.[9,10]

N-type calcium channels are located presynaptically, and although they are also sensitive to cadmium blockade they are resistant to dihydropyridines. N-type channel activities are inhibited by several neuroactive substances, including acetylcholine, GABA, norepinephrine, opioids, and 5-hydroxytryptamine.[11] Compared to L-type channels, N-type channels are activated by slightly weaker depolarizations, have a faster inactivation rate, and have decreased conductances.

T-type calcium channels are situated in the thalamus, pontine reticular formation, and inferior olive[12] where they are thought to mediate intrinsic oscillatory activity. They are also located on hippocampal pyramidal cells.[13] T-type channels are activated by weak depolarizations, rapidly inactivate, and have relatively small conductances. Unlike N- or L-type channels, they are not affected by dihydropyridines or divalent cations.

35.2.3 Intracellular Calcium-Dependent Second Messenger Systems

In excitatory neurotransmission neuron depolarization is associated with calcium spikes, transient elevations of intracellular calcium. This activity usually leads to activation of numerous enzymes including protein kinases, proteases, phosphodiesterases, phosphatases, phospholipases, and regulatory proteins, such as synapsin. In addition, there is direct and indirect modulation of a number of calcium-dependent ion channels. Some of the metabolic cascades initiated by the calcium influx are tightly regulated and rarely impinge on other calcium-activated pathways. Other cascades are comprised of less specific messengers, creating the opportunity for substantial "cross-talk" between signal transduction pathways.

The cytosolic calcium binding protein calmodulin serves as an important intermediary in neuronal signal transduction. During calcium influxes into the neuron, calmodulin becomes activated as it binds calcium ions. It then becomes a messenger and may successively activate a variety of calmodulin-dependent processes. One of the most important is the activation of type II calcium/calmodulin-dependent kinase (CaM Kinase II).

35.2.4 Calmodulin-Kinase II

Calmodulin-Kinase II (CaM Kinase II) has been implicated in the mediation of several second messenger effects of calcium in neurons, both pre- and postsynaptically. Reported roles include modulation of neurotransmitter release through vesicle-membrane interaction, synaptic function, cytoskeletal architecture, and neuronal excitability.[14,15] CaM Kinase II is highly enriched in mammalian brain, constituting 0.8 to 1% of total brain protein. In regions such as the hippocampus, which is important in learning and memory functions as well as in the development of seizures, CaM Kinase II is even more concentrated, constituting almost 2% of total protein weight. At the cellular level, CaM Kinase II is concentrated at the synapse.

The most prominent substrates that CaM Kinase II phosphorylates are microtubule-associated protein II (MAP II) and synapsin I. MAP II is a cytoskeletal component involved in intracellular transfer, and its phosphorylation state determines the extent of its polymerization. Presynaptic synapsin links neurotransmitter vesicles with the cytoskeleton. After the cell is depolarized, phosphorylation by CaM Kinase II reduces the affinity between synapsin and neurotransmitter vesicles, causing the vesicles to orient toward the cell membrane, and increasing the probability of vesicle-membrane fusion and neurotransmitter release.[16]

CaM Kinase II can regulate its own enzymatic function independently of calcium and calmodulin. It can undergo autophosphorylation, ultimately inhibiting its own enzymatic activity.[17] This intrinsic mechanism allows for initial signal amplification followed by signal termination. As with other phosphoproteins, the phosphorylation state of CaM Kinase II is also regulated by the activity of phosphatases.

35.2.5 Protein Kinase C and Phospholipase C

Protein Kinase C (PKC) refers to a family of isoenzymes which are activated by the simultaneous stimulation of diacylglycerol (DAG) and elevated cytosolic calcium. When phospholipase C cleaves phosphatidylinositol 4,5-bisphosphate, DAG and inositol polyphosphates such as IP_3 are generated. IP_3 in turn can independently liberate calcium from the endoplasmic reticulum. Phospholipase C isoforms are also linked by a G protein to a number of receptors, including metabotropic glutamate and muscarinic cholinergic receptors, and these receptors exert their effects via this pathway.[18] PKC, which is activated at the end of this cascade, has been implicated in regulating membrane conductances in hippocampal pyramidal cells.[19] PKC activation may also stimulate the release of neurotransmitters, including glutamate.[20]

35.2.6   Other Calcium-Dependent Pathways

In addition to the above-described calcium-dependent activities, many other intracellular processes depend on calcium. For instance, activated calmodulin can in turn activate adenylate cyclase, which leads to elevated cyclic AMP (cAMP) and activation of cAMP-dependent protein kinase. This protein kinase may play a role in the phosphorylation and activation of L-type calcium channels. Activated calmodulin may also play a role in desensitizing these channels, by activating the phosphatase calcineurin.[3,9] Calcium flux can activate calmodulin-stimulated phosphodiesterase, which degrades cAMP and other cyclic nucleotides.

Elevated intracellular calcium is associated with increased phospholipase A_2 activity. This enzyme generates arachidonic acid and its metabolites — prostaglandins and leukotrienes — both of which have been shown to modulate certain ion conductances. Arachidonic acid inhibits glial cell uptake of glutamate, and thereby may intensify excitatory amino acid effects.[21] In addition to these complex metabolic pathways, several membrane ion channels are regulated in a calcium-dependent fashion, including two types of potassium channels which can modulate neuronal excitability.

Finally, calcium-dependent processes may initiate a series of events leading to the expression of immediate early genes (IEG) and alterations in transcriptional and posttranscriptional programs for gene regulation. Calcium may therefore be involved in long-term alterations in neural function and adaptive mechanisms involved in neuronal plasticity. This subject will be further discussed later in this chapter.

35.3   CALCIUM SYSTEMS IN SEIZURES AND EPILEPSY

Seizures and interictal bursts occur as a result of increased excitation or diminished inhibition, and the mechanisms for effecting these changes include a variety of calcium-dependent processes.[1] Epilepsy is distinguished from seizures in that it is a syndrome of recurrent seizures, implying a permanent change in neuronal excitability. Epileptogenesis may occur through localized enrichments of certain effectors of the calcium systems.

Interictal discharges are the electrographic signature of epilepsy, and occur when a population of neurons synchronously generates burst discharges. At the individual cell level, the paroxysmal, high-voltage depolarization with superimposed fast and slow action potentials is termed a depolarizing shift (DS). Depolarizing shifts are terminated by hyperpolarization of the cell, and loss of this hyperpolarization occurs in the transition from the interictal to the ictal state.[22]

35.3.1   Influx of Extracellular Calcium

Prior to or during seizure onset, extracellular calcium decreases and extracellular potassium increases.[2] These shifts of extracellular calcium to the cell interior occur through the NMDA receptor ionophore and through voltage-regulated calcium channels, particularly the L-type. However, these channels allow calcium entry only after the cell membrane has been partially depolarized. Dichter[22] postulated that initial depolarization at the glutamatergic synapse may occur via sodium influx through the AMPA receptor cation channel, thereby relieving the NMDA receptor blockade by magnesium. Calcium influx could then contribute to neuronal depolarization and the late component of the excitatory postsynaptic potential (EPSP), in turn optimizing conditions for deactivation of L-type calcium channels. An alternative hypothesis is that high-frequency interictal bursting might lead to summation of non-NMDA receptor-mediated EPSPs, partially depolarizing the membrane, and thereby unblocking NMDA receptor-coupled channels.

35.3.2 Neuronal Excitability and Epileptogenesis

While the sequence of events described above applies to interictal bursting mechanisms and the occurrence of seizures, the process of epileptogenesis implies a sustained alteration in neuronal excitability that outlasts individual seizures. Long-term potentiation (LTP) is a model of the hyperexcitable state based on increased synaptic efficacy.

It is likely that there are differences between the mechanisms responsible for the induction and for the maintenance of LTP. In hippocampal pathways that rely on EEA neurotransmitters, nonpotentiating EPSPs can be blocked by nonselective EAA antagonists, while induction of LTP with high-frequency stimulation can be blocked by specific NMDA antagonists. After LTP has been induced, specific NMDA antagonists do not block generation of the potentiated response.[23] However, a specific antagonist of the AMPA receptor could depress potentiated responses after LTP has been established.[24] The activation of the NMDA receptor during LTP may lead to positive modulation of an AMPA subtype, and this modulation may occur via calcium as a second messenger.

There is a body of evidence supporting the theory that calcium is an essential second messenger in the sequence of events that leads to enhanced synaptic efficacy. After EGTA, a calcium chelator, was injected into postsynaptic cells, LTP could no longer be established.[25] Further studies have shown that at least two kinases, CaM Kinase II and PKC, are involved postsynaptically in the induction of LTP, while PKC is involved in the maintenance of LTP, possibly presynaptically.[26,27]

35.3.3 Direct Evidence That the Induction of Epileptogenesis Is Calcium Dependent

Primary hippocampal neuronal cultures have been utilized to investigate the properties of neuronal excitability and synaptic organization. A powerful *in vitro* model of "epilepsy" has been recently developed in populations of hippocampal neurons in culture.[28] Populations of neurons can be caused to develop recurrent seizures for the life of the neurons in culture. This *in vitro* model of epilepsy was used to investigate the processes of epileptogenesis in this preparation.

The induction of epileptogenesis in hippocampal neurons in culture was accomplished by a brief (3-h) treatment of the culture with media free of magnesium. Following return of the culture to normal media the neurons manifest recurrent seizure activity. Recent studies in this preparation have demonstrated that the induction of epileptogenesis is calcium dependent.[28]

These studies provide the first direct evidence that some forms of epileptogenesis are calcium dependent. Identification of the calcium-dependent second messenger systems mediating the long-lasting changes in neuronal function that underlie the process of epileptogenesis can be investigated in this *in vitro* model of epilepsy.

35.3.4 Calcium, Seizures, and Epilepsy in the Intact Brain

Kindling is an *in vivo* model of altered neuronal excitability and epilepsy. It involves the repeated administration of an initially subconvulsive stimulus, that ultimately elicits typical electrographic ictal discharges and clinical seizures. After the kindled state has been induced, the change in neuronal excitability is permanent.

Kindling and LTP share a number of similarities. McNamara et al. have found that MK-801, an NMDA receptor antagonist, retards the kindling process.[29] However, MK-801 only partially attenuates the kindled response after it has already been established, making it an ineffective anticonvulsant. In addition, activation of NMDA receptors mediates the loss of GABAergic inhibition, as measured by paired-pulsed depression of population spikes, an index of inhibitory efficacy.[30] This decrease in GABAergic inhibition is a long-lasting effect. The sequence of events that ultimately couples activation of the NMDA receptor with altered GABA-mediated inhibition has not yet been established. However, the conductance of hippocampal $GABA_A$ receptors is inhibited by a calcium-dependent process, such as the CaM-dependent phosphatase calcineurin, and is enhanced through phosphorylation by a kinase.

CaM Kinase II is another calcium/calmodulin-dependent system that is modulated in kindling. Several anticonvulsants (phenytoin, carbamazepine, and diazepam) inhibit CaM Kinase II activity *in vitro*, suggesting that this enzyme is involved in models of altered neuronal excitability.[14,15] However, the lack of a selective CaM Kinase II inhibitor suitable for *in vivo* applications has frustrated attempts to clarify the importance of this protein kinase in either the kindling process or in the generation of seizures in a kindled subject.

The loss of CaM Kinase II activity has also been associated with experimentally induced status epilepticus. Following 90 min of hippocampal stimulation — a model which culminates in limbic status epilepticus — CaM Kinase II activity was diminished.[31] The decrease in enzyme activity correlated with the intensity and complexity of electrographic discharges during the stimulation protocol, while the concentration of the enzyme itself remained stable. In this status epilepticus model, it is not clear whether loss of CaM Kinase II activity relates to increased neuronal excitability, or to excitotoxic processes which may already have been initiated at the time of tissue sampling. In either case, loss of CaM Kinase II activity could interfere with its capacity for autoregulation, as well as with the signal transduction pathways it subserves.

35.3.5　Calcium and Excitotoxicity

EAAs exert toxic effects on neurons in several ways. Sustained depolarization leads to massive influx of sodium ions, and this ionic shift can lead to the osmotic influx of water, and ultimately to swelling and osmotic lysis. This process of osmotic damage is associated with early cell death.

There is also evidence that calcium is involved in the neurotoxicity of EAAs.[32–34] When extracellular calcium is limited, there is improved survival of neurons after exposure to pharmacological levels of EAAs.[35] EAA neurotoxicity is also prevented when intracellular calcium is chelated.[36] Conversely, excessive calcium produces late excitotoxicity, or delayed cell death. Given the numerous calcium-dependent pathways, calcium-mediated delayed cell death is likely due to cellular dysregulation due to disruption of normal calcium homeostasis. It has been shown that kindled seizures generate transient but significant increases in cytoplasmic calcium,[37] and in experimental models of status epilepticus calcium accumulates in the cytoplasm and mitochondria.[38]

35.3.6　Calcium Binding Proteins and Calcium Buffering

The buffering capacity of calcium binding proteins helps to protect cells involved in kindling from excitotoxicity. The hippocampus is implicated in the origin of temporal lobe seizures, and is highly enriched in both NMDA receptors[39] and in CaM Kinase II.[40] After sustained perforant path stimulation of the rat hippocampus, cell loss is significantly greater among cells with low levels of the calcium buffering proteins parvalbumin and calbindin.[41] Dentate gyrus cells devoid of these proteins were protected from signs of deterioration when a selective calcium chelator was administered intracellularly.[36]

In kindling models of epilepsy, calcium buffering capacity of surviving neurons is impaired. Dentate gyrus granule cells taken from kindled animals show subnormal levels of calbindin-like activity. In addition, calcium currents in these cells inactivate more rapidly than in cells from control animals, and this effect is normalized by intracellular injection of a selective calcium chelator. It is not clear whether this finding constitutes a mechanism of, or a response to, increased excitability. However, in any situation in which the calcium buffering capacity of the neuron is overwhelmed, the calcium signal may activate other calcium-responsive pathways.

Calcium-dependent ion channels are another potential target for aberrant functioning of calcium-regulated second messengers. For example, after CaM Kinase II was injected into cultured spinal cord neurons, there was loss of the after-hyperpolarization that usually followed each spike.

Conversely, injection of a monoclonal antibody against CaM Kinase II was associated with enhanced hyperpolarization.[42] The loss of hyperpolarization could be mimicked by blocking two types of calcium-dependent potassium channels. This finding raises the possibility that these potassium channels are modulated by a calcium-mediated phosphorylation event, and that their dysregulation may play a role in the transition from interictal to ictal activity.

35.3.7   Calcium and Genetic Changes in Epilepsy

In addition to calcium's actions as a second messenger, it also has an important role in the regulation of gene expression. It has become increasingly apparent that the physiologic, and especially the pathologic, stimulation of neurons can initiate cascades of events that culminate in the transcription and translation of gene products that in turn influence the regulation of genes involved in normal neuronal functioning.[43] The ramifications for long-lasting neuronal changes are profound.

Immediate early genes (IEGs) were initially identified in dysregulated neoplastic tissue, and were described as protooncogenes. The protein products of these genes recognize motifs in the non-coding regulatory regions of genes, and promote or repress DNA transcription. An individual IEG may have a variable effect on transcription, depending on its interaction with other IEGs and the positional context of the recognition motifs along the DNA. In addition to the IEG c-*fos*, a number of additional IEGs have been induced in a variety of models of altered neuronal excitability.[44] In several seizure models (metrozol, picrotoxin, kainate) and in the kindling model of epilepsy, the IEG c-*fos* message is transiently elevated following an ictal event, especially in the dentate gyrus and pyriform cortex. This seizure-induced rise in c-*fos* is blocked by the NMDA antagonist MK-801.[45] The relationship between c-*fos* activity alone and neuronal plasticity is unclear. In the kindling model, c-*fos* levels are transiently elevated after seizures elicited during the kindling process as well as after spontaneous seizures in the kindled subject.[44]

Another group of agents that influence gene expression are the cAMP response element binding proteins (CREBs), which modulate transcription by interacting with recognition motifs known as cAMP response elements.[46] CREBs are indirectly activated by calcium influx, which initiates a CaM-mediated metabolic cascade, culminating in an increased production of cAMP and activation of PKA. PKA activates CREBs, leading to increased transcription of CREB-inducible genes. After PKA phosphorylation, CREB activity can be increased by phosphorylation by PKC, a kinase whose activity is also stimulated in the setting of calcium influx. CREBs may also be phosphorylated by CaM Kinase II.[47] CREBs regulate a number of genes.[46] The peptide products of these genes may act as neuromodulators, involved in seizure propagation.[48] The sustained accumulation of intracellular calcium may initiate these mechanisms of neuronal reprogramming. The ultimate importance of IEGs and transcriptional and posttranslational regulatory mechanisms derives from their capacity to control gene expression, and thereby impart long-term functional changes to a neuron's physiology.

It has been demonstrated that long-lasting changes in gene expression are observed in association with altered neuronal excitability in the kindling model of epilepsy.[49] These results provide some of the first direct evidence that conditions that induce long-term changes in brain excitability can produce enduring changes in the expression of specific gene products in neurons. The evidence also indicates that changes in gene expression associated with epilepsy are calcium induced. Thus, calcium may play a major role as a second messenger in modulating long-term changes in gene expression. These changes in genetic regulation may underlie several neurological disease processes such as epilepsy.

35.4 TRACE ELEMENTS AND EPILEPSY

In addition to calcium's direct effects, other ions can alter calcium-dependent aspects of neuronal function, such as neurotransmitter release and reuptake. Heavy metals such as lead, cadmium, and mercury inhibit acetylcholine release, possibly by reducing voltage-gated calcium entry into pre-synaptic nerve terminals.[50,51] Neurotransmitter uptake is inhibited by lead, copper, and mercury.[52,53]

These effects of metals have been observed *in vitro* and the exact mechanism of their neurotoxicity is not clear. Manganese, however, has reproducibly been shown to be associated with altered neuronal function *in vivo*. It is essential to many biologic functions, including glucose metabolism and dopaminergic activity in the brain. In experimental animals with manganese deficiency, clinical signs included neurologic problems as well as impaired growth and decreased reproductive capacity. In 1963, Hurley demonstrated that pregnant rats maintained on a manganese-deficient diet gave birth to offspring with increased susceptibility to electroshock-induced convulsions.[54] This susceptibility to seizures was reduced in the young rats after manganese supplementation.

Although manganese deficiency is rare in human beings, several researchers have demonstrated that people with convulsive seizure disorders tend to have lower levels of blood manganese.[55–57] Papavasiliou further showed an association between decreased blood manganese and seizure frequency, although this finding was not reproducible in a later study.[55]

It is not clear whether the low manganese concentration associated with epilepsy occurs because of seizure activity itself, or because of genetic factors associated with epilepsy. In studies of genetically epilepsy-prone rats, chronic seizures caused a decrease in whole-blood manganese levels, but not brain manganese concentrations. Brain glutamine synthetase activity, which requires Mn^{2+} or Mg^{2+} for activation, was unaffected by chronic seizures. However, its activity was significantly lower in genetically epilepsy-prone rats than in controls, suggesting that the abnormalities in manganese-dependent enzymes in epilepsy-prone rats are independent of seizure activity.[58]

It is possible that other divalent cations besides calcium may bind calmodulin, and therefore may have similar effects. Sutoo et al. found that serum levels of magnesium and zinc were lower and bone levels were higher in genetically epilepsy-prone mice compared to controls.[59] However, metal ion levels in whole brain were not significantly different between the two groups. This finding suggests the presence of other buffering systems for magnesium and zinc.

35.4.1   Metal Ions and Excitatory Amino Acids

The capacity of metal ions to produce epilepsy has been known and exploited for many years. Topical application of metallic substances to the brain is used in a number of seizure models. The substances used include cobalt,[60] alumina cream,[61] iron,[62] zinc,[63] and copper.[64] These substances induce electrographic epileptiform discharges in experimental animals, and sometimes induce spontaneous seizures. In other animal models of epilepsy, removal or reduction of metal ions has a protective effect. For example, chronic treatment with D-penicillamine, a metal chelator, protects photosensitive baboons from induced convulsions,[65] and zinc deficiency inhibits electrically induced seizures in mice.[66]

As with calcium, an excess of divalent metal ions can inhibit enzymes involved in the synthesis and degradation of neurotransmitters. After intense neuronal activity, metal ions are released into the extracellular space.[67–69] These ions may in turn affect the release or reuptake of neurotransmitters. Finally, metal ions in the extracellular space of the synapse may prolong the action of excitatory neurotransmitters[70] or antagonize the effect of GABA.[71]

Metal ions act as catalysts for some enzymes, and may play a regulatory role as a cofactor for other enzymes. Two of the enzymes essential in neurotransmitter synthesis are glutamate decarboxylase and glutamine synthetase, and both are inhibited by metal ions. Glutamate decarboxylase is the rate-limiting enzyme in the production of GABA, and requires Zn^{2+}. Glutamine synthetase requires Mg^{2+} or Mn^{2+} for activation, and is inhibited by Zn^{2+}. Therefore, it is likely that alteration in the concentrations of these metal ions can affect excitatory or inhibitory tone in the nervous system.

35.4.2 Metal Ions and Pyridoxal Phosphate

Pyridoxal-5^1-phosphate (PLP) is involved in a number of catalytic functions in amino acid metabolism, including aspartate transaminase, GABA transaminase, and glutamate decarboxylase. Pyridoxal phosphokinase is activated by Zn^{2+}, as well as by Mn^{2+}, Co^{2+}, Ni^{2+}, and Fe^{2+}.[72] One would expect that an increase in brain levels of PLP, either by administration or by stimulating the activity of pyridoxal kinase, would decrease GABA levels and enhance glutamate.[73] However, because numerous other PLP-dependent metal-sensitive enzymes are present, it is unlikely that the *in vivo* situation is that simple. The evidence suggests that the balance between GABA and glutamate may be influenced by metal ions and PLP, with excesses of these substances shifting the balance towards glutamate accumulation[64,74] and increased excitatory tone.

35.4.3 Release of Metal Ions Into Extracellular Space

It has been shown that metal ions are not always compartmentalized within neurons in the nervous system, and that under certain circumstances they are extruded into the extracellular space. For example, zinc ions in hippocampal tissues are released with electrical stimulation,[68] depolarization with high K^+,[67] and during kainate-induced seizures.[67,68] Transient increases in the extracellular concentration of metal ions could profoundly affect neuronal function, interfering with reuptake or release of neurotransmitters as well as with their postsynaptic actions. In addition, very high concentrations of metal ions could alter the pH of the microenvironment, leading to cellular damage.[73]

35.4.4 Metal Ions and Neurotransmitters in the Synapse

Divalent metal ions potentiate the release of excitatory amino acids. Spontaneous and veratrine-induced synaptosomal glutamate release is enhanced by Cu^{2+}, Fe^{2+}, and to a lesser extent, Zn^{2+}. Potassium-stimulated release is similarly influenced by the divalent cations.[73] On the other hand, metal ions also inhibit the uptake of neurotransmitters. They inhibit the uptake of catecholamines by adrenal medulla[75] and receptor binding of acetylcholine[76] and GABA.[77] Divalent cations similarly inhibit uptake of other neurotransmitters. Cu^{2+}, Zn^{2+}, and Fe^{2+} each inhibit the high-affinity uptake mechanisms of glutamate and GABA.[73] The effects of larger shifts in the concentrations of these divalent cations may therefore shift the balance of excitatory and inhibitory neuronal function.

35.4.5 Postsynaptic Receptors

The interaction between postsynaptic receptors and metal ions is being investigated. Smart and Constanti demonstrated that bath application of Zn^{2+} caused a prolongation of the excitatory postsynaptic potential, and irregular oscillation in the membrane potential.[70] In another study, the spontaneous firing rate increased following iontophoretic application of Zn^{2+}.[78] These studies demonstrate that zinc ions may modulate postsynaptic responses, but it is not clear whether the observed responses in these studies were due to direct action of the metal ions on the postsynaptic receptors, or due to effects on neurotransmitter release and reuptake, as previously discussed.

Clearly, the role of trace elements in neuronal function holds great interest, and future research may elucidate their role in seizures and epilepsy.

35.5 CONCLUSIONS

Calcium plays a crucial role not only in triggering seizures but also in the development of epilepsy. The systems discussed in this chapter contribute to the preservation of calcium homeostasis and maintenance of coordinated neuronal functioning. In these complex cascades of calcium-dependent events lies the potential for aberrations that may result in neuronal hyperexcitability. However, those sites of potential vulnerability are also potential sites for pharmacologic intervention. Anticonvulsants that are currently in use exploit some of these opportunities, and it is likely that future treatment strategies will utilize known calcium mechanisms both to prevent the development of epilepsy in high-risk patients (i.e., those with head injury or stroke) as well as to improve control of established seizure disorders. Greater knowledge of the many roles of calcium as a central second messenger in neuronal signal transduction will give us insight into the pathophysiological mechanisms of epilepsy, and broaden our understanding of the cellular and molecular basis of neuronal plasticity.

ACKNOWLEDGMENTS

We would like to thank Leslie Robinson for her work on preparing the manuscript. This work was supported by an NINDS Jacob Javits award (R01-NS23350) and Epilepsy Program Project (P01-NS25630) to RJD and the Sophie and Nathan Gumenick Neuroscience and Alzheimer Research Fund.

REFERENCES

1. Perlin JB and DeLorenzo RJ: Calcium and epilepsy. In: Pedley TA, Meldrum BS (eds): *Recent Advances in Epilepsy.* New York, Churchill Livingstone, 1992, pp. 15–36.
2. Meyer FB: Calcium, neuronal hyperexcitability and ischemic injury. *Brain Res. Rev.,* 1989, 14:227–243.
3. Kennedy MB: Regulation of neuronal function by calcium. *Trends Neurosci.,* 1989; 12:417–420.
4. Mayer ML and Westbrook GL: The physiology of excitatory amino acids in the vertebrate central nervous system. *Prog. Neurobiol.,* 1987; 28: 197–276.
5. Watkins JC: The NMDA receptor concept: origins and development. In: Watkins JC and Collingridge GL (eds.): *The NMDA Receptor.* IRL Press, Oxford, 1989, pp. 1–17.
6. Watkins JC, Krogsgaard-Larsen, and Honore T: Structure activity relationships in the development of excitatory amino acid agonist and competitive antagonists. *Trends Pharmacol. Sci.,* 1990; 11:25–33.
7. Boulter J, Hollmann M, O'Shea-Greenfield A, et al: Molecular cloning and functional expression of glutamate receptor subunit genes. *Science,* 1990; 249:1033–1037.
8. Sommer B, Keinanen K, Verdoorn TA, et al: Flip and flop: a cell-specific functional switch in glutamate-operated channels of the CNS. *Science,* 1990; 249:1580–1585.
9. Chad JE and Eckert R: An enzymatic mechanism for calcium current inactivation in dialysed *Helix* neurones. *J. Physiol. (London),* 1986; 378:31–51.
10. Yue DT, Backyx PH, and Imredy JP: Calcium-sensitive inactivation in the gating of single calcium channels. *Science,* 1990; 250:1735–1738.
11. Hess P: Calcium channels in vertebrate cells. *Annu. Rev. Neurosci.,* 1990; 13:337–356.
12. Coulter DA, Huguenard JR, and Prince DA: Calcium currents in rat thalamocortical relay neurones: kinetic properties of the transient low-threshold current. *J. Physiol. (London),* 1989; 414:587–604.
13. Tsien RW, Lipscombe D, Madison DV, et al: Multiple types of neuronal calcium channels and their selective modulation. *Trends Neurosci.,* 1988; 11:432–438.
14. DeLorenzo RJ: The calmodulin hypothesis of neurotransmission. *Cell Calcium,* 1981; 2:365–385.
15. DeLorenzo RJ: A molecular approach to the calcium signal in brain: relationship to synaptic modulation and seizure discharge. *Adv. Neurol.,* 1986; 44:435–464.
16. Llinas R, McGuiness TL, Leonard CS, et al: Intraterminal injection of synapsin I or calcium/calmodulin-dependent protein kinase alters neurotransmitter release at the squid giant axon. *Proc. Natl. Acad. Sci. U.S.A.,* 1985; 82:3035–3039.
17. Lou LL and Schulman H: Distinct autophosphorylation sites sequentially produce autonomy and inhibition of the multifunctional Ca^{2+}/calmodulin-dependent protein kinase. *J. Neurosci.,* 1989; 9:2020–2032.
18. Huang KP: The mechanism of protein kinase C activation. *Trends Neurosci.,* 1989; 12:425–432.
19. Baraban JM, Snyder SH, and Alger BE: Protein kinase C regulates ionic conductance in hippocampal pyramidal neurons: electrophysiological effect of phorbol esters. *Proc. Natl. Acad. Sci. U.S.A.,* 1985; 82:2538–2542.
20. Malenka RC, Madison DV, and Nicoll RA: Potentiation of synaptic transmission in the hippocampus by phorbol esters. *Nature,* 1986; 321:175–177.
21. Nicholls D and Atwell D: The release and uptake of excitatory amino acids. *Trends Pharmacol. Sci.,* 1990; 11:462–468.

22. Dichter MA: Cellular mechanisms of epilepsy and potential new treatment strategies. *Epilepsia*, 1989; 30 (Suppl. 1): S3–S12.

23. Collingridge GL, Kehl SJ, and McLennan H: The action of an *N*-methyl-aspartate antagonist on synaptic processes in the rat hippocampus. *J. Physiol. (London)*, 1983; 334:33–46.

24. Errington ML, Lynch MA, and Bliss TVP: Long-term potentiation in the dentate gyrus: induction and increased glutamate release are blocked by D(-)aminophosphonovalerate. *Neuroscience*, 1987; 20:279–284.

25. Lynch G, Larson J, Kelso S, et al: Intracellular injections of EGTA block induction of hippocampal long-term potentiation. *Nature*, 1983; 305:719–723.

26. Reymann KG and Matthies H: 2-Amino-4-phosphobutyrate selectively eliminates late phases of long-term potentiation in rat hippocampus. *Neurosci. Lett.*, 1989; 98:166–171.

27. Malinow RC, Schulman H, and Tsien RW: Inhibition of postsynaptic PKC or CaMKII blocks induction but not expression of LTP. *Science*, 1989; 245:862–866.

28. Sombati S and DeLorenzo RJ: Recurrent spontaneous seizure activity in hippocampal neuronal networks in culture. *J. Neurophysiol.*, 1995; 73(4):1706–1711.

29. McNamara JO, Russell RD, Rigsbee LC et al: Anticonvulsant and epileptogenic actions of MK-801 in the kindling and electroshock models. *Neuropharmacology*, 1988; 27:563–588.

30. Kapur J and Lothman EW: NMDA receptor activation mediates the loss of gabaergic inhibition induced by recurrent seizures. *Epilepsy Res.*, 1990; 5:103–111.

31. Perlin JB, Churn SB, Lothman EW et al: Continuous hippocampal stimulation decreases calcium-dependent protein kinase II activity. *Soc. Neurosci. Abstr.*, 1989; (Abstr. 181.10) 15:454.

32. Choi DW: Glutamate neurotoxicity in cortical cell cultures is calcium dependent. *Neurosci. Lett.*, 1985; 58:293–297.

33. Garthwaite G and Garthwaite J: Amino acid neurotoxicity: intracellular sites of calcium accumulation associated with the onset of irreversible damage to rat cerebellum neurons in vitro. *Neurosci. Lett.*, 1986; 71:53–58.

34. Siesjo BK: Calcium, excitotoxins and brain damage. *NIPS*, 1990; 5:120–125.

35. Rothman SM and Olney JW: Excitotoxicity and the NMDA receptor. *Trends Neurosci.*, 1987; 10:301–304.

36. Scharfman HE and Schwartzkroin PA: Protection of dentate hilar cells from prolonged stimulation by intracellular calcium chelation. *Science*, 1989; 246:257–260.

37. Kamphuis W, Huisman E, Wadman WJ, et al: Transient increase of cytoplasmic calcium concentration in the rat hippocampus after kindling-induced seizures: an ultrastructural study with the oxalate-pyro-antimonate technique. *Neuroscience*, 1989; 29:667–674.

38. Meldrum BS: Cell damage in epilepsy and the role of calcium in cytotoxicity. *Adv. Neurol.*, 1986; 44:849–855.

39. Collingridge GL and Lester RAJ: Excitatory amino acid receptors in the vertebrate central nervous system. *Pharmacol. Rev.*, 1989; 40:143–210.

40. Schulman H: The multifunctional Ca^{2+}/calmodulin-dependent kinase. In: Greengard P and Robison GA (eds): Advances in Phosphoprotein and Second Messenger Research. Raven Press, New York, 1988, pp 39–112.

41. Sloviter RS: Calcium-binding protein (calbindin-D_{28K}) and parvalbumin immunocytochemistry: localization in the rat hippocampus with specific reference to the selective vulnerability of hippocampal neurons to seizure activity. *J. Comp. Neurol.*, 1989; 280:183–196.

42. Sombati S, Forman RR, Attema BL, et al: Intracellular injection of CaM kinase II reduces spike after hyperpolarizations in cultured spinal cord neurons. *Soc. Neurosci. Abstr.*, 1988; (Abstr. 438.13) 14:1090.

43. DeLorenzo RJ: The challenging genetics of epilepsy. *Genet. Strategies Epilepsy Res.*, 1991; (Epilepsy Res. Suppl. 4): 3–17.

44. Dragunow M, Currie RW, Faull RLM, et al: Immediate-early genes, kindling, and long-term potentiation. *Neurosci. Biobehav. Rev.*, 1989; 13:301–313.

45. Labiner DM, Hosford DA, Cheolsu S, et al: An *N*-methyl-D-aspartate channel blocker, MK-801, inhibits seizure-induced rise in c-*fos* mRNA in hippocampus of kindled rats. *Epilepsia*, 1989; (Abstr. F-3) 30:697.

46. Montminy MR, Gonzalez GA, and Yamamoto KK: Regulation of cAMP-inducible genes by CREB. *Trends Neurosci.*, 1990; 13:184–188.

47. Sheng M, McFadden G, and Greenberg ME: Membrane depolarization and calcium induce c-Fos transcription via phosphorylation of transcription factor CREB. *Neuron*, 1990; 4:571–582.

48. Perlin JB, Lothman EW, and Geary WA, II: Somatostatin augments the spread of limbic seizures from the hippocampus. *Neurology*, 1987; 21 475–480.

49. Perlin JB, Gerwin CM, Panchision DM, et al: Kindling produces long-lasting and selective changes in gene expression of hippocampal neurons. *Proc. Natl. Acad. Sci. U.S.A.*, 1993; 90:1741–1745.

50. Cooper GP, Suszkiw JB, and Manalis RS: Heavy metals: effects on synaptic transmission. *Neurotoxicology*, 1984; 5:247–266.

51. Atchison WA, Clark AW, and Narahashi T: Presynaptic effects of methylmercury at the mammalian neuromuscular junction. In: Narahashi T, ed., *Cellular and Molecular Neurotoxicology*. New York, Raven Press, 1984, 23–42.

52. Komulainen H and Tuomisto J: Effects of heavy metals on monoamine uptake and release in brain synaptosomes and blood platelets. *Neurobehav. Toxicol. Teratol.*, 1982; 4:647–649.

53. Winder C and Kitchen I: Lead neurotoxicity: a review of the biochemical, neurochemical and drug-induced behavioral evidence. *Prog. Neurobiol.*, 1984; 22:59–87.

54. Hurley LS, Woolley DE, Rosenthal F, et al: Influence of manganese on susceptibility of rats to convulsions. *Am. J. Physiol.*, 1963; 204:493–496.

55. Papavasiliou PS, Kutt H, Miller ST, et al: Seizure disorders and trace metals: manganese tissue levels in treated epilepticus. *Neurology*, 1979; 29:1466–1473.

56. Dupont CL and Tanaka Y: Blood manganese levels in children with convulsive disorder. *Biochem. Med.*, 1985; 33:246–255.

57. Carl GF, Keen CL, Gallagher BB, et al: Association of low blood manganese concentrations with epilepsy. *Neurology*, 1986; 36:1584–1587.

58. Carl GF, Blackwell LK, Barnett FC, et al: Manganese and epilepsy: brain glutamine synthetase and liver arginase activities in genetically epilepsy prone and chronically seizured rats. *Epilepsia*, 1993; 34(3):441–446.

59. Sutoo D, Akiyama K, and Takita H: The relationship between metal ion levels and biogenic amine levels in epileptic mice. *Brain Res.*, 1987; 418:205–213.

60. Ross SM and Craig CR: γ-Aminobutyric acid concentration, L-glutamate-1-decarboxylase activity, and properties of the γ-aminobutyric acid postsynaptic receptor in cobalt epilepsy in the rat. *J. Neurosci.*, 1981; 1: 1388.

61. Heinemann U, Konnerth A, and Lux HD: Stimulation induced changes in extracellular free calcium in normal cortex and chronic alumina cream foci of cats. *Brain Res.*, 1981; 213:246.

62. Willmore LJ, Hurd RW, and Sypert GW: Epileptiform activity initiated by pial iontophoresis of ferrous and ferric chloride on rat cerebral cortex. *Brain Res.*, 1978; 152:406.

63. Itoh M and Ebadi M: Selective inhibition of hippocampal glutamic acid decarboxylase in zinc-induced epileptic seizures. *Neurochem. Res.*, 1982; 7:1287.

64. Chung SH and Johnson MS: Experimentally induced susceptibility to audiogenic seizure. *Exp. Neurol.*, 1983; 82:89.

65. Alley MC, Killam, EK, and Fisher GL: The influence of D-penicillamine treatment upon seizure activity and trace metal status in Senegalese baboon, *Papio papio, J. Pharmacol Exp. Ther.*, 1981; 217:138.

66. Tokuoka S, Takeshi F, Hiraoka H, et al: Neurochemical consideration on the alleviating effect of caudal resection of pancreas on epileptic seizures: relationship of zinc metabolism to brain excitability. *Bull. Yamaguchi Med. School*, 1967; 14:1.

67. Assaf SY and Chung SH: Release of endogenous Zn^{2+} from brain tissue during activity. *Nature*, 1984; 308:734.

68. Ben-Airi Y, Charton G, Leviel V, et al: Spontaneous and evoked release of Zn^{2+} in the Ammon's horn of the anesthetized rat. *J. Physiol.*, 1984; 357:40P.

69. Howell GA, Welch MG, and Frederickson CJ: Stimulation-induced uptake and release of zinc in hippocampal slices. *Nature*, 1984; 308:736.

70. Smart TG and Constanti A: Pre- and postsynaptic effects of zinc on *in vitro* prepyriform neurones. *Neurosci. Lett.*, 1983; 40:205.

71. Smart TG and Constanti A: A novel effect of zinc on the lobster muscle GABA receptor. *Proc. R. Soc. Lond. B*, 1982; 215:327.

72. McCormick DB, Gregory ME, and Snell EE: Pyridoxal phosphokinases. I. Assay, distribution, purification, and properties. *J. Biol. Chem.*, 1961; 236:2076.

73. Chung SH, Gabrielsson B, and Norris DK: Transition metal ions in epilepsy: an overview. *Adv. Exp. Med. Biol.*, 1986; 203:545–555.

74. Norris DK, Murphy RA, and Chung SH: Alteration of amino acid metabolism in epileptogenic mice by elevation of brain pyridoxal phosphate. *J. Neurochem.*, 1985; 44:1403.

75. Kirshner N: Uptake of catecholamines by a particulate fraction of the adrenal medulla. *J. Biol. Chem.*, 1962; 237:2311.

76. Hulme EC, Berrie CP, Birdsall NJM, et al: Regulation of muscarinic agonist binding by cations and guanine nucleotides. *Eur. J. Pharmacol.*, 1983; 94:59.

77. Baraldi M, Caselgrandi E, and Santi M: Effect of zinc on specific binding of GABA to rat brain membranes. In: Frederickson CJ, Howell GA and Kasarskis EJ (eds): *The Neurobiology of Zinc*. Alan R. Liss, New York, 1984, p. 59.

78. Wright DM: Zinc: effect and interaction with other cations in the cortex of the rat. *Brain Res.*, 1984; 311:343.

Chapter 36

A Possible Role of Calcium on Neurotransmitters and Biogenic Amines in Central Nervous System

Kiichiro Ota and Yukiharu Okamoto

CONTENTS

36.1 INTRODUCTION

Ca ($Ca^{2+)}$ ions play an essential role to regulate membrane excitability, enzyme activation, hormone and neurotransmitter secretion, and gene expression. Calcium channels, which open in response to membrane depolarization, are virtually ubiquitous in vertebrate cells. Ca^{2+} modulates the intracellular activity as a second messenger in nervous cells. In addition to calcium channels, *N*-methyl-D-aspartate (NMDA) receptor-associated ion channels are involved in the influx of Ca^{2+}, which affects signal transmission in the nerve cells.

One of the most important achievements of recent research in this field has been the recognition that different subtypes of voltage-operated calcium channels (VOCCs) are present not only in cells from different tissues, but also in individual cells.[1]

The development of new drugs and the discovery of natural toxins that act on calcium channel subtypes have given greater encouragement to pharmacologists, biologists, and even clinicians in projects aimed at characterizing these channels and studying the possible role of these molecules under physiological and pathological conditions.[2-4]

This brief review will describe the role of Ca^{2+} on neurotransmitters and biogenic amines in several disorders of the central nervous system (CNS).

36.2 CALCIUM-RELATED METABOLISM IN CNS TISSUES

Ca^{2+} ions trigger biological metabolic activities which relate directly or indirectly to the functions of the CNS. For example, following exposure to oxygen-free radicals generated by adding ferrous ions to the incubation mixture, Ca^{2+} uptake is increased to a greater extent in mitochondria from patients with Alzheimer's dementia than in controls.[4,5] These observations suggest that fibroblast mitochondria in Alzheimer's disease patients have impaired Ca^{2+} transport and increased sensitivity to oxygen-free radicals in CNS degeneration.[6] Thus, change in the concentration of intracellular

Ca^{2+} seems to be an important signaling mechanism which serves to communicate or act as a messenger between extracellular and intracellular events. As shown in Figure 36.1, glutamate binds to its receptor and activates phospholipase — this activity producing inositol triphosphate (IP_3) from polyphosphoinositides (PPIs). IP_3 production leads to an increment in cytoplasmic Ca^{2+} from release of the Ca^{2+} store (endoplasmic reticulum). Elevated cytosolic Ca^{2+} activates various proteases, lipases, and endonucleases in postsynaptic neurons, inducing neuronal death by an excitotoxic mechanism.

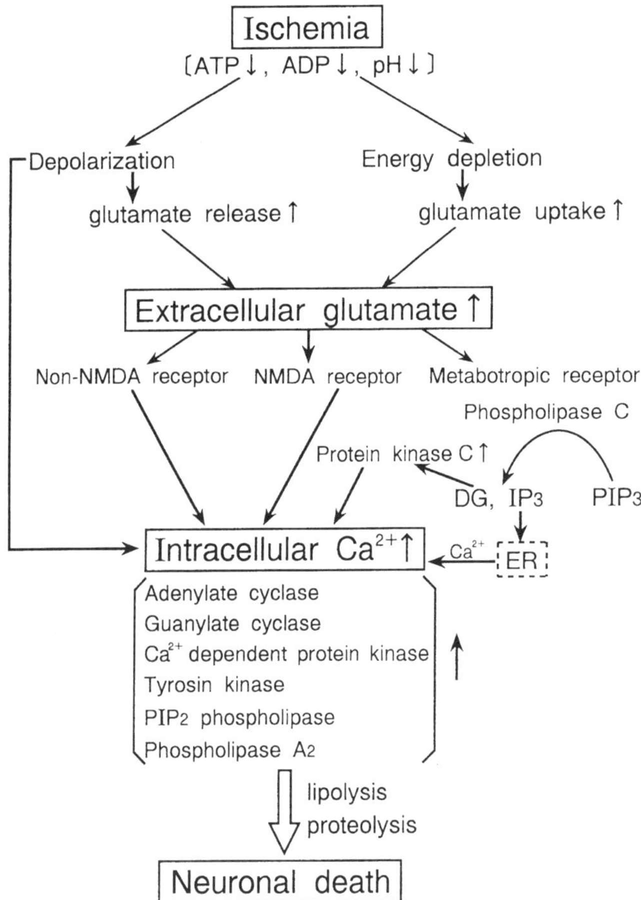

FIGURE 36.1 The tentative mechanism of neuronal death in nerve cell induced by ischemia. NMDA: *N*-methyl-D-asparate, DG: diacylglycerol, ER: endoplasmic reticulum, IP_3: inositol triphosphate.

A 113-ps tetraethylammonium (TEA)-sensitive K^+ channel is consistently absent from Alzheimer's disease (AD) fibroblasts, while it is often present in young and aged control fibroblasts. A second (166-ps) K^+ channel is present in all groups. Elevated external potassium raises intracellular Ca^{2+} concentration in young and aged control fibroblasts but not in AD fibroblasts. The invariable absence of a 113-ps TEA-sensitive K^+ channel dysfunction is seen in AD fibroblasts.

The proposed causes and/or predisposing factors for AD have ranged from defects in β-amyloid protein metabolism and environmental factors to abnormal Ca^{2+} homeostasis and/or Ca^{2+}-activated kinase levels or activity. Since the TEA-elicited Ca^{2+} signal essentially never occurs in fibroblasts of patients with AD, with or without a family history of AD, it appears to distinguish between individuals suffering from AD and those showing the signs of normal aging.[5]

Abnormalities of energy sources such as with brain ischemia bring about reductions of creatine phosphate, ATP, and intracellular pH. An increase in Ca^{2+} influx induces necrosis of nerve cells.[6,7]

Therefore, the final common mechanism of nerve cell death, including degeneration induced by free radicals, likely is a rise in the intracellular concentration of Ca^{2+}, which activates proteases and provokes mitochondrial dysfunction.[8]

An ischemic penumbra exists in tissue around the infarcted region, along with abnormal ionic gradients in the intra- and extracellular spaces of nerve cells. The first changes in the ischemic area are reduced energy supplies, along with changes in ion balance including Ca^{2+}, as well as altered concentrations of neurotransmitters such as glutamate.

Pharmacological studies report that because 5-hydroxytryptamine and prostaglandin $F_2\alpha$ may be involved in the pathogenesis of cerebral vasospasm, dihydropyridines (nimodipine, nicardipine) and Mg^{2+} can be used in the pharmacological treatment of this disorder. However, dihydropyridines, particularly nicardipine, are more potent vasodilators than Mg^{2+}.[9]

Administration of α-difluoromethylornithin (DFMO) did not substantially affect the intracellular Ca^{2+} release, nor the timing of the opening and closing of plasma membrane Ca^{2+} channels. These findings support the notion that Ca^{2+} signaling may be a target for inhibitors of polyamine metabolism. *In vitro*, polyamines have been found to inhibit the ATPase activity of a plasma membrane Ca^{2+} pump. Polyamines also affect the activities of protein kinase C and calmodulin and interact with cytoskeletal proteins and membrane phospholipids. All of these substances regulate the activity of membrane-associated Ca^{2+} pumps or Ca^{2+} exchangers, and may be the basis for the effects of polyamines. A regulatory site for polyamines on the NMDA receptor and the Na^{2+}, K^{2+}, and Ca^{2+}-gated channel has been described. Both spermidine and spermine enhance NMDA-induced Ca^{2+} currents in cells that express this receptor, and DFMO inhibits the expression of NMDA-induced injury. The NMDA receptor ion channel also has a voltage-dependent block by Mg^{2+} when the receptor complex is activated by NMDA or glutamate.[9,10]

Experimental and epidemiological studies have demonstrated that a long-standing reduction in Ca^{2+} in the drinking water of animals and humans provokes Ca^{2+} deposition in brain accompanied by CNS degeneration due to the derangement of the appropriate distribution of Ca^{2+} between bone and brain tissues.[11,12]

36.3 NMDA AND EXCITATORY AMINES

The content of glutamate is rich in brain. High concentrations of extracellular glutamate act as a neurotoxin.[13,14] Neurotoxicity has been suggested to result from Ca^{2+} influx mediated via the NMDA-associated ion channel receptor in nerve cells.[13] The excitatory amines, such as glutamate, further increase intracellular Ca^{2+} by opening receptor-operated channels (ROCC) in postsynaptic cells.[15–17] Glutamate receptors are present in hippocampus and glutamate is an excitatory neurotransmitter in this region. Delayed neuronal death, such as in the CA1 area of hippocampus, may be partially responsible for the large increase of glutamate.

Through the above process, a decrease of intracellular pH and lack of ATP produce membrane depolarization which releases large amounts of neurotransmitter from the cell invoked by the influx of Ca^{2+} mediated via the action of VOCC.

In many instances the plasma membrane contains Ca^{2+}-permeable channels which can be activated by depolarization. Ca^{2+} influx through these pathways has been shown to trigger neurotransmitter and hormone release, muscle contraction, and a large number of other cellular functions.

Voltage-sensitive Ca^{2+} channels are also the sites of action for several clinically important drugs. There exist low-threshold Ca^{2+} currents mediated by T-type Ca^{2+} channels and two high-threshold channels which are named L- and N-types. L-channels are the locus of action of dihydropyridine (DHP) drugs as well as a number of other agents known to regulate the cardiovascular system. N-channels are blocked by ω-conotoxin GVIA (ω-CgTX), a peptide toxin derived from the venom of the piscivorous marine mollusk *Conus geographus*. Such channels are present in cerebellar Purkinje neurons and are generally called P-type channels. The β-subunits of the channel may form intrinsic parts of N-channels as well as the various types of L-channels in brain.[18,19]

Dihydropyridine (DHP) and phenylalkylamine (PA) binding sites associated with L-type Ca^{2+} channels are mostly preserved in AD and Parkinson's disease (PD) brains.

DHP binding to L-type channels is mostly located on neuronal cell bodies and dendrites and not on nerve terminals in rat brain.[17,20] Treatment with DHP- and/or PA-related drugs in neurological disorders may not be hindered by deficits in Ca^{2+} channel binding proteins in AD and PD brains. Cellular Ca^{2+} entry is regulated by multiple types of Ca^{2+} channels and neurotransmitter receptors (e.g., ion channel of the N-methyl-D-asparate receptor). Controlling the permeability of the plasma membrane is one of the main mechanisms by which cells can regulate the cytoplasmic Ca^{2+} of nerve cells.

36.4 CALCIUM-RELATED DEGENERATIVE DISORDERS IN THE CNS

Several studies have established that brain aging is accompanied by a decrease in Ca^{2+} in the extracellular compartment. Paradoxically, cytosolic Ca^{2+} levels increase significantly with age in the rat brain. Transneuronal Ca^{2+} concentrations are apparently increased to highly toxic levels in the aged mammalian brain, including humans, especially in chronic cerebrovascular disorders. Interestingly, perturbation of the normal Ca^{2+} metabolism is correlated with learning and memory deficits in aging rabbits and aged or lesioned rats. Recently, antibodies to L-type voltage-gated calcium channels have been reported to be present in the sera of patients with amyotrophic lateral sclerosis (ALS), and antibody titers correlated with the rate of disease progression. These results suggest a role for autoimmune mechanisms in the pathogenesis of sporadic ALS. The presence of serum antibodies to VGGCs is not specific to patients with ALS. High titers of antibodies to L-type Ca^{2+} channels were also seen in sera from patients with Lambert-Eaton syndrome, an autoimmune disease involving antibodies which are cross-reactive to N-type and L-type VGGCs as well as synaptotagamin.[21] Vulnerability to the toxin (1-methyl-4-phenyl-1,2,5,6-tetrahydropyridine, MPTP) is found to vary among species with age and sex.

Toxicity to the MPP^+ (1-methyl-4-phenylpyridinium) ion, converted from MPTP, appears to be related to the vulnerability of the cell. Its toxicity leads to a deficit of neuromelanin, thereby being a cause of PD. Although not all of the mechanisms responsible for Ca^{2+}-mediated cell death are known, it is clear that elevation of cytosolic Ca^{2+} has been implicated in the toxicity of a wide range of toxins.[18] Differentiated PC12 cells were reported to react to Ca^{2+} stress through a sequence of regulatory processes which appeared to be independent of the apoptotic pathway.[4] Increased intracellular Ca^{2+} has also been associated with neurodegenerative changes occurring after trauma or ischemia in normal aging and in various neuropathological conditions.[8,9]

Selective vulnerability in the striatum is altered by dopamine neuron and by hypoxemia. The activity of cholinesterase is low in the hippocampus of patients with vascular dementia, and glutamate decarboxylase (GAD) and acetylcholine transferase activities (CAT) are decreased in vascular dementia. Neurotoxicity of glutamate is implicated in brain damage following epilepsy, Huntington's disease and olivo-ponto-cerebellar atrophy, as well as following ischemia.[22,23] These disorders are responsible for dysregulation of Ca^{2+} ion fluxes in nerve cell membranes. An epidemiological survey of ALS in Guam and the Kii Peninsula has reported that low concentrations of Ca^{2+} and Mg^{2+} and high concentrations of aluminum are present in the drinking water of various islands. High contents of Ca^{2+} and aluminum are found in brains from affected patients. The abnormal accumulation of minerals in the CNS likely produces cell degeneration. We believe that Ca^{2+} excess in brain leads to neuronal death through interactions with other neurotoxic metals.[11,12]

36.5 CONCLUSION

Appropriate Ca^{2+} levels are essential to maintain and activate nerve cell functions, but derangements of Ca^{2+} flux in the extra- and intracellular spaces can induce CNS degeneration. The regulation of Ca^{2+} in nerve cells should be studied further to clarify the causal factors for CNS degeneration.

REFERENCES

1. Catterall WA: Structure and function of voltage-sensitive ion channels. *Science*, 1988; 242:50–61.
2. Hosey MM and Lanzdunski M: Calcium channels: molecular pharmacology, structure and regulation. *J. Memb. Biol.*, 1988; 104:81–105.
3. Bean BP: Classes of calcium channels in vertebrate cells. *Annu. Rev. Physiol.*, 1989; 51:367–384.
4. Kumar U, Dunlop DM, and Richardson JS: Mitochondria from Alzheimer's fibroblast show decreased uptake of calcium and increased sensitivity of free radicals. *Life Sci.*, 1994: 54:1855–1860.
5. Etcheberrigaray R, Ito E, Oka K, et al: Potassium channel dysfunction in fibroblasts identifies patients with Alzheimer disease. *Proc. Natl. Acad. Sci. U.S.A.*, 1993; 90:8209–8213.
6. Carbone C and Swandulla D: Neuronal calcium channels: kinetics, blockade and modulation. *Prog. Biophys. Mol. Bio.*, 1989; 54:31–58.
7. Astrup J, Symon L, and Siiesjo BK: Thresholds in cerebral ischemia — the ischemic penumbra. *Stroke*, 1981; 12:723–725.
8. Farber JL, Chien KL, and Mittnachat S Jr: The pathogenesis of irreversible cell injury in ischemia. *Am. J. Pathol.*, 1991; 102:271–281.
9. Alborach E, Salom JB, Perales AJ, et al.: Comparison of the anticonstrictor action of dihydropyridines (nimodipine and nicardipine) and Mg^{2+} in isolated human cerebral arteries. *Eur. J. Pharmacol.*, 1992; 229:83–89.
10. Paschen W: Polyamine metabolism in different pathological states of brain. *Mol. Chem. Neuropathol.*, 1992; 16:241–271.
11. Yasui M, Yase Y, and Ota K: Distribution of calcium in central nervous system tissues and bones of rats maintained on calcium-deficient diet. *J. Neurol. Sci.*, 1991; 105:206–210.
12. Yasui M, Yase Y, Ota K, and Garuuto RM: Aluminum deposition in the central nervous system of patients with amyotrophic lateral sclerosis. *Neurotoxicology*, 1991; 12:615–620.
13. Benveniste H, Drejer J, Schousboe A, and Diemer NH: Elevation of the extracellular concentrations of glutamate and aspartate in rat hippocampus during transient cerebral ischemia monitored by intracerebral microanalysis. *J. Neurochem.*, 1984; 43:1369–1374.
14. Choi DW: Ionic dependence of glutamate neurotoxicity. *J. Neurosci.*, 1987; 7: 369–379.
15. Siesjo BK and Bengstsson F: Calcium fluxes, calcium antagonist, and calcium-related pathology in brain ischemia, hypoglycemia and spreading depression; a unifying hypothesis. *J. Cereb. Blood Flow Metab.*, 1989; 9:127–140.
16. Eyer ML and Westbrook GL: Cellular mechanisms underlying excitoxicity. *Trend Neurosci.*, 1987; 10:59–61.
17. Prasad SA, Boksa P, and Quirion R: Brain calcium channel-related dihydropyridine and phenylalkylamine binding sites in Alzheimer's, Parkinson's, and Huntington' disease. *Brain Res.*, 1993; 611:216–221.
18. Swandulla D and Carbone E: Do calcium channel classifications account for neuronal calcium channel diversity? *Trends Neurosci.*, 1991; 14:46–51.
19. Torri TF, Passafaro M, Clementi F, et al: Presynaptic localization of ω-Conotoxin-sensitive calcium channels at the frog neuromuscular junction. *Brain Res.*, 1991; 547:331–334.
20. Miller RJ: Voltage-sensitive Ca^{2+} channels. *J. Biol. Chem.*, 1992; 267:1403–1406.
21. Smith RG, Hamilton S, Hofmann F, et al: Serum antibodies to L-type calcium channels in patients with amyotrophic lateral sclerosis. *N. Engl. J. Med.*, 1992; 327:1721–1728.
22. Michael PP, Vyas S, Anglade P, et al: Morphological and molecular characterization of the response of differentiated PC12 cells to calcium stress. *Eur. J. Neurosci.*, 1944; 6:677–686.
23. Choi DW: Cerebral hypoxia: some new approaches and unanswered questions. *J. Neurosci.*, 1990; 10: 2493–2501.

SECTION 12

IRON

Chapter 37

Alterations in Iron in Neurodegenerative Disorders: Implications and Possible Therapeutic Agents

David T. Dexter and Peter Jenner

CONTENTS

37.1 INTRODUCTION

In 1887, histological studies by Zaleski revealed that certain nuclei of the human brain contained concentrations of iron well above those typically found in body tissues outside of the liver. Subsequent systematic histological and biochemical studies have shown that iron within the human brain is unevenly distributed and is highly localized in the extrapyramidal brain regions (globus pallidus, substantia nigra, putamen, red nucleus, thalamus, and caudate nucleus). A similar distribution pattern was also observed in rodents and primates.

The metabolism of iron has been the most intensively researched of all of the trace metals. Iron is not only vital for oxygen transport and exchange but is an essential cofactor for many heme and nonheme enzymes and is an essential component for many metabolic processes including DNA,

RNA, and protein synthesis. However, almost all of the literature deals with iron metabolism and function in peripheral tissues. Despite the profound biochemical consequences that can occur due to iron deficiency or overload, until recently little attention had been paid to iron metabolism in the brain.

At birth iron levels are low, but during the first 2 decades of life iron accumulates rapidly in the human brain and then starts to plateau at approximately 30 years of age.[1] Some brain areas, e.g., globus pallidus and substantia nigra, reach maximal levels that are maintained after age 30, whereas other brain regions, e.g., caudate nucleus and putamen, do not reach maximal levels until later in life (50 to 60 years). In the adult, the blood-brain barrier tightly regulates brain iron metabolism. This restricts the turnover of iron and protects the brain against changes in peripheral iron levels such as those occurring in hemochromatosis.

One of the most puzzling aspects of iron metabolism is its regional distribution. No other trace metal has a similar or such a well-defined distribution. Another puzzling aspect is that concentrations of iron in some brain regions exceed that found in the liver.[1] In addition, there are marked species variations in iron levels with areas of the human brain, like the substantia nigra, containing tenfold higher levels of iron compared to rodents and twice the levels found in primates (Dexter, D.T., unpublished data, 1988). Why the human brain contains levels of iron above those required for normal cellular biochemistry is not known. However, recently it has been postulated that iron is essential for the normal functioning of certain neurotransmitter systems and key brain functions, like learning and memory.

Although trace metals like iron are required for the normal physiological functioning of cells, under certain conditions some trace metals can potentially stimulate the formation of a variety of free radical species. If the formation of free radicals is not checked by cellular antioxidant defense systems their formation can impair the functioning of the cell or lead to its death. Clearly, if such toxic processes occur in the brain or spinal cord it could have catastrophic consequences. The devastating actions of free radicals can be observed in the familial form of motor neurone disease, which has been associated with a defect on chromosome 21q encoding for Cu,Zn-superoxide dismutase, a free radical scavenging enzyme.[2]

Indeed, the toxic actions of a variety of trace metals have recently been implicated in a range of neurodegenerative disorders, including Hallervorden Spatz disease, Parkinson's disease, Huntington's chorea, and Alzheimer's disease. This chapter will review the involvement of iron in neurodegenerative disorders such as Parkinson's disease and will discuss evidence supporting its toxic effects. In addition, we will discuss the possible therapeutic use of metal chelators in preventing the progression of Parkinson's disease.

37.2 IRON-DEPENDENT FORMATION OF REACTIVE RADICALS: WHERE DOES THE IRON COME FROM?

Iron and other divalent trace metals are involved in at least three stages of oxidative stress.

1. Iron ions can catalyze lipid peroxidation in two ways. First, it catalyzes the formation of initiating (H-abstracting) species. Second, it stimulates lipid peroxidation by reacting with lipid hydroperoxides and decomposing them to peroxyl radicals and alkoxyl radicals, which stimulate further peroxidation. Products of these complex decomposition reactions include hydrocarbon gases and a wide range of toxic carbonyl compounds, including aldehydes such as the highly cytotoxic 4-hydroxy-2,3-*trans*-nonenal. These aldehydes can damage adjacent cells; membrane-bound enzymes and receptors can also be inactivated.[3]

2. Iron can facilitate the formation of the highly toxic hydroxyl radical ($\cdot$OH) from hydrogen peroxide (H_2O_2) and superoxide (O_2^-), by the Fenton reaction. Hydroxyl radicals react at great speed with almost every molecule found in the living cell, including DNA (causing strand breakage and chemical alterations of the deoxyribose and of the purine

and pyrimidine bases), membrane lipids, proteins, and carbohydrates. Hydroxyl radicals are also capable of initiating the process of lipid peroxidation by abstracting a hydrogen atom from a polyunsaturated acid side chain in a membrane lipid.[3]

3. Iron is also involved in the generation of H_2O_2 and O_2^- by accelerating the nonenzymatic oxidation of molecules such as dopamine and glutathione.[3]

An obvious question then arises: What forms of iron present *in vivo* are able to accelerate these radical reactions?

In humans, iron is absorbed from the gut and enters the plasma attached to the protein transferrin. Each transferrin molecule is able to bind two molecules of iron. Transferrin crosses the blood-brain barrier and enters cells by endocytosis after initially binding to a transferrin receptor. The pH of the vacuole containing the transferrin/transferrin-receptor complex is then lowered, facilitating the release of the iron from the protein. The unloaded transferrin (apotransferrin) is ejected from the cell, and the iron released from it enters the intracellular mobile iron pool where it is utilized for the synthesis of intracellular iron proteins. Excess iron is stored in the protein ferritin, which acts as an intracellular housekeeping protein, binding up to 4,500 molecules of iron. The synthesis of ferritin is controlled by the intracellular iron availability, hence in conditions where cellular iron entry is increased synthesis of ferritin is triggered.

Iron attached to the transport protein transferrin or to the iron storage protein ferritin is thought to be incapable of stimulating oxidative stress under normal physiological conditions. By contrast, nonprotein-bound iron which forms the intracellular mobile iron pool is capable of stimulating oxidative stress.[4] Understandably, the cell appears to keep the intracellular mobile pool small.[4] Thus, under normal physiological conditions, in the extracellular space no iron is available to stimulate oxidative stress. Inside the cell catalytic iron is only present in the intracellular mobile pool. Hence an important function of intracellular antioxidant defense enzymes is to scavenge O_2^- and H_2O_2 before they can come into contact with any available iron.

After ischemia or brain trauma iron may be released from its storage proteins intracellularly or into the extracellular space. In peripheral tissue transferrin is only partially saturated with iron and there is an abundance of antioxidants. Thus the iron released is quickly mopped up by transferrin and the free radicals generated by its release quickly neutralized by the antioxidants. However, not all extracellular fluids are rich in metal-binding antioxidants. Despite the high iron content of the brain, the cerebrospinal fluid surrounding the brain has no significant iron binding capacity because its transferrin levels are low, or close to saturation.[5] Hence rapid free radical reactions may occur as a consequence of brain injury after ischemia or trauma. This problem is compounded by three additional factors. First, membrane lipids are very rich in polyunsaturated fatty acid side chains, which are especially sensitive to free radical attack. Second, the brain is relatively deficient in antioxidants, with poor levels of catalase and only moderate levels of superoxide dismutase and glutathione peroxidase. Third, the brain contains a high concentration of ascorbic acid. The choroid plexus has a specific active transport system that raises ascorbate concentrations in the CSF to tenfold plasma levels. Neurones have a second transport system that further concentrates intracellular ascorbate.[4] Ascorbic acid in the absence of transition metal ions has well-established antioxidant properties. However, ascorbate/iron and ascorbate/copper mixtures generate free radicals.[6] Thus, if catalytic iron were generated in the CNS as a result of injury, ascorbic acid might then stimulate ·OH generation within the brain and CSF.

37.3 ALTERATIONS IN IRON LEVELS IN PARKINSON'S DISEASE

37.3.1 Parkinson's Disease

Parkinson's disease is a progressive neurodegenerative disorder characterized by a dysfunction in movement consisting of akinesia, rigidity, tremor, and postural abnormalities. The disease affects approximately 1 in 500 of the general population. Symptoms usually appear after the age of 40 years, but the young are not exempt. The prevalence of Parkinson's disease is age related such that 1 in 100 of those over 60 develops the illness. Men and women are equally affected and the disease occurs worldwide.

The pathology of Parkinson's disease is distinct, involving the degeneration of pigmented brain stem nuclei, particularly the dopaminergic neurones, of the substantia nigra pars compacta. Inside remaining neurones, characteristic dense eosinophilic inclusions termed Lewy bodies are found. These are the pathologic hallmark of the disease process and appear to be derived from the cytoskeleton. The loss of cells in the substantia nigra leads to a dramatic reduction in the dopamine content in the caudate nucleus and to a greater extent in the putamen, since both receive projections from the substantia nigra.[7] However, it is necessary to lose approximately 80% of nigral neurones and to deplete the striatum of a least 70% of its dopamine content for clinical symptoms of Parkinson's disease to appear. For lesser degrees of nigral damage compensatory mechanisms occur within the surviving neurones, where dopamine synthesis and release is increased in an attempt to overcome the initial cell loss.

Despite intensive research the cause of neuronal death in the substantia nigra in Parkinson's disease remains unknown. Many possibilities, including the involvement of a virus, altered dopamine metabolism, and a role for neuromelanin have been invoked but not proven. However, recent interest has centered on the involvement of an ongoing toxic process involving increased production of free radicals, due to the presence of reactive iron or a neurotoxin.

37.3.2 Alterations in Iron Levels

The finding that nigral cell loss in Parkinson's disease was associated with increased iron levels was first demonstrated by Lehermitte et al.[8] and Earle,[9] using formalin-fixed tissues. However, the process of fixation can alter metal ion concentrations since many metal complexes are water soluble. Hence, we subsequently conducted a study in 1987[10] utilizing frozen brain tissue collected from control subjects and parkinsonian patients. Using the highly sensitive technique of inductively coupled plasma (ICP) spectroscopy, we were able to detect a selective increase of 35% in the levels of total iron in the parkinsonian substantia nigra when compared to age- and post-mortem-matched control subjects, so confirming the initial studies. Using a variety of techniques to measure brain iron, other investigators have confirmed these findings[11] and subsequently shown that the increased iron content was confined to the substantia nigra pars compacta.[12,13] However, the reported increases in iron content in the parkinsonian substantia nigra have been variable, with different investigators reporting increases up to 77% above control levels. The reason for this discrepancy is unknown but may relate to differences in the severity of the parkinsonism and hence cell loss in the patient groups studied. In our initial study, like other investigators, we utilized severely affected parkinsonian patients, whereas other studies utilizing mildly affected parkinsonian patients have reported no significant increase in iron levels in the substantia nigra.[14]

Magnetic resonance imaging techniques using T_2-weighted images have also demonstrated an accumulation of iron in the nigro-striatal system in parkinsonian patients *in vivo*.[15] However, it is not known which complex form of iron brings about the alterations in bulk water proton spin-spin relaxation times observed with this technique.[16]

Biophysical techniques provide measurements of metal ions in pieces of brain tissue but provide no information on the cellular localization of the increased iron load in Parkinson's disease. Attempts to define the localization were initially made using the traditional Perl's stain for iron in histological sections. These showed increased iron levels mainly localized in macrophages, astrocytes, and reactive microglia in substantia nigra, with occasional staining in nonpigmented neurones.[17] Such studies were not ideal since Perl's stain only detects iron (III) and the technique may involve the loss of water-soluble iron complexes. Subsequently, others have developed X-ray microanalysis and LAMMA techniques (see Chapter 38) for the detection of trace metals at the cellular level. Such studies have not only confirmed the findings of increased iron levels in the parkinsonian nigra but have shown these to be present in pigmented dopaminergic neurones.[18]

Obviously these findings raise the question, "Are these changes in iron levels toxic and are they responsible for the nigral cell loss observed in Parkinson's disease"? If this is the case such changes in iron levels should be selective to Parkinson's disease.

37.4 ALTERATIONS IN IRON LEVELS IN OTHER NEURODEGENERATIVE DISORDERS AFFECTING THE BASAL GANGLIA

To investigate whether the alterations in iron levels are selective for Parkinson's disease, we have investigated three other motor disorders affecting the basal ganglia, namely: multiple system atrophy (MSA), progressive supranuclear palsy (PSP), and Huntington's disease. In MSA with strionigral degeneration, degeneration of the substantia nigra also occurs but, in addition, pathology affects the caudate and putamen, olives, pons, and cerebellum. Similarly, in PSP extensive neurodegeneration occurs in the substantia nigra but there is additional pathology in the caudate and putamen, subthalamic nucleus, brainstem, and dentate nucleus. Whereas in Huntington's disease the substantia nigra is hardly affected but there is profound degeneration of the caudate and putamen.

In PSP and MSA, where the substantia nigra is affected, total iron levels were also increased by 70 and 59%, respectively (see Figure 37.1). In addition, increases in total iron levels were also observed in the putamen in PSP and MSA (36 and 67%, respectively), as well as in the caudate nucleus (44%) in MSA when compared with control material. In other brain regions examined in PSP and MSA iron levels were unchanged compared with control values. In contrast, in Huntington's disease total iron levels in the substantia nigra were comparable to control levels. However, in the putamen and caudate nucleus, where there is marked pathology in Huntington's disease, total iron levels were increased by 44 and 56%, respectively.[12]

Hence, increased iron levels are not selectively associated with neurodegeneration in the substantia nigra in Parkinson's disease. Indeed, increased iron levels appear to be associated with the pathological processes occurring in a variety of neurodegenerative disorders. It seems unlikely that the changes in brain iron are the primary cause of the different pathological changes characterizing these conditions. The neurodegenerative processes involved in the various basal ganglia diseases are likely to be different in view of the varied distribution of their pathology and their different cellular pathological markers (e.g., Lewy bodies in Parkinson's disease, intranuclear inclusion bodies in MSA, and neurofibrillary tangles in PSP). Indeed, today, Parkinson's disease is joined by an ever-growing list of neurodegenerative disorders, e.g., Alzheimer's disease, amyotrophic lateral sclerosis, multiple sclerosis, Hallervorden Spatz syndrome, and tardive dyskinesia, where brain iron levels are increased.[19,20]

37.5 MOLECULAR NATURE OF THE INCREASED IRON LOAD IN NEURODEGENERATIVE DISEASES

As previously discussed, iron bound to either the transport protein transferrin or to the storage protein ferritin is unable to stimulate oxidative stress under physiological conditions. Hence, for the increased iron levels found in neurodegenerative disorders to be toxic, iron must exist in a nonprotein-bound form. The cell understandably tries to keep the levels of nonprotein-bound iron to a minimum, mainly by the actions of ferritin which is the intracellular housekeeping protein for iron metabolism. Ferritin consists of a spherical protein shell of 24 subunits surrounding a core of up to 4500 ferric iron atoms in an inorganic hydrous ferric-oxide-phosphate complex.[21] The protein subunits are of two types: light or L (19,000 Da) and heavy or H (21,000 Da). The proportions of L and H subunits present in the protein shell vary between tissues, e.g., L-rich isoferritins predominate in iron storage organs like the liver, while H-rich isoferritins are found in organs where iron utilization is rapid, e.g., heart. The brain was recently shown to contain approximately 60% H subunits and 40% of L subunits. The biosynthesis of ferritin is regulated by iron availability. The induction of ferritin synthesis is rapid and reversible, a cellular response crucial to the prevention

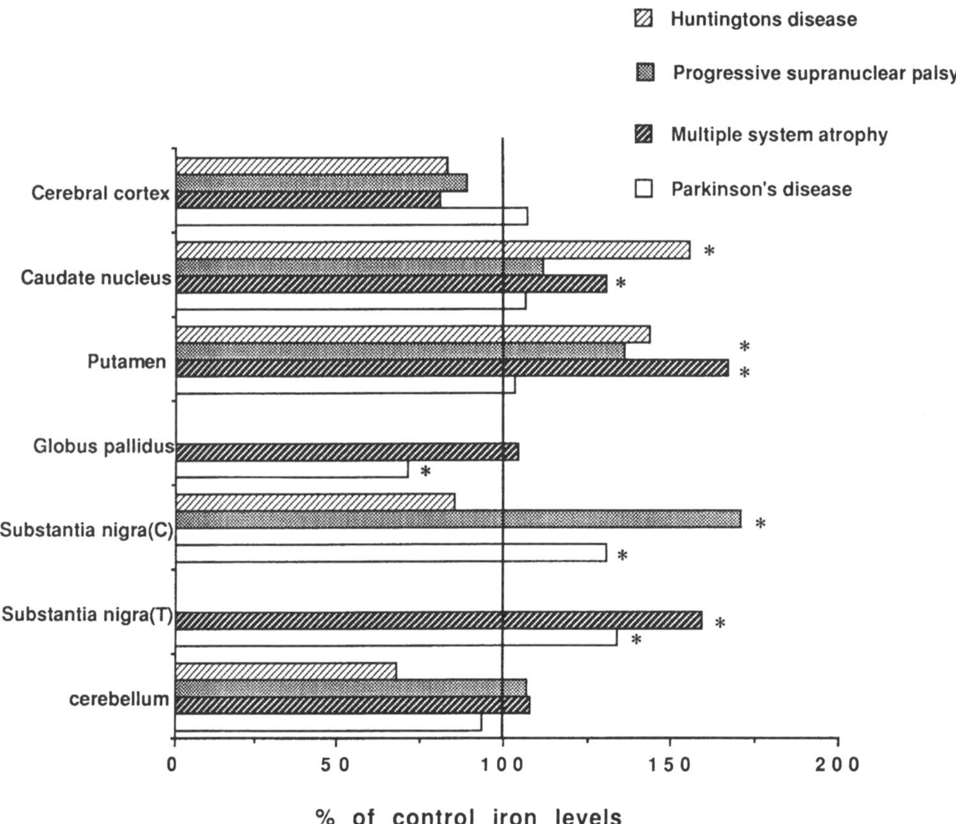

FIGURE 37.1 Total iron levels in Parkinson's disease, multiple system atrophy, progressive supranuclear palsy, and Huntingtons disease. Values are expressed as the percentage of control values.* = p <.05 compared with control subjects (Student's t test). C = compacta; T = total. (Taken from Dexter, D., Carayon, A., Javoy-Agid, F., et al., *Brain*, 114, 1953, 1991. With permission.)

of the toxic effects of iron. The presence of nonprotein-bound iron triggers the removal of an inhibitory molecule on mRNA encoding for ferritin, present in the ribonucleoprotein cell fraction. The activated mRNA migrates to polysomes where ferritin is synthesized. Consequently, the increased iron levels in neurodegenerative disorders may not be toxic if ferritin synthesis is induced.

In Parkinson's disease, there is conflicting evidence suggesting reduced, normal, or increased ferritin levels (for review see Reference 22). These discrepancies may relate to the disease severity, and hence iron loading, in the material examined. However, it is more likely due to differences in the specificity of the antibodies to ferritin used in the different studies. In the early studies antibodies raised to human liver or spleen ferritin were utilized, but with the recent discovery of a novel brain ferritin[23] such antibodies may not reflect the true brain ferritin content. Recently, Mann et al.[22] using an antibody raised against brain ferritin found no significant difference in ferritin levels in the substantia nigra when control subjects and patients with Parkinson's disease were compared.

Measurement of the ferritin levels in other neurodegenerative disorders affecting the basal ganglia reveals a different story. In PSP there is an increase in ferritin levels in the substantia nigra; similarly, in MSA there is a trend for a rise in the nigral ferritin content and an increase in the putamen. However, in Huntington's disease there is no alteration in ferritin levels in the substantia nigra or any other brain region examined.[12] In MSA and PSP, the increased iron levels in the substantia nigra are associated with an increase in ferritin synthesis and hence the increased iron load may be unreactive.

If the increased iron level in substantia nigra in Parkinson's disease is not associated with ferritin, it must exist in a nonprotein complex. Indeed recent X-ray microanalysis studies showed the increased iron levels present as a neuromelanin complex.[24] Neuromelanin, an oxidation product of catecholamines, is a poorly defined pigment that is located mainly in the cell bodies of catecholamine-producing neurones in the substantia nigra, locus ceruleus, and other brain stem nuclei. Neuromelanin has a high affinity for Fe^{3+}. Under normal physiological conditions melanin acts as a free radical scavenger.[25] However, under certain conditions, such as increased iron availability, it can potentiate the formation of oxygen-derived free radicals.[26] If neuromelanin granules selectively accumulate the increased iron content of the parkinsonian substantia nigra, this iron complex may be a strong candidate for the cytotoxic component of oxygen radical-induced selective degeneration of melanized dopaminergic cell populations.

37.6 EVIDENCE OF OXIDATIVE STRESS IN PARKINSON'S DISEASE

Under physiological conditions, there is a fine balance between the production of free radicals and their removal by antioxidant defense systems. Obviously, if the production of free radicals increases above the scavenging ability of the antioxidant defense systems, increased damage will occur to the cellular components. The term "oxidative stress" is often used when the balance is disrupted in favor of increased free radical production. If mild oxidative stress occurs, tissues often respond by increasing antioxidant defenses. On the other hand, severe oxidative stress can cause cell injury and death. Free radical-induced cell death can proceed as necrosis or apoptosis, with many antiapoptotic genes encoding for free radical scavengers.[27]

Although there is no direct proof of an increase in the formation of free radicals in Parkinson's disease, primarily due to their reactive nature, there is a considerable body of indirect evidence from post-mortem studies which support this concept.

37.6.1 Evidence of Increased Free Radical Attack on Cellular Components

Many free radical species e.g., OH·, can initiate lipid peroxidation by abstracting an atom of hydrogen from polyunsaturated fatty acid side-chains in membranes. This results in the oxidation of lipids, damage to proteins, leakage, and eventually, complete membrane breakdown.

Studies utilizing nigral tissue from patients with Parkinson's disease have revealed increased formation (by 35%) of malondialdehyde, a minor breakdown product in the lipid peroxidation process.[28] Examination of a more specific marker of lipid peroxidation, namely lipid hydroperoxides, revealed a tenfold increase in the substantia nigra in Parkinson's disease when compared to matched control subjects.[29] This suggests that in Parkinson's disease there is a marked increase in lipid peroxidation which is selective to the substantia nigra.

Although lipid peroxidation can result from increased free radical attack, it may also be a secondary product of cell death. However, further evidence from studies examining free radical attack on guanine in DNA, particularly by the OH· radical, have shown increased 8-hydroxy-2-deoxyguanosine levels in the parkinsonian substantia nigra but also in other brain areas.[30]

Inhibition of complex I of the mitochondrial respiratory chain can lead to increased free radical production.[31] Indeed, there is an inhibition of mitochondrial complex I activity in the substantia nigra in Parkinson's disease.[32] The cause remains unknown; there is no evidence of an $MPTP/MPP^+$-like toxin in the substantia nigra in Parkinson's disease, or evidence of complex I subunit alterations or of altered mitochondrial DNA encoding for complex I.

37.6.2 Alterations in Antioxidant Defense Systems

Post-mortem studies have revealed a specific increase in the activity of the manganese-dependent isoenzyme of superoxide dismutase (SOD)[33] in the substantia nigra in Parkinson's disease. SOD scavenges the superoxide radical as follows:

$$2O_2^- + 2H^+ \xrightarrow{\text{SOD}} H_2O_2 + O_2$$

One possible explanation for the increase in SOD activity is an induction in enzyme synthesis due to the increased level of oxidative stress in substantia nigra. Indeed, SOD synthesis is increased under conditions of increased oxidative stress in rodents.[34] However, the increase in SOD activity may itself be harmful through the production of increased amounts of hydrogen peroxide. Interestingly, there is an increase (by 25%) in monoamine oxidase (MAO-B) activity in the brain in Parkinson's disease.[35] Oxidative deamination of dopamine by MAO-B is the main catabolic pathway for dopamine within the nerve terminal, leading to the formation of 3,4-dihydroxyphenylacetic acid (DOPAC) and hydrogen peroxide. Thus there may be two sources of increased hydrogen peroxide formation in the parkinsonian brain. Although hydrogen peroxide itself is not very reactive, it can combine with reactive iron to form the highly toxic $\cdot$OH radical by the Fenton reaction as follows:

$$Fe^{2+} + H_2O_2 \longrightarrow Fe^{3+} + \cdot OH + OH^-$$

However, potential toxicity from hydrogen peroxide is prevented by the actions of two enzymes, catalase and glutathione peroxidase. Although the level of catalase is very low in human brain, its activity in substantia nigra in Parkinson's disease is reduced.[36] Glutathione peroxidase (GPX) is the main enzyme in brain that removes hydrogen peroxide by oxidizing the tripeptide glutathione (GSH) into its oxidized form (GSSG). GSH is also important for the maintenance of α-tocopherol and ascorbic acid in the reduced state. Normally a high ratio of reduced GSH to GSSG is maintained, but when a cell is oxidatively stressed the ratio is pushed more in favor of GSSG.

$$2GSH + H_2O_2 \xrightarrow{\text{GPX}} GSSG + 2H_2O$$

The activity of GPX has been reported to be decreased in substantia nigra in Parkinson's disease, but two other studies have shown no change.[37,38] More importantly the level of GSH is reduced by 40% in the nigra, pushing the GSH/GSSG ratio in favor of the oxidized form.[39] Such results are consistent with the involvement of oxidative stress in nigral cell loss in Parkinson's disease.

Therefore, in Parkinson's disease there is the potential for an increased capacity to produce hydrogen peroxide but a reduced ability to remove it. Due to the poorly reactive nature of hydrogen peroxide it can diffuse away from its site of formation, crossing cell membranes, to react with nonprotein-bound iron and produce the highly reactive OH$\cdot$ radical.

37.7 ARE ALTERATIONS IN IRON AND OXIDATIVE STRESS A PRIMARY OR SECONDARY EVENT IN PARKINSON'S DISEASE?

None of the previously discussed data indicates whether free radical involvement in Parkinson's disease is a primary or secondary event in nigral cell death or whether it occurs early or late in the disease process. This fact is very pertinent since 60 to 70% of nigral dopamine neurones are lost before clinical symptoms appear, suggesting that the disease has a long preclinical phase. If such factors are involved in the initial stages of cell death, then they should be present at the earliest stages of the illness. Indeed, they might be more evident in post-mortem tissue at the start of the illness when fewer nigral neurones have died. However, it is difficult to obtain brain tissue from patients at the onset of the illness to study this.

Examination of substantia nigra from control subjects without evidence of neurological disease in life has consistently revealed the presence of Lewy bodies in about 5 to 7% of the population.[40] Fearnley and Lees[41] found that subjects with incidental Lewy bodies have an average 27% loss of pigmented nigral neurones, in excess of that found in the normal ageing brain. Furthermore the topographical pattern of cell loss in the substantia nigra in incidental Lewy body disease was similar to that found in Parkinson's disease, but markedly different from that occurring in normal ageing. Subjects with incidental Lewy bodies in the substantia nigra therefore may be considered as showing the initial pathological indications of Parkinson's disease (presymptomatic Parkinson's disease).

To assess the significance of alterations in iron levels and oxidative stress in the substantia nigra in Parkinson's disease, we have measured total iron levels and markers of oxidative stress in incidental Lewy body disease. Nigral iron concentrations were unchanged in Lewy body-positive subjects when compared to control subjects.[42] Assuming that incidental Lewy body disease represents presymptomatic Parkinson's disease, these data suggest that changes in iron metabolism are unlikely to be the primary cause of neuronal cell death and that such changes are an important but secondary event. Similarly, Riederer and colleagues[14] found no increase in iron levels in patients with early Parkinson's disease and suggested that the accumulation of iron in nigra correlates with the severity of the disease. So even if an increase in reactive iron does not initiate nigral degeneration, it may contribute to a cascade of secondary events that accelerate the destruction of pigmented nigral neurones. Treatment strategies to reduce the impact of toxic iron could therefore delay progression of the disease.

However, a reduction in the levels of GSH was observed in the Lewy body-positive subjects without alteration in GSSG levels, resulting in a reduction in the ratio of GSH to GSSG. Such changes suggest that the nigra in patients with incidental Lewy body disease are exposed to some form of oxidative stress. What causes the oxidative stress at this stage of the illness is not known.

37.8 WHY DOES IRON ACCUMULATE IN BRAIN REGIONS UNDERGOING NEURODEGENERATION?

Why iron levels are increased in certain brain areas showing pathological changes is not known. Alterations in the permeability of the blood-brain barrier to iron have been proposed but the evidence is conflicting. For example, Pall and colleagues[43] demonstrated that iron levels, measured by electrothermal atomization/atomic absorption spectroscopy, in cerebrospinal fluid in Parkinson's disease patients are comparable to controls. Leenders,[44] using PET scanning techniques, demonstrated an increased uptake of $^{52}Fe^{3+}$ citrate into brain in Parkinson's disease. It seems unlikely that a generalized increase in iron uptake across the blood-brain barrier can explain the diversity of brain regions which accumulate iron in the different neurodegenerative diseases, although localized changes in the blood-brain barrier may occur as a result of the pathological process.

The brain material used to study neurodegenerative disorders inevitably comes from patients who will have been treated for long periods with a variety of drugs. Indeed some drugs, such as chlorpromazine and L-DOPA, are capable of binding metal ions and may influence their brain metabolism.[45] However, the initial studies demonstrating increased nigral iron levels in Parkinson's disease were carried out prior to the L-DOPA era.[8,9] Also, in recent studies we have conducted in chronic L-DOPA-treated primates, LAMMA techniques showed no effect on brain neuronal iron levels (Jenner, P., unpublished data, 1994).

Another possibility is that iron uptake via transferrin receptors may be increased in cells exposed to metabolic stress. Such a process may represent a protective mechanism related to increased cellular metabolism and an increased demand for iron. Support for this hypothesis has come from two studies. First, mice treated with the neurotoxin 1-methyl-4-phenyl-1,2,3,6-tetrahydropyridine (MPTP) initially showed an increased expression of neuronal transferrin receptors 7 days post MPTP treatment that correlated with the loss of dopamine nerve terminals.[46] However, there was a normalization in transferrin receptors in mice killed at later times perhaps reflecting an upregulation of transferrin receptors on surviving neurones. Second, in studies involving lesions of facial nerves there was a transient increase in transferrin receptors and an increase in iron uptake in regenerating motor neurones.[47]

37.9　IRON CHELATION AS A POSSIBLE THERAPY FOR PARKINSON'S DISEASE

Although the increase in iron levels appears to be a secondary phenomenon in Parkinson's disease, it may contribute to the neurodegenerative process and accelerate nerve cell death. Thus, treatment of Parkinson's disease patients with iron chelators may slow down the disease process. To assess the benefits of iron chelation therapy two recent studies using animal models of brain iron overload have been carried out.

Infusion of an iron solution into the rat substantia nigra produces a selective loss of dopaminergic neurones in the pars compacta of the substantia nigra.[48] This is accompanied by marked reductions in striatal levels of dopamine and its metabolites. Pretreatment with the iron chelator desferrioxamine reduced the loss of neurones in substantia nigra and decreased the loss of striatal dopamine.[48]

Invasive techniques which involve the puncturing of the blood-brain barrier cannot test whether iron chelators cross the blood-brain barrier and remove increased iron levels without adversely interfering with other essential systems, e.g., dopamine synthesis. To overcome these problems, we have recently developed a model of brain iron overload using a 3,5,5-trimethyl hexanoyl ferrocene derivative given in the diet or by gavage. Under normal circumstances increased dietary iron is bound by transferrin whose access to the brain is restricted by the blood-brain barrier. However, due to the highly lipid-soluble nature of the ferrocene iron complex it is readily absorbed; the intact complex then rapidly permeates cellular membranes, including the blood-brain barrier, resulting in the iron loading of the brain and peripheral organs. After 4 weeks of ferrocene treatment various brain regions, including substantia nigra, exhibit up to a 50% increase in iron levels,[49] similar to those observed in neurodegenerative disorders. Although such iron loading is not site specific, the blood-brain barrier remains intact allowing the full assessment of chelation therapy.

Desferrioxamine is the main iron chelator currently available; however, it lacks oral activity and generally has poor membrane permeability. Hence it may not be suitable as a therapeutic agent for neurodegenerative disorders. Recently, new classes of iron chelators which have high specificity for iron and which can penetrate bilayer lipid membranes readily have been developed, such as hydroxy pyridones. We have assessed the ability of desferrioxamine and two hydroxy pyridones (CP24 and CP94) for their ability to remove the increased iron load in the ferrocene-treated rat and for their effects on brain dopamine metabolism.

In normal animals, treatment for 2 weeks with CP24, CP94, or desferrioxamine did not affect total iron levels in a range of brain regions. However, in animals treated with the ferrocene complex for 4 weeks, subsequent administration of CP24 or CP94 for 2 weeks reversed the increase in brain iron (See Figure 37.2). Interestingly, similar results were observed with desferrioxamine despite its poor ability to cross lipid membranes.[49] So iron chelation therapy can remove excess iron levels after only short periods of treatment.

Administration of iron chelators did, however, have a deleterious effect on striatal dopamine metabolism. A decrease in striatal dopamine levels was observed at the higher doses of chelators and alterations in dopamine metabolite levels were observed at most doses (see Figure 37.3). Consequently iron chelation treatment not only removed the increased iron levels resulting from ferrocene treatment but also may have removed iron associated with key enzymes involved in dopamine metabolism. Obviously, in Parkinson's disease where the dopaminergic system is already compromised, iron chelators might exacerbate the clinical condition.

From this study, it seems unlikely that iron chelation therapy would be of benefit in Parkinson's disease. However, it remains to be seen whether lower doses of the chelator could be utilized which would remove the increased iron load over a longer period without affecting dopamine metabolism. Alternately, it may be possible to synthesize new chelators which remove iron from a specific pool, leaving essential iron bound to key enzymes. Indeed, we have recently tested new derivatives of desferriferrithiocin, a powerful iron chelator, in naive animals without any deleterious effect on

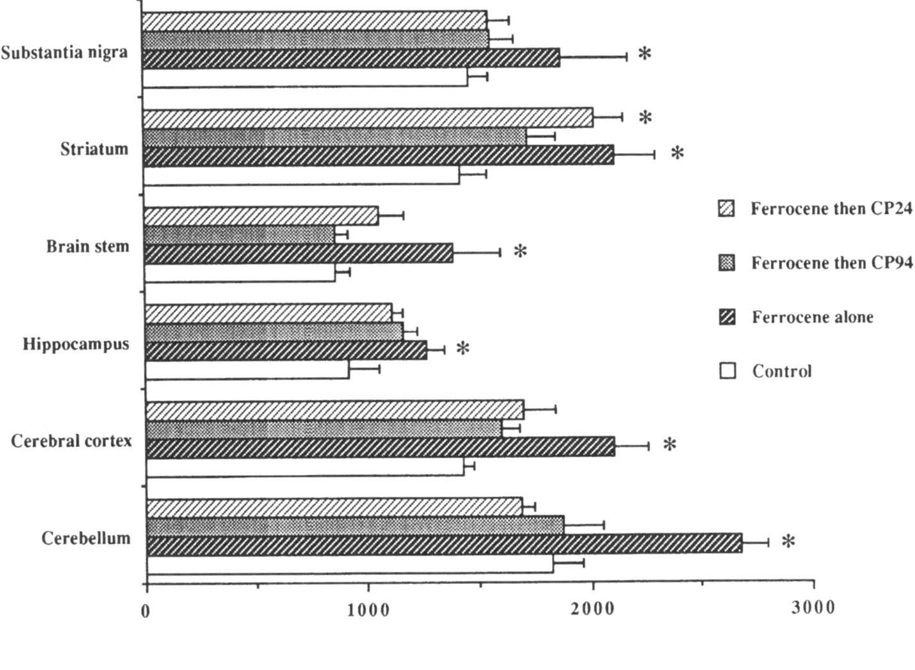

FIGURE 37.2 The effect of administration of the ferrocene derivative 3,5,5-trimethyl hexanoyl ferrocene for 4 weeks, with or without the subsequent administration of the hydroxypyridone chelators CP94 (100 mg/kg) and CP24 (30 mg/kg) for a further 2 weeks, on brain iron concentrations. Values shown are the mean $\pm$ 1 SEM, n = 6 per group. * = p <.05 compared with controls (Student's *t* test). (Modified from Ward RJ, Dexter DT, Florence A, et al: *Biochem. Pharmacol.*, 1995; 49:1821. With permission.)

dopamine metabolism (Dexter, D.T., unpublished data, 1994). Whether these new derivatives will have therapeutic potential remains to be seen.

37.10 CONCLUSIONS

There is at present only limited knowledge about the biochemical role of iron in normal brain. This lack of basic knowledge compounds the difficulty we face when interpreting the importance of increased iron levels in neurodegenerative disease. It seems clear from evidence collected so far that increased iron levels are associated with a variety of neurodegenerative disorders which have a variety of pathological origins. Therefore, accumulation of iron in neurodegenerative disorders appears to be nonspecific, but the reasons why it accumulates remain unknown.

However, in Parkinson's disease the increased iron levels may not be bound to the traditional iron storage proteins like ferritin, but rather to neuromelanin or some other low molecular weight chelate. In such complex forms, iron is capable of stimulating oxidative stress, supporting the involvement of free radicals in the disease process. The increases in iron levels are not observed in the early stages of the disease, indeed accumulation of iron may be associated with the disease progression. Consequently, in Parkinson's disease increased iron levels may not be the primary cause of nigral cell loss, but may act as an accelerating factor. Despite the lack of iron changes in early Parkinson's disease, evidence of oxidative stress can still be detected. The factors triggering oxidative stress at this point are not known but may have a dramatic impact on nigral cell survival and represent a potential target for therapeutic intervention.

In conclusion, iron and oxidative stress appear to play an important role in Parkinson's disease. Antioxidants and possibly iron chelators which can penetrate into the brain should be tested to

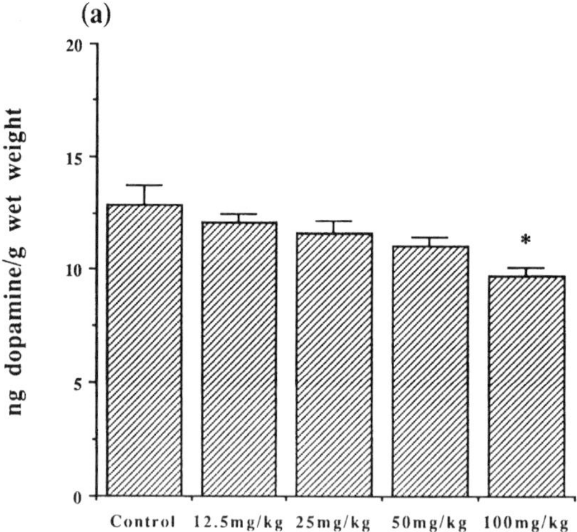

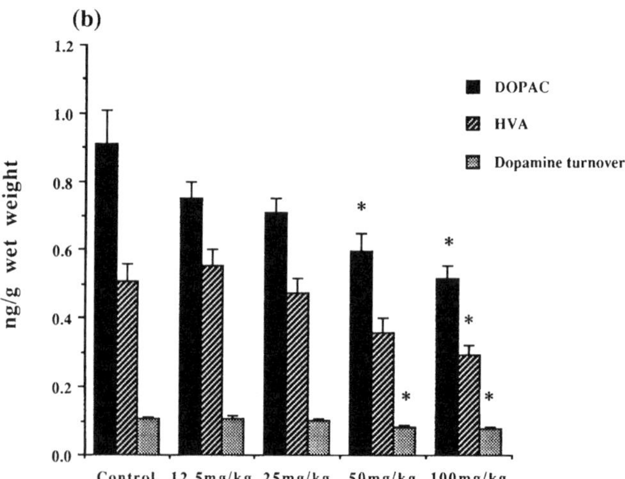

FIGURE 37.3 Effect of CP94 administration (12.5 - 100 mg/kg) in rats on (a) striatal dopamine levels and (b) striatal DOPAC, HVA, and dopamine turnover. Values shown are the mean $\pm$ 1 SEM, n = 6 per group. $* = p < .05$ compared with controls (Student's t test). (From Ward RJ, Dexter DT, Florence A, et al: *Biochem. Pharmacol.*, 1995; 49:1821. With permission.)

determine whether they can affect the progression of the disease. The success of such neuroprotective treatments may depend on the preclinical diagnosis of Parkinson's disease, since the majority of nigral cells are lost prior to the onset of clinical symptoms.

REFERENCES

1. Hallgren B and Sourander P: The effect of age on the non-haemin iron in human brain. *J. Neurochem.*, 1958; 3:41–51.
2. Robberecht W, Sapp P, and Viaene MK: Cu/Zn-superoxide dismutase activity in familial and sporadic amyotrophic lateral sclerosis. *J. Neurochem.*, 1994: 62:384–387.
3. Halliwell B: Reactive oxygen species and the central nervous system. In Packer L, Prilipko L, and Christen Y (eds): Free radicals in the brain; *Aging, Neurological and Mental Disorders*, Heidelberg, Springer-Verlag, 1992, 21–41.
4. Halliwell B: Reactive oxygen species and the central nervous system. *J. Neurochem.*, 1992; 59:1609–1623.
5. Halliwell B and Gutteridge JMC: Iron and free radical reactions: two aspects of antioxidant protection. *TIBS*, 1986; 11:372–375.

6. Halliwell B and Gutteridge JMC: Role of free radicals and catalytic metal ions in human disease: an overview. *Methods Enzymol.*, 1990; 186:1–85.

7. Duvoisin R: History of parkinsonism. *Pharm. Ther.*, 1987; 32:1–17.

8. Lehermitte J, Kraus W and McAlpine MA: On the occurrence of abnormal deposits of iron in the brain in Parkinson's disease with special reference to its location. *J. Neural. Psychopathol.*, 1924; 5:195–208.

9. Earle KM: Studies in Parkinson's disease including x-ray fluorescent spectroscopy of formalin fixed tissues. *J. Neuropathol. Exp. Neurol.*, 1967; 27:1–4.

10. Dexter DT, Wells FR, Agid F, et al: Increased nigral iron content in post-mortem parkinsonian brain. *Lancet*, 1987; ii:1219–1220.

11. Sofic E, Riederer P, Heinsen H, et al: Increased iron (III) and total iron content in post mortem substantia nigra of parkinsonian brain. *J. Neural Transm.*, 1988; 74:199–205.

12. Dexter DT, Carayon A, Javoy-Agid F, et al: Alterations in the levels of iron, ferritin and other trace metals in Parkinson's disease and other neurodegenerative diseases affecting the basal ganglia. *Brain*, 1991; 114:1953–1975.

13. Sofic E, Paulus W, Jellinger K, et al: Selective increase of iron in substantia nigra zona compacta of parkinsonian brains. *J. Neurochem.*, 1991; 56:978–982.

14. Riederer P, Sofic E, Rausch WD, et al: Dopaminforschung heute und morgen-L-DOPA in der Zukunft. In Riederer P and Umek P (eds), *L-DOPA-Substitution der Parkinson-Krankheit: Geschichte - Gegenwart - Zukunft*, Springer, New York, 127–144.

15. Olanow CW and Drayer B: Brain iron: MRI studies in Parkinson's syndrome. In Fahn S, Marsden CD, Calne D and Goldstein M (eds), *Recent Developments in Parkinson's Disease*, Vol. II, Macmillan, Florham Park, NJ, 135–143.

16. Chen JC, Hardy PA, Kucharczyk W, et al: MR of human post-mortem brain tissue: correlative study between T2 and assays of iron and ferritin in Parkinson and Huntington disease. *Am. J. Neuroradiol.*, 1993; 14:275–281.

17. Perl D and Good DF: Comparative techniques for determinating cellular iron distribution in brain tissues. *Ann. Neurol.*, 1992; Suppl. 32:S76–S81.

18. Hirsch EC, Brandet JP, Galle P, and Agid Y: Iron and aluminum increase in the substantia nigra of patients with Parkinson's disease. *J. Neurochem.*, 1991; 56:446–451.

19. Dedman DJ, Treffry A, Candy JM, et al: Iron and aluminum in relation to brain ferritin in normal individuals and Alzheimer's disease and chronic renal-dialysis patients. *Biochem. J.*, 1992; 287:509–514.

20. Riederer P and Youdim MBH: *Iron in Central Nervous System Disorders*. Springer-Verlag, New York.

21. Theil EC: Ferritin: structure, gene regulation and cellular function in animals, plants and microorganisms. *Annu. Rev. Biochem.*, 1987; 56:289–315.

22. Mann VM, Cooper JM, Daniel SE, et al: Complex I, iron and ferritin in Parkinson's disease substantia nigra. *Ann. Neurol.*, 1994; 36:876–881.

23. Dhar MS and Joshi JG: Detection and quantitation of the novel ferritin heavy chain message in human tissues. *Bio Factors*, 1994; 4: 147-149.

24. Jellinger K, Kienzl E, Rumpelmair G, et al: Iron-Melanin complex in substantia nigra of parkinsonian brains: An x-ray microanalysis. *J. Neurochem.*, 1992; 59:1168–1171.

25. Schwabe K, Lassmann G, Damerau W, and Naundorf H: Protection of melanome cells against superoxide radical melanin. *J. Cancer Res. Clin. Oncol.*, 1989; 115:597–600.

26. Ben-Shachar D, Riederer P and Youdim MBH: Iron-melanin interactions and lipid peroxidation. Implications for Parkinson's disease. *J. Neurochem.*, 1991; 57:1609–1614.

27. Sarafian TA and Bredesen DE: Is apoptosis mediated by reactive oxygen species? *Free Radical Res.*, 1994; 20:1–6.

28. Dexter DT, Carter CJ, Wells FR, et al: Basal lipid peroxidation in substantia nigra is increased in Parkinson's disease. *J. Neurochem.*, 1989; 52:381–389.

29. Dexter DT, Holley AE, Flitter WD, et al: Increased levels of lipid hydroperoxides in the parkinsonian substantia nigra. An HPLC and ESR study. *Movement Dis.*, 1994; 9:92–97.

30. Sanchez-Ramos JR, Overvik E, and Ames BN: A marker of oxyradical mediated DNA damage (8-hydroxy-2 deoxyguanosine) is increased in nigro-striatum of Parkinson's disease brain. *Neurodegeneration*, 1994; 3:197–204.

31. Cleeter MW, Cooper JM, and Schapira AHV: Irreversible inhibition of mitochondrial complex I by 1-methyl-4-phenylpyridinium: evidence for free radical involvement. *J. Neurochem.*, 1992; 58:786–789.

32. Schapira AHV, Cooper JM, Dexter DT, et al: Mitochondrial complex I deficiency in Parkinson's disease. *J. Neurochem.*, 1990; 54:823–827.

33. Saggu H, Cooksey J, Dexter DT, et al: A selective increase in particulate superoxide dismutase activity in parkinsonian substantia nigra. *J. Neurochem.*, 1989; 53:692–697.

34. Hass MA and Massaro D: Regulation of the synthesis of superoxide dismutase in rat lung during oxidant and hyperthermic stresses. *J. Biol. Chem.*, 1988; 263:776–781.

35. Riederer P, Sofic E, Rausch WD et al: Patho-biochemistry of the extrapyramidal system: a "short note" review. In Przuntek H and Riederer P (eds), *Early Diagnosis and Preventative Therapy in Parkinson's Disease*, Springer-Verlag, New York, 139–149.

36. Ambani LM, Van Woert MH and Murphy S: Brain peroxidase and catalase in Parkinson's disease. *Arch. Neurol.*, 1975; 32:114–118.

37. Kish SJ, Morito C and Hornykiewicz O: Glutathione peroxidase activity in Parkinson's disease brain. *Neurosci. Lett.*, 1985; 58:343–346.

38. Sian J, Dexter DT, Lees AJ, et al: Glutathione-related enzymes in brain in Parkinson's disease. *Ann. Neurol.*, 1994; 36:356–361.

39. Sian J, Dexter DT, Lees AJ, et al: Alterations in glutathione levels in Parkinson's disease and other neurodegenerative disorders affecting basal ganglia. *Ann. Neurol.*, 1994; 36:348–355.

40. Gibb WRG and Lees AJ: The relevance of the Lewy body to the pathogenesis of idiopathic Parkinson's disease. *J. Neurol. Neurosurg. Psychiatr.*, 1988; 51:745–752.

41. Fearnley JM and Lees AJ: Ageing and Parkinson's disease: substantia nigra regional selectivity. *Brain*, 1991; 114:2283–2301.

42. Dexter DT, Sian J, Rose S, et al: Indicies of oxidative stress and mitochondrial function in individuals with incidental Lewy body disease. *Ann. Neurol.*, 1994; 35:38–44.

43. Pall HS, Blake DR, Gutteridge JM, et al: Raised cerebrospinal-fluid copper concentrations in Parkinson's disease. *Lancet*, 1987; i:238–241.

44. Leenders KL: PET-studies in neurodegeneration. Symposium on neuroprotection in neurodegeneration, 21–24 April 1993, Wurzburg, Germany.

45. Weiner WJ, Nausieda PA, and Klawans HL: Effect of chlorpromazine on central nervous system concentrations of manganese, iron and copper. *Life Sci.*, 1977; 20:1181–1186.

46. Mash DC, Sanchez-Ramos J and Weiner WJ: Distribution and number of transferrin receptors in Parkinson's disease and in MPTP treated mice. *Ann. Neurol.*, 1990; 28:230–234.

47. Graeber MB, Raivich G and Kreutzberg GW: Increase of transferrin receptors and iron uptake in regenerating motor neurones. *J. Neurosci. Res.*, 1989; 23:342–345.

48. Ben-Shachar D and Youdim MBH: Intranigral iron injection induces behavioral and biochemical "Parkinsonism" in rats. *J. Neurochem.*, 1991; 57:2133–2135.

49. Ward RJ, Dexter DT, Florence A, et al: Brain iron in the ferrocene-loaded rat: its chelation and influence on dopamine metabolism. *Biochem. Pharmacol.*, 1995; 49:1821–1826.

Chapter 38

LAMMA Studies of Iron, Oxidative Stress, and Neuroprotective Strategies in Parkinson's Disease

Paul F. Good, C. Warren Olanow, and Daniel P. Perl

CONTENTS

38.1 INTRODUCTION

Neurodegenerative diseases, in particular those affecting the basal ganglia and motor neurons, are characterized by the selective degeneration and loss of specific neuronal populations. In virtually all of these diseases the neurons and circuits undergoing degeneration have been identified, yet the causes and mechanisms of such neurodegeneration remain largely unknown. While system-specific neurodegeneration encompasses many differing diseases and anatomic systems, the present article focuses on certain disorders affecting the basal ganglia and motor system: Parkinson's disease (PD), amyotrophic lateral sclerosis (ALS), and a variant of the two, amyotrophic lateral sclerosis/parkinsonism dementia complex (ALS/PDC) of Guam.

A number of recent studies have provided evidence for a novel concept of the mechanism of cell death in Parkinson's disease and other neurodegenerative disorders: that of oxidative stress. It is now thought that a state of intracellular oxidative stress results from factors specific to the neurons targeted by the disease process, leading to excess free radicals which results in selective cellular damage and subsequent neurodegeneration. One factor capable of producing oxidative stress is an excess of available intracellular iron. Oxidative stress may derive from different sources, and may be either an initiating cause of disease, or a consequence of other primary factors; nevertheless, in

379

either case, iron-induced oxidative stress represents a testable hypothesis in animal models as well as in therapeutic trials.

In humans, and probably all mammals, iron is normally found in the brain at concentrations approaching that of liver, especially in the globus pallidus and substantia nigra pars reticulata. Features of brain iron have been extensively reviewed in References 15, 16, 22, 26, 47, and 48 as well as in Chapter 37 of the present volume.

38.2 IRON AND OXIDATIVE STRESS

Oxidative stress is a state in which there is either an overproduction of reactive oxygen species (ROS), an underactivity of cellular protective mechanisms, or both, leading to uncontrolled oxidation of critical biological molecules. Under physiologic conditions an equilibrium exists determined by factors which promote or defend against the formation of free radicals. A change in equilibrium favoring the formation of ROS can result in damage to neighboring lipid membranes, proteins, and nucleic acids with consequent cell degeneration. Pushing cellular equilibria toward oxidative stress are oxidation reactions catalyzed by transition metals, such as iron, which can exist in more than one valence state.

Reactive oxygen species, superoxide, hydrogen peroxide, and hydroxyl radicals, can be produced in the sequence of reactions which occurs during oxidative phosphorylation, essential for the formation of high energy compounds:

$$O_2 \xrightarrow{+e^-} \cdot O_2^- \xrightarrow{+e^-} H_2O_2 \xrightarrow{+e^-} \cdot OH + OH^- \xrightarrow{+e^-} H_2O$$

$$\text{superoxide} \quad + 2H^+ \quad \underset{\text{peroxide}}{\text{hydrogen}} \quad \underset{\text{radical}}{\text{hydroxyl}} \quad + 2H^+$$

Iron can play a role in the production of ROS in a number of different ways, and an increase in iron concentration is associated with an increased likelihood that oxidation reactions will occur and that free radicals will be formed.[23,24] As iron changes between its Fe^{+2} and its Fe^{+3} valence states it can donate or accept an electron and thus promote redox reactions and the formation of cytotoxic radicals.

Perhaps the best known iron-catalyzed free radical reaction is the Fenton reaction:

$$H_2O_2 + Fe^{+2} \rightarrow \cdot OH + OH^- + Fe^{+3}$$

which readily converts hydrogen peroxide to the most cytotoxic of the reactive oxygen species, the hydroxyl radical. Recycling of iron between its oxidized and its reduced state allows iron to act as a catalytic agent, producing a cascade of free radicals. Recycling of iron to its reduced state can be mediated by tissue ascorbate, glutathione, or dopamine, as well as by superoxide in the Haber-Weiss reaction:

$$\cdot O_2^- + e^- + 2H^+ \rightarrow H_2O_2$$

$$\cdot O_2^- + Fe^{+3} \rightarrow O_2 + Fe^{+2}$$

$$H_2O_2 + Fe^{+2} \rightarrow \cdot OH + OH^- + Fe^{+3}$$

Nitrosylation or hydroxylation may occur through the mechanism of nitric oxide combining with peroxynitrite:

$$NO^{\cdot} + {}^{\cdot}O_2^- \rightarrow ONOO^-$$

nitric   peroxynitrite
oxide

Peroxynitrite, when protonated, can decompose into hydroxyl and nitrogen dioxide radicals:

$$ONOO^- + H^+ \rightarrow HONOO \rightarrow {}^{\cdot}OH + NO_2^{\cdot}$$

Nitration of tyrosine in the ortho position by nitronium ions, readily catalyzed by iron,[4] is a major product of peroxynitrite, resulting in damaged proteins.

All of the above reactions involve free radical attack on biomolecules as a final common pathway, producing oxidatively modified molecules including lipid hydroperoxides, carbonyl and nitrosylated modification of proteins, and DNA fragmentation. These mechanisms have been implicated in a variety of diseases including ischemia-reperfusion injury, arthritis, cataracts, heart disease, and cancer.[45]

A number of mechanisms have evolved to prevent or stop an oxidative cascade at differing points and thus reduce the likelihood of subsequent biochemical damage. These include ferritin, capable of storing iron in an unreactive form, detoxification of reactive oxygen species by enzymatic systems, such as superoxide dismutase (SOD), glutathione/glutathione peroxidase, and free radical scavengers such as ascorbate and α-tocopherol.

38.3 PARKINSON'S DISEASE

Parkinson's disease is caused by the selective degeneration of melanized dopaminergic neurons of the substantia nigra pars compacta. This degeneration results in the loss of dopaminergic neurotransmission to the basal ganglia, structures responsible for initiation and control of motor activity. While other clinical and pathological studies point to additional loci of pathological involvement in Parkinson's disease, understanding of the cause of degeneration of the dopaminergic neurons of the substantia nigra pars compacta remains a critical focus of Parkinson's disease research.

Substantia nigra neurons may be particularly at risk for damage by free radicals because hydrogen peroxide (H_2O_2) is generated by dopamine metabolism, either enzymatically by monoamine oxidase (MAO), or by autooxidation. Under normal circumstances H_2O_2 is cleared by glutathione in a reaction that is catalyzed by glutathione peroxidase. However, in the presence of reactive iron, H_2O_2 can generate highly reactive hydroxyl radicals.

38.3.1 Iron and Parkinson's Disease

Much of the current interest in the free radical hypothesis of PD is based on the finding, in studies employing many different methodologies, of increased iron in the SNc of PD patients, and on iron's known capacity to promote oxidant stress. Earle,[12] using X-ray fluorescent spectroscopy, initially reported that the brains of patients with PD had a two to three times higher concentration of iron when compared to normals. Subsequently, high field strength (1.5 Tesla) magnetic resonance imaging (MRI) studies demonstrated that signal hypointensity, consistent with iron accumulation, was increased in the substantia nigra of patients with PD and in the substantia nigra and striatum of patients with atypical parkinsonism.[11,35] Determinations of iron using bulk analytic methods confirmed these observations and noted that iron increase in PD was primarily confined to the SNc.[10,44] However, bulk histochemical and imaging techniques suffer from a number of drawbacks. The bulk analytic approach can provide precise data on regional differences, as has been shown for normal individuals[22] and patients with Parkinson's disease.[10,44] However, because of the brain's unique cellular diversity, one must carefully interpret bulk analytic studies as they relate to neurologic disorders since these techniques frequently miss alterations in the concentration of particular elements associated with specific cell types or focal pathologic processes present within the sample being analyzed. Bulk analysis is a technique that, of necessity, homogenizes the sample and consequently may dilute out focal accumulations which are characteristic of a given pathologic process. While bulk analyses may reveal gross alterations of elemental content, properly performed microprobe studies at the cellular and subcellular level of resolution are essential to reach an understanding of their possible pathogenetic role.

Histochemistry (Perls stain) is a relatively simple and inexpensive method that can localize iron in cellular compartments. However, it is relatively insensitive and only detects iron in the ferric form. The Perls stain procedure involves pretreatment with acid to liberate iron from loosely bound constituents prior to its reaction with potassium ferrocyanide to form the visible Prussian blue (ferric ferrocyanide). Iron in the ferrous form, which is more tightly bound, is not available for this reaction and does not stain. Hemosiderin will usually stain readily with Perls stain but iron in this form poses little threat to tissue because it is incorporated as ferric hydroxide and cannot be converted to Fe^{+2} necessary for participation in Fenton chemistry.[34] On the other hand, neuromelanin granules, which selectively accumulate iron[18] in a potentially critical site, do not stain with Perls stain in normal controls or in patients with PD.[30] Consequently, the use of Perls stain to monitor issues related to iron-mediated oxidant stress may be completely misleading.

38.4 MICROPROBE STUDIES OF TRACE ELEMENTS IN THE BRAIN

A determination of the precise cellular and subcellular distribution of iron in PD is of critical importance in evaluating its potential role in pathogenesis. Initial attempts to determine the subcellular distribution of iron in the SNc have employed relatively insensitive and imprecise methods. Using X-ray microanalysis and Perls stain, initial studies suggested that iron in PD is localized to areas devoid of neuromelanin.[44] Further studies using an X-ray microprobe have provided conflicting results, suggesting that iron both does and does not accumulate within neuromelanin granules.[27,29] X-ray microprobe analysis generally employs scanning electron microscopy (SEM) for imaging the specimen and identification of targets for analysis. In sectioned brain tissue the surface images provided by SEM do not provide sufficient cellular detail to enable definitive identification of a specific cell type, subcellular components, or pathological inclusions. X-ray spectrometry also has a relatively high minimum detection level for most elements of interest (approximately 200 to 300 ppm for iron) and thus may not detect lower concentrations of trace elements in many pathologic conditions. Additionally, because of surface irregularities of sectioned biologic tissues and variability in the depth of penetration of the electron beam, this technique does not provide consistent quantitative measures of the amount of element present in tissue samples. Accordingly, one cannot reliably compare the concentration of a particular element in one brain specimen with that found in another.

38.4.1 The Laser Microprobe

The laser microprobe mass analyzer, or LAMMA, which we employ in our laboratory for trace element analysis, provides precise histological detail through the use of semithin plastic-embedded sections, elemental detection limits on the order of 5 ppm, and the ability, with properly prepared standard samples, to quantitate the concentration of a given element of interest within the experimental tissue sample. The LAMMA incorporates a high resolution ($100\times$, N.A. 1.4) quartz objective which permits precise histologic identification of target sites measuring 1 µm in diameter. Elements are separated and detected by a time-of-flight mass spectrometer to produce a mass spectrum with simultaneous identification of elements throughout the entire periodic table. Additionally, we and others have found that the intensity of the spectral signal is proportional to the concentration of the element present within the specimen. Thus, relative concentrations can be determined by comparisons of peak intensity.[17,46] Alternatively, by using comparisons to assayed biological standards, quantitative estimates of intracellular concentration of individual elements can be determined.[19] In contrast to bulk analytic methods, the LAMMA collects elemental data from precisely defined subcellular targets thereby providing the means to identify accumulations of elements of interest and associate such accumulations with specific cellular sites.

The LAMMA provides analysis of a sample typically consisting of a plastic-embedded semithin (typically 0.75-μm-thick) histological section which is prepared from a tissue sample embedded as for electron microscopy. The section is stained with toluidine blue and mounted on a standard 3-mm electron microscopy grid. The grid is mounted on an X-Y stage traverse and placed in a high vacuum behind a quartz coverglass. The LAMMA is equipped with a high-power pulsed Nd:YAG laser that can be focused through the 100× objective of the light microscope to a spot on the specimen measuring approximately 1 μm in diameter. The high-energy laser is directed to the point of interest by a collinear continuous low-energy He:Ne laser which is visible as a red dot on the section. A brief (15 ns) pulse of the high-energy laser vaporizes and ionizes a small area of the tissue specimen, creating a minute perforation in the section. The ions produced by the laser perforation are attracted by a series of charged ion lenses into a 1.8-m-long time-of-flight mass spectrometry column where the ions are separated and detected according to their atomic mass number.

The tissues that are probed by LAMMA are examined as stained histologic sections with high-resolution light microscopy. In contrast to bulk analytic methods, the LAMMA collects elemental data by targeting and sampling specific cellular features. Accordingly, cells of interest are identified and targeted for analysis based on their light microscopic appearance. The size of the probe site (laser perforation) is sufficiently small that specific intracellular sites can be targeted and analyzed selectively. The site of microprobe analysis can be immediately confirmed by observing the location of the laser perforation which produced each individual spectrum.

38.4.2   Laser Microprobe Studies of Parkinson's Disease

Based on the findings of increased SNc iron in PD using MRI and bulk analytic methods, we examined the dopaminergic neurons of the substantia nigra pars compacta of PD and controls to determine the cellular and subcellular distribution of iron and other trace elements in PD. Brain specimens were derived from patients with Parkinson's disease through the University of South Florida Center for the Study of Neurodegenerative Disorders. The cases were chosen based on a classic clinical presentation of PD and a history of a good response to levodopa. Three age-matched controls were chosen for comparison based on normal extrapyramidal and cognitive functions during life and an absence of AD or PD changes on neuropathologic examination.

Intact neurons containing neuromelanin granules were targeted for LAMMA analysis. Laser probe sites were directed to neuromelanin granules, the nonmelanized portion of the cytoplasm, and to the neuropil adjacent to each selected neuron. Multiple laser shots were directed to each subcellular target and a total of ten neurons were examined in each case. The control cases were prepared and probed in a similar manner using identical instrumental parameters. Representative spectra are presented in Figure 38.1. The spectra for each target area were averaged and a multivariant ANOVA was performed to compare the intensity of aluminum and iron in the various target areas of the PD patients and the controls.

The mean integrated peak intensities for iron within probe sites directed to neuromelanin and adjacent neuropil for three PD patients and three controls are presented in Figure 38.2. Each data point represents the mean value of the total number of LAMMA acquisitions obtained for each target site. In specimens from control patients, intraneuronal neuromelanin granules contained prominent peaks related to the presence of iron in comparison to the nonmelanized neuronal cytoplasm or to the adjacent neuropil ($p < .05$). In patients with PD, the iron-related signal in the neuromelanin granules was significantly increased in comparison to controls ($p < .0001$). The iron-related peaks in the nonmelanized neuronal cytoplasm and in the adjacent neuropil were not significantly different between PD patients and controls. Interestingly, in two of three PD cases

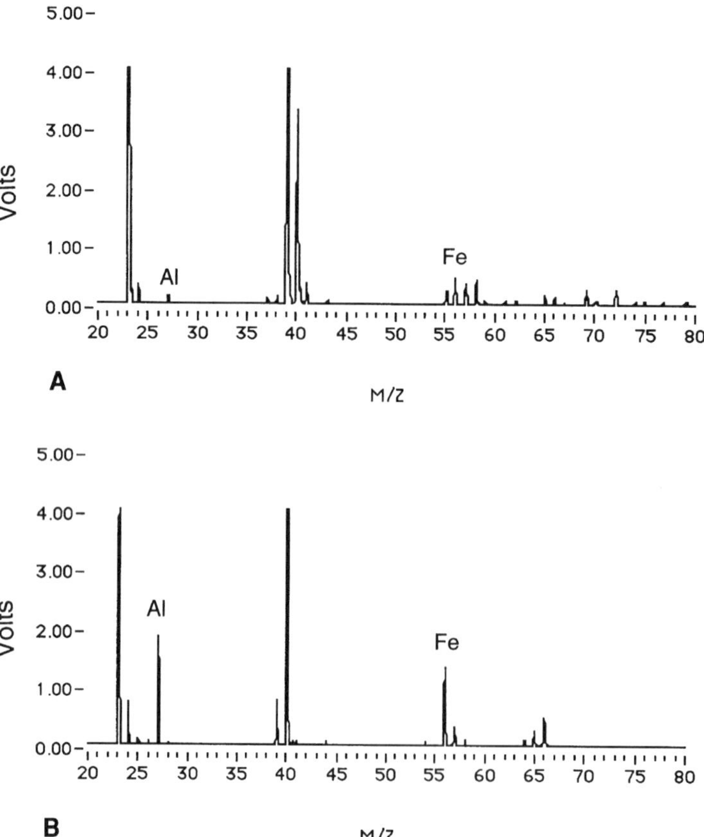

FIGURE 38.1 Representative partial mass spectra derived from a single probe site to neuromelanin of SNc neurons of (A) autopsy control, and (B) Parkinson's disease. Mass scale (m/z = mass to charge ratio) is indicated on the abscissa and peak height in volts on the ordinate. In the PD case note elevated peaks at mass 27 (aluminum) and 56 (major isotope of iron).

aluminum-related peaks were also dramatically increased in the neuromelanin granules in comparison to controls ($p < .005$).

The presence of aluminum colocalized with iron in neuromelanin is capable of producing a significant augmentation of pathology. While aluminum does not participate in Fenton chemistry because of its constant Al^{+3} oxidation state, it is a known neurotoxin and has been shown to increase iron-catalyzed lipid peroxidation by eightfold.[21] It was suggested by these authors that the effect of aluminum was mediated by a rearrangement of membrane lipids, facilitating peroxidative attack by iron.

38.5 AMYOTROPHIC LATERAL SCLEROSIS

Recent discoveries concerning familial ALS (FALS) suggest that neurodegeneration in ALS may also be a result of oxidative stress. In familial amyotrophic lateral sclerosis, a relatively uncommon form of the disease occurring in approximately 5% of all cases, multiple mutations in the gene encoding copper/zinc superoxide dismutase (Cu/Zn SOD) have been demonstrated. Rosen and colleagues[41] reported 11 different point mutations involving single amino acid substitutions in the cytosolic Cu/Zn superoxide dismutase (SOD1) gene in 15 families with FALS. These changes were not detected in more than 100 controls and were not considered to represent normal allelic variants. The development of FALS in association with 11 missense mutations in the SOD1 gene implies that an alteration in free radical defenses is critical to the development of FALS and is not simply a coexpressed marker. Superoxide dismutase is an important cellular mechanism which provides

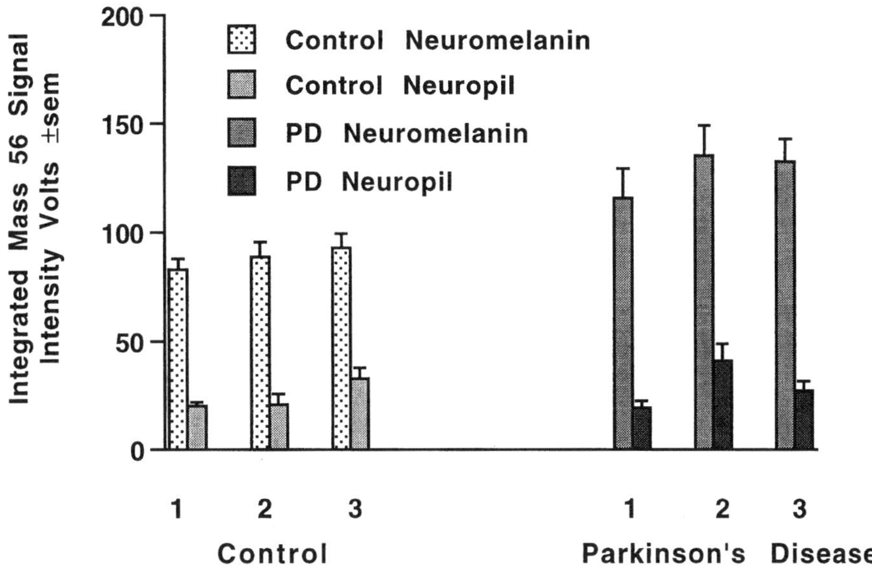

FIGURE 38.2   Mean integrated mass 56 signal intensity (± SEM) of probe sites to neuromelanin and adjacent neuropil of SNc neurons of autopsy controls and Parkinson's disease. In patients with PD, the iron-related signal in the neuromelanin granules was significantly increased in comparison to controls (p <.01 ANOVA and Student's t test). (From Good PF, Olanow CW, and Perl DP: *Brain Res.*, (1992) 593:343-346. With permission.)

protection against oxidative stress, and is responsible for detoxification of superoxide, producing hydrogen peroxide. That FALS is produced by defective SOD has been further demonstrated through the use of transgenic mice.[20,40] The phenotypic expression of the mutation in such mice is a selective degeneration of motor neurons with sparing of sensory neurons.[9,40] In addition, a second SOD1-mutation transgenic mouse model of ALS[40] has been produced which appears to more closely parallel the changes seen in sporadic ALS and which may now be tested for features of altered intracellular iron and oxidative damage to proteins.

Although oxidative stress is clearly involved in the production of neuronal death in FALS, the role it plays in sporadic ALS is less certain. The similar nature of the FALS and sporadic amyotrophic lateral sclerosis (SALS) phenotype[9] strongly argues that oxidative stress is, in fact, the underlying pathogenetic mechanism of SALS as well. While data regarding intraneuronal iron in ALS, in contrast to PD, have not as yet been forthcoming, the implication of free radical mechanisms in FALS provides strong impetus to investigate vulnerable neuronal systems in ALS for evidence of excess free iron. Whether increased iron or reduced defensive mechanisms are at the root of SALS must await further analysis of these brain specimens.

38.6 AMYOTROPHIC LATERAL SCLEROSIS/PARKINSONISM DEMENTIA COMPLEX OF GUAM

The island of Guam is the focus of an epidemic of neurodegenerative diseases with features of both ALS and parkinsonism, known as amyotrophic lateral sclerosis/parkinsonism dementia complex (ALS/PDC) of Guam. Although this disease complex occurs at an incidence of many times the rate of sporadic neurodegenerative disease elsewhere in the world, its etiology is unknown. Analysis of this epidemic has demonstrated that there is an extensive overlap among neuropathological changes found among clinically described cases of amyotrophic lateral sclerosis (ALS) and parkinsonism dementia complex (PDC). Post-mortem neuropathological examination of Guamanian ALS cases reveals system-specific neurodegeneration involving progressive loss of SNc neurons typical of Parkinson's disease (PD) in about 40% of cases, and similarly, motor neuron degeneration is seen in about 40% of parkinsonism dementia cases. These findings suggest that ALS/PDC of Guam is in fact a single disease entity caused by a single etiological agent with multiple modes of expression.

We hypothesize that oxidative stress represents the principal underlying mechanism of both the ALS and PDC manifestations, and is the common factor unifying the pathogenesis of these diseases. Additionally, in a recent study[14] of the Cu/Zn SOD gene from individuals diagnosed with ALS/PDC, no mutations were found, indicating that a genetic source for this disorder was unlikely. These concepts of oxidative stress are testable in the setting of ALS/PDC of Guam and have important implications for our understanding of sporadic neurodegenerative disease as seen elsewhere in the world.

We have recently investigated cases of ALS/PDC for the presence of increased iron in the substantia nigra of affected individuals. Figure 38.3 presents the results of this investigation and demonstrates that, as in sporadic PD, there is increased iron in the SNc of ALS/PDC as well. However, it is still not known whether increased iron is a primary factor or a secondary consequence of neuronal damage from other causes.

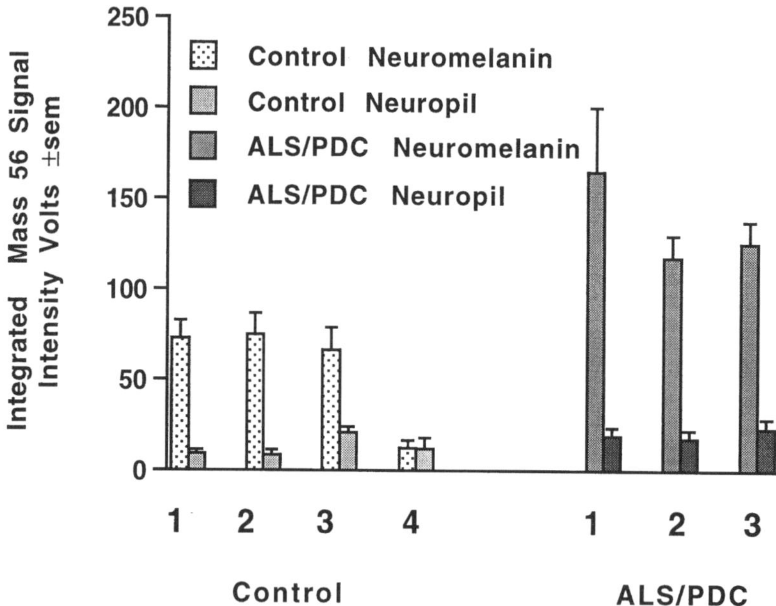

FIGURE 38.3   Mean integrated mass 56 signal intensity (± SEM) of probe sites to neuromelanin and adjacent neuropil of SNc neurons of U.S. autopsy controls and ALS/PDC of Guam. In patients with ALS/PDC, the iron-related signal in the neuromelanin granules was significantly increased in comparison to controls (p <.01 ANOVA and Student's t test).

38.7 ANIMAL MODELS OF IRON-INDUCED OXIDATIVE STRESS

Based on finding increased concentrations of iron within the SNc in PD, in an attempt to develop a model of PD based on iron-induced oxidant stress[42,43] we have infused low concentrations of iron citrate directly into the SNc of rats. The importance of iron as a mediator of neuronal damage in models of PD is illustrated by studies which show that iron chelators diminish neuronal toxicity associated with 6-hydroxydopamine administration.[5] We and others have demonstrated that SNc iron infusion induces focal neuronal degeneration and a dose-related decline in striatal dopamine.[6,43] These effects were associated with a transient increase in lipid peroxidation and were attenuated by coadministration of the iron-binding protein transferrin, suggesting that nigral damage occurs as a consequence of iron-induced oxidant stress. More recently, we have shown that a single infusion of a low concentration of iron placed directly into the SNc can induce progressive histologic, biochemical, and behavioral changes.[3] These findings suggest that iron infusion into the SNc induces a model of parkinsonism that more closely resembles PD than does 6-OHDA or MPTP, which tend to remain static or spontaneously improve. This iron infusion model in the rat also provides a system in which potential therapies designed to prevent degeneration or to rescue neurons from incipient degeneration may be tested.

38.8 THERAPEUTIC STRATEGIES

From a clinical perspective, the free radical-oxidative stress hypothesis in PD provides a scientific rationale for trials of antioxidants in the hope that they might be "neuroprotective" and modify the natural progression of the disease. The recently completed DATATOP study[36] demonstrated that the selective MAO-B inhibitor deprenyl delays development of clinical disability in otherwise untreated PD patients. There are a number of possible mechanisms which may explain this effect. First, the inhibition of MAO-B results in decreased metabolism of dopamine and the maintenance of dopaminergic neurotransmission. Secondly, the corollary to reduced oxidative metabolism of dopamine is a reduction in reactive oxygen species. However, deprenyl does not stop the progression of PD, and at the dose used in this study it does not prevent formation of peroxides generated by MAO-A or by autooxidation of dopamine. It is possible that more potent antioxidants could have a more profound neuroprotective effect and influence the natural progression of Parkinson's disease.

Recent evidence, however has demonstrated that the therapeutic action of deprenyl may be the result of mechanisms other than the inhibition of MAO-B. Tatton and co-workers[2] have demonstrated that axotomized facial motoneurons in the rat are rescued by deprenyl at concentrations too low to inhibit MAO-B and they speculate that the effects of deprenyl at these low concentrations is mediated by the same pathway as that of trophic factor-induced neuronal rescue. The significance of such findings is that axotomized neurons undergoing degeneration as a result of the loss of trophic factors do so by the mechanism of apoptosis. Apoptosis, often thought of as "cellular suicide" in contrast with necrosis or "cellular homicide", is postulated as a possible mechanism in neurodegenerative diseases and is thought to be mediated by free radical mechanisms.[28] While a clear link between Parkinson's disease and apoptosis has not been established, evidence demonstrating this possibility has been forthcoming[25,49] and indicates that deprenyl, neurotrophic factors, or other therapies designed to counter apoptotic events are worth pursuing.

38.9 REMAINING QUESTIONS

It is unclear why the brain, which has a limited capacity to maintain iron in a nonreactive state and a relatively deficient capacity for repair, should normally accumulate this potential toxin in concentrations which far exceed its known physiologic requirements. Many important questions about iron remain unresolved. How does excess iron gain access to vulnerable neuronal populations in pathologic conditions such as PD? Is it a consequence of a deficient blood-brain barrier or a redistribution of existing brain pools? Using ^{52}Fe PET, Leenders et al. recently reported an increase in iron transport across the blood-brain barrier in PD patients as compared to controls.[31] Moreover, in Alzheimer's disease, Connor et al.[7] have shown an increase in perivascular iron and ferritin immunoreactivity, suggesting an abnormality in the blood-brain barrier for iron transport.

Where precisely, at the cellular and subcellular level, and when in the course of the disease, does iron accumulate in PD? This information would provide critical knowledge essential to begin to understand the role of iron in PD and other neurodegenerative disorders. Iron accumulation within neuromelanin granules could explain the vulnerability of these cells. Excess iron within glial cells would suggest that they participate in the degenerative process and that a loss of trophic support may contribute to neuronal death. This might also account for the finding in PD of a profound reduction in glutathione, a free radical defense, which is primarily localized within glial cells. Accumulation of iron (and particularly aluminum) in a perivascular distribution would suggest that a breakdown in the blood-brain barrier is an important feature in PD. Finally, a determination that iron accumulation prior to the emergence of clinical or pathological features, in areas destined to undergo degeneration would provide evidence in favor of iron playing a primary role in the cell degeneration that characterizes PD. This fundamental information is necessary in order to evaluate the role of iron and other trace elements in the pathogenesis of PD.

Is iron that accumulates in PD in a reactive form? Studies of ferritin levels in Parkinson's disease as compared with controls have generated differing results. Most published reports suggest that ferritin fails to rise in response to an increase in iron[8,33] and indicate that iron may be unbound and capable of inducing oxidant stress, while a finding of increased ferritin in Parkinson's disease[39] has been interpreted as representing an increase of available iron.

Is iron a primary etio/pathogenetic factor in PD? Findings of point mutations in the SOD1 gene in FALS suggest that free radical damage may contribute to a sequence of progressive neurodegenerative changes, leading to the classic phenotypic expression of ALS. Such damage may arise as a result of impaired defenses as in FALS or from iron-induced increased free radical production in PD. Many studies have shown that iron can accumulate in association with a number of neurodegenerative processes, including multisystem atrophy, Huntington's disease, and in MPTP-induced parkinsonism. Studies of other disorders have also shown iron levels to be increased in parallel with pathological damage. Thus, in Hallevorden-Spatz disease, multiple sclerosis, and spastic paraplegia there are reports of increased iron deposition. The association between pathological damage to substantia nigra and the deposition of iron is clearly demonstrated by the increased iron levels detected within the brains of MPTP-treated primates.

Accordingly, changes in iron content are not selectively associated with nigral cell damage or with Parkinson's disease. Therefore, at present it is not possible to determine if increased iron is a primary event or follows an initial insult in the pathogenesis of parkinsonism. This does not mean that increases in iron in the substantia nigra are not an important component of Parkinson's disease. A common mechanism involving iron deposition may effectively act as an accelerator of pathological damage without regard to the primary cause of the specific disease.

Understanding the pathogenesis of Parkinson's disease is of extreme importance in view of the fact that PD is a debilitating and inexorably progressive chronic illness that is currently estimated to affect 1 million Americans. With increasing numbers of individuals reaching advanced age, it is estimated that by the year 2040 some 250,000 new cases of PD will be diagnosed each year.[32] At present there are numerous avenues of research that it is hoped will lead to more effective therapies for this disease.

REFERENCES

1. Anderson FH, et al: Neurofibrillary degeneration on Guam. Frequency in Chamorros and non-Chamorros with no known neurological disease. *Brain*, (1979) 102:65–77.

2. Ansari KS, et al: Rescue of axotomized immature rat facial motoneurons by R(-)-deprenyl: stereospecificity and independence from monoamine oxidase inhibition. *J. Neurosci.*, (1993) 13:4042–53.

3. Arendash GW, Olanow CW, and Sengstock GJ: Intranigral iron infusion in rats: A progressive model for excess nigral iron levels in Parkinson's disease?, in *Iron in Central Nervous System Disorders*, P. Riederer and M. Youdim, Editors. 1993, Springer-Verlag: Vienna. p. 87–101.

4. Beckman JS, et al: Kinetics of superoxide dismutase- and iron-catalyzed nitration of phenolics by peroxynitrite. *Arch. Biochem. Biophys.*, (1992) 298:438–445.

5. Ben-Shachar D, et al: The iron chelator desferrioxamine (Desferal) retards 6-hydroxydopamine-induced degeneration of nigrostriatal dopamine neurons. *J. Neurochem.*, (1991) 56:1441–4.

6. Ben-Shachar D and Youdim MB: Intranigral iron injection induces behavioral and biochemical "parkinsonism" in rats. *J. Neurochem.*, (1991) 57:2133–5.

7. Connor JR, et al: A histochemical study of iron, transferrin, and ferritin in Alzheimer's diseased brains. *J. Neurosci. Res.*, (1992) 31:75–83.

8. Connor JR, et al: A quantitative analysis of isoferritins in select regions of aged, parkinsonian, and Alzheimer's diseased brains. *J. Neurochem.*, (1995) 65:717–724.

9. Dal Canto MC and Gurney ME: Development of central nervous system pathology in a murine transgenic model of human amyotrophic lateral sclerosis. *Am. J. Pathol.*, (1994) 145:1271–1279.

10. Dexter DT, et al: Increased nigral iron content and alterations in other metal ions occurring in brain in Parkinson's disease. *J. Neurochem.*, (1989) 52:1830–6.

11. Drayer B, et al: MRI of brain iron. *AJNR*, (1986) 147:103–110.

12. Earle KM: Studies on Parkinson's disease including x-ray fluorescent spectroscopy of formalin fixed brain tissue. *J. Neuropathol. Exp. Neurol.*, (1968) 27:1–14.

13. Elizan TS, et al: Amyotrophic lateral sclerosis and parkinsonism-dementia complex of Guam. Neurological reevaluation. *Arch. Neurol.*, (1966) 14:356–368.

14. Figlewicz DA, et al: The Cu/Zn superoxide dismutase gene in ALS and parkinsonism-dementia of Guam. *Neuroreport*, (1994) 5:557–60.

15. Fishman JB, et al: Receptor-mediated transcytosis of transferrin across the blood-brain barrier. *J. Neurosci. Res.*, (1987) 18:299–304.

16. Francois C, Nguyen-Legros J, and Percheron G: Topographical and cytological localization of iron in rat and monkey brains. *Brain Res.*, (1981) 215:317–322.

17. Good PF, Katz RN, and Perl DP: Calculation of intraneuronal concentration of aluminum in neurofibrillary tangle (NFT) bearing and NFT free hippocampal neurons of Alzheimer's disease using laser microprobe mass analysis (LAMMA). *J. Neuropathol. Exp. Neurol.*, (1991) 50:300.

18. Good PF, Olanow CW, and Perl DP: Neuromelanin-containing neurons of the substantia nigra accumulate iron and aluminum in Parkinson's disease: a LAMMA study. *Brain Res.*, (1992) 593:343–346.

19. Good PF and Perl DP: A quantitative comparison of aluminum concentration in neurofibrillary tangles of Alzheimer's disease and parkinsonian dementia complex of Guam by laser microprobe mass analysis. *Neurobiol. Aging*, (1994) 15:S28.

20. Gurney ME, et al: Motor neuron degeneration in mice that express a human Cu,Zn superoxide dismutase mutation. *Science*, (1994) 264:1772–1775.

21. Gutteridge JMC, et al: Aluminum salts accelerate peroxidation of membrane lipids stimulated by iron salts. *Biochem. Biophys. Acta*, (1985) 835:441–447.

22. Hallgren B and Sourander P: The effect of age on the non-haemin iron in the human brain. *J. Neurochem.*, (1958) 3:41–51.

23. Halliwell B and Gutteridge JMC: Oxygen toxicity, oxygen radicals, transition metals and disease. *J. Biochem.*, (1984) 219:1–14.

24. Halliwell B and Gutteridge JMC: Oxygen free radicals and iron in relation to biology and medicine; some problems and concepts. *Arch. Biochem. Biophys.*, (1986) 246:501–514.

25. Hartley A, et al: Complex I inhibitors induce dose-dependent apoptosis in PC12 cells: relevance to Parkinson's disease. *J. Neurochem.*, (1994) 63:1987–90.

26. Hill JM and Switzer RC, The regional distribution and cellular localization of iron in the rat brain. *Neuroscience*, (1984) 11:595–603.

27. Hirsch EC, et al: Iron and aluminum increase in the substantia nigra of patients with Parkinson's disease: an X-ray microanalysis. *J. Neurochem.*, (1991) 56:446–51.

28. Hockenbery DM, et al: Bcl-2 functions in an antioxidant pathway to prevent apoptosis. *Cell*, (1993) 75:241–51.

29. Jellinger K, et al: Iron-melanin complex in substantia nigra of parkinsonian brains: an x-ray microanalysis. *J. Neurochem.*, (1992) 59:1168–71.

30. Jellinger K, et al: Brain iron and ferritin in Parkinson's and Alzheimer's diseases. *J. Neural Transm. Park Dis. Dement. Sect.*, (1990) 2:327–40.

31. Leenders KL, et al: Brain iron uptake in patients with Parkinson's disease (PD) measured by positron emission tomography (PET). *Ann. Neurol.*, (1993) 43: A389.

32. Lilienfeld DE and Perl DP: Projected neurodegenerative disease mortality in the United States, 1990-2040. *Neuroepidemiology*, (1993) 12:219–28.

33. Mann VM, et al: Complex I, iron, and ferritin in Parkinson's disease substantia nigra. *Ann. Neurol.*, (1994) 36:876–881.

34. O'Connell M, et al: Formation of hydroxyl radicals in the presence of ferritin and hemosiderin: is hemosiderin formation a biological protective mechanism? *Biochem. J.*, (1986) 234:727–731.

35. Olanow CW: Magnetic resonance imaging in parkinsonism. *Neurol. Clin.*, (1992) 10:405–20.

36. Parkinson Study Group, Effects of tocopherol and deprenyl on the progression of disability in early Parkinson's disease. *N. Engl. J. Med.*, (1993) 328:176–183.

37. Perl DP: Amyotrophic lateral sclerosis/parkinsonism-dementia complex of Guam: two distinct clinical entities or a single disease process with a spectrum of neuropathological and clinical expression, in *Adv. Res. Neurodegen.*, Y.E.A. Mizuno, Editor. 1994, Birkhauser: Boston. p. 209–219.

38. Perl DP, et al: Changes in the outbreak of ALS/parkinsonism-dementia complex: neuropathological studies of asymptomatic Chamorros. *J. Neuropathol. Exp. Neurol.*, (1995) 54:416.

39. Riederer P, et al: Transition metals, ferritin, glutathione, and ascorbic acid in parkinsonian brains. *J. Neurochem.*, (1989) 52:515–20.

40. Ripps ME, et al: Transgenic mice expressing an altered murine superoxide dismutase gene provide an animal model of amyotrophic lateral sclerosis. *Proc. Natl. Acad. Sci. U.S.A.*, (1995) 92:689–693.

41. Rosen DR, Siddique T, and Patterson D: Mutations in Cu/Zn superoxide dismutase gene are associated with familial amyotrophic lateral sclerosis. *Nature*, (1993) 362:59–62.

42. Sengstock GJ, et al: Iron induces degeneration of nigrostriatal neurons. *Brain Res. Bull.*, (1992) 28:645–649.

43. Sengstock GJ, et al: Infusion of iron into the rat substantia nigra: nigral pathology and dose-dependent loss of striatal dopaminergic markers. *J. Neurosci. Res.*, (1993) 35:67–82.

44. Sofic E, et al: Selective increase of iron in substantia nigra zona compacta of parkinsonian brains. *J. Neurochem.*, (1991) 56:978–82.

45. Stadtman ER and Oliver CN: Metal-catalyzed oxidation of proteins. Physiological consequences. *J. Biol. Chem.*, (1991) 266:2005–2008.

46. Verbueken AH, et al: Laser microprobe mass spectrometry, in *Inorganic Mass Spectrometry*, F. Adams, R. Gibels, and R. Van Grieken, Editors. 1988, John Wiley & Sons, New York. p. 173–256.

47. Youdim MBH: Brain iron metabolism: biochemical and behavioural aspects in relation to dopaminergic neurotransmission, in *Handbook of Neurochemistry*, A. Lajha, Editor. 1985, Plenum Press: New York. p. 637–670.

48. Youdim MBH: ed. *Brain Iron: Neurochemical and Behavioural Aspects*. 1988, Taylor & Francis: London. 1–148.

49. Ziv I, et al: Dopamine induces apoptosis-like cell death in cultured chick sympathetic neurons — a possible novel pathogenetic mechanism in Parkinson's disease. *Neurosci. Lett.*, (1994) 170:136–40.

SECTION 13

COPPER

Copper/Zinc Superoxide Dismutases in Familial Amyotrophic Lateral Sclerosis: Tyrosine Nitration Catalysis as a Possible Mutational Gain in Function

Jacinda B. Sampson, John P. Crow, Michael J. Strong, and Joseph S. Beckman

CONTENTS

39.1 INTRODUCTION

Mutations to copper zinc superoxide dismutase (Cu,Zn SOD) were first linked in 1993 to a subset of dominantly inherited familial amyotrophic lateral sclerosis (FALS).[1] No other linkage to antioxidant enzymes or defenses has yet been found in amyotrophic lateral sclerosis (ALS) in spite of extensive search.[2] Recent experimental evidence strongly suggests that the mutations to Cu,Zn SOD cause some gain in enzymatic function rather than a loss of antioxidant function.[3–5]

The sporadic and familial forms are clinically indistinguishable, raising hope that study of ALS-associated mutations in SOD will illuminate the etiology of the sporadic form. We have previously proposed that the gained function of the SOD mutations may be an increased catalysis of tyrosine nitration by peroxynitrite.[6] We will discuss the interaction of peroxynitrite with Cu,Zn SOD and the nitration of tyrosines and suggest their relation to the pathogenesis of ALS in this review.

39.2 AMYOTROPHIC LATERAL SCLEROSIS

The familial form of ALS, which is usually inherited as an autosomal dominant trait, accounts for approximately 10% of all ALS cases. In about 12% of FALS families, a structural mutation in Cu,Zn SOD can be identified.[1] Carriers express the mutated gene throughout life and are generally healthy until disease onset. After diagnosis the disease progresses rapidly, with death occurring in an average of 5 years.

The clinical hallmark of ALS is an insidious progression of muscle wasting and weakness that may begin in virtually any voluntary muscle group. The majority of patients progress to death by respiratory failure. Clinical, electrophysiological and neuropathological studies have clearly demonstrated that ALS is primarily (though not exclusively) a disorder of both upper and lower motor neurons. Demyelination of the posterior columns which was not clinically evident has been found upon autopsy in some sporadic and familial cases.[7] Motor neurons innervating the sphincteric and extraocular muscles are typically spared, but may become involved late in the disease in those patients maintained on ventilators.

Although ALS is not generally considered a disease of inflammation, reactive gliosis, neuronophagia, and lymphocytic infiltrates are found in regions of degenerating motor neurons.[8] Lymphocytic infiltrates in ALS are predominantly cytotoxic T cells, and are not seen in control patients with Wallerian (anterograde) degeneration of upper motor neurons due to stroke.[9]

Apart from the neuropathological evidence of a selective loss of motor neurons, early autopsy cases of ALS demonstrate a number of intraneuronal inclusions consisting of interwoven skeins or arrays of 10-nm filaments which are either neurofilament or neurofilament degradative products. Some inclusions also stain for ubiquitin, a posttranslational modification which flags proteins for turnover.[10] Under normal conditions, the neurofilament subunit proteins NFH (200 kDa by SDS-PAGE), NFM (160 kDa), and NFL (68 kDa) self-assemble into a triplet protein which is transported slowly down the axon, where the fully assembled intermediate filament is thought to maintain axonal calibre.[11] Neurofilaments are extensively modified following synthesis, with the primary modification being phosphorylation of the carboxy-terminus region. Phosphorylation occurs in a somatofugal fashion, with little phosphorylated neurofilament found in the neuronal perikarya under normal conditions. Perikaryal and proximal phosphorylated neurofilament aggregates are observed in ALS, suggesting that the modification of neurofilament synthesis, transport, or processing plays an integral role in the pathogenesis of the disease.[12–15]

The theory that neurofilament accumulation could disrupt axonal transport and initiate disease is strongly supported by several transgenic mouse models. The overexpression of NFL or NFH in transgenic mice is accompanied by perikaryal and axonal aggregates of neurofilament in motor neurons and dorsal root ganglia. Clinically, mice develop a hindlimb paralysis and muscle atrophy reminiscent of human ALS.[16,17] In mice overexpressing the human NFH gene, an inhibition of axonal transport by the neurofilamentous aggregates is thought to induce axonal degeneration and muscle atrophy.[18] The introduction of a proline residue in the C-terminal domain of an NFL mutant construct will induce motor neuron degeneration with neurofilamentous aggregates.[19] Although the mechanism is unknown, the proline mutation may mimic the effect of oxidatively modified NFL on the neuron. Based upon the average length of a motor axon and the rate of slow axonal transport, Lee et al.[19] propose that NFL has a half life of 1 to 2 years. The abundance and longevity of neurofilaments would allow them to accumulate oxidative damage. This vulnerability of neurofilaments may be the reason why motor neurons are unusually susceptible in ALS.

Chronic aluminum exposure will induce motor neuron degeneration in New Zealand white rabbits in which neurofilamentous aggregation is prominent.[20] In this paradigm, aluminum alters the phosphorylation state of NFH through a posttranslational modification.[13,14,20]

In each scenario, the overexpression or disruption of neurofilament is sufficient to induce motor neuron degeneration. These models are given added weight by the recent discovery of an allelic variant of NFH in humans. This NFH variant was linked to ALS in 5 out of 356 sporadic ALS patients, but was not found in any of 306 normal controls.[21]

39.3 Cu,Zn SUPEROXIDE DISMUTASE

There are three isoforms of superoxide dismutase in mammals: a mitochondrial form (manganese SOD), an extracellular copper-zinc form (extracellular SOD), and a cytosolic copper-zinc form (Cu,Zn SOD). Cu,Zn SOD is a 32,500-Da cytosolic homodimeric protein that averages 0.5% of total cell protein. Pardo et al. have found that its distribution in neuronal populations of the brain and spinal cord is not uniform.[22] Pyramidal cells in the cortex stain strongly for Cu,Zn SOD, as do interneurons, motor neurons, ependymal cells, and some astrocytes of the spinal cord. Cu,Zn SOD is distributed in the soma, dendrites, and axons of the motor neuron.

The Cu,Zn SOD protein was discovered well before the dismutase activity was found by McCord and Fridovich in 1969.[23] In the early 1980s, considerable controversy occurred over whether the primary function of Cu,Zn SOD was dismutation.[24] However, the crystal structure later revealed how exquisitely adapted Cu,Zn SOD was to superoxide dismutation. The dismutation reaction is as follows:

$$E \sim Cu^{+2} + O_2^- \rightarrow E \sim Cu^{+1} + O_2$$

$$E \sim Cu^{+1} + O_2^- + H^+ \rightarrow E \sim Cu^{+2} + HO_2^-$$

$$HO_2^- + H^+ \rightarrow H_2O_2$$

The rate of this reaction is remarkably fast (2×10^9 $M^{-1}s^{-1}$), and is 10,000 times faster than uncatalyzed dismutation. If 1% of oxygen consumed by mitochondrial respiration is assumed to form superoxide, than micromolar levels of superoxide will accumulate in the absence of catalyzed dismutation. In the presence of physiological concentrations of Cu,Zn SOD, this is reduced to picomolar levels. Only 1 in 10^{10} superoxide molecules will escape Cu,Zn SOD to undergo uncatalyzed dismutation. The concentration of the enzyme Cu,Zn SOD will actually be 1 million times higher than the concentration of its substrate *in vivo*.

The remarkable reaction rate of Cu,Zn SOD is due to the electrostatic guidance of the charged substrate (O_2^-) into the active site by conserved charged amino acid residues.[25] The amino acids involved in electrostatic attraction, metal binding, and catalysis are all highly conserved between species.[26]

Hydroxyl radical (OH·) has been proposed to be the mediator of superoxide toxicity via metal-catalyzed Fenton chemistry. Extreme reactivity limits the toxicity of the hydroxyl radical; it can only diffuse about 30 Å before reacting in the near vicinity of a Fenton metal.[27,28] For most biological targets superoxide anion is a poor oxidant, since the negative charge hinders abstraction of yet another electron. The toxicity of superoxide anion may be more related to an attraction to the positively charged, electron-rich transition metals of iron-sulfur enzymes such as NADH dehydrogenase and aconitase.[29]

Superoxide is not the only anion that reacts with Cu,Zn SOD. The wild-type enzyme will also react with peroxynitrite (ONOO−), and catalyze the nitration of tyrosines.[30,31] Peroxynitrite is formed by the diffusion-limited reaction of superoxide and nitric oxide at a rate of 6.7×10^9 $M^{-1}s^{-1}$, three times faster than the reaction of superoxide with Cu,Zn SOD.[32] The reaction of wild-type Cu,Zn SOD with peroxynitrite is fast ($k = 10^5$ $M^{-1}s^{-1}$), but limited to 9% of available peroxynitrite due to its complex chemistry.[30]

At present, 32 missense mutations in 25 different loci have been identified in Cu,Zn SOD in familial ALS families.[1,33–41] Mutation nomenclature names the normal amino acid, its position, and its missense replacement — for example, an alanine to threonine mutation at amino acid 4 would be written A4T. Two mutations involve metal ligands: H46R[33] and H48Q,[42] and there is one truncation mutation.[40] The SOD mutations cause disease in a dominantly inherited pattern in the presence of one wild-type allele of SOD.

Most mutations occur in or near the loops of the β-barrel, far from the active site or metal binding regions, or near the dimer interface as shown in Figure 39.1. A faster decline is correlated with some mutations (A4V, A4T), but no correlation has been found between a mutation and the age of onset. One mutation directly affecting the active site (H46R) curiously retains 80% normal activity and results in "milder" disease with slower progression.[33]

39.4 SUPEROXIDE DISMUTASE IN FAMILIAL AMYOTROPHIC LATERAL SCLEROSIS

The autosomal dominant inheritance pattern suggests that either a partial loss of activity is deleterious or that disease is caused by a gain rather than a loss of enzymatic function.[6] If one assumes that the mutant enzyme is an inactive dismutase but does not affect the normal allele of SOD, then cellular superoxide levels would double. Superoxide concentrations would remain in the picomolar range due to scavenging by the remaining wild-type Cu,Zn SOD as well as the manganese SOD and extracellular SOD forms. In reality, the mutant enzyme is not necessarily inactive. Transfected COS cells expressed active ALS-SOD with activities ranging from 0% (G85R) to 150% (G37R) that of wild-type, with an average of about 50% normal activity (A4V, G41D, G93C, I113T). Immortalized leukocytes from patients carrying the G37R mutation show 60 to 80% of normal activity.[43]

Dismutase activity in red blood cell lysates of FALS patients and carriers is roughly half that of normal controls or sporadic ALS patients.[5,36] This may be due to the decreased half-life of the mutant SOD rather than its inactivity. An increased turnover of the ALS-SOD has been seen in transfected COS cells, yeast, and in G37R patient lymphocytes.[43,44] Reports of SOD activity in FALS patients vary, depending on the tissue source, method of assay, and type of mutation.

Since mutant and wild-type subunits can form heterodimers, and mutant subunits have a shorter half-life, the mutant subunits could shorten their wild-type heterodimer partner's lifetime by targeting them for faster proteolysis. Another possible gain in function would be an inhibition of wild-type activity in wild-type/mutant heterodimers. Borchelt et al. addressed both questions in transfection experiments of monkey COS and murine N2a cells using G41D and G85R.[45] These missense mutations introduce a change in charge which allows homo- and heterodimers to be distinguished on gels. By pulse-chase and activity assays, the protein half-life and activity of wild-type subunits were not altered.[45]

Transgenic mice containing wild-type human and ALS-SOD have been developed in several labs.[3,46,47] These mice have their own normal murine SOD and overexpress the introduced human gene, resulting in up to three times normal SOD activity in the brain. Mice overexpressing ALS-SOD develop a progressive motor neuron disease, while mice overexpressing wild-type SOD do not,[3,46] ruling out a protein overload effect. All also have greater than normal SOD activity, which contradicts loss of dismutase activity or superoxide damage as a mechanism. Duplication of this experiment rules out an artifact from random gene integration. These mice support the theory that the mutants have a *gain in function* mutation, which is a deleterious property unrelated to its dismutase activity.[3,6]

The gained function might not be an enzymatic one. It may be analogous to sickle-cell anemia, where the physical presence of the abnormal protein, rather than its understood activity (oxygen transport), initiates the disease. Alternatively, the gain in function could be enzymatically related: nitration catalysis, peroxidase activity, or some yet-unknown reaction.[48] Because both dismutation and nitration require metal catalysis, understanding the role of copper and zinc in these mechanisms may give some insight into how the ALS-SODs may differ from the wild-type enzymes.

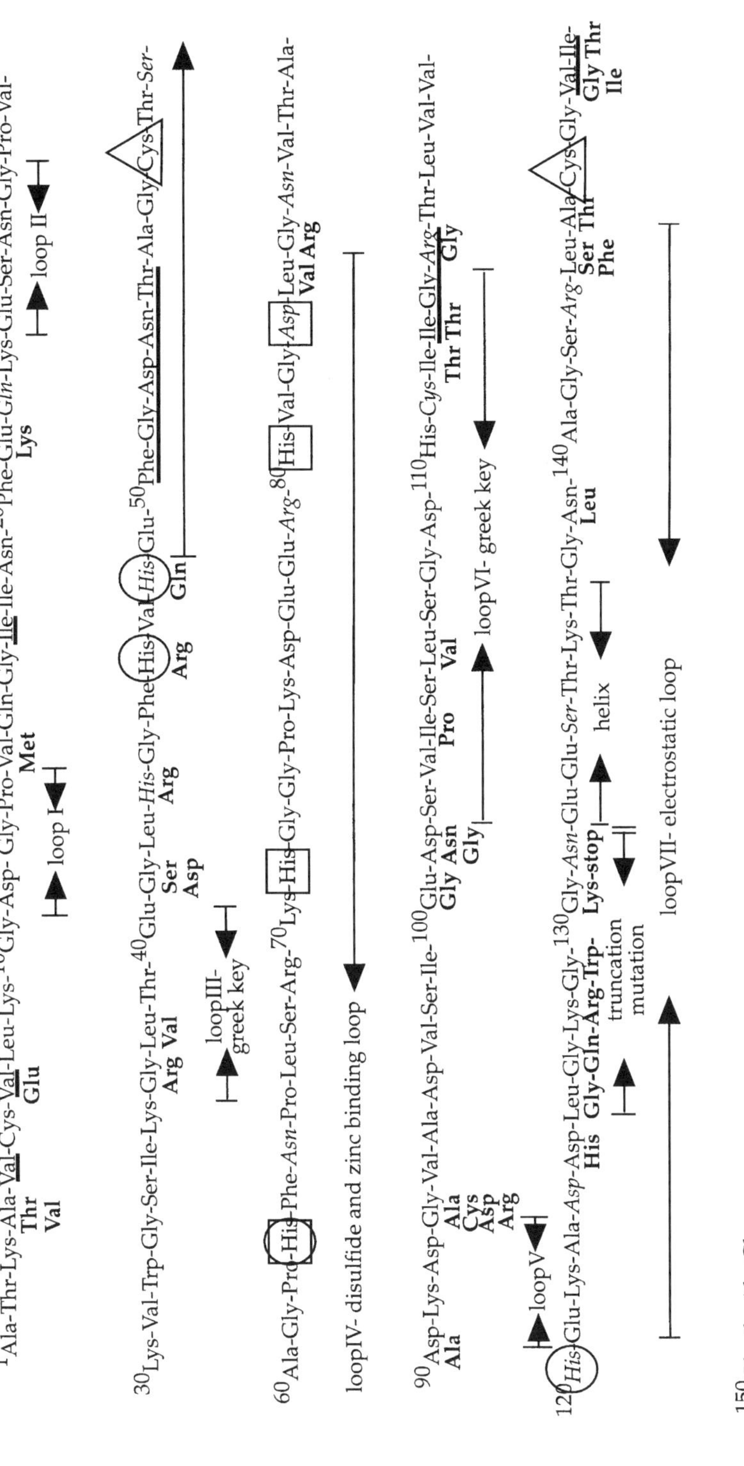

FIGURE 39.1 Locations of the known familial ALS mutations and structural motifs in the human Cu/Zn SOD sequence: circles indicate copper ligands, squares indicate zinc ligands, and triangles represent cysteines involved in a sulfhydryl bond. Underlined residues are involved in dimer contacts. Italicized residues indicate side chains involved in structurally important hydrogen bonds. Boldface residues below the sequence represent the mutant missense mutations. Arrows delineate the loops of the β-barrel and the one missense mutation, and are labeled. Sequence, structural, and mutational information were collated from the following references: 1, 24, 33–41, 51.

39.5 METALS IN THE DISMUTATION REACTION

Copper and zinc play important roles in both the dismutation and nitration reactions. Dismutation is performed by the copper, which requires two superoxides to complete a cycle. The copper is anchored to the active site by four histidines in a distorted square planar geometry, leaving one coordination site free to bind substrate.[49] Zinc is bound by one aspartate and three histidines. Copper and zinc share one histidine ligand (H63). The histidine is then thought to donate a proton to form hydrogen peroxide. The second proton required is provided by water in rapid equilibrium with the bulk phase.[26] The zinc plays no known catalytic role beyond stabilization of the bridging histidine; structurally, it is buried in the protein and anchors the zinc loop (49-84) to the β-barrel.[24] An electrostatic loop (121-144) bears lysines (122-136) involved in electrostatically routing superoxide into the active site channel.[50] The copper is ligated to the surface of the β-barrel and is encircled by the zinc loop and the electrostatic loop which arise from opposite sides of the β-barrel.[51] The highly hydrogen-bonded, compact β-barrel structure gives the wild-type SOD its remarkable resistance to heat denaturation, detergents, and denaturants.

Zinc binding is pH dependent, and is readily lost below pH 4.5.[52] Copper binding is also lost below pH 3, and coincides with denaturation.[52] Apoenzyme (E_2E_2SOD) is more susceptible to denaturation at acidic pH than holoenzyme (Cu_2Zn_2SOD).[53] Reversible removal of copper is most easily done with competitive chelators at acidic pH.[24] At pH 6.25 the copper site has a 1000-fold higher affinity for copper than zinc; the zinc site, however, has nearly equal affinity for copper or zinc.[54] The zinc site has little specificity for what metal it binds — copper, cobalt, cadmium, or mercury can all be substituted for zinc *in vitro*.[24] Another pH peculiarity is the complete rearrangement of Cu_2E_2SOD to $Cu_2Cu_2SOD + E_2E_2SOD$ at pH 9.5.[55] Zinc-deficient enzymes or enzymes with copper or cobalt substituting for zinc still have dismutase activity.[56]

Removing the copper eliminates both dismutation and nitration activities of Cu,Zn SOD.[31] However, superoxide dismutation activity can be inactivated by alkaline hydrogen peroxide, which leaves the copper intact.[57] This method oxidizes His118, a copper ligand, to 2-oxo-histidine.[58] Cu,Zn SOD inactivated in this manner is still able to catalyze tyrosine nitration.[31] While both dismutation and nitration require copper, their mechanisms are distinct.

39.6 MECHANISM OF PEROXYNITRITE NITRATION CATALYSIS

Peroxynitrite ($ONOO^-$) is not a radical, but an anion formed *in vivo* from the radical-radical reaction of nitric oxide and superoxide (O_2^-). The terminal oxygens can share electron density, forming a relatively unreactive ring-like *cis* conformation.[59] Peroxynitrite is a far better oxidant than either nitric oxide or superoxide and is comparable to hydroxyl radical and hypochlorous acid. It is capable of damaging many different cellular components by causing strand breaks in DNA, cross-linking proteins, oxidizing sulfhydryls and lipids, and attacking zinc-thiolate structures.[60–63] However, the reaction of peroxynitrite with superoxide dismutase ($k = 10^5$ $M^{-1}s^{-1}$) is rapid in comparison with its reactions with sulfhydryls ($k = 2.6$ to 5.9×10^3 $M^{-1}s^{-1}$).[30,63] Thus, SOD-catalyzed nitration can still be seen in the presence of alternative targets of peroxynitrite oxidation, as are present in tissue homogenates.

The peroxynitrite anion can protonate and isomerize from the relatively stable *cis* conformation (*cis*ONOOH) to the *trans* conformation (*trans*ONOOH), which is able to decay to nitrate.[30] Peroxynitrous acid has a pK_a of 6.8[61] so at neutral pH there is a rapid interchange from the *cis* to the *trans* conformation as the *trans* acid flows toward decay (nitrate). However, *cis*ONOOH is 1 to 2 kcal/mol more stable than *trans*ONOOH,[30] so reversion from *trans* back to *cis* is possible. These kinetics are quite complex. (64)

$$cis\text{ONOO}^- + H^+ \leftrightarrow cis\text{ONOOH}$$

$$cis\text{ONOOH} \leftrightarrow trans\text{ONOOH} \rightarrow HNO_3 \text{ (decay)}$$

In the *trans* anion conformation, the terminal peroxy oxygens of ONOO⁻ bear a structural and electronic resemblance to one of its parental compounds: superoxide anion (O_2^-). *Trans*-peroxynitrite may be electrostatically guided into the active site of SOD, where the copper can withdraw electron density, leaving an NO_2^+-like moiety exposed to the solvent (Figure 39.2).[30] What is observed is an enhanced rate and yield of phenolic nitration in the presence of physiological SOD levels. Additionally, SOD trapping of *trans*-peroxynitrite favors the nitrating activity over the hydroxyl radical-like reactivity.[30] Neither the dismutation nor the nitration activity of Cu,Zn SOD is inactivated by peroxynitrite.

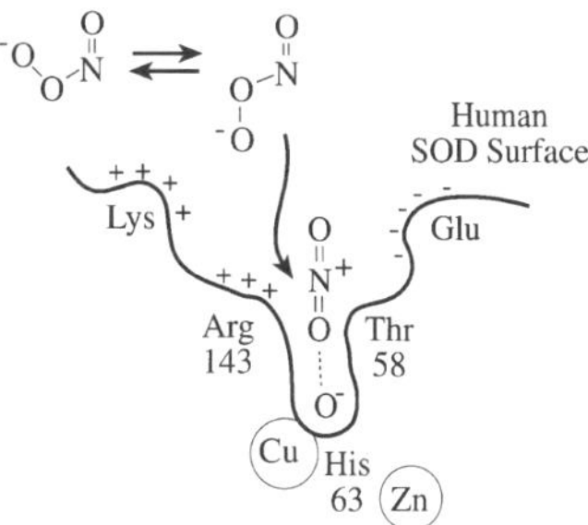

FIGURE 39.2 Peroxynitrite interaction with the active site of Cu,Zn SOD.

One peculiarity of the short half-life and conformational changes of peroxynitrite is the hyperbolic curve and plateau in the SOD dose-response curve (Figure 39.3). Nitration is seen without SOD at about half the rate of maximum SOD-catalyzed nitration.[30] Increasing amounts of SOD compete with the uncatalyzed nitration and decay pathways for the *trans*-peroxynitrite. The plateau occurs when the SOD concentration exceeds that of *trans*-peroxynitrite that can fit in the active site; at that point peroxynitrite isomerization, rather than reaction with the enzyme, is rate limiting. Only 9% of peroxynitrite can fit into the wild-type SOD active site under "plateau" conditions.[30] An increase in the nitration plateau suggests a greater percentage of peroxynitrite conformers have access to the active site.

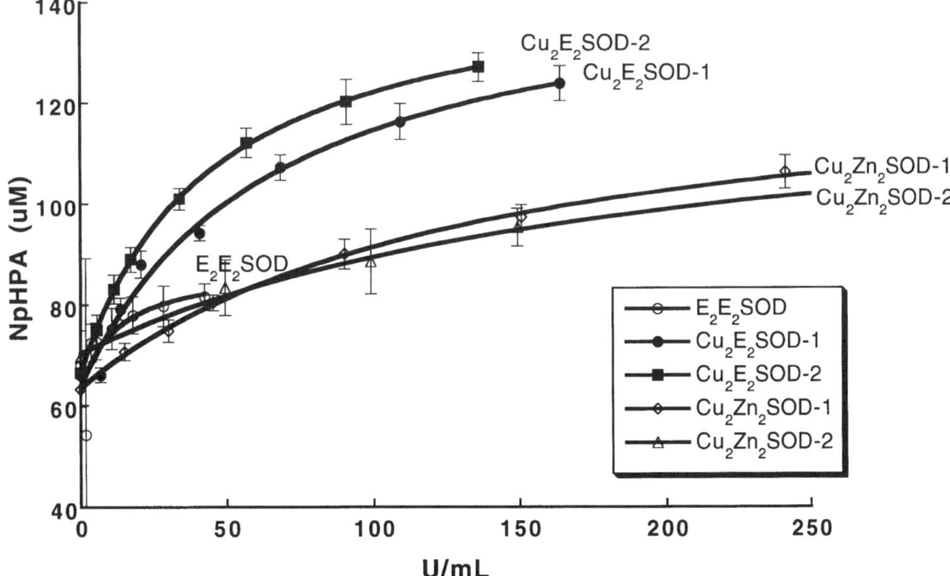

FIGURE 39.3 Nitration by Cu_2E_2 and Cu_2Me_2 bovine SOD: 1 mM 4-hydroxyphenylacetic acid was used as a tyrosine analog and nitration target. The nitration buffer contained 50 mM KP_i/100 mM NaCl/100 mM glucose/100 μM diethylene-triaminepentaacetic acid; 1 mM peroxynitrite was added by rapid mixing.

If wild-type Cu,Zn SOD has already been shown to catalyze dismutation, then what distinguishes mutant SOD from wild type? The shortened half-life of mutants in tissue culture models argues that the gain of function activity must be significant considering the smaller pool of mutant protein.

39.7 METALS IN THE MUTANT SUPEROXIDE DISMUTASES

We propose that the difference between the mutant and wild-type is that the mutant SODs have a lower affinity for zinc. Wild-type SOD is known for its stability; it retains activity after exposure to strongly denaturing conditions (8 M urea, 2% SDS).[65] Mutant SOD is not as stable, which may account for its decreased half-life *in vivo*. We have been expressing recombinant human ALS SODs in *E. coli* and purifying the proteins by anion exchange and HPLC. Under mildly denaturing conditions (2 M urea), both mutant and wild-type release zinc faster than copper. The zinc is also released to a greater extent when metal binding is in equilibrium with the chelator, 4-(pyridoazo)resorcinol. The mutants have a lower affinity for zinc than wild-type.[66]

Zinc-deficient SOD (Cu_2E_2SOD) is still capable of dismutation, though it is less active (30 to 50% of Cu_2Zn_2SOD).[24] This would result in a slight increase in superoxide concentration, which could then react with nitric oxide to form more peroxynitrite. Zinc deficiency nearly doubles the efficiency of wild-type SOD as a catalyst of nitration compared to holoenzyme (Cu_2Zn_2SOD) (Figure 39.3). This is true at any concentration on the dose-response curve, regardless of how SOD dose is expressed (protein concentration, activity, or copper content) (not shown).

Occupancy of the zinc site with zinc, copper, or cobalt returns nitration activity to that of the holoenzyme.[66] Nitration activity of the fully metallated, HPLC purified mutant SODs was indistinguishable from wild-type Cu,Zn SOD. Yet, with its decreased zinc affinity, maintaining zinc binding by the mutant SOD during purification is difficult. Zinc loss would release the zinc loop forming one rim of the active site. The increased conformational mobility of the zinc loop may explain the multiple peaks of zinc-deficient SOD on HPLC; these peaks resolve to a single peak when incubated with zinc.[66] Increased conformational freedom opens access to the active site to more conformers of peroxynitrite, and results in the increased nitration catalysis observed in zinc-deficient SOD.

The half-life of mutant SODs may be different in motor neurons than in tissue culture or red blood cells, leading to an accumulation of the zinc-deficient, highly nitrating species of SOD. In SOD, metal loss has been shown to precede proteolysis.[67] If the loss of metals is ordered, as has been seen in our experiments, the zinc-deficient, highly nitrating species will exist before denaturation.

39.8 NEUROFILAMENTS AS A NITRATION TARGET

Tyrosine nitration of neurofilament could be an epiphenomenon; the proteins of inclusion bodies escaping proteolysis could accumulate oxidative modifications including tyrosine nitration as a result of prolonged exposure. Functionally, nitration alters the pK_a of tyrosine from 10 to 7.5, making it negatively charged at neutral pH; this charge change could have as much of a functional impact as phosphorylation. NFL is an excellent nitration target, with 20 tyrosines per subunit in a domain-critical assembly. A charge change there could disrupt both lateral and longitudinal assembly. Neurofilament must be disassembled in order to act as a nitration target. Addition of nitrated NFL to normal recombinant NFL has recently been shown to disrupt *in vitro* assembly.[66]

39.9 CONCLUSION: MORE QUESTIONS

If and where the pathogenic mechanisms of the sporadic and familial forms of ALS intersect is not known. Finding a link between a gene and a disease does not necessarily find the initiator of the disease. The cause of the late onset of ALS is still unknown. Something "upstream" of expression of mutant Cu,Zn SOD protein is likely to initiate the disease, perhaps related to the environment or aging. Initiation of the disease might be related to an increase in peroxynitrite production due to increases in nitric oxide or superoxide or nitric oxide synthase induction. Motor neuron injury is known to induce nitric oxide synthase[68] which contributes to their death.[69] Superoxide production may also be increased in aging.[70] Superoxide is also a product of damaged, uncoupled nitric oxide synthase.[71] Because nitration has been detected in sporadic ALS tissue, and the zinc-deficient wild-type SOD has increased catalysis of nitration, the production of nitric oxide in or near motor neurons may be a common link between the pathogenesis of the familial and sporadic forms of ALS.

REFERENCES

1. Rosen DR, Siddique T, Patterson D, Figlewicz DA, et al: Mutations in Cu/Zn superoxide dismutase gene are associated with familial amyotrophic lateral sclerosis. *Nature*, 1993; 362:59–62.
2. Parboosingh J, Rouleau G, Meninger V, McKenna-Yasek D, et al: Absence of mutations in the Mn superoxide dismutase or catalase genes in familial amyotrophic lateral sclerosis. *Neuromusc. Disord.*, 1995; 5:7–10.
3. Gurney ME, Pu H, Chiu AY, Dal Corto MC, et al: Motor neuron degeneration in mice that express a human Cu,Zn superoxide dismutase mutation. *Science*, 1994; 264:1772–1775.
4. Rabizadeh S, Gralla EB, Borchelt DR, Gwinn R, et al: Mutations associated with amyotrophic lateral sclerosis convert superoxide dismutase to a proapoptotic gene: studies in yeast and neural cells. *Proc. Natl. Acad. Sci. U.S.A.*, 1995; 92:3024–3028.
5. Bowling A, Barkowski E, McKenna-Yasek D, Sapp P, et al: Superoxide dismutase concentration and activity in familial amyotrophic sclerosis. *J. Neurochem.*, 1995; 64:2366–2369.
6. Beckman JS, Carson M, Smith CD, and Koppenol WH: ALS, SOD and peroxynitrite. *Nature*, 1993; 364:584.
7. Engel WK, Kurland LT, and Klatzo I: An inherited disease similar to amyotrophic lateral sclerosis with a pattern of posterior column involvement. An intermediate form? *Brain*, 1959; 82:203–220.
8. Troost D, van den Oord J, de Jong J, and Swaab D: Lymphocytic infiltration in the spinal cord of patients with amyotrophic lateral sclerosis. *Clin. Neuropathol.*, 1989; 8:289–294.
9. Troost D, Van den Oord J, and Vianney de Jong J: Immunohistochemical characterization of the inflammatory infiltrate in amyotrophic lateral sclerosis. *Neuropathol. Appl. Neurobiol.*, 1990; 16:401–410.
10. Lowe J: New pathological findings in amyotrophic lateral sclerosis. *J. Neurol. Sci.*, 1994; 124 (S): 38–51.
11. Nixon R, Paskevich P, Sihag R, and Thayer C: Phosphorylation on carboxyl terminus domains of neurofilament proteins in retinal ganglion cell neurons in vivo: influences on regional neurofilament accumulation, interneurofilament spacing, and axon caliber. *J. Cell Biol.*, 1994; 126:1031–1046.
12. Strong M: Aluminum neurotoxocity: an experimental approach to the induction of neurofilamentous inclusions. *J. Neurol. Sci.*, 1994; 124 (S): 20–26.
13. Strong M and Jakowec D: 200 kDa and 160 kDa neurofilament protein phosphatase resistance following in vivo aluminum chloride exposure. *Neurotoxicology*, 1994; 15:799–808.
14. Strong M, Mao K, Nerurkar V, Wakayama I, et al: Dose-dependent selective suppression of light (NFL) and medium (NFM) molecular weight neurofilament gene transcription in acute aluminum neurotoxicity. *Mol. Cell Neurosci.*, 1994; 5:319–326.
15. Troost D, Sillevis Smitt P, de Jong J, and Swaab D: Neurofilament and glial alterations in the cerebral cortex in amyotrophic lateral sclerosis. *Acta Neuropathol.*, 1992; 84:664–673.
16. Xu Z, Cork L, Griffin J, and Cleveland D: Increased expression of neurofilament subunit NF-L produces morphological alterations that resemble the pathology of human motor neuron disease. *Cell*, 1993; 73:23–33.
17. Cote F, Collard J-F, and Julien J-P: Progressive neuronopathy in transgenic mice expressing the human neurofilament heavy gene: a mouse model of amyotrophic lateral sclerosis. *Cell*, 1993; 73:35–46.
18. Collard J, Cote F, and Julien J-P: Defective axonal transport in a transgenic mouse model of amyotrophic lateral sclerosis. *Nature*, 1995; 375:61–64.
19. Lee M, Marszalek J, and Cleveland D: A mutant neurofilament subunit causes massive, selective motor neuron death: implications for the pathogenesis of human motor neuron disease. *Neuron*, 1994; 13:975–988.
20. Strong M, Gaytan-Garcia S, and Jakowec D: Reversibility of neurofilamentous inclusion formation following repeated sublethal intracisternal inoculums of $AlCl_3$ in New Zealand white rabbits. *Acta Neuropathol.*, 1995; 90:57–67.
21. Figlewicz D, Krizus A, Martinoli M, Meininger V, et al: Variants of the heavy neurofilament subunit are associated with the development of amyotrophic lateral sclerosis. *Hum. Mol. Genet.*, 1994; 3:1757–1761.
22. Pardo CA, Xu Z, Borchelt DR, Price DL, et al: Superoxide dismutase is an abundant component in cell bodies, dendrites, and axons of motor neurons and in a subset of other neurons. *Proc. Natl. Acad. Sci. U.S.A.*, 1995; 92:954–958.
23. McCord JM and Fridovich I: Superoxide dismutase: an enzymic function for erythrocuprein (hemocuprein). *J. Biol. Chem.*, 1969; 244:6049–6055.

24. Valentine JS, Panteliano MW: Protein-Metal Ion Interactions in Cuprozinc Protein (Superoxide Dismutase). In Spiro TG (ed.): *Copper Proteins*. New York, Wiley Interscience, 1981, pp 291–358.

25. Koppenol WH and Margoliash E: Faster superoxide dismutase mutants designed by enhancing electrostatic guidance. *Chemtracts, Biochem. Mol. Biol.*, 1992; 3:330–333.

26. Fridovich I: Superoxide Dismutases. *Adv. Enzymol. Rel. Aspects Mol. Biol.*, 1986; 58:61–97.

27. Hutchinson F: The distance that a radical formed by ionizing radiation can diffuse in a yeast cell. *Radiat. Res.*, 1957; 7:473–483.

28. Fridovich I: Superoxide dismutases: an adaptation to a paramagnetic gas. *J. Biol. Chem.*, 1989; 264:7761–7764.

29. Boveris A and Cadenas E: Production of superoxide radicals and hydrogen peroxide in mitochondria. In Oberley L (ed.): *Superoxide Dismutase*. Boca Raton, FL, CRC Press, 1982, pp 15–30.

30. Beckman JS, Ischiropoulos H, Zhu L, van der Woerd M, et al: Kinetics of superoxide dismutase and iron catalyzed nitration of phenolics by peroxynitrite. *Arch. Biochem. Biophys.*, 1992; 298:438–445.

31. Ischiropoulos H, Zhu L, Chen J, Tsai HM, et al: Peroxynitrite-mediated tyrosine nitration catalyzed by superoxide dismutase. *Arch. Biochem. Biophys.*, 1992; 298:431–437.

32. Huie RE and Padmaja S: The reaction rate of NO with superoxide. *Free Radical Res. Commun.*, 1993; 18:195–199.

33. Aoki M, Ogasawara M, Matsubara Y, Narisawa K, et al: Mild ALS in Japan associated with novel SOD mutation. *Nature Genet.*, 1993; 5:323–324.

34. Aoki M, Abe K, Houi K, Ogasawara M, et al: Variance of age at onset in a Japanese family with amyotrophic lateral sclerosis associated with a novel Cu/Zn superoxide dismutase mutation. *Ann. Neurol.*, 1995; 37:676–679.

35. Elshafey A, Lanyon W, and Conner J: Identification of a new missense point mutation in exon 4 of the Cu/Zn superoxide dismutase (SOD-1) gene in a family with amyotrophic lateral sclerosis. *Hum. Mol. Genet.*, 1994; 3:363–364.

36. Esteban J, Rosen D, Bowling A, Sapp P, et al: Identification of two novel mutations and a new polymorphism in the gene for Cu/Zn superoxide dismutase in patients with amyotrophic lateral sclerosis. *Hum. Mol. Genet.*, 1994; 3:997–998.

37. Hirano M, Fujii J, Sonobe M, Okamoto K, et al: A new variant Cu/Zn superoxide dismutase (Val7 → Glu) deduced from lymphocyte mRNA sequences from Japanese patients with familial amyotrophic lateral sclerosis. *Biochem. Biophys. Res. Commun.*, 1994; 204:572–577.

38. Jones C, Shaw P, Chari G, and Brock D: Identification of a novel exon 4 SOD1 mutation in a sporadic amyotrophic lateral sclerosis patient. *Mol. Cell Probes*, 1994; 8:329–330.

39. Kawamata J, Hasegawa H, Shimohama S, Kimura J, et al: Leu106 → Val(CTC → GTC) mutation of superoxide dismutase-1 gene in patient with familial amyotrophic lateral sclerosis in Japan. *Lancet*, 1994; 343:1501.

40. Pramatarova A, Goto J, Nanba E, Nakashima K, et al: A two basepair deletion in the SOD1 gene causes familial amyotrophic lateral sclerosis. *Hum. Mol. Genet.*, 1994; 3:2061–2062.

41. Suthers G, Laing N, Wilton S, Dorosz S, et al: "Sporadic" motoneuron disease due to familial SOD1 mutation with low penetrance. *Lancet*, 1994; 244:1773.

42. Enayat Z, Orrell R, Claus A, Ludolph A, et al: Two novel mutations in the gene for copper zinc superoxide dismutase in UK families with amyotrophic lateral sclerosis. *Hum. Mol. Genet.*, 1995; 4:1239–1240.

43. Borchelt DR, Lee MK, Slunt HS, Guarnieri M, et al: Superoxide dismutase 1 with mutations linked to familial amyotrophic lateral sclerosis possesses significant activity. *Proc. Natl. Acad. Sci. U.S.A.*, 1994; 91:8292–8296.

44. Nishida CR, Gralla EB, and Valentine JS: Characterization of three yeast copper-zinc superoxide dismutase mutants analogous to those coded for in familial amyotrophic lateral sclerosis. *Proc. Natl. Acad. Sci. U.S.A.*, 1994; 9906–9910.

45. Borchelt DR, Guarnieri M, Wong PC, Lee MK, et al: Superoxide dismutase 1 subunits with mutations linked to familial amyotrophic lateral sclerosis do not affect wild-type subunit function. *J. Biol. Chem.*, 1995; 270:3234–3238.

46. Avraham K, Sugarman H, Rotshenker S, and Groner Y: Down's syndrome: morphological remodelling and increased complexity in the neuromuscular junction of transgenic CuZn-superoxide dismutase mice. *J. Neurocytol.*, 1990; 20:208–215.

47. Ripps ME, Huntley GW, Hof PR, Morrison JH, et al: Transgenic mice expressing an altered murine superoxide dismutase gene provide an animal model of amyotrophic lateral sclerosis. *Proc. Natl. Acad. Sci. U.S.A.*, 1995; 92:689–693.

48. Hodgson E and Fridovich I: The interaction of bovine erythrocyte superoxide dismutase with hydrogen peroxide: chemiluminescence and peroxidation. *Biochemistry*, 1975; 14:5299–5303.

49. Beem KM, Richardson DC, and Rajagopalan KV: Metal sites of copper-zinc superoxide dismutase. *Biochemistry*, 1977; 16:1931–1937.

50. Policelli F, Battistoni A, Bottaro G, Carri M, et al: Mutation of Lys-120 and Lys-134 drastically reduces the catalytic rate of Cu,Zn superoxide dismutase. *FEBS Lett.*, 1994; 352:76–78.

51. Parge HE, Hallewell RA, and Tainer JA: Atomic structures of wild-type and thermostable mutant recombinant Cu,Zn superoxide dismutase. *Proc. Natl. Acad. Sci. U.S.A.*, 1992; 89:6109–6113.

52. Pantoliano MW, McDonnell PJ, and Valentine JS: Reversible loss of metal ions from the zinc binding site of copper-zinc superoxide dismutase: the low pH transition. *J. Am. Chem. Soc.*, 1979; 101:6454–6456.

53. Fee JA and Phillips WD: The behavior of holo- and apo- forms of bovine superoxide dismutase at low pH. *Biochim. Biophys. Acta*, 1975; 412:26–38.

54. Hirose J, Yamada M, Hayakawa C, Nagao H, et al: Selective binding behavior of zinc (II) and copper (II) ions to their native sites of apo-bovine superoxide dismutase. *Biochem. Int.*, 1984; 8:401–408.

55. Valentine JS, Pantoliano MW, McDonnell PJ, Burger AR, et al: pH-dependent migration of copper (II) to the vacant zinc-binding site of zinc-free bovine erythrocyte superoxide dismutase. *Proc. Natl. Acad. Sci. U.S.A.*, 1979; 76:4245–4249.

56. Pantoliano MW, Valentine JS, Mammone RJ, and Scholler DM: pH dependence of metal binding to the native zinc site of bovine erythrocuprein (superoxide dismutase). *J. Am. Chem. Soc.*, 1982; 104:1717–1723.

57. Glaser AN, Delange RJ, and Sigman DS; Chemical modification of proteins: selected methods and analytical procedures; New York, Elsevier, 1975.

58. Uchida K and Kawakishi S: Identification of oxidized histidine generated at the active site of Cu,Zn superoxide dismutase exposed to H_2O_2. *J. Biol. Chem.*, 1994; 269:2405–2410.

59. Tsai J: A Theoretical and Experimental Study of the Conformation of Peroxynitrite (ONOO⁻) and Its Stability [Ph.D. Thesis]: University of Alabama at Birmingham, 1994.

60. Koppenol WH, Moreno JJ, Pryor WA, Ischiropoulos H, et al: Peroxynitrite: a cloaked oxidant from superoxide and nitric oxide. *Chem. Res. Toxicol.*, 1992; 5:834–842.

61. Crow J, Beckman J, and McCord J: Sensitivity of the essential zinc-thiolate moiety of yeast alcohol dehydrogenase to hypochlorite and peroxynitrite. *Biochemistry*, 1995; 34:3544–3552.

62. Radi R, Beckman JS, Bush KM, and Freeman BA: Peroxynitrite-induced membrane lipid peroxidation. The cytotoxic potential of superoxide and nitric oxide. *Arch. Biochem. Biophys.*, 1991; 288:481–487.

63. Radi R, Beckman JS, Bush KM, and Freeman BA: Peroxynitrite oxidation of sulfhydryls: the cytotoxic potential of superoxide and nitric oxide. *J. Biol. Chem.*, 1991; 266:4244–4250.

64. Crow JP, Spruell C, Chen J, Gunn C, et al: On the pH-dependent yield of hydroxyl radical products from peroxynitrite. *Free Radical Biol. Med.*, 1994; 16:331–338.

65. Steinman H: Superoxide dismutases: protein chemistry and structure-function relationships. In Oberley L (ed.): *Superoxide Dismutase*. Boca Raton, FL, CRC Press, 1982, pp 11–68.

66. Crow J, Strong M, Sampson J, Kowall N, et al: Unpublished data, 1996.

67. Salo DC, Pacifici RE, Lin SW, Giulivi C, et al: Superoxide dismutase undergoes proteolysis and fragmentation following oxidative modification and inactivation. *J. Biol. Chem.*, 1990; 265:11919–11927.

68. Wu W: Expression of nitric-oxide synthase (NOS) in injured CNS neurons as shown by NADPH diaphorase histochemistry. *Exp. Neurol.*, 1993; 120:153–159.

69. Wu W and Li L: Inhibition of nitric oxide synthase reduces motoneuron death due to spinal root avulsion. *Neurosci. Lett.*, 1993; 153:121–124.

70. Sohal R and Brunk U: Mitochondrial production of pro-oxidants and cellular senescence. *Mutat. Res.*, 1992; 275:295–304.

71. Pou S, Pou W, Bredt D, Snyder S, et al: Generation of superoxide by purified brain nitric oxide synthase. *J. Biol. Chem.*, 1992; 267:24173–24176.

Chapter 40

The Role of Ceruloplasmin for Iron and Copper Metabolism in Central Nervous System Disorders

Hiroshi Morita and Nobuo Yanagisawa

CONTENTS

40.1 INTRODUCTION

The neurotoxic effects of copper and iron on the central nervous system are well known but have yet to be clearly defined. Iron metabolism in the basal ganglia is an important topic in neurology. Parkinson's disease has been the focus of iron metabolism research. As for copper metabolism disturbances, the best-known disorder is Wilson disease. Ceruloplasmin, a marker for the diagnosis of Wilson disease, is a ferroxidase;[1] but iron metabolism disturbance is not a major problem, copper deposition in the tissues being the main pathological finding. Moreover, recent studies have shown that Wilson disease is not directly linked to a molecular defect in ceruloplasmin.[2,3]

In his study of the relationship between iron and copper metabolism, Owen showed that copper-deficient rats accumulated large amounts of iron in the liver and that the serum ceruloplasmin concentration was decreased in copper- and iron-deficient rats.[4] Since his work, there has been little investigation of this relationship. Our group and some others recently have reported cases of hereditary ceruloplasmin deficiency with hemosiderosis, providing a new viewpoint on the relationship between copper and iron metabolism in the central nervous system.[5–9] Taking into account the relationship between ceruloplasmin and iron metabolism, we discuss the pathological significance of iron and copper metabolism in the central nervous system disorders, including Wilson disease and hereditary ceruloplasmin deficiency with hemosiderosis.

40.2 IRON METABOLISM AND CERULOPLASMIN

Ceruloplasmin, an α_2-globulin that has six copper atoms, is an acute reactant substance.[1] It not only transports copper but scavenges free radicals, thereby preventing metal-catalyzed free-radical tissue damage and lipid peroxidation.[10] Another important physiological activity of ceruloplasmin is its function as a ferroxidase, and since the 1960s many researchers have focused on this activity.[11] Ceruloplasmin has oxidative activity for Fe^{2+} to Fe^{3+}-transferrin, and oxidase activity for aromatic amines, as well as having a copper transport function.[1] Therefore, it is very likely that a ceruloplasmin deficiency disturbs iron transport from the cell to the plasma due to the abnormal intracellular accumulation of iron. Roeser et al. showed that the plasma iron level increased after an infusion of ceruloplasmin to swine made hypoceluroplasminic by copper deprivation.[12]

FET3 protein of the yeast *Saccharomyces cerevisiae* is notable. It is a multicopper oxidase that acts as a ferrireductase and ferrous iron transporter. Whereas ceruloplasmin is a serum protein, FET3 is a plasma membrane protein, but their physiological functions are almost the same. Both proteins oxidize Fe^{2+} to Fe^{3+}. Results of recent studies show that *S. cerevisiae* with the mutant FET3 protein suffers from an iron transport disturbance.[13,14] In summary, it can be expected that iron metabolism disturbance is induced in congenital ceruloplasmin deficiency.

40.3 PATHOGNOMONIC MECHANISM OF WILSON DISEASE AND METAL METABOLISM

One of the most important findings in Wilson disease is that the serum ceruloplasmin levels are reduced. Deposition of copper in the liver and brain is the main pathological change seen in this disease, but some studies also have found an iron metabolism disturbance.[12,15] The question is why the iron metabolism is not significantly disturbed in Wilson disease. Although an iron deficiency or a low plasma iron concentration sometimes is associated with anemia, especially for the patients whose serum ceruloplasmin content was less than 5% of normal, no iron deposition has been found in Wilson disease.[12,15] The serum level of ceruloplasmin is low, but not zero, in patients with Wilson disease. The gene of Wilson disease is reported to be a copper-transporting ATPase, its mutation or deletion has been reported in Wilson disease, and the dysfunction of this ATPase causes a copper-incorporation disturbance in apoceruloplasmin.[2,3] This gene is quite different from the one that encodes ceruloplasmin. There is, additionally, no observation that indicates a direct relationship between the decrease of ceruloplasmin and copper deposition in tissues in Wilson disease.

Considering these studies, the previous hypothesis that the small amount of ceruloplasmin may have enough ferroxidase activity to prevent iron deposition seems correct.[12,15] As compared with the following new disease entity, we assume that this explanation would be true.

40.4 HEREDITARY CERULOPLASMIN DEFICIENCY WITH HEMOSIDEROSIS

Hereditary ceruloplasmin deficiency is a good model for investigating the relationship of ceruloplasmin to copper and iron metabolism in humans.[5–9] It is a new disease entity, and we have shown ceruloplasmin gene abnormality in patients with this deficiency.[16] Details of our study of a family with this deficiency are reported elsewhere.[6,9,16] Our findings are described briefly below.

40.4.1 Case Record

Seven Japanese siblings of consanguineous parents (first cousins) and their families were studied. Four of the siblings are homozygotes. The others, as well as their father and one of their daughters, are heterozygotes for ceruloplasmin gene mutation. Three of the four homozygotes show neurological abnormalities, and all have received insulin infusions for diabetes mellitus.

The most severely affected sibling died of bronchopneumonia at age 60. Diabetes mellitus had been diagnosed at age 40, and insulin therapy was begun at age 47. Forgetfulness and motor disturbances were noted when she was 51 years old, and these gradually worsened. At age 55, there was no Kayser-Fleischer ring, but mild pigmentary retinal degeneration was present. She presented with an ataxic speech, torticollis, and facial grimacing, but without flapping tremor. Deep tendon reflexes were hyperactive, and her gait was ataxic. Her WAIS full-scale IQ was 80; verbal IQ 80, performance 72. Laboratory examinations showed mild microcytic hypochromic anemia, hyperglycemia, and an increase in the protein content of the cerebrospinal fluid. No serum ceruloplasmin or apoceruloplasmin was detected. Serum ferritin was increased markedly, whereas serum iron was decreased, and TIBC and transferrin levels were normal. Serum and urinary copper contents were also decreased (Table 40.1).

Table 40.1 Comparison of Wilson Disease and Hereditary Ceruloplasmin Deficiency With Hemosiderosis

		Hereditary ceruloplasmin deficiency with hemosiderosis		
	Wilson disease	**Miyajima et al.[5,7,17]**	**Logan et al.[8]**	**Morita et al.[6,9,16]**
Serum ceruloplasmin	Decrease	Absent	Absent	Absent
Serum ferritin	Normal	Increase	Increase	Increase
Serum copper	Decrease	Decrease	Decrease	Decrease
Urine copper	Increase	Increase	Normal	Decrease-normal
Serum iron	Normal	Decrease	Decrease	Decrease
Diabetes mellitus	—	—	+	+
Hemoglobin	Normal-decrease (mild)	Decrease	Decrease (mild)	Normal-decrease (mild)
Liver copper	Increase	Increase (mild)	Normal	Increase (mild)
Liver iron	Normal	Increase	Increase	Increase
Liver cirrhosis	Common	—	—	—
Dementia	Common	—	+	+
Extrapyramidal symptoms	Common	Mild (blepharospasms)	—	Severe
Low signal intensity in MRI[a]	Rare	+	+	+

[a] For both T1 and T2 weighted images of the brain.

Brain CT showed high density areas in the basal ganglia, thalamus, and dentate nucleus of the cerebellum. The basal ganglia and dentate nucleus had low signal intensity areas in both the T1 and T2 weighted MR images. The center of the caudate nucleus showed a low signal in the T1 weighted image and a high signal in the T2 weighted one, indicative of cystic degeneration (Figure 40.1). The liver showed diffuse high density in the CT scan, and low signal intensity in both the T1 and T2 weighted MR images.

Her symptoms gradually worsened, and she became bedridden with severe dystonic posture when 58 years old. D-penicillamine had no effect for her symptoms. She died of bronchopneumonia at age 60.

40.4.2 Pathological Findings

Varying degrees of iron deposition were seen in the visceral organs. The liver and pancreas were the most severely affected, the heart, kidney, spleen, thyroid gland, and retina being mildly affected. There was no iron deposition in other tissues such as skeletal muscles, peripheral nerves, or skin.

The liver showed coarse iron granules in the cytoplasm of many hepatocytes and Kupffer's cells (Figure 40.2A). There was no cirrhosis. Quantitative metal analysis indicated a large amount of iron but little increase of copper in the liver. Electron microscopy detected several electron-dense deposits in the cytoplasm of hepatocytes with diameters of 0.5 to 2.5 μm and small electron-dense bodies in lysosome-like structures with a single limiting-membrane, larger bodies having irregular membrane-unbound shapes. The presence of iron in these electron-dense materials was confirmed by energy-dispersive X-ray microanalysis. In the pancreas, iron deposition was more pronounced

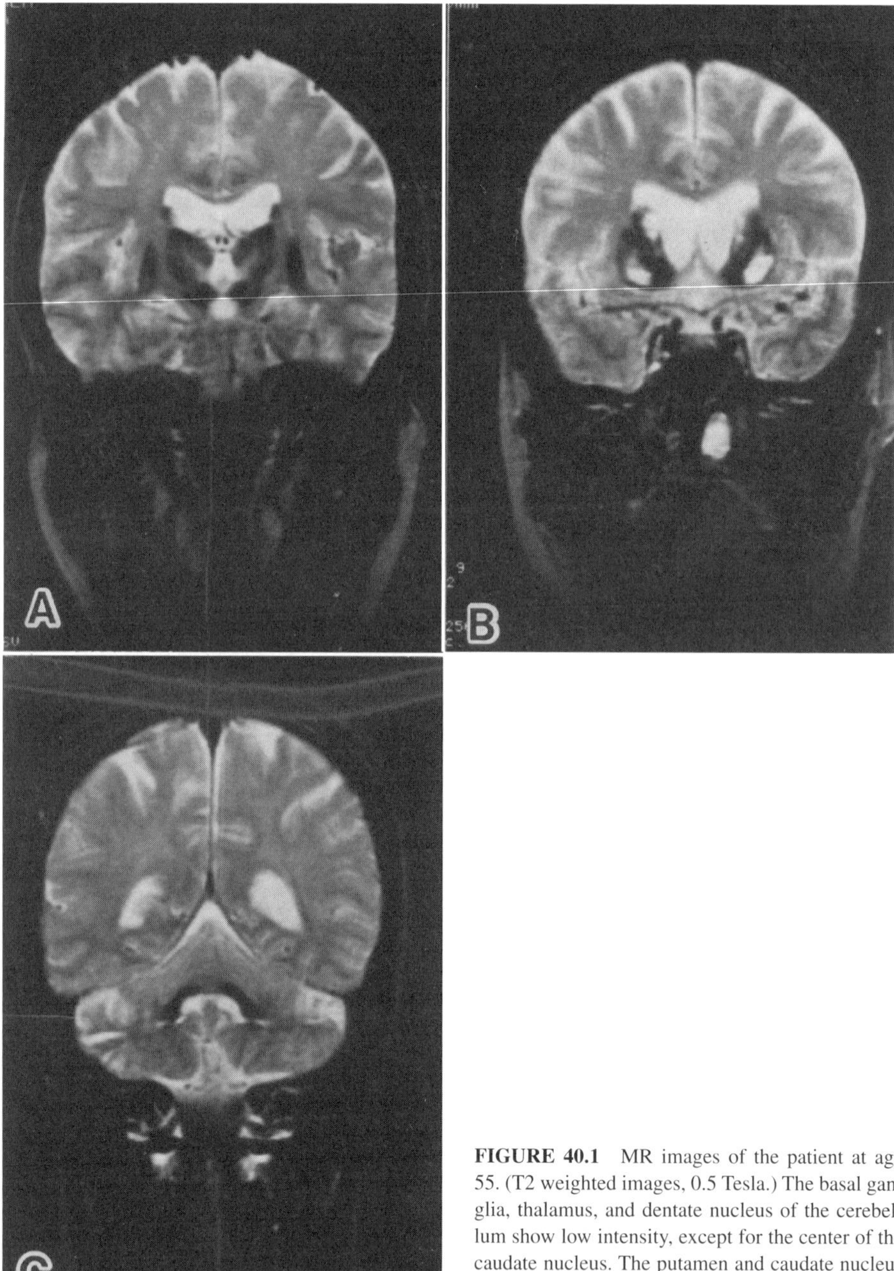

FIGURE 40.1 MR images of the patient at age 55. (T2 weighted images, 0.5 Tesla.) The basal ganglia, thalamus, and dentate nucleus of the cerebellum show low intensity, except for the center of the caudate nucleus. The putamen and caudate nucleus show cavitation.

in the exocrine than the endocrine cells. Many myocardial fibers showed iron deposition (Figure 40.2B). In the kidney, iron pigment was limited mainly to the podocyte of the glomerulus.

In the central nervous system, iron deposition was found in glial and nerve cells in association with decreased numbers of neurons in many parts of the brain. The caudate nucleus and putamen had the most striking lesions, both showing spongy softening with severe neuronal loss and little evidence of reactive scarring (Figure 40.2C). Iron pigment was diffuse in the remaining neurons and glial cells, and there were many siderophages around vessels. In the other basal ganglia and dentate nucleus, pathological changes were similar to those in the striatum, but less severe (Figure

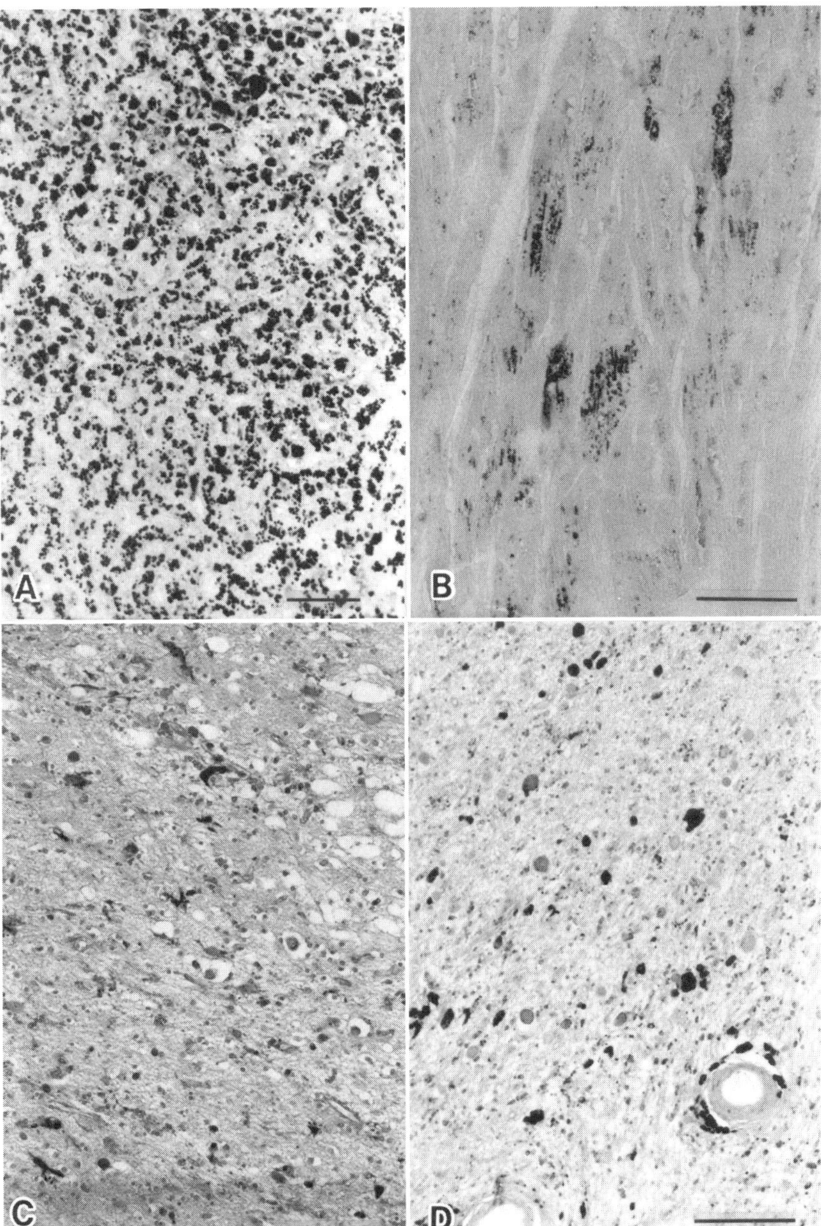

FIGURE 40.2 Pathological findings for the patient at autopsy (Prussian-blue stain). A: The liver has many iron granules both in the hepatocytes and the reticular system. Note that there is no cirrhosis (bar = 50 μm). B: The myocardium has iron granules, mainly around the perinuclear lesion (bar = 50 μm). C: The caudate nucleus shows severe spongy softening with little gliosis. Most of the neuronal and glial cells have been lost (bar = 100 μm). D: The dentate nucleus of the cerebellum shows essentially the same pathological changes as the caudate nucleus, but they are less severe. Many siderophages are present around the vessels (bar = 100 μm).

40.2D). The cerebellar cortex showed a marked loss of Purkinje cells, but the granular cell layer appeared to be unaffected.

An electron microscopic examination showed that most nerve and glial cells in the basal ganglia had many cytoplasmic inclusions of high electron density and pleomorphic shape. The nuclei of these cells appeared to be normal. The cytoplasms of the macrophages were filled with this type of electron-dense material. Energy-dispersive X-ray microanalysis confirmed that iron was the main component of the electron-dense inclusions in the nerve and glial cells. Quantitative analysis of the metal components in the brain tissues showed a greatly increased iron content in all the tissues examined. As for copper content in the brain and liver, it remained normal in this case; however, we must direct attention to the effect of D-penicillamine for this result. Before the treatment, the copper content of the liver was slightly increased, and that was similar to other reports (Table 40.1).[6,8,9]

40.4.3   Gene Analysis

We searched for the mutation in the ceruloplasmin gene in patients. The methodological details are described elsewhere.[16]

By sequencing of RT-PCR amplified cDNA from the proband, we identified a 5-base deletion (CACAG nucleotide position 3019-3023). Sequencing of genomic DNA confirmed that this deletion resulted from a G to A transition at the splice acceptor site of the upstream intron. We concluded that this mutation eliminated the consensus AG dinucleotides at the splice acceptor site and caused a 5-base deletion by defective splicing. Consequently, it introduced a premature termination codon (TAG) at the amino acid position 991:

....ttaag ^{3019}CAC AGG GGA GTT.... → ....ttaa**a**cacag GGG AGT TTA TAG...

(substituted adenine in bold type, 5-base deletion underlined).

A screening method for this mutation has been established; *Dra I* restriction site analysis of amplified genomic DNA. This screening showed that all the family members with the complete ceruloplasmin deficiency were homozygotes, and those with partial ceruloplasmin deficiency were heterozygotes for this mutation.

40.5   CLINICOPATHOLOGICAL SIGNIFICANCE OF HEREDITARY CERULOPLASMIN DEFICIENCY WITH HEMOSIDEROSIS

The clinical manifestations shown by our three patients were progressive extrapyramidal symptoms and cerebellar ataxia, both of which manifested in middle age. Two of them showed dementia. Biochemically, the characterizations were a deficiency of serum ceruloplasmin and an increase in serum ferritin. All the patients had been treated with insulin for diabetes mellitus. The severities of the symptoms varied for the four homozygotes; but the radiological findings of their brains and livers were very similar, with the exception of cavitation of the basal ganglia in the autopsy subject. Histological examinations including liver biopsies of the three homozygotes with neurological abnormalities showed systemic hemosiderosis, mainly in the liver, pancreas, and brain. Surprisingly there was no liver cirrhosis, which differs markedly from findings for Wilson disease and hereditary hemochromatosis (Table 40.1). In the brain in particular, iron deposition with severe neuronal loss occurred mainly in the basal ganglia and cerebellum, which corresponds to the clinical pictures of the patients. These pathological changes are very different from Wilson disease, hereditary hemochromatosis, or Hallervorden-Spatz disease.[9]

Some of the clinical investigations in the literature probably are those of our disease entity. Miyajima et al. reported a woman with familial apoceruloplasmin deficiency who had iron deposition in the liver and probable iron deposition in the brain.[5,7] The CT and MR findings for her liver and brain were similar to those of our patients, suggesting that their patient also had accompanying iron deposition in the brain. Her clinical symptoms were blepharospasm and retinal degeneration, and the progress of her disease was much slower than in our patients. Recently they described a 5-bp insertion in exon 7 resulting in a truncated open reading frame.[17]

In 1994, a family that very closely resembles ours was reported by Logan et al.[8] The symptoms were diabetes mellitus and dementia. MR images of the family members were very similar to those of our patients; moreover, the members of that family probably have an abnormality in the ceruloplasmin gene. We therefore conclude that hereditary ceruloplasmin deficiency with hemosiderosis is a novel disease entity (Table 40.1). Although only three families have been reported in the literature, there must be more cases around the world (personal communications).

Previous biochemical and physiological studies suggested that ceruloplasmin acted as ferroxidase in cells, and ceruloplasmin deficiency would show an iron metabolism disturbance that was not observed in Wilson disease. This disease proved the hypothesis and provided a key for understanding the function of ceruloplasmin as ferroxidase. Future investigations will provide the functional significance of the copper site of ceruloplasmin, and the functional relationship between copper, iron, and related proteins.

ACKNOWLEDGMENT

We thank Dr. Shu-ichi Ikeda and Dr. Kunihiro Yoshida of our department for their help with the pathological and genetic investigations reported in our study.

REFERENCES

1. Cousins R: Absorption, transport, and hepatic metabolism of copper and zinc: special reference to metallothionein and ceruloplasmin. *Physiol. Rev.*, 1985; 65:238–309.
2. Bull PC, Thomas GR, Rommens JM, et al: The Wilson disease gene is a putative copper transporting P-type ATPase similar to the Menkes gene. *Nature Genet.*, 1993; 5:327–337.
3. Tanzi RE, Petrukhin K, Chernov I, et al: The Wilson disease gene is a copper transporting ATPase with homology to the Menkes disease gene. *Nature Genet.*, 1993; 5:344–350.
4. Owen CA Jr: Effects of iron on copper metabolism and copper on iron metabolism in rats. *Am. J. Physiol.*, 1973; 224:514–518.
5. Miyajima H, Nishimura Y, Mizoguchi K, et al: Familial apoceruloplasmin deficiency associated with blepharospasm and retinal degeneration. *Neurology*, 1987; 37:761–767.
6. Morita H, Inoue A, and Yanagisawa N: A case with ceruloplasmin deficiency which showed dementia, ataxia and iron deposition in the brain. *Rhinshoshinkeigaku (Tokyo)*, 1992; 32:483–487 (in Japanese).
7. Miyajima H, Kaneko G, Kotani K, Maekawa M, and Kanno T: Excessive iron deposition and lipid peroxidation in familial ceruloplasmin deficiency. *Shinkeinaika (Tokyo)*, 1994; 41:48–54 (in Japanese).
8. Logan JI, Harveyson KB, Wisdom GB, et al: Hereditary ceruloplasmin deficiency, dementia and diabetes mellitus. *Q. J. Med.*, 1994; 87:663–670.
9. Morita H, Ikeda S, Yamamoto K, et al: Hereditary ceruloplasmin deficiency with hemosiderosis: a clinicopathological study of a Japanese family. *Ann. Neurol.*, 1995; 37:646–656.
10. Samokyszyn VM, Reif DW, Miller DM, and Aust SD: Effects of ceruloplasmin on superoxide-dependent iron release from ferricin and lipid peroxidation. *Free Radical Res. Commun.*, 1991; 12–13 Pt 1: 153–159.
11. Osaki S, Johndon FA, and Frieden E: The possible significance of the ferrous oxidase activity of ceruloplasmin in normal human serum. *J. Biol. Chem.*, 1966; 241:2746-2751.
12. Roeser HP, Lee GR, Nacht S, and Cartwright GE: The role of ceruloplasmin in iron metabolism. *J. Clin. Invest.*, 1970; 49:2408-2417.
13. Dancis A, Yuan DS, Haile D, et al: Molecular characterization of a copper transport in *S. cerevisiae*: an unexpected role for copper in iron transport. *Cell*, 194; 76:393–402.
14. Askwith C, Eide D, van Ho A, et al: The FET3 gene of *S. cerevisiae* encodes a multicopper oxidase required for ferrous iron uptake. *Cell*, 1994; 76:403–410.
15. O'Reilly S, Pollycove M, and Bank WJ: Iron metabolism in Wilson disease. Kinetic studies with iron[59]. *Neurology*, 1968; 19:634–644.
16. Yoshida K, Furihata K, Takeda S, et al: A ceruloplasmin gene mutation causes hereditary systemic hemosiderosis in humans. *Nature Genet.*, 1995; 9:267–272.
17. Harris ZL, Takahashi Y, Miyajima H, et al: Aceruloplasminemia. Molecular characterization of this disorder of iron metabolism. *Proc. Natl. Acad. Sci. U.S.A.*, 1995; 92:2539–2543.

Chapter 41

Treatment of Wilson Disease and Metal Metabolic Interactions

George J. Brewer

CONTENTS

41.1 INTRODUCTION TO THE DISEASE

Wilson disease is a disorder of copper accumulation which is inherited in an autosomal recessive manner.[1–3] Copper is an essential nutrient and the diet typically contains somewhat more than is required. The mechanism for ridding the body of the excess copper involves hepatic excretion into the bile for loss in the stool.[4] This mechanism has been rendered ineffective in Wilson disease due to the genetic mutation.[5,6]

Due to the defect in getting rid of excess copper, copper accumulates and causes toxicity of an oxidant nature, primarily in the liver and the brain. This results in a clinical presentation, usually in the second to fourth decade of life, of one of three types. One type is hepatic. The excess copper is stored in the liver and eventually exceeds the storage capacity of the liver and causes damage. The patient may present in childhood through the early twenties, sometime even later, with hepatic disease of one of several types. One type is hepatic failure, which may range from mild to severe and fulminant failure. Another type produces a hepatitis picture, which may be chronic, leading to a diagnosis of chronic active hepatitis. Or the patient may have chronic cirrhosis, and if the patient uses alcohol, they may be incorrectly labeled as having alcoholic cirrhosis.

In other patients the liver undergoes damage but the liver disease does not declare itself clinically. These patients may go on to develop neurologic or psychiatric disease, often in their late teenage years or early twenties. As the capacity of the liver to store copper is exceeded, copper accumulates in other parts of the body and the next most sensitive organ appears to be the brain.

Portions of the brain that are particularly sensitive are those that coordinate movement, such as the basal ganglia. Thus the patient may develop a movement disorder involving increasing problems with speech, swallowing, and coordination of fine, and later coarse, movements.[7] In some patients tremor is a major problem.

Some patients go through a prolonged period of psychiatric disturbance before developing neurologic symptoms. The psychiatric symptoms may include depression, changes in temperament, various kinds of neurotic behavior, and even suicidal ideation.[8] This type of behavioral abnormality is often mislabeled as substance abuse. Patients may have psychiatric symptoms for several years prior to the development of the neurologic signs and symptoms which lead to the diagnosis.

41.2 DIAGNOSIS OF WILSON DISEASE

Early diagnosis of Wilson disease is critical because it is a disorder which can be treated and further damage prevented. The later in the course of the disease therapy is instituted, the more likely that permanent damage of both the liver and the brain will occur. When the disease is suspected on clinical grounds, a series of studies can be instituted to verify the diagnosis (Table 41.1). These would normally include assay of serum ceruloplasmin, which is generally low in Wilson disease, but in about 10% of patients may be normal. Measurement of 24-h urine copper is a very valuable screening procedure.[1] It is always elevated in symptomatic Wilson disease. The only situation in which it may be elevated, and the patient not have Wilson disease, is in obstructive liver disease. It may be intermediately elevated in presymptomatic patients.[9] Examination of the eyes by slit lamp for the presence of copper deposits in the cornea, called Kayser-Fleischer (KF) rings, is a useful, noninvasive procedure as well. Kayser-Fleischer rings normally occur in almost all patients who have neurologic or psychiatric presentation of Wilson disease. They occur in about 50% of patients who present with liver disease.

Table 41.1 Useful Procedures for the Diagnosis of Wilson Disease

Procedure	Normal	Wilson Disease
Assay of serum ceruloplasmin	20–35 mg/dl	Usually very low. May be normal in about 10% of patients.
24-h Urine copper	25–50 μg	Always over 100 in symptomatic patients. May also be elevated in obstructive liver disease. May be intermediate in presymptomatic patients.
Slit lamp exam for KF rings	Negative	Positive in all patients with neurologic or psychiatric symptoms. Negative in about 50% of patients with liver presentation.
Liver biopsy for quantitative copper	25–50 μg/g dry weight	Always over 200 in untreated patients. May also be elevated in obstructive liver disease.

The gold standard for diagnosis is the use of hepatic biopsy and quantitative measurement of hepatic copper. This is always diagnostically elevated in Wilson disease.[1] The only situation in which elevated levels of hepatic copper may occur in the absence of Wilson disease is with obstructive liver disease. It is important not to rely on a copper stain. Such stains are only positive if the copper has become sequestered. If the copper is diffusely cytoplasmic, as is often the case, copper stains may be negative.

It is of course important to work-up the family of a newly diagnosed patient. Every full sibling is at 25% risk for the disease and can be treated prophylactically once the diagnosis is made.[10,11] Diagnosis is made by using the same kinds of screening technology mentioned in the previous paragraphs. Generally, measurement of 24-h urine copper and ceruloplasmin are quite helpful. However, at the presymptomatic stage not all patients show diagnostic elevation of urinary copper. The level may be intermediate between normal and diagnostically elevated levels. In these patients, liver biopsy is required for ruling in or out the diagnosis.

Very recently the gene for Wilson disease has been identified and sequenced.[12–14] This gene codes for a membrane-bound, copper-binding ATPase-type protein which probably acts as a copper pump across either intracellular or intercellular membranes. So far, it appears that there are a large number of mutations in this gene that produce clinical Wilson disease. This will make it more difficult to develop DNA tests to facilitate making the diagnosis in isolated patients. However, simpler and more direct tests are needed for patients who present with liver disease and psychiatric disorders, and so such tests should be developed.

41.3 NATURE OF COPPER TOXICITY IN WILSON DISEASE

The primary cause of tissue damage in Wilson disease is clearly copper toxicity.[1–3] Tissue damage is related to elevated levels of copper, and cessation of tissue damage is related to lowering copper levels through any one or more of a variety of therapeutic approaches.

It has been shown that excess copper causes cell injury and death (evidence reviewed in Reference 1). However, it is also clear that certain ligands mitigate or eliminate the toxicity of high levels of copper. Natural ligands include ceruloplasmin, metallothionein, and albumin. The potentially toxic copper is ionic copper and copper that is freely dissociable from other ligands.

It appears that copper toxicity is mediated through oxidant mechanisms (evidence reviewed in Reference 1). Copper produces H_2O_2,[15] probably derived from superoxide.[16] In red cells, Heinz bodies, a hallmark of oxidant damage, are produced[17] and levels of reduced glutathione, a major intracellular protection against oxidation, are lowered.[17,18] The critical damage may be in cell membranes or cytoskeletons.[19,20]

41.4 TREATMENT

41.4.1 Classification of Treatment

It is useful to classify the treatment of Wilson disease into several categories because the drug or drugs of choice for various categories are different. Table 41.2 presents such a classification. It consists first of initial treatment, broken down into whether the clinical presentation is neurologic or hepatic. Initial treatment may be defined as that period during which copper levels are brought down to subtoxic levels, usually a period of 2 to 6 months.

Table 41.2 Classification of Treatment in Wilson Disease

Category of treatment	Treatment of choice	Other treatments
Initial		
Neurologic	Tetrathiomolybdate	Zinc, penicillamine, trientine
Hepatic	Trientine plus zinc	Penicillamine
Maintenance	Zinc	Trientine, penicillamine
Presymptomatic from the beginning	Zinc	Trientine, penicillamine
Pregnant	Zinc	Trientine, penicillamine

The next category of treatment is maintenance therapy. This is the period after copper levels are brought down to subtoxic levels extending for the duration of the patient's life. Anticopper treatment must be taken during the rest of the patient's life in order to prevent the reaccumulation of copper and the recurrence of symptoms.

Another category of treatment is the treatment of the presymptomatic patient. Because this is a genetic disease of a recessive nature, each sibling of a newly diagnosed patient is at 25% risk for the disease. Work-up should be pursued aggressively on such siblings to see which of them have Wilson disease at the presymptomatic stage. After the diagnosis is established, such patients should go on prophylactic therapy for the rest of their lives.

The last category of treatment is the pregnant patient. Because of the teratogenic nature of some of the drugs used for Wilson disease, some women have stopped therapy during pregnancy. This has led to disastrous consequences in many reported cases.[1] It is very important that the pregnant woman continues therapy during her pregnancy and therefore it is useful to consider this category of patient separately.

41.4.2 Therapeutic Agents

Penicillamine was developed in 1956[21] and has been available since as a treatment for Wilson disease. It works by chelation and causing excretion of copper in the urine. Since this drug has been around for so long, and during much of that period was the only treatment available, it became the standard therapy for this disease for several decades.

Penicillamine is usually taken in doses of 250 mg four times a day, separated from food. It is clear that the drug is completely effective in terms of putting the patient into a negative copper balance and bringing copper down to subtoxic thresholds.

The main problem with penicillamine is its toxicity. It has a high risk of immediate toxicity of a hypersensitivity type which may require cessation of therapy or at least temporary withdrawal.[1,2] Oftentimes, the drug can be reinstituted under coverage of steroids.

In addition to hypersensitivity reactions, penicillamine has a very long list of potential side effects.[1] These include bone marrow depression, proteinuria, a number of immunological disorders, a tendency to produce abnormalities of connective tissue, and many others. The connective tissue effects may result in wrinkling of the skin, a cosmetic problem, or may cause risk of damage to the connective tissue of blood vessels which may lead ultimately to aneurysms, strokes, and the like.

Because of the large number of side effects with penicillamine, an alternative chelator drug, called trientine, has been developed.[22] This drug is approved for use in patients who are intolerant of penicillamine. The mechanism of action is believed to be similar and the dose is similar to that used with penicillamine. As with penicillamine, trientine appears to be 100% effective in producing a negative copper balance and reducing the copper to subtoxic levels.

Trientine also has side effects, such as proteinuria and autoimmune problems, but the frequency of these seem to be less than with penicillamine.

Zinc has been developed into a practical treatment of Wilson disease by our group[1,11,23–27] in the U.S. and by Hoogenraad's group[28,29] in the Netherlands. We use zinc acetate, whereas Hoogenraad uses the zinc sulfate salt. Zinc acts by a completely different mechanism. It induces intestinal cell metallothionein which then causes a mucosal block of copper absorption.[27] Zinc is completely effective in producing a negative copper balance and reducing copper to subtoxic thresholds.

Zinc is taken in doses of 50 mg of the element, three times a day, separated from food and beverages other than water. The big advantage of zinc over the other agents, is its almost complete lack of toxicity.[1] No side effects from zinc are known, other than an occasional complaint of gastric intolerance, particularly with the first morning dose if taken prior to breakfast. In such patients, taking the first morning dose between breakfast and lunch usually mitigates this complaint.

Tetrathiomolybdate (TM) is another drug that we are developing, in this case into an initial therapy for neurological Wilson disease.[30] TM acts by forming a complex of copper and protein and renders that copper unavailable for absorption if the complex forms in the intestinal tract, or for cellular uptake if the complex forms in the blood.

We have used TM for the initial treatment of patients with Wilson disease who present with neurologic symptoms.[30] We use the drug in initial doses of 20 mg, six times a day; 20 mg three times a day with meals in order to prevent absorption of food copper and reabsorption of endogenously secreted copper and the other three doses are given between meals to accomplish absorption and titration of the potentially toxic copper of the blood into an albumin TM-complex. The between-meals dose is escalated until there is a stoichiometric equilibrium of molybdenum and copper.

TM has been completely effective in gaining control of copper toxicity without loss of neurologic function.[30] Its toxicity seems to be limited to an occasional anemia from suppression of red cell production, which is reversible. This may be a result of localized copper deficiency in the bone marrow since TM is such a potent anticopper agent, resulting in a failure of heme production as copper is required for heme synthesis.

41.4.3 Treatment Recommendations

Our treatment recommendations are listed in Table 41.2. For the initial treatment of patients who present with neurologic disease, we use TM for a period of 8 weeks. The purpose of this therapy during this period is to gain control of copper metabolism without losing neurologic function. The patient is then transitioned to zinc therapy for maintenance.

The disadvantages of penicillamine in terms of hypersensitivity and side effects have already been discussed briefly. In addition, in the setting of a new patient presenting with neurological disease, penicillamine may produce neurological worsening of the disease, probably due to mobilization of hepatic copper and temporary flushing of the brain with additional copper.[1] This risk appears to be about 50%. About a half of that 50%, or 25% of the original sample, never recover to baseline values.

The use of trientine for initial treatment of neurologically affected patients has not really been explored to any significant extent. Theoretically it carries the same risk of worsening as does penicillamine.

We believe that zinc is rather slow acting for this kind of patient. It takes zinc about 4 to 6 months to gain control of copper and bring it below subtoxic thresholds and during this period the disease may progress.

The initial treatment of patients who present with hepatic disease has not been well worked out. Certainly those patients who present in acute fulminant failure may require hepatic transplantation to save their lives. For patients who present with mild hepatic failure or with hepatitis, medical therapy is indicated and works well. Empirically we use a combination of trientine and zinc. The trientine is to produce a relatively brisk negative copper balance. We choose trientine over penicillamine for this purpose because of its greater safety. Zinc is added because, in addition to the induction of intestinal cell metallothionein, zinc will induce hepatic metallothionein. Metallothionein in the liver will sequester copper and render it nontoxic. Thus zinc has a theoretical potential of having an additive effect to that of trientine. We treat such patients for 4 to 6 months with the combination and then discontinue the trientine and the patients continue on zinc for their maintenance therapy.

For maintenance therapy, we believe that zinc is the treatment of choice. All three drugs, penicillamine, trientine, and zinc are equally effective. The big advantage of zinc over the other two is the almost complete lack of side effects. We have treated 125 patients with zinc maintenance therapy. The longest has been treated for 13 years and we have treated 62 patients for 5 years or longer. Zinc is at the present time undergoing a review of a new drug application (NDA) by the U.S. Food and Drug Administration. Once it is approved, it should take its place as the treatment of choice for maintenance therapy.

We treat the presymptomatic patients as if they were in the maintenance phase from the beginning. Thus we treat them with zinc. We have treated 15 presymptomatic patients with zinc from the beginning, with complete success.[11]

The pregnant patient is at some risk of teratogenic effects from either penicillamine or trientine. For that reason we recommend zinc, which has specifically shown not to be teratogenic. We have treated 11 patients through pregnancy with zinc, without problems.[1]

41.5 RECOVERY AND PROGNOSIS

As part of our protocol for the study of TM in the initial treatment of Wilson disease, we have developed techniques for quantitating neurologic function.[30] These include a quantitative neurologic exam and a quantitative speech exam. We also quantitate the abnormalities on brain magnetic resonance image (MRI).

From our studies we have good data of the recovery of function in patients after initial treatment. Generally speaking, no recovery takes place until 6 months after therapy has begun. Recovery then takes places between about 6 and 24 months after initiation of therapy. That disability which remains after 2 years is usually permanent. As compared to the improvement in neurologic and speech function being limited to the first 2 years, the brain MRI shows further recovery over about a 4-year period before leveling out.

Generally speaking, patients with mild to moderate neurologic disability recover very well. They may be left with a mild speech defect or other stigma of the disease, but usually they recover to a very good functional state. The one exception to this is tremor. Tremor seems to show less recovery, in our experience, than the other abnormalities in Wilson disease. Patients with severe neurologic disability at the time treatment is initiated have a poorer prognosis. While some recovery occurs, the patient is often left with significant permanent disability.

Recovery of hepatic function follows a similar curve to that of neurologic function. Usually the liver shows significant repair by about 6 months and repair is completed by 2 years. Patients with mild hepatic failure usually recover to normal hepatic function although they are always left with underlying cirrhosis. However the functional status of their liver, in our experience, remains constant over many years of follow-up, so far up to 10 years.[1] Such patients often show significant portal hypertension as reflected in hypersplenism, thrombocytopenia, and leukopenia. However, these things do not progress. The only real risk in these patients is the possibility of bleeding from esophageal or gastric varices.[1]

Patients who are treated in the presymptomatic category usually never show any ill effects on their health. Some of these do have significant cirrhosis and portal hypertension as reflected by thrombocytopenia and leukopenia, but this never becomes clinically important and we have never seen varicele bleeding in such a patient.

41.6 METAL INTERACTIONS IN TREATMENT

It is interesting that two of the four treatment agents are themselves metals. Thus the treatment of Wilson disease involves metal-metal interaction.

In the case of zinc, this interaction occurs through the commonality of one protein, metallothionein. In other words, zinc is an excellent inducer of metallothionein but the metallothionein, once induced, has a higher affinity for copper. This is taken advantage of in two ways in treatment. In maintenance therapy, zinc induces metallothionein in the intestinal cell, which blocks absorption of copper. In initial treatment of the hepatic presentation of Wilson disease, zinc therapy induces hepatic metallothionein, which sequesters potentially toxic copper of the liver in a nontoxic form.

The interaction of copper and molybdenum is somewhat more direct although it requires a third agent. In this case, copper and tetrathiomolybdate are part of a complex with the third agent, protein. This complex is rendered both unabsorbable and in a state where it is not taken up by cells. Again, this is taken advantage of in two ways in treatment. Tetrathiomolybdate given with food blocks the absorption of food copper and copper in endogenous secretions. Tetrathiomolybdate given between meals is absorbed into the blood and sequesters the potentially toxic copper of the blood and renders it nontoxic.

41.7 CONCLUSION

Metal interactions can be used to great advantage in the treatment of Wilson disease. In fact, the best maintenance therapy is zinc, which interacts with copper through a protein that both metals interact with, metallothionein. Second, a new agent, tetrathiomolybdate, is an excellent initial treatment for patients presenting with neurologic disease who are otherwise at serious risk of neurologic worsening of their disease. Tetrathiomolybdate acts by complexing copper with protein, making the copper unavailable for cellular uptake, and therefore nonabsorbable and nontoxic.

41.8 SUMMARY

Wilson disease is a recessively inherited disease of copper accumulation which causes liver and/or brain damage in older children and young adults. One of the biggest problems is recognition of the disease, since it can masquerade as hepatitis, cirrhosis, or liver failure from other causes, or as some type of psychiatric disorder. When the disease presents in its neurologic form, as a movement disorder, it is more easily recognizable.

When Wilson disease is suspected, there are a series of screening procedures and tests which can be used to determine or exclude the diagnosis. One of the best is a 24-h urine copper determination. Urine copper is always diagnostically elevated in symptomatic Wilson disease. Kayser-Fleischer ring evaluation by slit lamp examination is extremely useful when positive, but can't be used to exclude the diagnosis when negative. Ceruloplasmin assay in serum can be used to guide the index of suspicion, because it is usually low, but it is normal in 10% of patients and can also be low in the heterozygous carrier. Quantitative copper assay of liver biopsy tissue remains the gold standard of diagnosis. The gene has recently been cloned, and ultimately DNA tests may become available. Since the disease is inherited, it is very important to work up full siblings for possible Wilson disease, since they are at 25% risk and the disease is preventable with anticopper therapy.

Wilson disease is very treatable with anticopper agents. For maintenance therapy, treatment of the presymptomatic patient from the beginning, and treatment of the pregnant patient, we recommend zinc. For the initial treatment of the patient presenting with neurological disease who is at great risk of neurologic worsening if treated with penicillamine, we use tetrathiomolybdate for 8 weeks and then a transition to zinc. For the initial treatment of the patient presenting with liver disease, we use a combination of trientine and zinc for 4 to 6 months and then transition to zinc alone.

Patients show clinical recovery over the 6- to 24-month period after anticopper treatment is initiated. For patients only mildly to moderately affected, recovery of both liver and neurologic function is usually quite good.

REFERENCES

1. Brewer GJ and Yuzbasiyan-Gurkan V: Wilson Disease. *Medicine*, 1992; 71:139–164.
2. Schienberg IH and Sternlieb I: *Wilson Disease*; Smith L H, Jr., ed.; *Major Problems in Internal Medicine*; Philadelphia; W. B. Saunders, 1984.
3. Danks DM: Disorders of copper transport; Scriver CR, Beaudet AL, Sly WS, and Valle D, eds.; *Metabolic Basis of Inherited Diseases*; 6th ed., New York; McGraw-Hill, 1989; 1411–31.
4. Cartwright GE and Wintrobe MM: Copper metabolism in normal subjects. *Am. J. Clin. Nutr.*, 1964; 14:224–32.
5. Frommer DJ: Defective biliary excretion of copper in Wilson disease. *Gut*, 1974; 15:125–29.
6. Gibbs K and Walshe JM: Biliary excretion of copper in Wilson disease. *Lancet*, 1980; 2:538–39.
7. Starosta-Rubinstein S, Young AB, Kluin K, et al: Clinical assessment of 31 patients with Wilson disease. Correlations with structural changes on magnetic resonance imaging. *Arch. Neurol.*, 1987; 44:365–70.
8. Akil M, Schwartz J A, Dutchak D, et al: The psychiatric presentations of Wilson disease. *J. Neuropsychiatry*, 1991; 3:377–82.
9. Yuzbasiyan-Gurkan V, Johnson V, and Brewer GJ: Diagnosis and characterizations of presymptomatic patients with Wilson disease and the use of molecular genetics to aid in the diagnosis. *J. Lab. Clin. Med.*, 1991; 118:458–65.
10. Sternlieb I and Schienberg IH: Prevention of Wilson disease in asymptomatic patients. *N. Engl. J. Med.*, 1968; 278:352–59.
11. Brewer GJ, Dick RD, Yuzbasiyan-Gurkan V, et al: Treatment of Wilson disease with zinc. XIII. Therapy with zinc in presymptomatic patients from the time of diagnosis. *J. Lab. Clin. Med.*, 1994; 123:849–58.
12. Bull P C, Thomas G R, Rommens J M, et al: The Wilson disease gene is a putative copper transporting P-type ATPase similar to the Menkes gene. *Nature Genet.*, 1993; 5:327–37.

13. Tanzi RE, Petrukhin K, Chernov I, et al: The Wilson disease gene is a copper transporting ATPase with homology to the Menkes disease gene. *Nature Genet.*, 1993; 5:344–50.

14. Yamaguchi Y, Heiny ME, and Gitlin JD: Isolation and characterization of a human liver cDNA as a candidate gene for Wilson disease. *Biochem. Biophy. Res. Commun.*, 1993; 197:271–7.

15. Metz EN and Sagone AL: The effect of copper on the erythrocyte hexose monophosphate shunt pathway. *J. Lab. Clin. Med.*, 1972; 80:405–13.

16. Rifkind JM: Copper and the autooxidation of hemoglobin. *Biochemistry*, 1974; 13:2475–81.

17. Deiss A, Lee GR, and Cartwright GE: Hemolytic anemia in Wilson disease. *Ann. Intern. Med.*, 1970; 73:413–18.

18. Todd JR and Thompson RH: Studies on chronic copper poisoning. II. Biochemical studies on blood of sheep during haemolytic crisis. *Br. Vet. J.*, 1963; 119:161–73.

19. Hochstein P, Kumar KS, and Forman SJ: Lipid peroxidation and the cytotoxicity of copper. *Ann. N.Y. Acad. Sci.*, 1980; 355:240–48.

20. Sokol RJ, Devereaux MW, Traber MG, et al: Copper toxicity and lipid peroxidation in isolated rat hepatocytes: effect of vitamin E. *Pediatr. Res.*, 1989; 25:55–62.

21. Walshe JM: Penicillamine. A new oral therapy for Wilson disease. *Am. J. Med.*, 1956; 21:487–95.

22. Walshe JM: Treatment of Wilson disease with trientine (Triethylene tetramine) dihydrochloride. *Lancet*, 1982; 1:643–47.

23. Brewer GJ, Hill GM, Prasad AS, et al: Oral zinc therapy for Wilson disease. *Ann. Intern. Med.*, 1983; 99:314–320.

24. Hill GM, Brewer GJ, Prasad AS, et al: Treatment of Wilson disease with zinc. I. Oral zinc therapy regimens. Hepatology, 1987; 7:522–28.

25. Brewer GJ, Hill GM, Dick RD, et al: Treatment of Wilson disease with zinc. III. Prevention of reaccumulation of hepatic copper. *J. Lab. Clin. Med.*, 1987; 109:526–31.

26. Brewer GJ, Yuzbasiyan-Gurkan V, Lee D-Y, et al: Treatment of Wilson disease with zinc. VI. Initial treatment studies. *J. Lab. Clin. Med.*, 1989; 114:633–38.

27. Yuzbasiyan-Gurkan V, Grider A, Nostrant T, et al: The treatment of Wilson disease with zinc. X. Concurrent induction of intestinal metallothionein and suppression of copper absorption by zinc therapy. *J. Lab. Clin. Med.*, 1992; 120:380–86.

28. Hoogenraad TU, Koevoet R, and De Ruyter Korver EGWM: Oral zinc sulfate as long-term treatment in Wilson disease (hepatolenticular degeneration). *Eur. Neurol.*, 1979; 18:205–11.

29. Hoogenraad TU, Van Hattum J, and Van den Hamer CJA: Management of Wilson disease with zinc sulfate. Experience in a series of 27 patients. *J. Neurol. Sci.*, 1987; 77:137–46.

30. Brewer GJ, Dick RD, Johnson V, et al: Treatment of Wilson disease with ammonium tetrathiomolybdate. I. Initial therapy of 17 neurologically affected patients. *Arch. Neurol.*, 1994; 51:545–54.

Chapter 42

Neuropsychological Characteristics in Wilson Disease

Harald Hefter

CONTENTS

42.1 INTRODUCTION

The clinical features of Wilson disease (WD) were already known at the end of the nineteenth century;[1,2] the pathophysiology of this disease entity, however, remained unclear until now. In 1912, S.A.K. Wilson realized that the described clinical features were part of a "familial nervous disease associated with cirrhosis of the liver" which he called "progressive lenticular degeneration."[3] An etiological role of copper in WD was postulated in 1948.[4] Progress in the understanding of possible pathophysiological mechanisms was made recently. In 1985, it was established that the gene for WD resides on the long arm of chromosome 13.[5] In 1993, evidence was presented that the WD gene encodes a P-type ATPase.[6] Some linkage and haplotype studies suggested that relevant mutations would be quite limited.[7] However, in a recent study on 58 WD-families at least 23 mutations, occurring with various frequency, were found differing in the age of onset of symptoms.[8] Furthermore, comparison of cDNA prepared from brain and liver tissue revealed alternative splicing.[9] Alternative splicing may represent a tissue-specific mechanism to regulate the amount of mRNA encoding the full-length, functionally active protein. This mechanism could supplement the apparent tissue-specific transcriptional regulation of the WD gene expression.[6] Alternatively, the variant transcripts could encode proteins with distinct cellular functions; for example, the truncated forms of the WD protein with intact metal-binding sites might serve as copper storage proteins which shuttle copper ions inside the cell.[9] How the variability of the genetic defect helps to understand the variable phenotype has to be worked out during the next few years.

42.2 NEUROPSYCHIATRIC SYMPTOMS

42.2.1 Neuropathological and Neuroimaging Abnormalities

The phenotype or the clinical outcome of a WD patient critically depends on the amount and duration of copper intoxication. Copper is an essential trace element. Since copper ions have high cell toxicity,[10] a careful copper homeostasis is necessary but is disturbed in WD because of an impaired excretion of copper via the bile. In WD copper is stored throughout the entire body causing a broad spectrum of clinical symptoms arising from the affection of various organs including heart, kidneys, bones, blood, and skin.[11] In that aspect, terms like "hepatolenticular" or "progressive lenticular degeneration" are misleading to characterize WD. However, the dominant clinical features arise from an affection of the liver and brain.

Neuropathological analyses of the brain copper content in WD patients have demonstrated elevated levels throughout the entire gray matter.[12] Correspondingly, **PET** and **MRI** studies,[13,14] (especially when using the new high-Tesla machines) revealed a reduced glucose metabolic rate and altered spin resonance behavior, respectively, not only in the basal ganglia, the thalamus, the brainstem, and the cerebellum, but also in the cerebral cortex. The reduction of the glucose metabolism correlates with various clinical features;[13] the correlation between structural alterations detected by MRI scanning with clinical symptoms, however, is only weak.[14] No correlative study between neuropsychiatric symptoms and neuroimaging abnormalities has been presented so far.

42.2.2 Statistical Analysis of Clinical Symptoms

Recent **cluster analysis**[15] of clinical symptoms in 400 WD patients linked together out of four larger series of patients revealed four different clusters. An hepatic and a neurological subgroup were detected in all four series analyzed. Additional clusters were an asymptomatic and a mixed hepatic and neurological subgroup. Transitions between clusters may occur since this type of analysis heavily relies on the sensitivity of the discriminating variables. In one series, age of onset was helpful to split up the neurological subgroup into two different clusters, probably corresponding to the classification of Denny-Brown who distinguished between a "pseudosclerotic" and a "progressive lenticular" type with different mean ages of onset.[16] In the largest series of 195 patients[17] a neuropsychiatric and a neurological subgroup could be distinguished from each other and from the rest of the patients.

Factor analysis was also performed to find out individual symptoms characterizing the different clusters. Clinical factors highly loading the neurological factor were dysphagia, dystonia, rigidity, dysarthria, tremor, KF rings present, age of onset, outcome score, and duration of admission. This is in agreement with the ranking of neurological symptoms in other larger series of patients[10,11,18] (see Table 42.1).

Clinical features loading the psychiatric factor were personality change, irritability, cognitive impairment, and depression. In our series, mood disturbances and personality changes were the leading psychiatric symptoms[19] (see Table 42.1).

42.2.3 Characteristic Psychiatric Features

In about 90% of 60 WD patients we observed **mood disturbances** ranging from major depression to mania, although those extremes were very rare. Relatives remarked about the aggressiveness and impulsiveness of the patients, who themselves noticed their own emotional lability without being able to control their behavior. In addition, **personality changes** are present in about 50% of

Table 42.1 Neuropsychiatric Symptoms in Wilson Disease

Structure impaired	Neurological symptoms	Frequency of occurrence	Response to therapy	Disorders of	Psychiatric symptoms	Frequency of occurrence	Response to therapy
Basal ganglia	Dysarthria	+++	+	Affect	Mood disturbance (emotional lability)	+++	+
	Bradykinesia	+++	+				
	Dystonia	++	++		Depression	++	+
	Action tremor	++	++		Euphoria	+	+
	Hypersalivation	+	++		Mania	(+)	?
Cerebellum	Gait ataxia	+	++	Personality	Impulsiveness	++	+
	Intention	+	+++		Irritability	++	+
	tremor				Lack of judgment	+	+
Brain stem	Slow saccad.	+	?		Antisocial behaviour	+	++
	Abnormal breathing	+	?				
Cortex	Focal neuropsych. deficits as apraxia or dyslexia, etc.	(+)	?	Cognition	Mental slowing	+	+
					Memory deficits	+	?
					Delusional disord.	(+)	+

Note: +++ = frequent; + = rare; +++ = excellent response to therapy; + = some response to therapy.

our patients, making psychosocial interaction rather difficult in about 20% of the patients. Lack of judgment, poor work performance, in addition to mental slowing may lead to drug abuse, loss of driving permission, and unemployment. Extremely abnormal behavior such as permanent thievery as well as wasting money occurred in two patients.

In some patients with moderate to severe neurological impairment a considerable mental slowing also may be present. Especially in combination with a change in personality, those patients may appear as being demented. Therefore it has been claimed that **dementia** is a typical finding in WD. However, during psychometric testing, especially when tests are used that do not rely on motor function, these handicapped patients produce surprisingly good and often fairly normal results. Therefore "treated WD should not be considered a cause of dementia."[20,21]

Memory deficits are claimed by about half of the WD patients; but in comparison to patients with neurodegenerative diseases such as Huntington's chorea[21] or multisystem atrophy[22] of comparable age and neurological status, the memory deficits in WD patients are mild.

Whether there is an association between **schizophrenia** and WD remains a matter of debate. Because of a high interest in this controversy it is overrepresented in the literature compared to its clinical relevance. Single cases have been described with improvement of psychotic episodes by copper-trapping therapy.[23] We have come across only 2 patients with clear-cut schizophrenia-like episodes out of 60 patients; Dening describes only 3 out of 195,[17] and Oder et al. 3 out of 45 patients.[18] "When fairly rigorous diagnostic criteria are used, such states are uncommon, but they do occur, and probably at a rate higher than that of ideopathic schizophrenia in the general population."[24]

In his earlier papers, Wilson doubted that mental symptoms were an integral part of the clinical picture.[3] Consistent correlations between psychiatric and neurological symptoms, however, especially the association between disorders of conduct and personality with dysarthria and the "dystonic type of neurological WD," suggest an organic basis for the mental affection of WD patients.[20] A common affection of the parallel (cortico-putamo-pallido-thalamo-cortical) motor and the (cortico-caudato-thalamo-cortical) cognitive loop[25] may account for this correlation.

42.3 PSYCHOMETRIC TESTING

The first attempts to apply psychometric tests were made in the late 1960s. Application of the Wechsler Adult Intelligence Scale (WAIS) in 19 WD patients confirmed that, in general, WD patients were not demented.[26] With the use of Rorschach patterns and Figure Drawing Styles it became clear "that Wilson disease occurs in such a variety of personalities that these patients represent a sample from the general population, and not a particular type."[26] Of the 49 patients, 30 had either neurological or psychiatric symptoms, 25 patients improved under therapy, and 14 also revealed an improvement of their psychiatric disturbance.[26]

About 20 years later the results of psychometric testing in 10 WD patients with acute or subacute WD without overt neuropsychiatric signs or symptoms of hepatic encephalopathy were described. The test battery contained 11 subtests: 5 of these subtests (Token, Dominant Tapping, Animal Naming, Digit Span Forward, Digit Span Backward) did not yield significant differences to a group of 20 controls matched in age and education level. Since in six subtests (Symbol Digit, Trails A and B, Nondominant Tapping, Block Design, Digit Supraspan) significant differences were found (after one-sided testing without Bonferoni adjustment) the authors concluded that subacute or acute WD may be associated with cerebral dysfunction.[27] However, this conclusion lacks good statistical support because after two-sided testing at a significance level of 5% and alpha adjustment, none of these tests would have been significant.

A more detailed psychometric analysis of 19 neurologically impaired and 12 neurologically asymptomatic WD patients as well as 15 controls used a large ensemble of tests, including the Wechsler Adult Intelligence Scale (revised form, WAIS-R), the Wechsler Memory Scale (WMS), the Dementia Rating Scale (DRS), the Wisconsin Card Sorting Test (WCST), the Boston Naming Test, the Trail Making Test, and the Animal Naming Test. The neurologically impaired patients were significantly worse in some timed visuomotor tasks than the asymptomatic patients, probably because of interference with their impaired motor function. As the only nonmotor cognitive deficit a mild memory impairment was found.[28] More detailed testing (Rey Auditory Verbal learning test) of memory function in the same patients[29] suggests that the memory deficit in WD is not a learning or storage problem but arises from difficulties to recall stored material, similar to the retrieval problem in patients with Huntington's disease. However, as in the previous study the significance level after alpha adjustment was not reached.[29]

This is different in another detailed psychometric analysis of 17 patients where the following tests were used: 3 subtests of the WAIS (Digit Span, Mathematical Abilities, Picture Arrangement), the Multiple Choice Vocabulary Test (form B, MWT-B), the Standard Progressive Matrices Test (Raven), the Short Term Memory, Attention, and Perceptual Speed (German: Syndromkurztest; SKT), the Benton Test, the Achievement Assessment System (German: Leistungs-Prüf-System; LPS), the Intelligence Structure Test (IST), Psychiatric assessment (German: documentation system of the "Arbeitsgemeinschaft für Methodik und Dokumentation in der Psychiatrie;" AMDP-scale; comparable to the American DSM III-scale). In spite of highly restrictive statistical analysis with alpha adjustment, a significant reduction of perceptual speed could be detected in 17 WD patients. These authors emphasize that "all statistically different subtests or those close to it were time-limited. Differences were more pronounced when a speed and a visuo-spatial factor were combined."[21] This finding that WD patients reveal a poor performance in time-limited tests has been mentioned previously and was confirmed later on.[19]

The most sophisticated analysis of WD patients by standardized neurological, psychiatric, and psychological tests so far included the Webster Parkinson's Disease Rating Scale, the North-Western University Disability Scale, the Frenchay Dysarthria Scale, the Comprehensive Psychopathological Rating Scale, the General Health Questionnaire, the Personal Assessment Schedule, the Mini-Mental State (MMS), the Benton Test, the Spot-the-Word Test, the Signal Detection Memory Test, the Global Assessment Scale, the Drug Risk Number, and the analysis of spontaneous eye blink rates. As in previous studies, the leading neurological symptoms were bradykinesia, tremor, gait disturbances, and dysarthria (see Table 42.1). Whereas tremor was not related to psychopathology or cognition, bradykinesia was clearly associated with psychopathology and gait disorders were related to depression and cognition and to abnormal personality. Cognitive impairments were mild. After correction for premorbid intelligence, there was no correlation between cognition and neurological or psychiatric variables. Disorders of behavior and personality correlated with dysarthria, bradykinesia, and rigidity. Interestingly, a correlation between depressive symptoms and hepatic dysfunction was not described before.[30]

In summary, neuropsychological testing presents evidence that most of the treated WD patients are not demented, have only mild cognitive deficits such as low perceptual speed, and a mild memory dysfunction. However, in tests assessing motor skill and which have to be performed in a time-limited way or including visuospatial discrimination tasks, neurologically impaired WD patients reveal a significantly worse performance than asymptomatic WD patients or controls.

42.4 ELECTROPHYSIOLOGICAL TESTING OF COGNITIVE IMPAIRMENT

42.4.1 P300 Event-Related Potentials

The latency of the P300 component, which can be elicited by rare, task-relevant target stimuli embedded into sequences of otherwise irrelevant stimuli, contains information about the time course of stimulus evaluation. P300 latency is prolonged in demented patients. In patients with Parkinson's disease slowing of P300 latency was demonstrated that paralleled fluctuating changes of global mental status. Especially in those parkinsonian patients with poor results in the subtests of the MMS testing recent memory and constructional ability, the P300 latency was prolonged.[31] In more than 75% of consecutively recruited patients with manifest Huntington's disease and 25% of clinically unimpaired subjects at risk, P300 latencies were prolonged. P300 latency was correlated with psychometric testing of selective attention (d2-test), recent memory (SKT, Benton), vulnerability to distraction (d2-test, SKT), but not with depression or psychosis scores.[32] The best correlation was between P300 latency and subtests of the SKT testing attention and visual memory under time pressure. Multiple regression analysis demonstrated that P300 latency is highly correlated with tests requiring speeded processing of visually presented material and is less associated with language-related abilities. P300 amplitude correlated with verbal and nonverbal subtests of the WAIS. Both parameters correlated with the overall disability score in Huntington patients.

In WD only a few patients underwent P300 testing. In a case study, improvement of auditory-elicited P300 maps with therapy was described (Figure 42.1). The initially dominating frontal amplitudes decreased during treatment whereas the parietal field increased in amplitude.[33] This single observation was confirmed by our systematic analysis of standard P300 recordings in an auditory oddball paradigm in 19 consecutively recruited WD patients and in an age- and sex-matched control group. P300 testing was compared to psychometric testing (MWT-b; Raven, SKT, AMDP-scale). None of the earlier N1,N2,P2-components revealed a significant latency prolongation. In 8 patients the P300 component was completely abolished; in the remaining 11 WD patients P300 latency was significantly prolonged and the P300 amplitude significantly reduced. The correlation between P300 latency and the SKT just failed to be significant, the reduction of P300 amplitude was significantly inversely related to the clinical score (details in Reference 19). Thus, the analysis of P300 event-related potentials in WD yields similar results to those in patients with Huntington's and Parkinson's diseases. The abolition of the P300 component in a fairly large percentage of WD patients (8 out of 19) on the one hand demonstrates its vulnerability, but its correlation with the clinical score underlines its potential usefulness for monitoring cognitive improvement under therapy, on the other hand. Therefore more sophisticated event-related studies should be performed on larger groups of WD patients in the future.

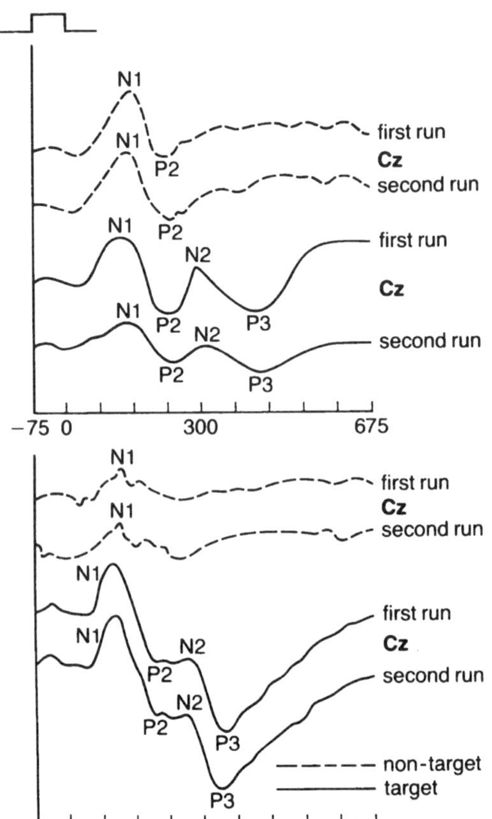

FIGURE 42.1 Auditory event-related P300 potentials recorded by means of an auditory oddball paradigm in a WD patient (upper part) in comparison to an age- and sex-matched control of equal educational level. The potentials resulting from hearing the frequent tones (dashed lines) have the same shape in the patient and in the control. The late P3 component (solid lines) which results from counting the rare tones has a longer peak latency and a reduced amplitude in the WD patient in comparison to its control. (Note that the curves were smoothed considerably before presentation).

42.4.2 Time Estimation

A circumscribed cognitive deficit in WD patients became apparent when we tested the ability to estimate the duration of short time intervals. For that purpose sequences of five tones were presented via earphones. The intervals between the first four tones were constant of a given length (100, 200, 300, 400 ms). The length of the interval between the fourth and fifth tone was varied randomly in seven steps around the basic interval length. For each sequence of tones WD patients or controls had to decide whether the last interval was longer or shorter than the previous ones. More than 200 runs per basic interval length were tested in each subject. Thereafter the cumulative distribution for the probability to detect that the last interval was longer was determined. This is demonstrated in Figure 42.2 for the base interval 300 ms. After z-transformation the correlation between the seven interval lengths tested and the z-transformed probabilities was determined (compare Figure 42.2). For the controls (n = 15) the mean correlation coefficient lies close to 0.9 (Figure 42.3: left side). The 10 WD patients had considerable difficulties in estimating the durations of intervals in the 200 ms range (Figure 42.3; right side). The lengths of intervals in the 100-, 300-, and 400-ms range, however, were estimated correctly. Since the mean cycle duration of the maximal tapping rate in these WD patients lay in the same range as the interval durations which were estimated significantly worse than the rest of the intervals, and close to the mean cycle duration of their hand tremor, this finding may suggest that tremor does not only disturb motor activity but also cognitive processes. A speculative interpretation is that tremor disturbs the function of the parallel motor and cognitive loop (mentioned above) in a similar way.[34] This hypothesis is stimulating to analyze neuropsychological impairment in WD in more detail in comparison to other patient groups with tremor.

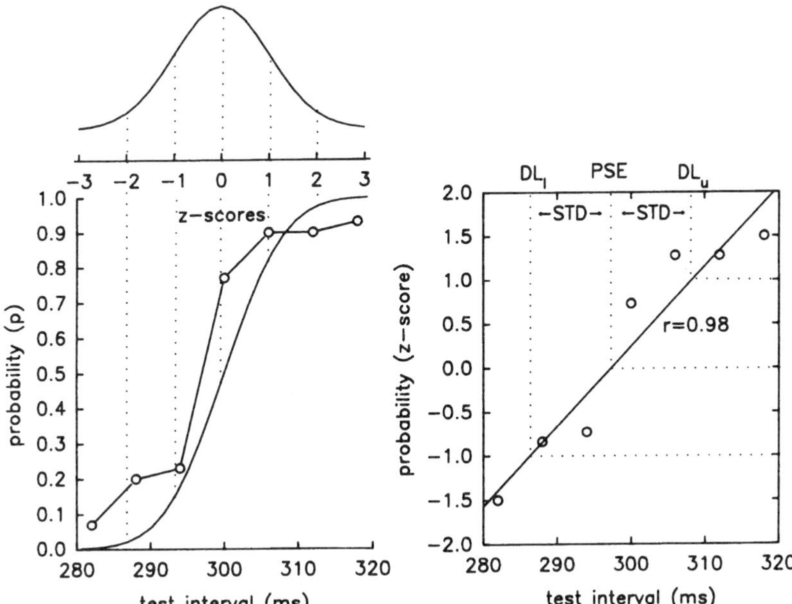

FIGURE 42.2   (Left side): The cumulative probabilities that the interval between the last two tones of a sequence of five tones was estimated to be longer than the previous ones was plotted against the seven test intervals varying around the base interval, which is 300 ms in the presented example. In a second step, the correlation between the test interval lengths and the z-transformed probabilities was calculated (right side) for each base interval and each subject.

42.5 INFLUENCE OF TREATMENT ON NEUROPSYCHIATRIC SYMPTOMS

When copper-trapping therapy is started in WD, in most patients a considerable improvement is found. This has been reported for clinical and laboratory findings, motor testing, evoked potentials,

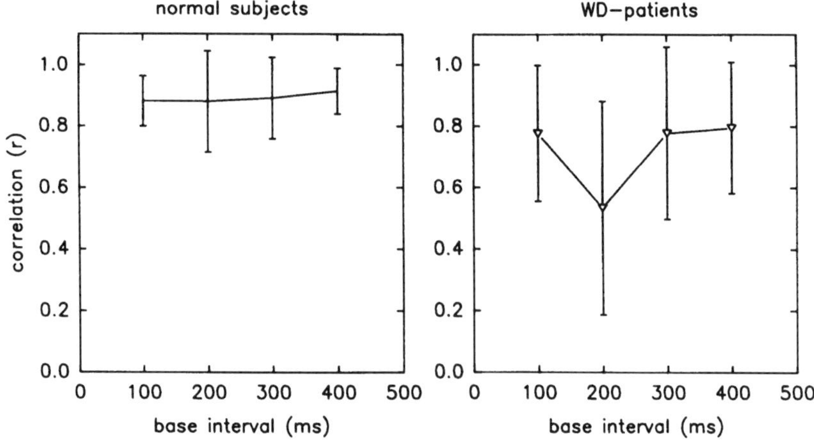

FIGURE 42.3   In normal subjects the mean correlation coefficients between test interval lengths and z-transformed cumulative probabilities were close to 0.9 for each base interval (left side). In the WD patients (right side) the mean correlation coefficient for the 200-ms base interval was significantly lower than in the normals. For the other base intervals, no significant difference could be detected between the WD patient and the control group. Thus WD patients present with the very special cognitive deficit that they cannot estimate time intervals correctly with a duration close to their tremor cycle length.

as well as CT, MRI, SPECT, and PET scans. It has also been reported for neuropsychiatric symptoms, P300 recordings, and some subtests of the German WAIS (picture arrangement, block design, object assembly, for a review see Reference 11). In a few patients worsening of neurological symptoms, especially dysarthria, has been reported.[35]

Therefore the question arises why in cross-sectional studies a considerable percentage of long-term-treated WD patients still present with neurological and neuropsychiatric symptoms in spite of a good response to drug treatment. Some light was shed on this problem when in 12 WD patients the cerebral glucose metabolism of the basal ganglia was analyzed (using PET scanning) in dependence on the duration of drug treatment. In patients who were treated for less than 6 to 7 years, a significant increase of the basal ganglia glucose metabolism was found; but for those patients who had been treated for more than 7 years, a significant decline of the glucose metabolic rate in the basal ganglia was found. Since this decline was inversely related to the 24-h urinary copper excretion, this finding suggests that too low a maintenance dose was administered. This is certainly one of the reasons why neuropsychiatric deficits persist in long-term-treated WD patients.[36]

Little is known about improvement of major psychiatric symptoms with therapy.[21] Intellectual functioning at the onset of therapy has been compared to its outcome after 21 to 34 years of treatment in 7 WD patients.[37] In spite of considerable improvement of hepatic and neurological symptoms no significant changes between the initial and later testing was found. However, since all seven patients did not show relevant deficits in the initial testing this study only demonstrates that intellectual functioning does not deteriorate under treatment. In a previous series, improvement of neuropsychiatric symptoms is mentioned, but no statistical support has been presented. A mild improvement was described in a prospective study on 31 cases; in single patients a clear improvement may be found.[38] A larger longitudinal study on the improvement of neuropsychiatric symptoms and psychometric tests with duration of therapy is not available so far.

42.6 CONCLUSIONS

In contrast to Wilson initial opinion, there is common agreement now that neuropsychiatric symptoms form an integral part of WD. Some psychiatric features correlate with neurological symptoms strongly, suggesting an organic origin. About 10 to 15% of the WD patients present with psychiatric symptoms as the first clinical manifestation of WD. Psychometrical test analysis reveals that in spite of considerable neurological deficits, neuropsychological impairment is usually mild in WD. However, when present, in spite of clear improvement at the onset of therapy neuropsychiatric deficits may persist even under sufficient long-term treatment. This implies that WD should be diagnosed as early as possible and treated with a sufficiently high dose of copper chelating agents. It may very well be that in the future gene analysis will help to diagnose WD early in life, so that most of the WD patients will be treated before severe brain damage occurs and will be kept in an asymptomatic state.

ACKNOWLEDGMENTS

This study as been supported by grants from the DFG (SFB 194, A5) and the German Ministry of Research and Technology BMFT 01/KL 9004.

REFERENCES

1. Westphal C: Über eine dem Bilde der cerebrospinalen grauen Degeneration ähnlichen Erkrankung des zentralen Nervensystems ohne anatomischen Befund, nebst einigen Bemerkungen über paradoxe Kontraktionen. *Arch. Psychiatr. Nervenk.*, 1883; 14:87–134.
2. Strümpell A: Ein weiterer Beitrag zur Kenntnis der sogenannten Pseudosklerose. *Dtsch. Z. Nervenheilk.*, 1899; 14:348–355.
3. Wilson SAK: Progressive lenticular degeneration. A familial nervous disease associated with cirrhosis of the liver. *Brain*, 1912; 34:295–509.
4. Cumings JN: The copper and iron content of brain and liver in the normal and in hepatolenticular degeneration. *Brain*, 1948; 71:410–415.
5. Frydman F, Bonné-Tamir B, Farrer A, et al: Assignment of the gene for Wilson disease to chromosome 13: Linkage to the esterase D locus. *Proc. Natl. Acad. Sci. U.S.A.*, 1985; 82:1819–1821.
6. Tanzi RE, Petrukhin K, Chernov I, et al: The Wilson disease gene is a copper transporting ATPase with homology to the Menkes disease gene. *Nature Genet.*, 1993; 5:344–350.
7. Bowcock AM, Tomfohrde J, Weissenbach J, et al: Refining the position of Wilson disease by linkage disequilibrium with polymorphic microsatellites. *Am. J. Hum. Genet.*, 1994; 54:79–87.
8. Thomas GR, Forbes JR, Roberts EV, et al: The Wilson disease gene: spectrum of mutations and their consequences. *Nature Genet.*, 1995; 9:210–217.
9. Petrukhin K, Lutsenko S, Chernov I, et al: Characterization of the Wilson disease gene encoding a P-type copper transporting ATPase: genomic organization, alternative splicing, and structure/ function predictions. *Hum. Mol. Genet.*, 1994; 3:1647–1656.
10. Brewer GJ and Yuzbasiyan-Gurkan Y: Wilson Disease. *Medicine*, 1992; 71,3:139–164.
11. Hefter H: Wilson disease. Review of pathophysiology, clinical features and drug treatment. *CNS Drugs*, 1994; 2:26–39.
12. Walshe JM and Gibbs KR: Brain copper in Wilson disease. *Lancet*, 1987; 2:1030.
13. Kuwert T, Hefter H, Scholz D, et al: Regional cerebral glucose consumption measured by positron emission tomography in patients with Wilson disease. *Eur. J. Nucl. Med.*, 1992; 19:96–101.
14. Grimm G, Prayer L, Oder W, et al: Comparison of functional and structural brain disturbances in Wilson disease. *Neurology*, 1991; 41:272–276.
15. Dening TR and Berrios GE: Wilson disease: clinical groups in 400 cases. *Acta Neurol. Scand.*, 1989; 80:527–534.
16. Denny-Brown D: Hepatolenticular degeneration (Wilson disease). Two different components. *N. Engl. J. Med.*, 1964; 270:1149–1156.
17. Dening TR and Berrios GE: Wilson disease: psychiatric symptoms in 195 cases. *Arch. Gen. Psychiatr.*, 1989; 46:1126–1134.
18. Oder W, Grimm G, Kollegger H, et al: Neurological and neuropsychiatric spectrum of Wilson disease: a prospective study of 45 cases. *J. Neurol.*, 1991; 238 281–287.
19. Arendt G, Hefter H, Stremmel W, and Strohmeyer G: The diagnostic value of multi-modality evoked potentials in Wilson disease. *Electromyogr. Clin. Neurophysiol.*, 1994; 34:137–148.
20. Dening TR: The neuropsychiatry of Wilson disease: A review. *Int. J. Psychiatry, Med.*, 1991; 21(2):135–148.
21. Lang C, Müller D, Claus D, and Druschky KF: Neuropsychological findings in treated Wilson disease. *Acta Neurol. Scand.*, 1990; 81:75–81.
22. Wenning GK, Ben Shlomo Y, Magalhaes M, et al: Clinical features and natural history of multiple system atrophy. An analysis of 100 cases. *Brain*, 1994; 117:835–845.
23. Modai I, Karp L, Liberman UA, and Munitz H: Penicillamine therapy for schizophreniform psychosis in Wilson disease. *J. Nervous Mental Dis.*, 1985; 173:698–701.

24. Dening TR: Psychiatric aspects of Wilson disease. *Br. J. Psychiatry*, 1985; 147:677–682.

25. Hoover JE and Strick PL: Multiple output channels in the basal ganglia. *Science*, 1993; 259:819–821.

26. Scheinberg H, Sternlieb I, and Richman J: Psychiatric manifestations in patients with Wilson disease. *Am. J. Psychiatry*, 1968; 124:85–87.

27. Tarter RE, Switala J, Carra J, et al: Neuropsychological impairment associated with hepatolenticular degeneration (Wilson disease) in the absence of overt encephalopathy. *Int. J. Neurosci.*, 1987; 37:67–71.

28. Medalia A, Isaacs-Glabermann K, and Scheinberg H: Neuropsychological impairment in Wilson disease. *Arch. Neurol.*, 1988; 45:502–504.

29. Isaacs-Glabermann K, Medalia A, and Scheinberg H: Verbal recall and recognition abilities in patients with Wilson disease. *Cortex*, 1989; 25:353–361.

30. Dening TR and Berrios GE: Wilson disease: a prospective study of psychopathology in 31 cases. *Br. J. Psychiatry*, 1989; 155:206–213.

31. Hayashi R, Hanyu N, Shindo M, et al: Event-related potentials, reaction time and cognitive state in patients with Parkinson's disease. In Narabayashi H, Nagatsu T, Yanagisawa N, Mizuno Y (eds): *Advances in Neurology*, Vol. 60, New York, Raven Press, 1993, pp 429–433.

32. Hömberg V, Hefter H, Granseyer G, et al: Event-related potentials with Huntington's disease and relatives at risk in relation to detailed psychometry. *Electroenceph. Clin. Neurophysiol.*, 1986; 63:552–569.

33. Maurer K, Ihl R, and Dierks Th: Topographie der P300 in der Neuropsychiatrischen Pharmakotherapie. III. Kognitive P300-Felder beim organischen Psychosyndrom (Morbus Wilson) vor und während einer Therapie mit D-Penicillamin. *Z. EEG EMG*, 1988; 19:62–64.

34. Brown RG and Marsden CD: Cognitive function in Parkinson's disease: from description to theory. *TINS*, 1990; 13(1):21–29.

35. Brewer GJ, Carol AT, Aisen AM, et al: Worsening of neurologic syndrome in patients with Wilson disease with initial penicillamine therapy. *Arch. Neurol.*, 1987; 44:490–493.

36. Hefter H, Kuwert T, Herzog H, et al: Relationship between striatal glucose consumption and copper excretion in patients with Wilson disease treated with D-penicillamine. *J. Neurol.*, 1993; 241:49–53.

37. Medalia A and Scheinberg IH: Intellectual functioning in treated Wilson disease. *Ann. Neurol.*, 1991 29:573–574.

38. Rosselli M, Lorenzana P, Rosselli A, and Vergara I: Wilson disease, a reversible dementia: case report. *J. Clin. Exp. Neuropsych.*, 1987; 9,3:399–406.

Chapter 43

Copper Localization in the Central Nervous System of Wilson Disease

Masayuki Yasui and Kiichiro Ota

CONTENTS

43.1 INTRODUCTION

Extrapyramidal motor disorder is well known to occur by excessive copper (Cu) accumulation in the brain of Wilson disease (hepatolenticular degeneration) or by manganese (Mn) in manganism. This chapter will examine two kinds of hepatocerebral disease, Wilson disease and portal-systemic encephalopathy (PSE), chiefly in terms of copper and manganese contents in the CNS. It will discuss three main points.

1. Although the textbooks suggest that copper distribution in the brain in Wilson disease is localized chiefly in the basal ganglia, neutron activation analysis (NAA) showed instead a high-level diffuse distribution throughout the cerebral regions.
2. Although Wilson disease and PSE look very similar clinically, their etiology is different. They can be easily differentiated in terms of the copper content in the brain and liver by NAA.
3. Although the manganese content in the brain in Wilson disease and PSE is similar, there is a correlation in the case of Wilson disease but none in the case of PSE. NAA also makes this clear.

Additionally, this chapter gives data on the manganese and aluminum contents of biopsied or autopsied liver tissues from 12 patients, 8 with Wilson disease and 4 with liver cirrhosis, and then discusses the biochemical interaction among these metals in the CNS and liver.

43.2   NEUTRON ACTIVATION ANALYSIS FOR CNS SAMPLES OF WILSON DISEASE AND PSE

The CNS tissues for the simultaneous determination of copper and manganese were obtained from three cases of Wilson disease and two cases of PSE and five controls. Wet mass weight of each sample ranged between 50 to 150 mg. Samples were dried at 105°C until a constant weight (10 to 30 mg) was obtained (results were expressed as micrograms per gram dry weight). Samples were irradiated in a thermal flux of 2.3×10^{13} neutron/cm^{2}s for 5 min.[1]

The following 20 CNS anatomical regions were analyzed: (A) gray matter (frontal, parietal, temporal, occipital and cerebellar cortex, hippocampus); (B) white matter (frontal, parietal, occipital and cerebellar white matter, capsula interna); (C) basal ganglia group (thalamus, caudate nucleus, globus pallidus, putamen, substantia nigra); (D) brain stem (pons, olivary nucleus and medulla); (E) spinal cord.

43.3   COPPER DISTRIBUTION IN THE WILSON DISEASE BRAIN

The mean contents of copper in 20 brain regions from 3 Wilson disease, 2 PSE cases, and 5 controls are shown in Figure 43.1. Copper content shows a significant difference between the Wilson disease cases and the PSE cases. The copper distribution in the brain of Wilson disease cases is significantly higher, and more diffuse in distribution. Copper was not only localized in the basal ganglia but found in all gray and white matter (Figure 43.2a). This finding strongly indicates that copper may affect cerebral cortexes at a functional level rather than classical morphological changes at the basal ganglia.

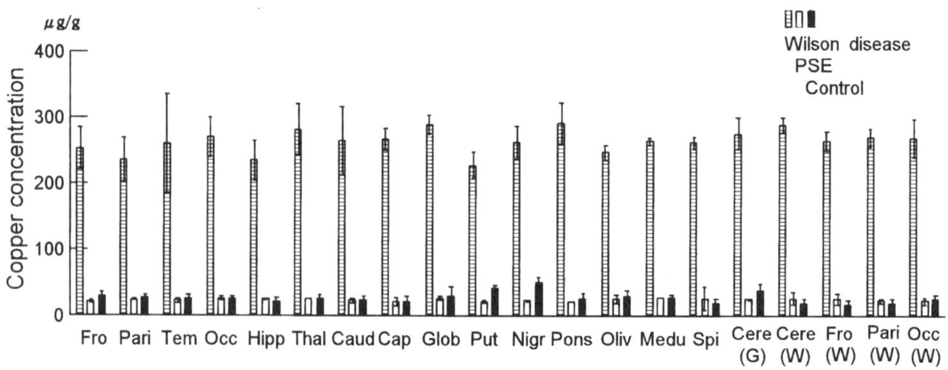

FIGURE 43.1   Copper concentrations (μg/g dry weight, mean ± SD) in 20 anatomical regions of the CNS of Wilson disease (3 cases), portal systemic encephalopathy (2 cases), and controls (5 cases).

Manganese content was of no specific distribution pattern, but two of three Wilson disease cases showed a positive correlation between copper and manganese ($p < .01$) (Figure 43.2b). No correlation was seen between them in either PSE cases or controls.

43.4 COPPER, MANGANESE, AND ALUMINUM IN THE LIVER

Table 43.1 shows copper, manganese, and aluminum contents of biopsied or autopsied tissues simultaneously analyzed by NAA. Two Wilson disease cases showed extremely high aluminum content and a third moderately high aluminum content in the liver besides high copper content as compared with the liver cirrhosis cases. Two of the Wilson disease livers (Cases 1 and 4) were biopsied before the D-penicillamine treatment started.

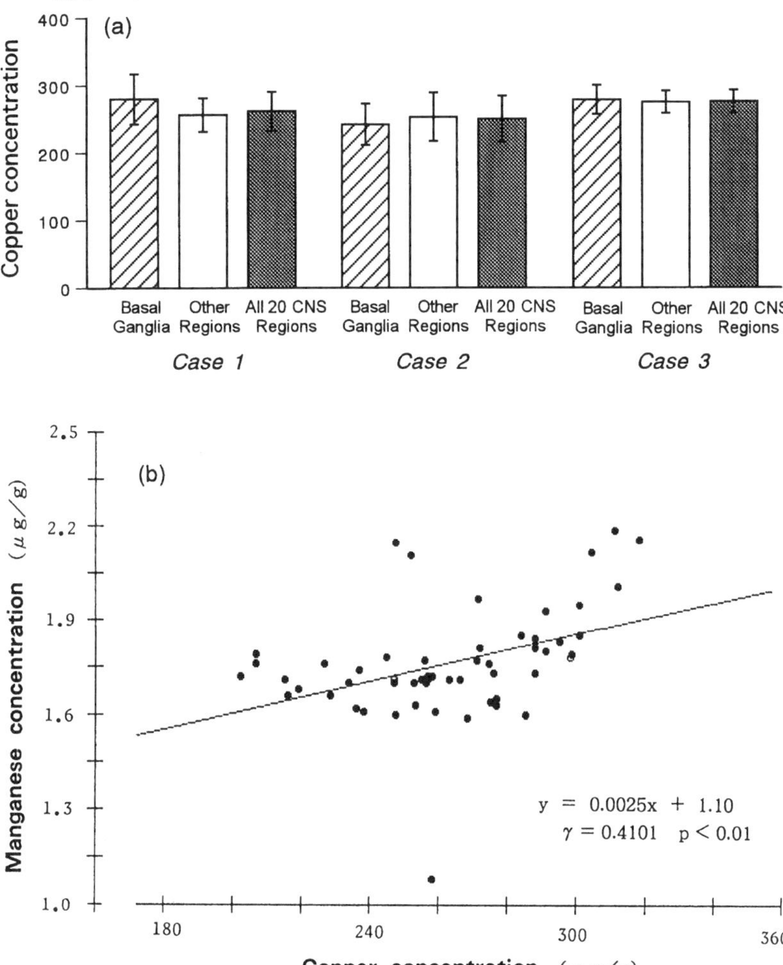

FIGURE 43.2 (a) Copper concentrations (μg/g dry weight, mean ± SD) in basal ganglia, other regions, and all 20 CNS regions of the 3 cases with Wilson disease. (b) Correlation between copper and manganese in the CNS of the Wilson disease cases.

43.5 DISCUSSION

The simple determination of copper and other metals of CNS tissue in Wilson disease cases has been done by many authors, but reports on a systematic search for trace metals and their interaction in the body are rare. In normal controls, copper is more abundant in gray matter than in white matter of CNS tissue[2,3] and is especially rich in the substantia nigra and the locus ceruleus, being three to eight times the white matter level.[4] Therefore, it may be significant to determine the content of various metals simultaneously for the same region of CNS tissue of Wilson disease or PSE by NAA or other compatible methods.

There is no difference in copper contents between gray and white matter in Wilson disease and also in PSE, although a higher copper value was obtained in the CNS tissue of Wilson disease and a normal value in PSE as compared with the controls. It is interesting to point out that the copper content in the brain of Wilson disease is diffuse in distribution, without significantly higher value in the basal ganglia, despite the fact that neuropathologic changes are mainly located in the basal ganglia. This may indicate that the basal ganglia is very susceptible to copper in the term of

Table 43.1 Copper, Manganese, and Aluminum Contents in Liver[a]

	Cu	Mn	Al
Biopsy			
Wilson-1	987.1	4.0	54.1
Wilson-2	208.8	2.5	51.2
Wilson-3	210.0	2.3	1345.6
Wilson-4	548.0	2.7	416.0
Wilson-5	237.0	1.2	156.0
Liver cirrhosis-1	10.7	3.5	42.3
Liver cirrhosis-2	15.2	3.3	38.7
Autopsy			
Wilson-4	588.6	1.6	74.4
Wilson-5	1065.6	3.5	88.6
Wilson-6	708.5	2.6	52.3
Liver cirrhosis-3	17.8	4.8	51.4
Liver cirrhosis-4	31.7	1.2	53.0
Immature infant-1	156.1	3.5	84.2
Immature infant-2	228.9	2.7	72.2
Controls (n=5)	25.9 ± 14.1	3.2 ± 0.7	29.1 ± 6.0

[a] Micrograms per gram dry weight, mean ± SD; by NAA.

neuronal degeneration. The white matter and cerebral cortex may be involved in less than 10% of the patients of Wilson disease. As the copper content of the CNS and the liver tissue in PSE cases shows a normal value, it is possible to differentiate Wilson disease from PSE by the determination of copper content in the organ.

Iyengar et al.[5] hypothesized that the "regulatory copper" (they use the term to describe the way the liver regulates copper balance by excretion of copper) is contained in ceruloplasmin, and a high molecular weight copper-containing fragment found in bile-enriched duodenal secretions from normal controls cross-reacts with ceruloplasmin antibodies and contains an appropriate amount of copper (about 0.2 mg per day) to regulate copper balance.[5] Since the Wilson disease gene and the ceruloplasmin gene are on different chromosomes (13 and 3, respectively), the ceruloplasmin gene and linked regulatory elements are not responsible for Wilson disease. Suggesting that transcription and translation of the ceruloplasmin are not the problem, the antigenic cross-reacting material to ceruloplasmin in the liver of patients with Wilson disease was found.[5] Thus, the Wilson disease gene might be involved in some posttranslational modification of apoceruloplasmin, which is required for its subsequent secretion into blood and bile. This theory clearly accounts for the uniformly low rate of secretion of ceruloplasmin into the blood in Wilson disease, and also for its failure to appear in the form of cross-reacting material in the bile of these patients.[6] Based on the etiopathologies of both Wilson disease and PSE, it is certain that the different amounts of copper in CNS and liver between Wilson disease and PSE are due to different mechanisms of copper regulation *in vivo*.

On the other hand, copper content in the liver of Wilson disease, ranging from 200 to 1000 µg/g, is much higher than that in CNS tissue. As higher copper content is found in the liver of the untreated than the treated cases, the chelating agent of copper appears to be more effective in the liver tissue than in the brain tissue. It was previously observed that while four patients with Wilson disease undergoing long-term D-penicillamine treatment revealed a large visually evoked potential amplitude of the late component in the initial stage, the amplitude of the late component became smaller in two cases along with disappearance of the Kayser-Fleischer rings in one case and a reduction in motor disturbance, but without change in the Kayser-Fleischer rings or in clinical laboratory data in the other.[7] Our study indicated that serial visually evoked potential determinations are important in monitoring the effectiveness of D-penicillamine.

Along with D-penicillamine, oral administration of zinc in Wilson disease is reported to prevent reaccumulation of copper in the liver.[8] Induction of the intestinal copper-binding protein metallothionein is the proposed mechanism. Based on experimental studies in rats,[9,10] the mechanism of the anticopper action by zinc leads to the induction by zinc of metallothionein in the intestinal cells. Brewer et al.[8] reported that metallothionein binds copper with high affinity and prevents the transfer of copper into the blood, and the metallothionein-bound copper is excreted in the stool when the intestinal cell is sloughed. Blockade of the uptake of copper, derived both from the diet and from saliva and gastric juice, etc. allows zinc to produce a negative copper balance. As the high levels of liver copper are stable during zinc therapy, they hypothesized that the extensive previous penicillamine therapy could have caused the elimination of the available copper of the liver, and rendered it impossible to mobilize the remaining copper.[8]

Butt et al.[11] earlier reported that there was no statistically significant difference in manganese content as seen by the mean value of the contents in brain from eight Wilson disease cases (2.6 µg/g) and controls (2.4 µg/g). It can be argued that Butt's method, some 40 years ago, was not sensitive enough for manganese, for it is well known that manganese plays an important role in the occurrence of parkinsonian symptoms, one of the same extrapyramidal diseases which include Wilson disease. The possible triggering role of manganese in the pathogenetic process might be considered in Wilson disease and PSE cases not by a simple role of manganese but rather by an interaction of this metal with others. Excess copper is known to have harmful effects on mitochondria and perioxisomes of liver cells in patients with Wilson disease.[12,13]

Many enzymes involved in the oxidant defense systems such as glutathione reductase and acetylcholine esterase are inhibited by copper or aluminum.[6,14] Copper has an important role as the H_2O_2 may be derived from superoxide (O_2^-) produced by the enhanced oxidation of hemoglobin resulting from exposure to copper.[15] The evidence favors the hypothesis that copper toxicity occurs because of generation of oxidant radicals, probably beginning with superoxide. Also, the mechanism of cellular protection is the binding and detoxifying of copper by metallothionein as inhibitory effects of copper. The evidence that the aluminum content was increased in the liver of Wilson disease strongly suggests that many metals are involved and interact in the process of neurodegenerative disorders in the brain as well as systematically through the liver.

Manganese has an acute as well as chronic effects on the CNS, producing similar symptoms as seen in Parkinson's disease.[16] Manganese-poisoned brain has the microscopic extrapyramidal structures similar to those found in Parkinson's disease.[17] In Parkinson's disease the accumulation of iron and aluminum,[14,18] which are known to promote oxidant stress, may account for the selective degeneration of neuromelanin-containing neurons.[18] Divalent trace metals, such as copper and manganese, also have prooxidant capabilities. Dexter et al.[19] reported that copper levels were decreased in the substantia nigra in Parkinson's disease but total iron content was increased in the same area, and there were no consistent alterations of manganese levels in basal ganglia structures. Certain metals are implicated in the neurodegenerative disorders such as copper in Wilson disease, iron in Hallervorden's disease, aluminum in Alzheimer's disease, Guamanian parkinsonism-dementia, and dialysis encephalopathy.[20–22] Although it is interesting to note that copper and manganese are essential metals for enzyme activities in the CNS, and their excessive bioavailability and subsequent absorptions by the brain make it susceptible to deposition to its preferred brain regions like the lenticular nuclei, our data on Wilson disease do not support this observation.

The accumulation of aluminum, manganese, and copper seem to induce the inhibition of Cu/Zn superoxide dismutase (SOD) which is a homodimeric metalloenzyme that catalyzes the dismutation of the superoxide anion to oxygen and hydrogen peroxide. Excessive oxygen radicals levels, such as superoxide anion, have been implicated in neuronal injury and the formation of more reactive oxygen species as seen in hydroxyradicals.[23]

43.6 CONCLUSION

It has been firmly established that abnormal copper metabolism is the pathogenic basis for development of extrapyramidal syndrome in genetically susceptible family members. However, from our study as well as others, interactions of trace metals at molecular levels remain to be elucidated, particularly in the areas of bioavailability, the roles in triggering free radicals in apoptosis of the dopaminergic neurons, and genetic susceptibility.

ACKNOWLEDGMENTS

Our gratitude is extended to Dr. K.-M. Chen, Guam Memorial Hospital, for his many helpful comments in preparing this manuscript and to Mr. Leonard Lundmark for his support.

REFERENCES

1. Yasui M, Yase Y, Ota K, et al: Aluminum deposition in the central nervous system of patients with amyotrophic lateral sclerosis. *Neurotoxicology*, 1991; 12:615–620.
2. Warren PJ, Earl CJ, and Thompson RHS: The distribution of copper in human brain. *Brain*, 1960; 83:709–717.
3. Courville CB, Nusbaum RE, and Butt EM: Changes in trace metals in brain in Huntington's chorea. *Arch. Neurol.*, 1963: 8; 481–489.
4. Cumings JN: Trace metal in the brain and in Wilson disease. *J. Clin. Pathol.*, 1967; 21:1–7.
5. Iyengar V, Brewer GJ, Dick RD, et al: Studies of cholecystokinin-stimulated secretions reveal a high molecular weight copper-binding substance in normal subjects that is absent in patients with Wilson disease. *J. Lab. Clin. Med.*, 1988; 111:267–274.
6. Brewer GJ and Yuzbasiyan-Gurkan V: Wilson disease. *Medicine*, 1992; 71:139–164.
7. Yasui M: Electrophysiological study on Wilson disease. Changes of visual evoked potential and EEG by D-penicillamine treatment. *Wakayama Med. Rep.*, 1978; 21:49–58.
8. Brewer GJ, Hill GM, Rick RD, et al: Treatment of Wilson disease with zinc. III. Prevention of reaccumulation of hepatic copper. *J. Lab. Clin. Med.*, 1987; 109:526–531.
9. Hall AC, Young BW, and Bremner I: Intestinal metallothionein and the mutual antagonism between copper and zinc in the rat. *J. Inorg. Biochem.*, 1979; 11:57–66.
10. Menard MP, McCormic CC, and Cousins RJ: Regulation of intestinal metallothionein biosynthesis in rats by dietary zinc. *J. Nutr.*, 1981; 111:1353–1361.
11. Butt EM, Nusbaum RE, Gilmour TC, et al: Trace metal patterns in disease state. *Am. J. Clin. Pathol.*, 1958; 30:479–497.
12. Sternlieb I: Mitochondria and fatty changes in hepatocytes of patients with Wilson disease. *Gastroenterology*, 1968; 5:354–367.
13. Sternlieb I and Quintana N: Abnormalities of human hepatocellular peroxisomes. *Ann. N.Y. Acad. Sci.*, 1982; 386:530–533.
14. Yasui M, Kihira T, and Ota K: Calcium, manganese and aluminum concentrations in Parkinson's disease. *Neurotoxicology*, 1992; 13:593–600.
15. Hochstein P, Kumar KS, and Forman SJ: Lipid peroxidation and the cytotoxicity of copper. *Ann. N.Y. Acad. Sci.*, 1980; 355:240–248.
16. Calne DB, Chu N-S, Huang C-C, et al: Magnesium and idiopathic parkinsonism: Similarities and differences. *Neurology*, 1994; 44:1583–1586.
17. Wennberg A, Iregren A, Struwe G, et al: Manganese exposure in steel smelters a health hazard to the nervous system. *Scand. J. Work Environ. Health*, 1991; 17:255–262.
18. Good PF, Olanow CW, and Perl DP: Neuromelanin-containing neurons of the substantia nigra accumulate iron and aluminum in Parkinson's disease. A LAMMA study. *Brain Res.*, 1992; 593:343–346.
19. Dexter DT, Carayon A, Javoy-Agid F, et al: Alterations in the levels of iron, ferritin and other trace metals in Parkinson's disease and other neurodegenerative diseases affecting the basal ganglia. *Brain*, 1991; 114:1953–1975.
20. Lukiw WJ, Kruck TPA, and McLachlan DRC: Alternation in human linker histone-DNA binding in the presence of aluminum salts in vitro and in Alzheimer's disease. *Neurotoxicology*, 1987; 8:291–302.
21. Garruto RM, Fukatsu R, Yanagihara R, et al: Imaging of calcium and aluminum in neurofibrillary tangle-bearing neurons in parkinsonism-dementia of Guam. *Proc. Natl. Acad. Sci. U.S.A.*, 1984; 81:1875–1879.
22. Alfrey AC, Mishell JM, Burks J, et al: Syndrome of dyspraxia and multifocal seizures associated with chronic hemodialysis. *Trans. Am. Soc. Artif. Intern. Organs*, 1972; 18:257–261.
23. Ocorr KA, Walters ET, and Byrne JH: Associative conditioning analog selectively increases cAMP levels of tail sensory neurons in Aplysia. *Proc. Natl. Acad. Sci. U.S.A.*, 1985; 82:2548–2552.

Chapter 44

Menkes Kinky Hair Disease

Tsunekazu Yamano and Morimi Shimada

CONTENTS

44.1 INTRODUCTION

Menkes kinky hair disease (MKHD) is an X-linked recessive trait characterized by growth retardation, progressive neural deterioration, convulsions, and abnormal hair (Figure 44.1a,b).[1,2] The underlying biochemical abnormality is a derangement of copper absorption and transport.[3] Several groups of investigators have isolated a candidate gene whose mutation is responsible for this disease.[4–6] However, the pathogenesis of MKHD has not been completely clarified.

Model mice of MKHD have been reported.[7,8] They have contributed greatly to research on the pathogenesis of MKHD. In Japan, a mutant strain named the macular mouse was discovered by Nishimura in 1973.[8] We have studied this mutant mouse intensively since it bears a close clinical, biochemical, and neuropathological resemblance to human MKHD.[9] The macular mouse is also presented here to elucidate the pathogenesis of MKHD.

44.2 MACULAR MUTANT MOUSE AS A MODEL OF MENKES KINKY HAIR DISEASE

This mutant mouse shows an X-linked recessive inheritance of the MKHD-like trait. Individual pups in the litter are of the same weight and size at birth. At approximately 3 days of age, the fur of the pups is clearly white in the hemizygote, a white and agouti mosaic in the heterozygote, and brownish black in the normal mouse. The hemizygote has curly whiskers, and shows a gradual weight loss after 9 days of age. It also suffers from frequent tonic seizures and ataxia, and finally becomes inactive and dies at about 15 days of age in an emaciated condition.[9] However, exogenous copper therapy for this mutant mouse strain is quite successful; the administration of cupric chloride improves these clinical manifestations.[10]

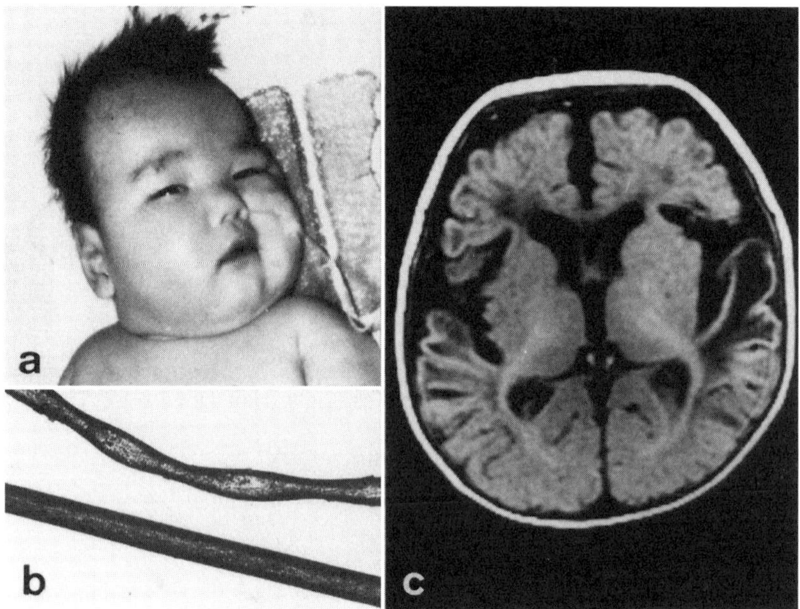

FIGURE 44.1   a: A patient with MKHD. Note pudgy cheeks, sagging jowl, and abnormal eyebrows. b: Abnormal hair (pili torti). c: T1-weighted MRI of a 1.5-year-old patient with MKHD shows brain atrophy, intense infarction at bilateral temporal lobes, and hypomyelination at anterior lobes.

44.3 BIOCHEMISTRY

Copper accumulates in fibroblasts isolated from an MKHD patient when these cells are cultured in a copper-enriched medium, since the efflux of copper is defective.[3,11] This accumulation also increases the intracellular metallothionein concentration. The proliferation of these fibroblasts is inhibited, and some of them become detached and float in the medium. This toxic effect may result from excess "free" copper, which does not bind to metallothionein. In these cells, copper localizes mainly to the organelle-free cytoplasm.[11] The fibroblasts from the macular mutant mouse also reveal the same defects.[12]

In patients with MKHD, the copper concentration increases significantly in the intestinal wall and the kidneys, but decreases in the brain and liver.[13,14] In the mutant mouse, the pattern of copper distribution is similar to the MKHD patient (Table 44.1). The copper content in the intestinal wall and kidneys of these mice is about four times the levels in their normal littermates.[15] The affected cells are epithelial cells and vascular endothelial cells in the intestinal wall, and the proximal tubular cells in the kidney.[16] Copper also localizes specifically to the organelle-free cytoplasm of these cells like in the cultured human fibroblasts.[16] These results suggest that MKHD may result from an abnormality in copper transport from the cytosol to the organelles, in addition to the efflux of copper. Recently, three groups have successfully isolated a candidate gene for MKHD. This gene is mapped to band Xq 13, and encodes a copper-transporting ATPase. The complete structure of this gene product will yield a great deal of information for understanding iron transport and the role of heavy metals in cellular functions.[4–6] The abnormalities of copper transport responsible for this disease will be elucidated in the near future.

In the MKHD patient, the oral administration of copper does not elevate serum copper and ceruloplasmin levels because it must be absorbed by the intestinal epithelial cells. Therefore, the serum copper levels are always low in these patients. To reach the neurons in the central nervous system, the copper has to pass through the blood-brain barrier which consists of endothelial cells and the end-foot processes of astrocytes. Even in these cells, the efflux of copper is defective and copper accumulates.[17] Therefore, the intestinal wall and the blood-brain barrier make use of oral

Table 44.1 Copper Contents in Various Organs of the Macular Mouse on Day 14[a]

	Normal littermate	Hemizygote	Treated hemizygote
Cerebrum	7.0 ± 0.7	1.9 ± 0.4[b]	3.7 ± 0.6[b]
Cerebellum	11.3 ± 3.3	3.8 ± 1.0[b]	6.0 ± 2.2[b]
Brainstem	8.6 ± 1.7	2.5 ± 0.6[b]	4.8 ± 0.9[b]
Liver	29.1 ± 10.4	9.3 ± 2.5[b]	16.0 ± 5.6[b]
Kidney	9.4 ± 1.2	35.5 ± 10.9[b]	188.8 ± 28.9[b]
Intestine	6.1 ± 1.3	24.3 ± 7.7[b]	40.0 ± 8.8[b]

[a] Micrograms per gram dry weight, mean ± SD.

[b] Significantly different from normal littermate ($p < .001$).

and parenteral copper administration as therapeutic modalities ineffective in the MKHD patient. In 14-day-old macular mutant mice, the copper content in the cerebrum, cerebellum, and brain stem were only 9, 29, and 38%, respectively, of the value in the corresponding CNS of the normal littermate (Table 44.1). Copper deficiency is more pronounced in neurons because the copper is retained by the blood-brain barrier.

Copper ions are vital cofactors for several enzymes. Copper deficiency in the brain, liver, and other organs, except for the kidneys and intestinal wall, reduces the activity of copper-dependent enzymes such as cytochrome oxidase, monoamine oxidase, tyrosinase, superoxide dismutase, xanthine oxidase, cytochrome c reductase, ascorbic acid oxidase, δ-aminolevulinic acid dehydrase, dopamine-β-oxidase, ceruloplasmin, galactose oxidase, and lysyl oxidase. Reduction in the activity of these enzymes plays an important role in the clinical manifestations of MKHD.

44.4 PATHOLOGY

Characteristic findings for MKHD are seen on the arterial vessels, bone, and the nervous system.[18–20] Since arterial and bone abnormalities contribute to the clinical manifestations, they are discussed in the following section.

The brain is severely atrophic in MKHD patients, and meningeal vessels are frequently dilated. Microscopically, common findings are patchy necroses and neuronal cell loss with widespread fibrillary gliosis.[18,19] In some cases, hypomyelination is also observed. Pathologic changes in the cerebellum are particularly severe.[19] Many granule cells and Purkinje cells are destroyed and the remaining Purkinje cells show the following characteristic abnormalities. Their dendritic arborization is considerably delayed, and somal spines and sprouts persist without regression. This appearance is observed in immature Purkinje cells from normal animals; thus, the surviving Purkinje cells in MKHD are developmentally arrested and immature. Their dendrites may be swollen and may have cactus-like projections forming a "weeping willow"-shaped dendritic tree. Electron microscopic examination reveals that neurons in the cerebral cortex and the Purkinje cells contain many abnormal mitochondria.[19] Histochemical studies indicate that activity of cytochrome c oxidase (CCO), a copper-dependent enzyme, is decreased in the cerebrum and cerebellum.[21] During the final stage of MKHD, brain infarction and hemorrhage are produced by arterial lesions (Figure 44.1c).

In the macular mutant mouse, the neuropathological findings are similar to those in the MKHD patient except for arterial abnormalities.[9,10,22] In addition, the activities of NADH diaphorase (NADH) and succinic dehydrogenase (SDH) are increased, especially in cerebellar Purkinje cells and hippocampal pyramidal neurons.[23] CCO, NADH, and SDH are restricted to the mitochondria and function in high-energy metabolism. NADH and SDH are copper-independent enzymes that function upstream of the CCO stage in the electron transport system. Copper deficiency produces an excess of abnormal mitochondria with CCO deficiency, and thus the elevation of NADH and SDH activities in the macular mouse may be a compensatory phenomenon that attempts to produce more energy in the neurons. Although hypomyelination has not been confirmed neuropathologically in the brains of mutant mice, 2′,3′-cyclic nucleotide 3′-phosphodiesterase (CNP) activity is reduced in the brains of mutant mice.[24] A close correlation between the developmental increase in myelin and CNP activity has been confirmed. Therefore, this reduction in CNP activity suggests that myelination is probably affected in the macular mouse. Some investigators suggested that the neuropathological changes seen in MKHD patients are induced by vascular insufficiency. Since no vascular abnormalities were recognized in the mutant mice, these neuropathological changes probably resulted from the copper deficiency rather than vascular insufficiency due to arterial abnormalities. Most of these neuropathological changes can be reversed by parenteral copper therapy in the macular mouse.[10,22]

44.5 CLINICAL MANIFESTATIONS

During the neonatal period, a patient with MKHD manifests hypothermia and hyperbilirubinemia with/without any abnormal hair.[20] Hypothermia is probably due to the reduction in CCO activity. Around 3 months of age, the patient begins to exhibit developmental delays and convulsions. Soon thereafter, mental retardation becomes severe and the convulsions often become intractable.[1,20] The skin is hypopigmented and abnormal hair is also apparent at this stage. The hair is coarse, tangled, and lusterless, showing a grayish color due to the hypopigmentation. Microscopic studies of the hair shafts reveal periodic narrowing (monilethrix), twisting (pili torti), and fragmentation (trichorrhexis nodosa) (Figure 44.1b). Hypopigmentation may also be caused by the lowered activity of tyrosinase due to the copper deficiency. The patient has a striking face characterized by pudgy cheeks, sagging jowls, and abnormal eyebrows (Figure 44.1a).

X-ray findings of the skeletal system reveal osteoporosis, flaring of the ribs, and metaphyseal spurring, which may be caused by the lowered activity of ascorbic acid oxidase due to the copper deficiency, since it contributes to normal bone metabolism. These bone abnormalities may result in pathologic fractures, and rib fractures are very common.

Arteriograms show elongation, tortuosity, and variable caliber in the major arteries of the brain, viscera, and limbs. These lesions may cause arterial rupture and thrombosis. Consequently, the MKHD patient frequently suffers from intracranial hematoma and brain infarction (Figure 44.1c). Microscopically, these arteries are composed of structural fragmentation and reduplication of the internal elastic lamina, with intimal thickening.[25] Copper deficiency reduces the activity of lysyl oxidase, which contributes to the formation of collagen and elastin and thus causes these arterial lesions.

MKHD patients frequently suffer from recurrent urinary tract infections due to diverticula of the bladder or ureters, and also have diverticula of the gastrointestinal tract.[26] These diverticula are believed to be formed by abnormalities in collagen and elastin metabolism.

44.6 DIAGNOSIS

The diagnosis of MKHD is easily made by the clinical manifestations of neural deterioration and abnormal hair, and radiologic findings of osteoporosis and metaphyseal spurring. Serum copper and ceruloplasmin levels can be normal in the first 2 to 3 weeks of life, and then decrease. They do not elevate, even if copper sulfate is given orally. These laboratory changes are also helpful for making a diagnosis.

For a prenatal diagnosis, amniotic cells and/or chorionic villi cells can be cultured in a copper-enriched medium. Copper retention is then examined, since copper accumulates in these cells if they are from an affected or heterozygous fetus. It is difficult to distinguish an affected fetus from the heterozygous fetus, because the cells from the heterozygote show similar abnormalities as an affected hemizygous fetus. Therefore, it is important for a prenatal diagnosis to determine the sex of the fetus, since MKHD shows X-linked recessive inheritance. When the copper content increases in these cells from a male fetus, it is diagnosed as having MKHD. In the near future, a genetic test based on the cloned gene responsible for MKHD will facilitate the prenatal diagnosis.

In the macular mutant mouse, the copper contents double in the placenta of a hemizygous fetus vs. a normal fetus.[27] Histochemically, the copper accumulates in the chorion of the placenta. Thus, it may be useful for the diagnosis to measure the copper levels in biopsied chorionic villi.

44.7 TREATMENT

Several reports have been published on copper therapy for patients with MKHD. The parenteral administration of copper after birth produces no improvement in the neural degeneration of patients with MKHD.[14,28] Other treatments recently attempted include parenteral copper therapy in combination with either D-penicillamine, chelators, or vitamin E, or treatment with vitamin C and treatment with parenteral copper-histidinate. However, none of these resulted in any significant improvement in the clinical course.

Copper therapies have also been used in the macular mutant mouse. The macular mutant mouse is injected intraperitoneally 4 times with 10, 20, 20 and 30 µg of cupric chloride on days 4, 6, 8, and 10, respectively. The mouse thus treated survives, and shows an almost normal growth pattern without any clinical neural abnormalities.[10] The copper content in the cerebrum, brain stem, and cerebellum of treated mutants also increases in comparison with nontreated mutants, and are about 53, 56 and 53%, respectively, of their normal littermates on day 14 (Table 44.1). This copper therapy also reduces the number of abnormal mitochondria, normalizes CCO, NADH, and SDH activities, and causes the regression of somal sprouts and spines in the Purkinje cells.[10,22] In the macular mutant mouse, copper therapy not only improves the clinical manifestations, but also biochemical and neuropathological changes. From the point of histogenesis in the central nervous system, the brain of the mutant mouse, including the blood-brain barrier, is much more immature in comparison to human MKHD patients treated with copper. This difference may depend on whether the copper therapy is beneficial or not. However, the parenteral administration of copper to the pregnant mouse as a therapy for the affected fetus does not work, because most of the copper is taken up by the placental villi without moving across to the affected fetus.[27]

REFERENCES

1. Menkes JH, Alter M, Steigleder G, et al: A sex-linked recessive disorder with retardation of growth, peculiar hair, and cerebral and cerebellar degeneration. *Pediatrics*, 1962; 29:764–779.
2. Danks DM, Campbell PE, Stevens BJ, et al: Menkes kinky hair syndrome. An inherited defect in copper absorption with widespread effects. *Pediatrics*, 1972; 50:188–201.
3. Danks DM, Cartwright E, Stevens BJ, and Townley RRW: Menkes kinky hair disease: Further definition of the defect in copper transport. *Science*, 1973; 179:140–1142.
4. Vulpe C, Levinson B, Whitney S, et al: Isolation of a candidate gene for Menkes disease and evidence that it encodes a copper-transporting ATPase. *Nature Genet.*, 1993; 3:7–13.
5. Chelly J, Tumer Z, Tonnesen T, et al: Isolation of a candidate gene for Menkes disease that encodes a potential heavy metal binding protein. *Nature Genet.*, 1993; 3:14–19.
6. Mercer JFB, Livingston J, Hall B, et al: Isolation of a partial candidate gene for Menkes disease by positional cloning. *Nature Genet.*, 1993; 3:20–25.
7. Hunt DM: Primary defect in copper transport underlies mottled mutants in the mouse, *Nature*, 1974; 249:852.
8. Nishimura M: A new mutant mouse, Macular(Ml) (in Japanese). *Exp Anim. (Tokyo)*, 1975; 24:185.
9. Yamano T, Shimada M, Kawasaki H, et al: Clinico-pathological study on macular mutant mouse. *Acta Neuropathol.*, 1987; 72:256–260.
10. Kawasaki H, Yamano T, Iwane S, et al: Golgi study on macular mutant mouse after copper therapy. *Acta Neuropathol.*, 1988; 76:606–612.
11. Kodama H, Okabe I, Yanagisawa M, and Kodama Y: Copper deficiency in the mitochondria of cultured skin fibroblasts from patients with Menkes syndrome. *J. Inher. Metab. Dis.*, 1989; 12:386–389.
12. Katsura T, Yamano T, and Shimada M: Effect of copper on cultured fibroblasts from macular mouse as a model of Menkes kinky hair disease. *Congenital Anomalies*, 1987; 27:251–258.
13. Danks DM, Stevens BJ, Campbell PE, et al: Menkes kinky-hair syndrome. *Lancet*, 1972; 1:1100–1103.
14. Nooijen JL, De Groot CJ, Van den Hamer CJA, et al: Trace element studies in three patients and a fetus with Menkes disease. Effect of copper therapy. *Pediatr. Res.*, 1981; 15:284–289.

15. Katura T, Kawasaki H, Yamano T, and Shimada M: Copper contents and pathological changes in various organs of macular mouse. *Congenital Anomalies*, 1988; 28:85–92.

16. Kodama H, Abe T, Takama M, et al: Histochemical localization of copper in the intestine and kidney of macular mice. Light and electron microscopic study. *J. Histochem. Cytochem.*, 1993; 41:1529–1535.

17. Kodama H, Meguro Y, Abe M, et al: Genetic expression of Menkes disease in cultured astrocytes of the macular mouse. *J. Inher. Metab. Dis.*, 1991; 14:896–901.

18. Aguilar MJ, Chadwick DL, Okuyama K, and Kamoshita S: Kinky hair disease. I. Clinical and pathological features. *J. Neuropathol. Exp. Neurol.*, 1966; 25:507–522.

19. Hirano A, Llena JF, French JH, and Ghatak NR: Fine structure of the cerebellar cortex in Menkes kinky hair disease. X-chromosome-linked copper malabsorption. *Arch. Neurol.*, 1977; 34:52–56.

20. Baerlocher K and Nadal D: Das Menkes-Syndrom. *Ergeb. Inn. Med. Kinderheilkd.*, 1988; 57:77–144.

21. Maehara M, Ogasawara N, Mizutani N, et al: Cytochrome c oxidase deficiency in Menkes kinky hair disease. *Brain Dev.*, 1983; 5:533–540.

22. Yamano T, Shimada M, Onaga A, et al: Electron microscopic study on brain of macular mutant mouse after copper therapy. *Acta Neuropathol.*, 1988; 76:574–580.

23. Kumode T, Yamano T, and Shimada M: Histochemical study of mitochondrial enzymes in cerebellar cortex of macular mutant mouse, a model of Menkes kinky hair disease. *Acta Neuropathol.*, 1994; 87:313–316.

24. Sasahara A, Yamasaki S, Tatiiri T, et al: Biochemical study on the brain of the macular mutant mouse as a model of Menkes kinky hair disease. *Brain Dev.*, 1988; 10:54–56.

25. Oakes BW, Danks DM, and Campbell PE: Human copper deficiency. Ultrastructural studies of the aorta and skin in a child with Menkes syndrome. *Exp. Mol. Pathol.*, 1976; 25:82–98.

26. Hara K, Oohira A, Nogami H, et al: Kinky hair disease. Biochemical, histochemical, and ultrastructural studies. *Pediatr. Res.*, 1979; 13:1222–1226.

27. Xu GQ, Yamano T, and Shimada M: Copper distribution in fetus and placenta of the macular mutant mouse as a model of Menkes kinky hair disease. *Biol. Neonate*, 1994; 66:302–310.

28. Garnica AD: The failure of parenteral copper therapy in Menkes kinky hair syndrome. *Eur. J. Pediatr.*, 1984; 142:98–102.

Index

Iron, 5
  absorption, 367
  aging and, 301
  aluminum and, 12, 84, 92, 93
  animal models, 374
  axonal transport, 63, 66, 67
  beneficial low-level exposure, 23
  blood-brain barrier and, 366, 367, 373, 374,
      387–388
  brain trauma and, 367
  cadmium interactions, 230
  ceruloplasmin, 406
    aceruloplasminemia, 9
    hereditary ceruloplasmin deficiency, 405,
        407–410
  chronic overload, 6
  CNS distribution, 6, 365–366, 380, 408–410
  copper interactions, 9, 405
  CSF, 367
  deficiency, 6
  dopamine receptor function and, 6
  genetic disorders, 6–7, See also Hemochromatosis
  Hallervorden-Spatz disease, 7
  histochemistry, 382
  lead displacement of, 22
  liver, hereditary ceruloplasmin deficiency, 408
  manganese interactions, 299
  NAA
    neurologically diseased CNS tissue, 57
    normal range of controls, 56
  neurodegenerative diseases, 388
    ferritin, 369–370
    neuromelanin complex, 371
    reasons for accumulation, 373
  neuroleptics and, 151
  neuromelanin, 387
  neuronal morphology and, 67
  neurotransmitter uptake/release, 353
  non-parkinsonian neurodegenerative disorders affecting
      the basal ganglia, 369
  normal brain distribution, 380
  oxidative processes, 7, 366–368, 379–381
    animal models, 386–387
    Fenton reaction, 324, 366, 372, 380
  Parkinson's disease and, 7, 367–368, 381–382,
      405
    analytical methods, 381–384
    cellular localization, 368
    CSF levels, 373
    early Parkinson's disease, 373
    ferritin studies, 370, 388
    future research problems, 387–388
    neuromelanin, 382, 383–384
    oxidative stress model, 368
  Perl's Prussian blue reaction, 63, 66, 67
  plasma, 367
  poisoning, 6
  seizures and, 352
  storage, 367
  transport, 367, See also Transferrin
Iron chelation therapy for Parkinson's disease, 374

Ischemia
  calcium and, 330, 358–359
  iron release, 367
Isotopic tracers, 62, 81–82

J

Jakai, 28

K

Kainate receptor, 346
Kainic acid, 207
Kayser-Fleischer rings, 8, 414, 419, 434
Keshan disease, 10
Kidney
  aluminum elimination, 87, 92, 128
  cadmium distribution, 230
  magnesium elimination, 220
  mercury toxicity, 171
Kii Peninsula endemic neuropathies, 28, 107–110
  aluminum and, 120
  soil and water analyses, 329–330
Kindling model, 349–351
Kinesin, 62
Kraepelin, Emil, 114

L

Lactate dehydrogenase, 150
  aluminum toxicity, 28–30
  selenium treatment, 22
Lambert-Eaton syndrome, 360
LAMMA, 368, 382–384
LAP–4 receptors, 346
Laser microprobe mass analyzer (LAMMA), 368, 382–384
Lead, 5, 243–245, 263, See also Lead neurotoxicity
  abasement, 280, 289–290
  absorption and transport, 11
  ALS, 68
  axonal transport, 63, 68
  bioaccumulation, 19
  blood-brain barrier, 245
  blood levels, 18, 245, 263–264, 276, 286
  bone, 19, 276
  cadmium interactions, 230
  calcium interactions, 11, 22, 243, 246–248, 258–259,
      267
  calmodulin and, 258–259
  chemical properties, 254
  CNS distribution, 11, 264
  cognitive deficits, 237
  cytoplasmic concentrations, 245
  distribution, 11
  effective ionic radius, 254
  excretion, 11
  fetal exposure, measurement, 277
  in gasoline, 10, 19, 286
  in hair, 237
  hippocampal neurofibrillary tangles and, 151
  hyperactivity in animals, 268